STRUCTURE FORMATION IN ASTROPHYSICS

Understanding the formation of objects at all scales in the universe, from galaxies to stars and planets, is a major issue in modern astrophysics, and one of the most exciting challenges of twenty-first century astronomy. Even though they are characterized by different scales, the formation of planets, stars and galaxies share many common physical processes and are rooted in the same underlying domains of physics.

This unique reference for graduate students and researchers in astrophysics is the first to cover structure formation on various scales in one volume. This book gathers together extensive reviews written by world experts in physics and astrophysics working in planet, star and galaxy formation, and related subjects. It addresses current issues in these fields and describes the recent observational status and state-of-the-art theoretical and numerical methods aimed at understanding these problems.

GILLES CHABRIER is Head of the Theoretical Astrophysics group at the Ecole Normale Supérieure de Lyon. He has recently won the Johann Wempe prize and the silver medal of the Centre National de la Recherche Scientifique (CNRS). His research interests include dense matter physics, compact objects, stellar and planetary physics, as well as galactic implications.

STRUCTURE FORMATION
IN ASTROPHYSICS

Edited by

GILLES CHABRIER

Ecole Normale Supérieure, Lyon

CAMBRIDGE
UNIVERSITY PRESS

CAMBRIDGE UNIVERSITY PRESS
Cambridge, New York, Melbourne, Madrid, Cape Town, Singapore,
São Paulo, Delhi, Dubai, Tokyo, Mexico City

Cambridge University Press
The Edinburgh Building, Cambridge CB2 8RU, UK

Published in the United States of America by Cambridge University Press, New York

www.cambridge.org
Information on this title: www.cambridge.org/9780521182744

© Cambridge University Press 2009

First published 2009
First paperback edition 2010

A catalogue record for this publication is available from the British Library

Library of Congress Cataloguing in Publication data
Structure formation in astrophysics / edited by Gilles Chabrier.
p.   cm.
1. Astrophysics.   2. Large scale structure (Astronomy)   1. Chabrier, Gilles.   11. Title.
QB461.S82   2008
523.01–dc22
2008019587

ISBN 978-0-521-88779-3 Hardback
ISBN 978-0-521-18274-4 Paperback

Contents

Contributors

Tom Abel, KIPAC, Stanford University, USA

Yann Alibert, Physikalisches Institut University of Benn, Switzerland

Philippe André, Service d'Astrophysique, CEA Saclay, France

Robi Banerjee, ITA, University of Heidelberg, Germany

Isabelle Baraffe, Ecole Normale Supérieure de Lyon, CRAL, France

Shantanu Basu, Department of Physics and Astronomy, University of Western Ontario, Canada

Willy Benz, Physikalisches Institut University of Bern, Switzerland

Ian Bonnell, SUPA, School of Physics and Astronomy, University of St Andrews, United Kingdom

Volker Bromm, Department of Astronomy, University of Texas at Austin, USA

Greg Bryan, Department of Astronomy, Columbia University, USA

Gaspard Duchêne, Laboratoire d'Astrophysique de Grenoble, France

Cornelis Dullemond, Max Planck Institut für Astronomie, Heidelberg, Germany

Richard Durisen, Astronomy Department, Indiana University, USA

Richard Ellis, Department of Astrophysics, Oxford, UK

Andrea Ferrara, SISSA/International School for Advanced Studies, Trieste, Italy

Tristan Guillot, OCA, France

Alexander Heger, Los Alamos National Laboratory, USA; Department of Astronomy and Astrophysics, UCSC, USA; School of Physics, University of Minnesota

Fabian Heitsch, Department of Astronomy, University of Michigan, Ann Arbor, USA

Patrick Hennebelle, Ecole Normale Supérieure, Paris, France

Ahmad Hujeirat, ZAH – Center for Astronomy, University of Heidelberg, Germany

Shu-Ichiro Inutsuka, Department of Physics, Graduate School of Science, Kyoto University, Japan

Richard Klein, Department of Astronomy, University of California, Berkeley and Lawrence Livermore National Laboratory, USA

Ralf Klessen, ZAH/ITA, Universität Heidelberg, Universität, Germany

Mark Krumholz, Department of Astrophysical Sciences, Princeton University, USA

Gregory Laughlin, University of California, Santa-Cruz, USA

Emmanuel Lévêque, Laboratoire de physique, Ecole normale supérieure de Lyon, France

Zhi-Yun Li, University of Virginia, USA

Mordecai-Mark Mac Low, American Museum of Natural History, New York, USA

Thomas Megeath, Department of Physics, University of Toledo, USA

Francois Menard, Laboratoire d'Astrophysique de Grenoble, France

Subhanjoy Mohanty, Harvard University, USA

James Muzerolle, Steward Observatory, University of Arizona, USA

Aake Nordlund, Niels Bohr Institute, Denmark

Rachid Ouyed, Department of Physics and Astronomy, University of Calgary, Canada

John Papaloizou, DAMT, Centre for Mathematical Sciences, University of Cambridge, United Kingdom

Ralph Pudritz, Physics and Astronomy Department, McMaster University, Hamilton, Canada

Edwin Salpeter, Space Sciences, Cornell University, USA

Wolfram Schmidt, Institut für Theoretische Physik und Astrophysik, Universität Würzburg, Germany

Joseph Silk, Department of Physics, University of Oxford, United Kingdom

James Stone, Department of Astrophysical Sciences, Princeton University, USA

Romain Teyssier, Service d'Astrophysique CEA Saclay, France

Stephane Udry, Observatoire de Genéve, Switzerland

Enrique Vazquez-Semadeni, Universidad Nacionál Autónoma de México, Mexico

Gerhard Wurm, Institut für Planetologie, Münster, Germany

Preface

Understanding the formation of gravitationally bound structures at all scales in the universe is one of the most fascinating challenges of modern astronomy. It is now realized that the initial building blocks of galaxies were small collapsing dark matter halos, produced by the primordial fluctuations. These blocks then merged and were assembled into progressively larger galaxies, a scheme generally described as the hierarchical model of galaxy formation. The modern understanding of star formation involves large-scale turbulent motions producing local overdensities which eventually collapse and form prestellar cores under the action of gravity. The most likely scenario for planet formation is the collapse of a vast gaseous envelope onto a central dense core formed from the aggregation of millimetre-size grains in the original protoplanetary nebula, although disk fragmentation could remain an alternative scenario in some situations. The detailed processes responsible for the formation of these structures, however, remain poorly understood. Many important issues remain unsettled, so the robustness of these general paradigms is still ill determined. All these scenarios for the formation of galaxies, stars and planets, although involving vastly different scales, share many underlying physical mechanisms. They all involved hydrodynamical processes, generally leading to turbulent motions, but the very nature of these motions and their real role in structure formation remains unclear. The role of magnetic fields, in the collapse itself and in the generation of winds and jets, remains one of the major unknowns in the formation of structures. As a structure starts to collapse, its ability to cool determines the final bound objects and thus radiation, or more precisely radiation hydrodynamics is a key process in galaxy, star and planet formation. How the ubiquitous presence of gravity modifies the impact of all these processes remains a major issue. Only a comprehensive description of these complex physical mechanisms, and of their interplay, will enable us to fully assess the validity of the aforementioned scenarios for the formation of gravitationally bound structures in the universe.

It was indeed the aim of this conference, organized at the foot of Mont Blanc, in Chamonix, to bring together world experts in galaxy, star and planet formation, in order to address these issues and to share their expertise and problems over the course of a week. Since 'the star is always right', all the sessions started with an observational review, presenting our current understanding of the problem and the main questions to be answered. The theoretical talks addressed in detail the various physical problems encountered in the astrophysical situations of interest, as well as the numerical methods and challenges in the description of these complex processes. The talks were followed by lively discussions between the participants during the daily poster sessions, the lunches taken together on the site and the long morning and afternoon breaks, promoting interactions between the various communities.

I am deeply indebted to the different speakers for their outstanding reviews and for the remarkable review chapters they have written, which are included in the present volume. Bringing together various authors to write a common, genuine review was a challenge and I am really thankful to all the contributors for having accepted this task. I believe this has generated a volume which will be useful to young and less young researchers interested in or working in the exciting field of structure formation in astrophysics. My profound gratitude to all the members of the Scientific Organizing Committee (SOC),[1] whose input was essential in choosing the various topics and speakers, and in shaping up the scientific backbone of the conference. I am also deeply indebted to Cathy Meot and to the whole team at the Hotel Majestic for their kindness, their efficiency and their remarkable professionalism. Their role was essential in making this conference a success. Special thanks also to Jimmy Paillet, our talented artistic graduate student, who designed the logo of the conference, and to the city and the mayor of Chamonix, for hosting the conference in this unique site. Finally, it is important to mention that this conference could not have been organized without the essential financial support of various sponsors, namely the Ecole Normale Supérieure de Lyon, the Centre National de la Recherche Scientifique, the Ministère de la Recherche et de la Technologie, the Institut National des Sciences et de l'Univers, the Centre de Recherche Astrophysique de Lyon, the Région Rhône-Alpes and the Conseil Général de Haute-Savoie. My last thoughts will be for the IAU and the 'experts' of the various Divisions/Commissions, that the proposal had been submitted to, for deciding *not* to support the conference, for the general reason that it was unlikely that people coming from different scientific communities would be persuaded to listen to each other. This certainly strengthened my wish to take up the challenge!

---

[1] The SOC of the conference was composed of: Ph. André (Paris), I. Baraffe (Lyon), P. Bodenheimer (UCSB), V. Bromm (Austin), C. Cesarsky (ESO), G. Kauffmann (Garching), R. Klessen (Heidelberg), R. Larson (Yale), M. Mayor (Geneva), C. McKee (Berkeley), Ph. Myers (Harvard), J. Papaloizou (Cambridge), E. Salpeter (Cornell), F. Shu (UCSD) and J. Silk (Oxford).

Part I

Physical Processes and Numerical Methods Common
to Structure Formations in Astrophysics

1

The physics of turbulence

E. Lévêque

Turbulence in fluids is a topic of great interest. First and foremost, most flows in nature are turbulent and this is particularly true in the astrophysical context (Kritsuk & Norman 2004). Also, turbulence leads to very peculiar mechanics that still escapes to a great extent from our understanding. Since the pioneering works conducted by Osborne Reynolds at the end of the nineteenth century (around 1895), turbulence in fluids has become a rich and challenging research subject in which scientists from engineering, theoretical and experimental physics have been involved with many different perspectives. There is no doubt that bridging ideas from one field to another, and therefore stimulating new interdisciplinary approaches, should provide a fruitful means of gaining understanding on turbulence in the future.

In this chapter, the *background* physics of turbulence will be discussed spontaneously at a (very) basic level, i.e. without getting into details or precise formulation. The discussion will be limited to incompressible hydrodynamics governed by the Navier–Stokes (NS) equations. Firstly, general comments on turbulence (as a statistical-mechanical problem) will be made. Then, I shall attempt to provide some hints (rather than definite answers) to a series of questions: What is generally the source of turbulence? What are the main statistical features of turbulence? How to deal with turbulence? Much more elaborated developments and references may be sought in the following books (among many others) dealing with turbulence:

- a reference book on the physics of turbulence: *A first course in turbulence* by H. Tennekes and J. L. Lumley, MIT Press, Cambridge, USA (1972)
- a reference book on turbulence as a statistical-mechanical problem: *Turbulence: The Legacy of A. N. Kolmogorov* by U. Frisch, Cambridge University Press, Cambridge, UK (1995)
- a reference textbook on the modelling of turbulence: *Turbulent flows* by S. Pope, Cambridge University Press, Cambridge, UK (2000)

Structure Formation in Astrophysics. ed. G. Chabrier. Published by Cambridge University Press.

- a reference book on the numerical modelling of turbulence: *Large-eddy simulation for incompressible flows – An introduction* by P. Sagaut, Springer-Verlag, Scientific Computation series (2005).

1.1 General comments on turbulence

Turbulence is employed to label flows with the common characteristics of complexity and disorder. *Complexity* refers to the complicated swirling motion of the fluid, the ability to distort material fluid elements into complex convoluted geometries. *Disorder* is related to this dynamics being random (or unpredictable). In this respect, turbulence should be approached from the viewpoint of statistical mechanics (Monin & Yaglom 1975).

Complexity in turbulence has to do with the essential non-linearity arising from the advection term in the dynamical equations:

$$\partial_t \mathbf{u}(\mathbf{x}, t) + (\mathbf{u}(\mathbf{x}, t) \cdot \nabla) \mathbf{u}(\mathbf{x}, t) = forces\ by\ unit\ mass.$$

$\mathbf{u}(\mathbf{x}, t)$ denotes the velocity field. By recasting this term in the Fourier space (and assuming local homogeneity), one gets

$$\partial_t \widehat{u}_i(\mathbf{k}, t) = -ik_j \sum_{\mathbf{p}+\mathbf{q}=\mathbf{k}} \widehat{u}_i(\mathbf{p}, t)\widehat{u}_j(\mathbf{q}, t) + ...,$$

where the summation is over all allowed wavevectors $\mathbf{p}$ and $\mathbf{q}$. Thus, the time evolution of mode $\mathbf{k}$ is a priori driven by the triadic interactions with all modes such that $\mathbf{p} + \mathbf{q} = \mathbf{k}$. A specific feature of turbulence (as a dynamical system) lies in the impossibility to reduce this interaction to a restricted set of interacting modes $\mathbf{p}$ and $\mathbf{q}$. On the contrary, *strong interactions* with all triads of modes must be considered. Furthermore, long-range (phase) coherency between Fourier modes is expected to play an essential role; it is heuristically connected to the concentration of vorticity into intense thin fluid structures such as vortex filaments (Figure 1.1).

Disorder in turbulence has to do with a strong departure from absolute statistical equilibrium. From a theoretical viewpoint, the statistical problem of turbulence is a priori well posed if at initial time t_0 the mean velocity $U_i(\mathbf{x}, t_0)$ and the two-point correlation function

$$R_{ij}(\mathbf{x}, t_0; \mathbf{x}', t_0) \equiv \langle u_i(\mathbf{x}, t_0)u_j(\mathbf{x}', t_0) \rangle$$

are prescribed for all $\mathbf{x}$ and $\mathbf{x}'$ and if it is assumed that the (multivariate) distribution of the turbulent (fluctuating) velocity field $u_i(\mathbf{x}, t_0)$ is normal. However, at times $t > t_0$, it is observed that the multivariate distribution of $u_i(\mathbf{x}, t)$ strongly deviates from the normal distribution due to statistical correlations generated by

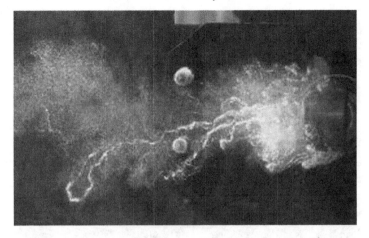

Fig. 1.1. Vortex filaments in a turbulent jet visualized by micro-bubbles. Courtesy of Olivier Cadot, École nationale supérieure de techniques avancées, Paris.

the hydrodynamical forces. Analytically, this departure from the normal distribution is governed by the entire infinite sequence of statistical equations (deduced from the NS equations) for all multi-point multi-time correlation functions

$$R_{ijk...}(\mathbf{x}, t; \mathbf{x}', t'; \mathbf{x}'', t''; ...) \equiv \langle u_i(\mathbf{x}, t) u_j(\mathbf{x}', t') u_k(\mathbf{x}'', t'') ... \rangle.$$

An identified fundamental issue resides in finding an appropriate closure condition to convert this infinite hierarchy into a closed (low-dimensional) subset of equations and eventually solve it.

In practice, turbulence is often investigated from a phenomenological standpoint, i.e. starting from hypotheses motivated by experimental and numerical observations. This line of study has yielded very fruitful results during the past half century and continues to expand nowadays (in absence of any successful statistical theory).

1.2 What is the source of turbulence?

Turbulence is generally triggered by the inhomogeneities of the flow (by the mean velocity field not being uniform in space). Boundary conditions are often responsible for such inhomogeneities (Figure 1.2).

In order to dissect this mechanism, let us consider the flow over a (solid) flat plate. Because of the no-slip condition at the boundary, the velocity field necessarily decreases to zero in the vicinity of the plate. This implies a strong gradient of the mean velocity in the direction perpendicular to the plate. From kinematic considerations, this strong gradient may be viewed as a vorticity sheet attached to the plate. This sheet is generically unstable and generates streaky structures that

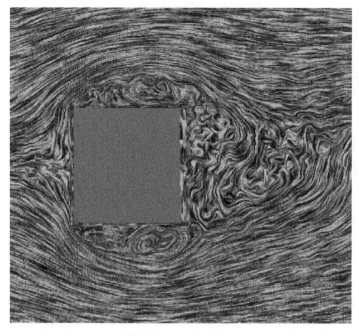

Fig. 1.2. Streamlines of a turbulent flow (from left to right) around an obstacle. It appears that turbulence originates in the vicinity of the obstacle.

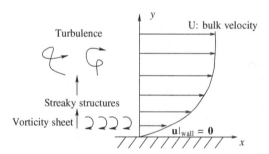

Fig. 1.3. A solid boundary may be viewed as a source of vorticity which is generically unstable, detaches from the boundary and sustains turbulence in the bulk.

eventually detach, interact and contribute to sustain turbulence in the bulk of the flow (Figure 1.3).

As rule of thumb, one may claim that a key ingredient involved in the generation of turbulence is the inhomogeneity of the mean flow, i.e. strong mean velocity gradients, and the instability of these gradients. This mechanism also tells us that it is important to learn about the mean flow before getting to the turbulent fluctuations.

1.3 What are the main statistical features of turbulence?

This section is devoted to the statistical features of turbulent motions, usually viewed as the net result of the interaction of a *gas of turbulent eddies*. By analogy with a molecule, an eddy may be seen as a glob of fluid of a given size, or (spatial) scale, that has a certain structure and life history of its own. This interaction is unpredictable in detail; however, statistically distinct properties can be identified and profitably examined. The need for a statistical description arises from both the *intrinsic* complexity of individual solutions and the instability of these solutions to infinitesimal perturbations (in the initial and boundary conditions). This makes it natural to examine ensemble of realizations rather than each individual realization and seek for robust statistical features insensitive to the details of perturbations.

It is characteristic of turbulence that turbulent eddies are distributed over a wide range of size scales and associated turn-over timescales. This range spreads from the *integral scale* L_S, which refers to the size of the largest eddies in the flow (triggered by the inhomogeneities of the mean flow as seen previously), to the *elementary scale* η, which nails down the size of the smallest eddies. The macroscale L_S can be estimated by equalling the timescale related to the local mean velocity gradient, $1/\|\nabla U\|$, and the turn-over time of the largest eddies, L_S/u, where u denotes the root-mean-squared turbulent velocity. Let us note that the norm of the mean velocity gradient is often referred to as the *shear* in the literature, and L_S is called the *shear length scale*. The elementary microscale η is the viscous cut-off scale. Formally, it is estimated by equalling the viscous timescale η^2/ν, where ν is the kinematic viscosity of the fluid, and the turn-over time of the smallest eddies. This latter must be modelled (see p. 9).

The range of excited scales

$$L_S \sim \frac{u}{\|\nabla U\|} \geq r \geq \eta$$

is called the *inertial range*. In that range, interactions between turbulent eddies result in an effective transfer of kinetic energy from the large scales (comparable to L_S), where energy is fed into turbulence, to the small scales (comparable to η), where this energy is dissipated by molecular viscosity. In 1941, Kolmogorov envisaged to describe this mechanism by a self-similar cascade of kinetic energy, which is local in scale and in which all statistical information concerning the large scales is lost (except for the mean energy-cascade rate ε). Kolmogorov's theory yields the celebrated *universal law* for the kinetic energy spectrum (in wavenumber k):

$$E(k) = C\varepsilon^{2/3}k^{-5/3},$$

where C is a universal constant (Kolmogorov 1941). It is worth noting that this law does not include any characteristic scale; this refers to the idea that the energy cascade is a self-similar process in scale.

The Kolmogorov's energy spectrum represents a form for the inertial range of wavenumbers, in which energy is transferred from small to large wavenumbers by the process of vortex stretching. Indeed, the chaotic nature of turbulence tends to separate any two fluid elements initially near to each other. Consequently, there is a tendency to stretch initial vorticity into ever-elongated and thinning structures, until viscosity stops the thinning (Figure 1.4). By Kelvin's theorem, if the cross section of a vortex structure decreases under stretching, the fluid in the vortex must spin faster. The combination of stretching and spin-up means a transfer of energy from lower to higher wavenumbers. The viscous cut-off scale η, which identifies the bottom scale of the energy cascade, is given (from dimensional arguments) by

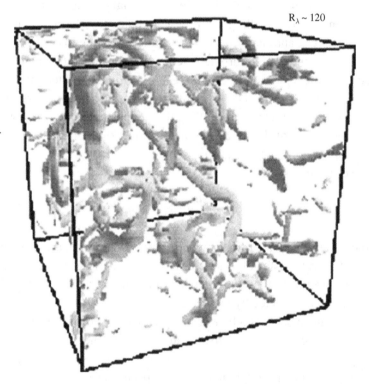

Fig. 1.4. Snapshot of high-enstrophy isosurfaces from a numerical simulation of three-dimensional turbulence; the local enstrophy is defined by $|\vec{\nabla} \times \vec{u}(\vec{x}, t)|^2$. The swirling activity of the flow concentrates into very localized elongated and thin fluid structures: the *vortex filaments* (E. Lévêque).

$$\eta = \left(\frac{\nu^3}{\varepsilon}\right)^{1/4}.$$

A large body of experimental and numerical measurements corroborate the Kolmogorov's energy spectrum (Figure 1.5). However, higher-order statistical correlations are not universal in the sense of Kolmogorov's hypothesis. These discrepancies are rooted in the fact that the cascade of energy is actually a highly non-uniform process in space and time (Figure 1.6). This feature is usually referred to as *intermittency* in the literature. From the viewpoint of statistical mechanics, intermittency implies that the macroscopic parameter ε is not sufficient to characterize the energy-cascade state of turbulence, but fluctuations of $\varepsilon(\mathbf{x}, t)$ should be taken into account. Once Kolmogorov's *mean field theory* is abandoned, a Pandora's box of possibilities is opened and a specific contact with the dynamics of turbulence (solution of the NS equations) must be achieved. Current models have not succeeded to establish this contact. They essentially rely on plausible hypotheses but fail to relate themselves to the actual dynamics. More recent works attempt to correlate turbulent high-order velocity correlations with the presence of highly

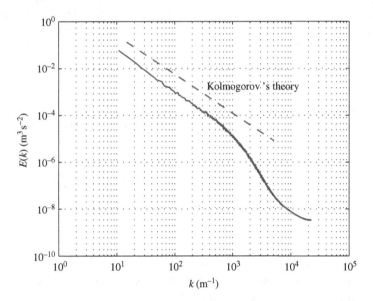

Fig. 1.5. The density (in wavenumber k) of turbulent kinetic energy $E(k)$ exhibits a universal $k^{-5/3}$ decrease (obtained from a laboratory experiment, courtesy of C. Baudet and S. Ciliberto, ENS-Lyon, France) in agreement with the Kolmogorov's theory. Turbulent motions are strongly damped by molecular viscosity at very large wavenumbers. The energy is not equally distributed among Fourier modes: $E(k) \sim k^2$ would be expected at statistical thermodynamic equilibrium. Turbulence is a far-from-equilibrium system.

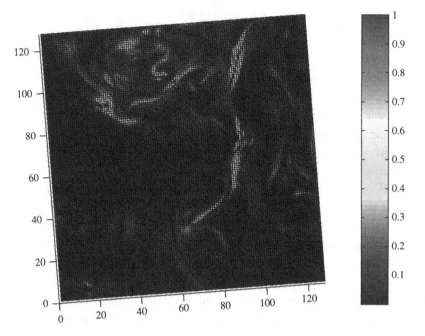

Fig. 1.6. Slice of a snapshot of the energy-dissipation rate $\varepsilon(\mathbf{x}, t)$ obtained from a numerical simulation of the NS equations (E. Lévêque). The magnitude of $\varepsilon(\mathbf{x}, t)$ is represented by a grayscale bar ranging from 0 to 1. Energy dissipation is concentrated on fine structures; it is not uniformly distributed.

coherent dynamical structures, the so-called *She–Lévêque model* (She & Lévêque 1994), for instance.

High-order multi-point correlations of turbulent velocity fluctuations are commonly investigated through the velocity structure functions, defined by

$$S_p(r) \equiv \left\langle |\mathbf{u}(\mathbf{x}, t) - \mathbf{u}(\mathbf{x} + \mathbf{r}, t)|^p \right\rangle$$

for $p = 1, 2, \ldots$ and the separation scale r within the inertial range. It is observed both experimentally and numerically that the $S_p(r)$'s exhibit power-law scalings:

$$S_p(r) \sim r^{\zeta_p}$$

and the scaling exponents ζ_p are found to be universal. The She–Lévêque model relates the set of scaling exponents ζ_p to the presence of vortex filaments and yields a formula without any adjustable parameter:

$$\zeta_p = \frac{p}{9} + 2\left[1 - \left(\frac{2}{3}\right)^{p/3}\right].$$

This model, which is found in very good agreement with experimental and numerical data, establishes a concrete link between the dynamics and the statistics of turbulence. Furthermore, its formulation is very general and can apply to a vast class of turbulent systems. In astrophysics, it has been successfully employed to relate the statistics to coherent dynamical structures in the interstellar medium, in the solar wind or in cosmic rays.

1.4 How to deal with turbulence?

1.4.1 The Reynolds-averaged Navier–Stokes equations

How to account for turbulence in the mean flow? This fundamental question was raised by Osborne Reynolds more than a century ago (Reynolds 1895) and remains mostly open today.

In the turbulent regime, the mean flow is solution of the NS equations complemented by a force which encompasses the exchange of momentum between the mean flow and the turbulent agitation:

$$\rho \frac{dU_i}{dt} = NS(U) - \rho \frac{\partial \overline{u_i u_j}}{\partial x_j}.$$

$-\rho \overline{u_i u_j}$ is termed the *Reynolds stress* (the overbar represents the statistical mean value). These equations are commonly called the *Reynolds-averaged Navier–Stokes* (RANS) equations. It is necessary to express the Reynolds stress in terms of the mean velocity field in order to close the RANS equations.

1.4.1.1 The concept of turbulent viscosity

The idea behind the introduction of a turbulent viscosity is to treat the deviatoric (traceless) part of the Reynolds stress like the viscous stress in a Newtonian fluid:

$$-\rho \overline{u_i u_j} + \frac{1}{3} \rho \overline{u_k u_k} \, \delta_{ij} = 2\rho \nu_{turb} \overline{S}_{ij},$$

where ν_{turb} is the (kinematic) turbulent viscosity and $\overline{S}_{ij}$ is the mean rate-of-strain tensor (the symmetric part of the velocity gradient tensor). The turbulent viscosity $\nu_{turb}(\mathbf{x}, t)$ depends on the position $\mathbf{x}$ in the flow and time t; it is a property of the flow not of the fluid. The introduction of a turbulent viscosity relies on the hypothesis of an internal friction (related to the turbulent agitation of the fluid) responsible for the diffusive transport of momentum from the rapid to the slow mean-flow regions.

Dimensionally, ν_{turb} is equivalent to the product of a velocity and a length. This suggests that (by analogy to the kinetic theory for gases)

$$\nu_{turb} = u \, \ell,$$

where u and ℓ would represent the characteristic velocity and length of the turbulent agitation. In the *mixing-length* model, u and ℓ are specified on the basis of the geometry of the flow. In two-equation models (the so-called $(k - \epsilon)$ model being the prime example), u and ℓ are related to the turbulent kinetic energy and the turbulent dissipation, for which modelled constitutive equations are explicated. See Wilcox (1993) for a comprehensive review.

1.4.1.2 The mixing-length model (zero-equation model)

In 1925, the so-called *mixing-length* model has been proposed by Prandtl for the two-dimensional flow over an horizontal wall. The turbulent viscosity reads

$$\nu_{\text{turb}} = \ell_{\text{m}}^2 \left| \frac{\partial U_x}{\partial y} \right|,$$

where y is the coordinate perpendicular to the wall (along the x axis).

The turbulent length is the mixing length and the velocity scale is locally determined by the mean velocity gradient:

$$\ell = \ell_{\text{m}} \quad \text{and} \quad u = \ell_{\text{m}} \left| \frac{\partial U_x}{\partial y} \right|.$$

Here, u is zero wherever the mean velocity gradient is zero. This may be problematic for bounded flows in some particular circumstances (on the centreline of a channel flow, for instance).

In practice, the mixing length $\ell_{\text{m}}(\mathbf{x}, t)$ must be specified accordingly to the geometry of the flow. This specification is usually empirical; it is derived from experiment and observation rather than theory. The mixing-length model, which is arguably the simplest modelling of turbulence, appears to be valuable in simple flows such as shear-layer or pressure-driven flows, but cannot account for the transport effects of turbulence. Indeed, it implies that the local level of turbulence depends on the local generation and dissipation rates. In reality, turbulence may be carried or diffused to locations where no turbulence is actually being generated at all.

1.4.1.3 One-equation model: the k-model

In 1942, independently, Kolmogorov and Prandtl suggested to base the turbulent velocity scale on the mean turbulent kinetic energy: $u \propto \bar{k}^{1/2}$. If the turbulent length scale is taken as the mixing length, the turbulent viscosity then becomes

$$\nu_{\text{turb}} = \ell_{\text{m}} \bar{k}^{1/2}.$$

In that case, a transport equation for $\bar{k}$ is required.

The (exact) dynamical equation for $\overline{k}$ (deduced from the NS equations) is as follows:

$$\frac{\overline{D}\,\overline{k}}{\overline{D}t} = -\frac{\partial}{\partial x_j}\left(\underbrace{\frac{1}{2}\overline{u_iu_iu_j} + \frac{1}{\rho}\overline{u_jp} + v\frac{\partial\overline{k}}{\partial x_j}}_{\text{turbulent transport}}\right) \underbrace{-\overline{u_iu_j}\,\overline{S_{ij}}}_{\text{production: }P} \underbrace{-v\overline{\left(\frac{\partial u_i}{\partial x_j}\right)^2}}_{\text{dissipation: }\varepsilon}$$

- $-\overline{u_iu_j} \approx 2v_{\text{turb}}\overline{S_{ij}} - 2/3\,\overline{k}\delta_{ij}$ with $v_{\text{turb}} = \ell_m\overline{k}^{1/2}$ according to the turbulent-viscosity hypothesis.
- $1/2(\overline{u_iu_iu_j}) + 1/\rho(\overline{u_jp}) \approx (v_{\text{turb}})/\sigma_k(\partial\overline{k})/\partial x_j$. This stems from the gradient-diffusion hypothesis. The empirical coefficient σ_k may be viewed as a turbulent Prandtl number (of order unity) for the kinetic energy.
- From a dimensional argument, the mean dissipation rate (per unit mass) $\varepsilon = C\overline{k}^{3/2}/\ell_m$, where C is a model constant.

Finally, the closed transport equation for the mean turbulent kinetic energy is as follows:

$$\frac{\overline{D}\,\overline{k}}{\overline{D}t} = -\frac{\partial}{\partial x_j}\left(\left(v + \frac{v_{\text{turb}}}{\sigma_k}\right)\frac{\partial\overline{k}}{\partial x_j}\right) + P - \varepsilon$$

with $v_{\text{turb}} = \ell_m\overline{k}^{1/2}$ and $\varepsilon = C\overline{k}^{3/2}/\ell_m$ together with the turbulent-viscosity hypothesis for $-\overline{u_iu_j}$ and the specification of the mixing length ℓ_m.

The k-model does allow for the transport of turbulence into regions where there is locally no generation. It is therefore inherently capable of simulating some phenomena more realistically than the mixing-length model. However, the mixing length remains an empirical parameter and knowledge is almost totally absent for recirculating and three-dimensional flows. This model is found useful in boundary-layer flows, where the mixing length is fairly well known.

1.4.1.4 Two-equation model: the $(k-\varepsilon)$ model

The $(k-\varepsilon)$ model belongs to the class of two-equation models, which are more efficient in the case of flows in complex geometry and higher Reynolds numbers (Jones & Launder 1972).

The two relevant macroscopic parameters in the $(k-\varepsilon)$ model are the turbulent mean kinetic energy and mean energy dissipation. From dimensional analysis, one gets that the turbulent length scale $\ell \sim \overline{k}^{3/2}/\varepsilon$ and the turbulent velocity scale $u \sim \overline{k}^{1/2}$. It follows that

$$v_{\text{turb}} = C_\mu\frac{\overline{k}^2}{\varepsilon},$$

where C_μ is an empirical constant.

The standard model equation for ε is viewed as

$$\frac{\overline{D}\varepsilon}{\overline{Dt}} = \frac{\partial}{\partial x_j}\left(\left(\nu + \frac{\nu_{\text{turb}}}{c\sigma_\varepsilon}\right)\frac{\partial\varepsilon}{\partial x_j}\right) - 2c_1\frac{\nu_{\text{turb}}}{k}\frac{\varepsilon}{}\overline{S_{ij}}\frac{\partial\overline{u}_i}{\partial x_j} - c_2\frac{\varepsilon^2}{k}.$$

The five empirical coefficients (C_μ, c_1, c_2, σ_k, σ_ε) are fixed by considering experimental results for simple-geometry flows.

1.4.1.5 A general comment

In principle, the concept of viscosity is properly applicable only when there is a significant separation between the length scale of inhomogeneity of the mean field and the mixing length of the agitation. This condition is not satisfied in turbulent flows where turbulent motions display a continuous distribution of scale sizes (turbulent motions do not occur at some characteristic mixing length). Consequently, the concept of a local transport has not much justification here, except in a very crude qualitative sense. A more realistic approach, mainly devoted to the numerical simulation of turbulent flows, consists of the large-scale modelling of turbulence dynamics (see Section 1.4.2).

1.4.2 The Large-eddy Simulation of turbulent flows

1.4.2.1 The general context

The prohibitive cost of the *direct numerical simulation* (DNS) of turbulent flows (that is, without calling to any simplification of the dynamics) motivates the elaboration of reduced models, requiring less computation but still relevant (to some degree) for reproducing the large-scale dynamics of the flow (Lesieur 1997). Indeed, on the one hand, the elementary scale η fixes the required grid spacing of the simulation. On the other hand, the integration domain must encompass the largest scales of motion, comparable to the macroscale L_S. Therefore, the number of grid points required to suitably resolve all spatial scales of motion is given by

$$\left(\frac{L_S}{\eta}\right)^3 \sim Re^{9/4},$$

where

$$Re = \frac{uL_S}{\nu}$$

denotes the *Reynolds number*. Here, it has been assumed that $\varepsilon \sim u^3/L_S$. Furthermore, when the Courant–Friedrich–Levy time-step constraint is factored in,

$$\frac{u\delta t}{\eta} < 1,$$

one ends up with a computational cost for the DNS which grows as Re^3. In real flows, the encountered Reynolds numbers are typically very large ($Re \sim 10^{10}$ in astrophysics), which definitively rules out this direct numerical approach. Reduced models of turbulence must be considered.

Roughly speaking, large-scale motions transport most of the kinetic energy of the flow. Their strength makes them the most efficient carriers of conserved quantities (momentum, heat, mass, etc.). On the contrary, small-scale motions are primarily responsible for the dissipation while they are weaker and contribute little to transport. From mechanical aspects, the large-scale (energy-carrying) dynamics are thus of particular importance, and the costly computation of small-scale dynamics should be avoided. Furthermore, while large-scale motions are strongly dependent on the external flow conditions, small-scale motions are expected to behave more universally. Hence, there is a hope that numerical modelling can be feasible and/or requires few adjustments when applied to various flows.

The RANS approaches (introduced in Section 1.4.1) lead to the estimation of mean quantities but do not resolve the turbulent fluctuations; they are based on questionable closure conditions and often appeal to numerous empirical parameters. But their computational cost is unbeatable. The *large-eddy simulation* (LES) offers a compromise between the DNS and the RANS approaches. In a LES, the large-scale dynamics of turbulence are explicitly integrated in time and the interaction with the unsolved small-scale motions is modelled. In order to separate the small-scale component and the large-scale component of the turbulent velocity field, an explicit filtering procedure is widely used.

A spatial filtering is conceptually introduced as

$$\widetilde{\phi}(\mathbf{x}, t) = \int \phi(\mathbf{x}', t) G_\Delta(\mathbf{x} - \mathbf{x}') d\mathbf{x}',$$

where the filter width Δ fixes the size of the smallest scales of variation retained in the flow variable $\phi(\mathbf{x}, t)$. In practical applications, Δ is chosen much larger than the cut-off spatial scale of $\phi(\mathbf{x}, t)$ so that $\widetilde{\phi}(\mathbf{x}, t)$ may be properly considered as the large-scale component of $\phi(\mathbf{x}, t)$. Applying the previous filtering procedure to the NS equations yields

$$\frac{\partial \widetilde{u}_i}{\partial t} + \widetilde{u}_j \frac{\partial \widetilde{u}_i}{\partial x_j} + \frac{\partial \tau_{ij}}{\partial x_j} = -\frac{\partial \widetilde{p}}{\partial x_i} + \nu \frac{\partial^2 \widetilde{u}_i}{\partial x_k \partial x_k} \quad \text{with} \quad \frac{\partial \widetilde{u}_i}{\partial x_i} = 0,$$

where $\widetilde{u}_i(\mathbf{x}, t)$ and $\widetilde{p}(\mathbf{x}, t)$ represent the large-scale velocity and pressure, respectively, and ν is the kinematic viscosity of the fluid.

These equations are amenable to numerical discretization with a grid spacing comparable to Δ, since $\widetilde{\mathbf{u}}(\mathbf{x}, t)$ is expected to vary smoothly over the distance Δ.

$$\tau_{ij}(\mathbf{x}, t) \equiv \widetilde{u_i(\mathbf{x}, t) u_j}(\mathbf{x}, t) - \widetilde{u}_i(\mathbf{x}, t) \widetilde{u}_j(\mathbf{x}, t)$$

is named the *subgrid-scale* (SGS) *stress tensor* and encompasses all interactions between the grid-scale component and the unresolved SGS component of $\mathbf{u}(\mathbf{x}, t)$. In LES, $\tau_{ij}(\mathbf{x}, t)$ needs to be expressed in terms of the grid-scale velocity field $\widetilde{\mathbf{u}}(\mathbf{x}, t)$ only, which is a difficult problem.

Eddy-viscosity models parameterize the SGS stress tensor as

$$\tau_{ij} - \frac{1}{3}\tau_{kk}\delta_{ij} = -2\nu_T \widetilde{S}_{ij},$$

where $\nu_T(\mathbf{x}, t)$ is the scalar eddy-viscosity and

$$\widetilde{S}_{ij}(\mathbf{x}, t) \equiv \frac{1}{2}\left[\frac{\partial \widetilde{u}_i}{\partial x_j}(\mathbf{x}, t) + \frac{\partial \widetilde{u}_j}{\partial x_i}(\mathbf{x}, t)\right]$$

is the resolved rate-of-strain tensor. This empirical modelization is rooted in the idea that SGS motions are primarily responsible for a diffusive transport of momentum from the rapid to the slow grid-scale flow regions. The theoretical basis for the introduction of an eddy-viscosity is rather insecure; however, it appears to be *workable in practice*. The eddy-viscosity $\nu_T(\mathbf{x}, t)$ is then primarily designed to ensure the correct mean drain of kinetic energy from the grid-scale flow to the SGS motions. Another important feature is that $\nu_T(\mathbf{x}, t)$ must vanish in laminar-flow regions (where there is no turbulence).

1.4.2.2 The Smagorinsky model

The Smagorinsky model is certainly the simplest and most commonly used eddy-viscosity model of turbulence (Smagorinsky 1963). The prescription for $\nu_T(\mathbf{x}, t)$ is

$$\nu_T(\mathbf{x}, t) = (C_S \Delta)^2 \|\widetilde{S}(\mathbf{x}, t)\|,$$

where $\|\widetilde{S}\| \equiv (2\widetilde{S}_{ij}\widetilde{S}_{ij})^{1/2}$ represents the magnitude of the resolved rate-of-strain and C_S is a non-dimensional coefficient (called the *Smagorinsky constant*). The major merits of the Smagorinsky model are its manageability, its computational stability and the simplicity of its formulation (involving only one adjusted parameter). All this makes it a very valuable tool for many applications. However, while this model is found to give acceptable results in the LES of homogeneous and isotropic turbulence (with $C_S \approx 0.2$), it is found to be too dissipative with respect to the resolved motions in strongly non-homogeneous flow regions, due to an excessive eddy-viscosity arising from the mean shear. Furthermore, the eddy-viscosity predicted by Smagorinsky is non-zero in laminar-flow regions; the model introduces spurious dissipation which has the effect of damping the growth of small perturbations and thus restrain the transition to turbulence.

1.4.2.3 The shear-improved Smagorinsky model

Recently, a shear-improved Smagorinsky model (SISM) has been introduced (Lévêque *et al.* 2007). In this refined modelling, it is proposed to subtract the magnitude of the shear from the magnitude of the instantaneous resolved rate-of-strain in the eddy-viscosity:

$$v_T(\mathbf{x}, t) = (C_S \Delta)^2 \left(\|\widetilde{S}(\mathbf{x}, t)\| - \mathcal{S}(\mathbf{x}, t) \right).$$

$\mathcal{S}(\mathbf{x}, t)$ denotes the shear at the position $\mathbf{x}$ and time t; C_S is the Smagorinsky constant and $\Delta = (\Delta x \Delta y \Delta z)^{1/3}$ is the local grid spacing. It is worth noting that the eddy-viscosity involved in the SISM implies some information concerning the mean flow: the shear $\mathcal{S}(\mathbf{x}, t)$. It is assumed that the flow is well enough resolved in the direction of the shear, so that

$$\mathcal{S}(\mathbf{x}, t) \approx \|\langle \widetilde{S}(\mathbf{x}, t)\rangle\|. \tag{1.1}$$

Angle brackets $\langle \rangle$ denote statistical (ensemble) average (in practice, space average over homogeneous directions and/or time average should be considered by invoking ergodic properties of the flow).

The main concerns of the SISM are, first, to take into account shear effects in the exchanges of momentum to the SGS motions (without any kind of adjustment) and, second, to make the eddy-viscosity automatically vanish in laminar-flow regions (without any ad hoc damping function). The SISM stems from analytical developments derived from the NS equations.

In flow regions where the fluctuating part of the rate-of-strain is much larger than the shear, turbulence can be considered as homogeneous and isotropic and the SISM naturally reduces to the original Smagorinsky model (which is known to perform reasonably well). In flow regions where shear effects are significant, the SISM yields a transfer of energy to SGS motions fully consistent with the exact energy budget obtained from the NS equations.

A stringent test of the SISM has been performed recently in the backward-facing step geometry (Toschi *et al.* 2006) (Figure 1.7). In addition to wall effects, the flow over a backward-facing step is strongly affected by the detached shear layer and the recirculating motions behind the step. An adequate resolution of the instability of the shear layer is required to predict correctly the location of the reattachment point. The computation of the skin-friction coefficient is displayed in Figure 1.8. The skin-friction coefficient changes sign from negative to positive at the reattachment point (indicated by the arrow). We observe that the prediction is in very good agreement with experimental and DNS results (at comparable Reynolds number). We also see that the SISM performs well with respect to the *dynamic Smagorinsky model* (Germano *et al.* 1991) (a reference eddy-viscosity model) but with the great advantage of being much less demanding in computational resources.

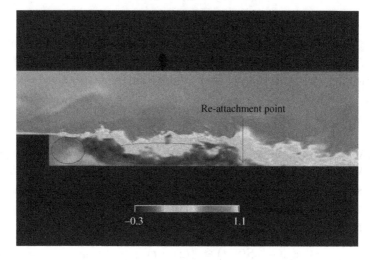

Fig. 1.7. Configuration of the backward-facing step flow.

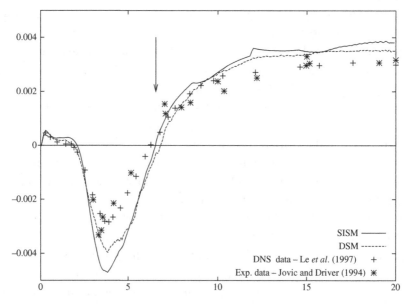

Fig. 1.8. For the backward-facing step flow at $Re = 4800$ (based on the step height), the prediction of the skin-friction coefficient, here displayed as a function of the distance to the step, is in good agreement with experimental data, DNS data and with a LES based on the *dynamic Smagorinsky model* (DSM) at comparable Reynolds number. Notice that the strength of the secondary recirculation bubble near the corner (a sensitive diagnostic) is also correctly predicted. (Courtesy of Federico Toschi.)

First tests indicate that the SISM possesses a good predictive capacity with a computational cost and a manageability comparable to the original Smagorinsky model. The implementation of the SISM in the astrophysics code ENZO (http://lca.ucsd.edu/software/enzo/) is currently under development and the first results are encouraging (Wolfram Schmidt, personal communication).

References

Germano, M., Piomelli, U., Moin, P. and Cabot, W. H., 1991, *Phys. Fluids A* **3**, 1760

Jones, W. O. and Launder, B. E., 1972, *Int. J. Heat Mass Tran.* **15**, 301

Jovic, S. and Driver, D., 1994, *NASA Tech. Mem.* 108807

Kolmogorov, A. N., 1941, *Dokl. Akad. Nauk SSSR* **30**, 299

Kritsuk, A. G. and Norman, M. L., 2004, *ApJ.* **601**, L55

Le, H., Moin, P. and Kim, J., 1997, *J. Fluid Mech.* **330**, 349–374

Lesieur, M., 1997, *Turbulence in Fluids* (Kluwer Academic Publishers Dordrecht, Boston, London)

Lévêque, E., Toschi, F., Shao, L. and Bertoglio, J.-P., 2007, *J. Fluid Mech.* **570**, 491

Monin, A. S. and Yaglom, A. M., 1975, *Statistical Fluid Mechanics* (MIT Press, Cambridge)

Reynolds, O., 1895, *Phil. Trans. Roy. Soc. Lond. Ser. A* **186**, 123

She, Z.-S. and Lévêque, E., 1994, *Phys. Rev. Lett.* **72**, 334

Smagorinsky, J., 1963, *Mon. Weather Rev.* **91**, 99

Toschi, F., Kobayashi, H., Piomelli, U. and Iaccarino, G., 2006, *Proceedings of the Summer Program 2006*, Center for Turbulence Research, Stanford, USA

Wilcox, D. C., 1993, *Turbulence Modeling for CFD* (DWC Industries Inc., La Canada, USA)

2

The numerical simulation of turbulence

W. Schmidt

Turbulence is a remarkable subject in physics. The underlying equations, which are in their simplest formulation the Euler equations, were published 250 years ago (Euler 1757). Yet a theoretical grasp of the phenomenology emerging from these equations had not been achieved before the mid-twentieth century, when Heisenberg (1923) and Kolmogorov (1941) obtained their first analytical results. Eventually, it took the capabilities of modern supercomputers to obtain a full appreciation of the complexity that is inherent to the Euler equations. Astrophysics is now at the very frontier of numerical turbulence modelling. Among the additional ingredients for making turbulence in astrophysics even more complex are supersonic flow, self-gravity, magnetic fields and radiation transport. In contrast, terrestrial turbulence is mostly incompressible or only weakly compressible. External gravity is, of course, an issue in the computation of atmospheric processes on Earth. Self-gravity, however, is only encountered on large, astrophysical scales. The dynamics of turbulent plasma has met vivid attention in research related to nuclear fusion reactors but, otherwise, is not encountered under terrestrial conditions.

In this chapter, I give an overview of the various approaches towards the numerical modelling of turbulence, particularly, in the interstellar medium (ISM). The discussion is placed in a physical context, i.e. computational problems are motivated from basic physical considerations. Presenting selected examples for solutions to these problems, I introduce the basic ideas of the most commonly used numerical methods. For detailed methodological accounts, the reader is invited to follow the references. Some important results and astrophysical implications are briefly outlined. Since turbulence, in a strict sense, is genuinely three-dimensional, I almost exclusively consider three-dimensional simulations. Furthermore, turbulence is a multi-scale phenomenon, and, in order to capture its properties correctly, sufficient numerical resolution is essential. This is why newer, higher-resolution computations are generally preferred as examples (as regards self-gravitating

Structure Formation in Astrophysics. ed. G. Chabrier. Published by Cambridge University Press.
© Cambridge University Press 2009.

turbulence, however, high-resolution simulations are appearing just now and are not included here).

To begin with, I briefly discuss the equations which are numerically solved in astrophysics and related theoretical aspects. The following sections deal with supersonic turbulence, self-gravitating turbulence and magnetohydrodynamic (MHD) turbulence. Naturally, there are intersections. Representative examples for numerical simulations are placed depending on their main objectives. Finally, I give a resume of what has been achieved and which challenges are likely to be met.

2.1 Fundamentals

In the following, we consider self-gravitating inviscid fluid subject to the action of external force fields and, possibly, magnetic fields. The evolution of the mass density ρ, the velocity v and the specific energy e of the fluid is given by the compressible Euler equations:

$$\frac{D}{Dt}\rho = -\rho\nabla v, \tag{2.1}$$

$$\rho\frac{D}{Dt}v = -\nabla P + \rho(f + g), \tag{2.2}$$

$$\rho\frac{D}{Dt}e + \nabla \cdot v P = \Gamma - \Lambda + \rho v \cdot (f + g), \tag{2.3}$$

where the Lagrangian time derivative is defined by

$$\frac{D}{Dt} = \frac{\partial}{\partial t} + v \cdot \nabla. \tag{2.4}$$

The total energy per unit mass is given by

$$e = \frac{1}{2}v^2 + \frac{P}{(\gamma - 1)\rho}, \tag{2.5}$$

where γ is the adiabatic exponent and the pressure P is related to the mass density and the temperature T via the ideal gas law:

$$P = \frac{\rho k_B T}{\mu m_H}. \tag{2.6}$$

The constants k_B, μ and m_H denote, respectively, the Boltzmann constant, the mean molecular weight and the mass of the hydrogen atom. The energy budget can be altered by heating (Γ) and cooling (Λ), by non-gravitational forces (ρf), as well as by gravity (ρg). Generally, heating and cooling processes are important for the dynamics of the ISM (Sections 2.3 and 2.4).

Non-gravitational forces supplying energy to the flow can be mechanical or magnetic. An example of a mechanical force would be the random driving force that is commonly used in turbulence simulations. This is an external force, i.e. it is independent of the dynamics of the fluid (Section 2.2). Quite the opposite holds for conducting fluids in the presence of magnetic fields. In the case of ideal magnetohydrodynamics, the fluid is dragged by the Lorentz force $\rho f = J \times B$, where the current $J = \nabla \times B$. The magnetic field B, in turn, depends on the flow via Faraday's law,

$$\frac{\partial B}{\partial t} = \nabla \times (v \times B). \tag{2.7}$$

The interaction between turbulent flow and magnetic fields alters properties of turbulence such as the fragmentation behaviour (Section 2.4).

The gravitational acceleration $g = -\nabla \cdot \phi$ arises, in part, from the self-gravity of the fluid and, depending on the boundary conditions, the gravitational field of external sources. In the case of periodic boundary conditions, gravity is solely produced by density fluctuations with respect to the global mean (Section 2.3). In this case, the gravitational potential is determined by the Poisson equation

$$\nabla^2 \phi = 4\pi G(\rho - \langle \rho \rangle), \tag{2.8}$$

where G is Newton's constant. The mean mass density $\langle \rho \rangle = \rho_0$ is a constant because of mass conservation.

Letting the curl operator $\nabla \times$ act upon Eq. (2.2), an evolutionary equation for the vorticity of the flow, $\omega = \nabla \times v$, is obtained:

$$\frac{D}{Dt}\omega = S \cdot \omega - d\omega + \frac{1}{\rho^2}\nabla\rho \times \nabla P + \nabla \times f. \tag{2.9}$$

The rate-of-strain tensor S is the symmetrized velocity gradient and the divergence $d = \nabla \cdot v$. Apart from the stirring of fluid due to rotational force components, vorticity can be generated by the baroclinic term $\nabla\rho \times \nabla P$, if the gradient of the mass density and the pressure gradient are not aligned. With the ideal gas Eq. (2.6), it follows that baroclinic vorticity generation occurs in non-isothermal gas. The term $S \cdot \omega - d\omega$ accounts for the stretching of vortices caused by strain, in addition to the expansion or contraction of vortices due to the compressibility of the fluid.

The time evolution of the divergence d is given by

$$\frac{D}{Dt}d = \frac{1}{2}\left(\omega^2 - |S|^2\right) - \frac{1}{\rho}\nabla^2 P$$
$$+ \frac{1}{\rho^2}\nabla\rho \cdot \nabla P - 4\pi G(\rho - \rho_0) + \nabla \cdot f. \tag{2.10}$$

This equation also follows from the conservation law (2.2) by applying the operator ∇ and substituting the Poisson equation for the gravitational potential (2.8). The norm of the rate of strain is defined by $|S| = (2S_{ij}S_{ij})^{1/2}$. From the first term, one can see that vorticity contributes to positive divergence, while strain is associated with negative divergence, i.e. converging flow. In the incompressible case, the first term on the right-hand side of Eq. (2.10) is cancelled by the Laplacian of the pressure divided by the density. Whereas vorticity cannot be generated by a gradient field such as gravity, an increasing rate of convergence will be caused by gravity, if the term $4\pi G(\rho - \rho_0)$ dominates. In this case, gravitational collapse of the gas ensues.

The numerical simulation of astrophysical turbulence has to account for all fluid dynamical processes which become manifest in the evolutionary equations of the vorticity (2.9) and the divergence (2.10). In many terrestrial applications, on the other hand, incompressible turbulence without gravity and magnetic fields is considered. Then Eqs (2.9) and (2.10) reduce to

$$\frac{D}{Dt}\omega = \mathbf{S} \cdot \omega + \nabla \times f,\tag{2.11}$$

$$\frac{D}{Dt}d = 0.\tag{2.12}$$

The only direct mechanism generating vorticity in this case is to stir the fluid by *solenoidal* (rotational) forces. Non-linear turbulent interactions stretch and fold the large-scale eddies produced by stirring into thin vortex filaments. This is the very essence of turbulent fluid dynamics. Here, the question arises whether the generation of turbulence in the ISM resembles the scheme of stirring, stretching and folding at all. Interestingly, this presumption underlies the application of solenoidal driving forces in most numerical simulations.

Incompressible fluid dynamics serves as an important reference case for which scaling properties of isotropic turbulence in the ensemble average can be derived analytically. Kolmogorov (1941) showed that the root mean square (RMS) velocity fluctuations $v'(\ell)$ at a certain length scale ℓ obey the so-called 2/3-law,

$$v'(\ell)^2 \propto \ell^{2/3},\tag{2.13}$$

provided that ℓ is small compared to the integral length scales L at which energy is supplied to the flow. The 2/3-law implies that the rate of kinetic energy dissipation ϵ is asymptotically constant, as $\epsilon \sim v'(\ell)^3/\ell \simeq$ const. in the limit $\ell/L \ll 1$. This result is known as the law of positive energy dissipation (Frisch 1995). Both the 2/3-law and the law of positive energy dissipation appear to be robust properties of turbulence as long as the flow becomes isotropic and nearly incompressible towards small length scales. This is not always the case. Magnetic fields, for instance, introduce small-scale anisotropy. Moreover, there is no consensus yet as to what extent the scaling properties of incompressible turbulence carry over to supersonic flow.

2.2 Supersonic turbulence

Turbulence is called supersonic if the RMS Mach number, $\mathcal{M}_{RMS} = \langle (v/c_s)^2 \rangle^{1/2}$, exceeds unity. Kinetic energy dissipation in supersonic turbulence caused by shock fronts propagating through the fluid dominates viscous dissipation. Since the resulting dissipative heating raises the speed of sound to a value comparable to the typical flow velocity within a dynamical time, supersonic turbulence cannot be maintained if the excess heat is not efficiently removed from the fluid. As an idealization, heating is exactly balanced by cooling such that the temperature remains constant. In this case, continuously driven turbulent flow settles into a steady state in which the RMS Mach number is asymptotically constant. In isothermal gas, solenoidal forcing is used to produce turbulence because the density gradient is aligned with the pressure gradient and, consequently, $\nabla \rho \times \nabla P = 0$. In this case, it can be seen from Eq. (2.9) that non-zero vorticity either results from discontinuities (shock fronts) or it is generated by the curl of the force field. As a consequence of the constant speed of sound, the properties of driven supersonic turbulence in isothermal gas depend on a single parameter only, namely, the RMS Mach number (provided that the forcing is purely solenoidal (Schmidt & Federrath 2008).

A lot of effort has gone into the numerical simulation of supersonic flow by means of finite-volume schemes of higher order, in particular, the piecewise parabolic method (PPM) (Colella & Woodward 1984). The applicability of PPM to turbulence was tested in simulations of adiabatic turbulence (Sytine *et al.* 2000) and nearly isothermal transonic turbulence. An example for the latter is the driven turbulence simulation by Porter *et al.* (2002). They utilized an explicit cooling function and, thereby, maintained an RMS Mach number ≈ 1.0. In Schmidt *et al.* (2006), the equation of state of degenerate electron gas was applied in simulations of forced transonic turbulence with PPM. Degeneracy entails an extremely large heat capacity. Since the isothermal case corresponds to the limit of infinite heat capacity, the gas in these simulations is effectively isothermal. One of the conclusions was that the numerical dissipation of PPM acts as an implicit subgrid scale model in the weakly compressible regime (at small scales).

A new approach is the application of adaptive mesh refinement (AMR) in turbulence simulations (Kritsuk *et al.* 2006; Schmidt *et al.* 2008). AMR is based on the finite-volume approach with a hierarchy of grid patches of different resolution. The basic idea is to represent relatively smooth flow regions on coarse grids only, whereas steep gradients of the velocity, density or any other field are computed with higher accuracy by dynamically inserting grids of higher resolution (Berger & Oliger 1984; Berger & Colella 1989). In astrophysics, AMR has been successfully used for the treatment of strong shocks or gravitational collapse in a variety of problems (O'Shea *et al.* 2004). On the contrary, the simulation of developed turbulence with AMR has been regarded as infeasible, because, in the picture of the turbulent cascade of eddies, small-scale features of the flow would be space-filling (Frisch 1995). For a particular flow realization, however, turbulent eddies occupy an ever decreasing fraction of the flow domain towards smaller length scales at any given instant of time. This follows from the intermittency of turbulence (Frisch 1995).

Picking up the intermittent picture of turbulence, Kritsuk *et al.* proposed that AMR would offer a computational advantage compared to static grids of fixed resolution (Kritsuk *et al.* 2006). This can be motivated from the Landau estimate of the number of degrees of freedom Landau and Lifshitz (1987),

$$N \sim Re^{9/4}, \tag{2.14}$$

where Re is the Reynolds number of the flow. Present-day supercomputers manage $N \sim 10^{10}$, which allows for $Re \sim 10^4$. Note that fully developed turbulence is known to set in at Reynolds numbers of a few thousand. On the contrary, invoking intermittency models of turbulence such as the β-model (Frisch 1995), we have

$$N \sim Re^{3D/(D+1)}, \tag{2.15}$$

where D is interpreted as the fractal dimension of dissipative structures. Assuming a value of $D \approx 2$ as in the Burger model of supersonic turbulence, it follows that $N \sim 10^8$. This figure is smaller than the Landau estimate (corresponding to a fixed-resolution grid) by two orders of magnitude. Of course, there is substantial computational overhead with AMR in comparison to static grids. Apart from that, more sophisticated intermittency models and numerical studies suggest that $D > 2$ in supersonic turbulence (Boldyrev 2002; Boldyrev *et al.* 2002). Nevertheless, AMR is expected to pay off if several levels of refinement are used. In particular, this applies to scenarios including gravitational collapse (see Section 2.3).

In Kritsuk *et al.* (2007), the effective resolution (i.e. the resolution corresponding to the most refined grids) of an AMR simulation of supersonic turbulence was raised to the yet unprecedented value of 2048^3. Nearly isothermal gas was maintained by setting the adiabatic exponent γ to a value that differs only by a small fraction from unity. This means that the internal energy $E_{int} = P/(\gamma - 1)\rho$ is artificially increased by a huge factor such that $E_{int} \gg \frac{1}{2}\rho v^2$, and, hence, it takes many dynamic timescales to heat the gas significantly by kinetic energy dissipation. In combination with a 1024^3 static grid simulation, a great wealth of results was obtained from the analysis of the numerical data. Figure 2.1 shows the projected mass density in logarithmic scaling for one snapshot of the AMR simulation. One can see voids in between high-density regions which display intricate turbulent structure. This is a tell-tale sign of the pronounced intermittency of supersonic turbulence.

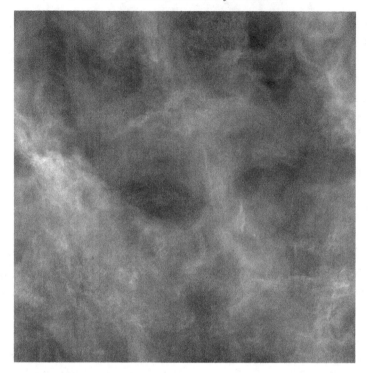

Fig. 2.1. Projected mass density in an AMR simulation of supersonic turbulence with effective resolution 2048^3 by Kritsuk *et al.* (2007), using Enzo. Bright regions contain gas of high density, while under-dense gas appears dark.

Further indications of intermittent properties come from the probability density function of the time-averaged mass density, which follows very closely a log-normal distribution (Padoan & Nordlund 2007), and the scaling exponents of the velocity structure functions. The definition of structure functions is based on spatial correlations of the velocity field:

$$S_p(\ell) = \langle |v(x + \ell) - v(x)| \rangle^p. \tag{2.16}$$

The square brackets denote the average over the whole domain of the flow. Structure functions probe the correlations of the velocity field at different spatial positions depending on the separation ℓ. It is both a theoretical prediction and an experimentally well-established fact that the structure functions of isotropic turbulence obey power laws

$$S_p(\ell) \propto \ell^{\zeta_p}. \tag{2.17}$$

The exponents ζ_p are called the scaling exponents. For incompressible turbulence, Kolmogorov found that $\zeta_p = p/3$. In the case of the second-order structure functions ($p = 2$), this result corresponds to the 2/3-law for the velocity fluctuations mentioned in Section 2.1. Calculating the scaling exponents from their numerical data, Kritsuk *et al.* were able to demonstrate significant deviations form the Kolmogorov relation in accordance with predictions of the intermittency model proposed by Boldyrev *et al.* (2002), which is a generalization of the She–Lévêque model for incompressible turbulence (She & Lévêque 1994). In particular, it was found that $\zeta_2 \approx 0.76$, whereas $\zeta_2 = 2/3$ in the Kolmogorov theory. Intriguingly, the analysis also revealed that the scaling exponents applicable to incompressible turbulence are obtained if the statistics is computed for the density-weighted variable $\rho^{1/3}\boldsymbol{v}$ in place of $\boldsymbol{v}$, a result that is not fully understood yet, but might bear important implications on the nature of supersonic turbulence.

2.3 Self-gravitating turbulence

According to the linear stability analysis of the Euler equations by Jeans (1902), a perturbation in isothermal gas of temperature T and uniform density ρ_0 becomes unstable against gravitational collapse, if its size exceeds the Jeans length

$$\lambda_J = c_s \sqrt{\frac{\pi}{G\rho_0}}, \tag{2.18}$$

where $c_s \propto T^{1/2}$ is the isothermal speed of sound. Beginning with Chandrasekhar's proposition to substitute c_s^2 by

$$c_{\text{eff}}^2 = c_s^2 + \frac{1}{3}\langle v^2 \rangle \tag{2.19}$$

in the above expression for the Jeans length, several attempts have been made to extend the Jeans stability analysis to turbulent gas of RMS velocity $\langle v^2 \rangle^{1/2}$ (Chandrasekhar 1951; Bonazzola *et al.* 1987, 1992). Equation (2.19) implies that the effective gas pressure is given by the sum of the thermal pressure P and the turbulence pressure $P_{\text{turb}} = (2/3)E_{\text{kin}}$, where E_{kin} is the mean kinetic energy density.

However, even the most advanced analysis put forward so far (Bonazzola *et al.* 1992) suffers from severe constraints. A perturbation analysis can be carried out for a statistically stationary equilibrium state only. Thus, it has to be assumed that the free-fall timescale $T_{\text{ff}} \sim (G\rho_0)^{-1/2}$ is much greater than the dynamical timescale of turbulent flow. From the divergence Eq. (2.10), one can see that this assumption implies that the self-gravity term remains small compared with the other terms at all times. Allowing for gravitational collapse, however, self-gravity eventually dominates and

$$\frac{D}{Dt}d \sim -G\rho. \tag{2.20}$$

A non-perturbative theory of the regime in which the free-fall timescale and the dynamical timescale of turbulence are comparable was suggested by Biglari and Diamond (1989). Combining scaling relations from the β-model of turbulence (Frisch 1995) with the assumption of energy equipartition between gravity and turbulence at all scales, they derived an intermittent hierarchical cloud model of self-gravitating turbulence.

At present, however, there is no general theory that would fully encompass the highly non-linear and non-stationary interplay between gravity and turbulence in the ISM. Numerical simulations, on the contrary, can aid in the understanding of self-gravitating turbulence, although one has to resort to artificial mechanisms of producing turbulence or imposing more or less arbitrary initial conditions. One technique is to apply a driving force as mentioned in Section 2.2. Alternatively, some random initial velocity field might be assumed as initial condition. This results in decaying turbulence. Yet another option is to consider gas in an unstable state and apply small perturbations. The most prominent example is the thermal instability (Kritsuk & Norman 2002; Elmegreen & Scalo 2004).

Numerical simulations of self-gravitating driven or decaying supersonic turbulence were initially performed with smoothed particle hydrodynamics (SPH) because of its ability to tackle variations of the mass density over many orders of magnitude (Monaghan 1992). SPH makes use of the Lagrangian framework of fluid mechanics and represents fluid parcels by particles. Examples for the application of SPH to self-gravitating turbulence are the simulations by Klessen *et al.* (2000), and Klessen (2000, 2001). In these simulations, the influence of self-gravity on probability density functions (Klessen 2000) and the fragmentation properties of the gas were investigated both for driven and for decaying supersonic turbulence (Klessen *et al.* 2000; Klessen 2001). Sink particles were introduced to capture collapsing regions beyond a certain density threshold in order to prevent the mass density from growing indefinitely. Figure 2.2 illustrates that sink particles are mostly formed within filamentary structures (which are also seen in cosmological simulations). Varying the Mach number, it was found that higher turbulence intensity slows down considerably the rate at which sink particles are produced (Figure 2.3). The major conclusion drawn by Klessen *et al.* was that supersonic turbulence inhibits gravitational collapse globally, as one would suspect from Eq. (2.19). The strong gas compression caused by shocks, on the contrary, can trigger gravitational collapse locally and initiate the formation of stars.

Jappsen *et al.* (2005) studied the mass spectrum produced by turbulent fragmentation in non-isothermal gas. They defined the state variables by piecewise polytropic relations, i.e. $P \propto \rho^{\gamma_{eff}}$ for a certain range of densities. If $\gamma_{eff} < 1$, the gas

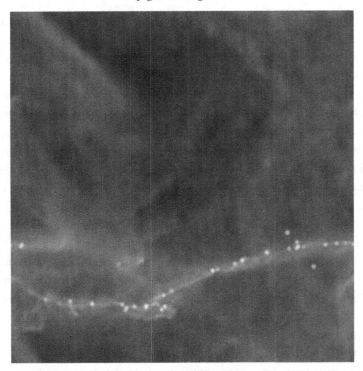

Fig. 2.2. Filamentary structure with sink particles in an SPH simulation of self-gravitating turbulence (courtesy of R. Klessen). In this plot, over-dense regions appear bright.

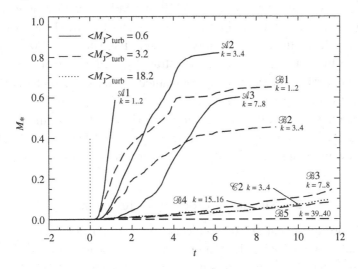

Fig. 2.3. Total mass fraction captured by sink particles as a function of time in SPH simulations with different Mach numbers and varying driving scale of turbulence (Klessen *et al.* 2000).

cools with increasing density. This effect enhances the compressibility of the gas and therefore supports gravitational collapse (negative $\nabla \rho \cdot \nabla T$ contribution adds to the gravity term in the divergence Eq. (2.10)). At some threshold density, however, the cooling regime ceases and γ_{eff} becomes greater than unity. From various simulations, it was found that the transition from the cooling regime to isothermal or nearly isothermal gas sets a characteristic mass scale of turbulent fragmentation. This mass scale was interpreted to correspond to the peak of the observed initial mass function. Vàzquez-Semadeni *et al.* (2007) included explicit heating and cooling in SPH simulations of colliding gas streams. They made use of a model for the cooling function Λ in the energy Eq. (2.3) that was proposed by Koyama and Inutsuka (2002). Due to the thermal instability (compression causes the gas to cool), the gas at the collision interface undergoes gravitational collapse and becomes increasingly turbulent. This is interpreted as possible mechanism of molecular cloud formation. Remarkably, approximate equipartition of gravitational and kinetic energy was found, although clearly no state of virial equilibrium was approached by the collapsing gas. Further applications of SPH, addressing problems such as the formation of brown dwarfs, binary star systems and stellar clusters via turbulent fragmentation, as well as the origin of the initial mass function, were presented by Bate *et al.* (2002a,b), Bate and Bonnell (2005), and Bonnell *et al.* (2003).

Despite the numerous important contributions to the understanding of the role of turbulence in star formation, the adequacy of treating turbulence with SPH has been a matter of debate. On the one hand, numerical studies of self-gravitating systems indicate basic agreement between AMR and SPH (Commercon *et al.* 2007). However, comparisons with grid-base codes suggest that SPH dissipates small-scale velocity fluctuations significantly stronger (Padoan *et al.* 2007; Kitsionas *et al.*, 2008). In particular, it was demonstrated that the steeper power spectra obtained from SPH simulations are accompanied by mass distributions which are not consistent with observations. Apart from that, it appears to be rather difficult to accommodate magnetohydrodynamics in the SPH formalism. Although there are ongoing attempts to overcome this shortcoming (Maron & Howes 2003), MHD has been treated with great success using finite-volume discretization.

2.4 Magnetohydrodynamic turbulence

In purely hydrodynamical (HD) turbulence, there is no preferred spatial direction for small-scale velocity fluctuation. The randomization of the flow due to non-linear turbulent interactions produces statistical isotropy towards small scales which can be interpreted as a symmetry of the ensemble average (Frisch 1995). This symmetry is broken by a magnetic field B in turbulent conducting fluid if the ratio of the thermal to the magnetic pressure,

$$\beta = \frac{P}{P_{\mathrm{M}}} = \frac{8\pi P}{B^2}, \tag{2.21}$$

is comparable to unity or less. If the curl of the Lorentz force, $\nabla \times (\boldsymbol{J} \times \boldsymbol{B})$, in the vorticity Eq. (2.9) is sufficiently strong, then the vorticity will mainly grow in the direction of $\boldsymbol{B}$. In developed MHD turbulence, this effect causes eddies to shrink perpendicular to the field lines. As a further implication, the anisotropy of MHD turbulence increases towards smaller length scales (Biskamp 2003).

The numerical simulation of MHD turbulence is rather challenging. Godunov-based schemes of higher order such as PPM become very complex upon including MHD. For this reason, simpler schemes have been adopted for MHD. Examples are the Zeus code (Stone & Norman 1992), the Stagger code (Nordlund & Galsgaard 1995) and the Ramses code which also features AMR (Teyssier 2002; Fromang *et al.* 2006). As an illustration, Figure 2.4 shows a simulation of the gravitational collapse of a magnetized cloud with up to six levels of refinement. One of the major problems that needs to be addressed when solving the MHD equations numerically is to keep the magnetic field divergence-free, i.e. $\nabla \cdot \boldsymbol{B} = 0$. In Ramses

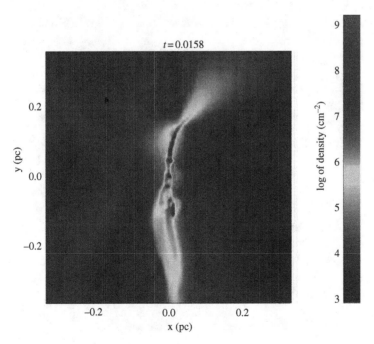

Fig. 2.4. Simulation of a collapsing gas cloud of 1000 solar masses with Ramses (courtesy of P. Hennebelle). The magnetic energy of the clouds is about the same as its kinetic energy.

and other codes, the constrained transport algorithm is employed to ensure this constraint (Evans & Hawley 1988).

A comparison of the properties of driven HD and MHD turbulence in high-resolution simulations was presented by Padoan *et al.* (2007). Whereas the energy spectrum functions, irrespective of magnetic fields, were found to follow closely the power law k^{-2} in the inertial subrange, the fragmentation properties of the gas appear to be markedly different for HD and MHD turbulence. In plots of the projected density fields for both cases (Figures 2.5 and 2.6), one can see that the gas is more concentrated in pronounced filaments under the action of magnetic fields. Utilizing a clump-find algorithm, it was demonstrated that a significantly steeper mass spectrum of dense cores (defined by a density threshold based on the Bonnor–Ebert mass) resulted from purely HD turbulence. Furthermore, they were able to reproduce the Chabrier initial mass function (Chabrier 2003) in the MHD case, although this result must be considered with care given the ambiguity of calculating mass spectra.

Fig. 2.5. Projected mass density in a simulation of hydrodynamic turbulence with the Stagger code (Padoan *et al.* 2007). For comparison with MHD turbulence, see Figure 2.6.

Fig. 2.6. Projected mass density in a simulation of MHD turbulence with the Stagger code (Padoan *et al.* 2007). For comparison with purely HD turbulence, see Figure 2.5.

Glover and Mac Low (2007) performed simulations of self-gravitating MHD turbulence including various thermal and chemical processes. They were able to show that the fast formation of molecular hydrogen within a few Myr, which is typically observed in molecular clouds, can be explained as an effect of turbulence. Hennebelle and Audit (2007) investigated thermally bistable MHD turbulence and found mass spectra similar to those inferred from CO observations of molecular clouds. Although first computed in only two dimensions, the setup was generalized to include self-gravity and computed in three dimensions as well. The ansatz by Hennebelle *et al.* is different from simulations of isothermal MHD turbulence as it does not rely on a driving force or an initial turbulent velocity field.

2.5 Perspectives

Up to now, most approaches to the numerical simulation of astrophysical turbulence have focused on particular aspects, be it a thorough understanding of supersonic turbulence, the study of the dynamics of self-gravitating turbulent gas or an emphasis on the effects of magnetic fields. While impressive progress has been made in the area of isotropic turbulence, both in the HD and in the MHD case, more complex scenarios remain elusive. The treatment of thermal and chemical processes is still fairly approximate. Our understanding of self-gravitating turbulence remains poor despite many ongoing efforts. Radiation HD and MHD in combination with adaptive methods are just beginning.

There is one aspect that should be highlighted whenever we are talking about the simulation of turbulence in the ISM. The huge range of length scales from the galactic disk down to proto-stellar cores does not allow for coverage by a single numerical simulation. In general, there are two numerical cutoffs: one towards larger scales and the other towards smaller scales. The former poses the question of initial as well as boundary conditions and whether simple models such as periodic boxes and random forcing can account for those in a self-consistent fashion. The small-scale cutoff would, in general, necessitate closures. These are considered as essential in atmospheric sciences but have met very little attention in astrophysics.

As regards the large scales, it certainly will not be feasible in the near future to run a simulation of a disk galaxy or even a piece of a galactic disks including all processes of turbulence production, the different phases of the ISM, the formation of molecular clouds and the collapse of gas giving rise to the birth of stars. If energy is injected via random forcing in a simulation of turbulence over some subrange of scales in the ISM, one faces the question of what the influence of the forcing might be. That the outcome can vary depending on the applied forcing is illustrated by a simple numerical experiment. Figure 2.7 shows projected mass densities obtained from 128^3 simulations, in which purely solenoidal (left) and compressive (right) forcing was applied. Although there is no fully developed turbulence at such low resolution, the plots suggest that the flow morphology is genuinely different, with markedly higher density contrasts and more pronounced intermittency in the case

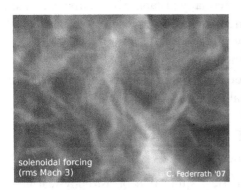
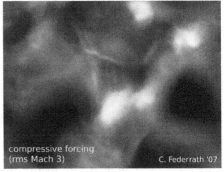

Fig. 2.7. Projected mass density in 128^3 simulations of turbulence driven by a solenoidal and a compressive force field, respectively. (Courtesy of C. Federrath.)

of compressive forcing. Indeed, this conclusion is confirmed by high-resolution simulations (Schmidt *et al.* 2008).

For the high-fidelity treatment of small scale effects, AMR emerges as the most promising tool, simply following the maxim to go to very high levels of refinement wherever the critical events take place. This was most impressively executed by Abel *et al.* (2002) in AMR simulations of the formation of the first stars in the Universe. However, circumstances are not that favourable when it comes to galactic star formation and it remains to be seen whether AMR by itself will hold its promises. An alternative approach was suggested by Niemeyer *et al.* (2005). The basic idea is to combine the techniques of AMR, which has been brought to great success in astrophysics, and subgrid scale modelling, which is commonly used in engineering. Furthermore, the incorporation of additional physics such as thermal processes, complex chemistry as well as radiation transport in AMR simulations are essential for realistic models of turbulence in the ISM.

References

Abel, T., Bryan, G. L. and Norman, M. L., 2002, *Science* **295**, 93

Ballesteros-Paredes, J., Gazol, A., Kim, J., Klessen, R. S., Jappsen, A.-K. and Tejero, E., 2006, *Astrophys. J.* **637**(1), 384

Bate, M. R. and Bonnell, I. A., 2005, *Mon. Not. R. Astron. Soc.* **356**(4), 1201

Bate, M. R., Bonnell, I. A. and Bromm, V., 2002a, *Mon. Not. R. Astron. Soc.* **332**, L65

Bate, M. R., Bonnell, I. A. and Bromm, V., 2002b, *Mon. Not. R. Astron. Soc.* **336**, 705

Berger, M. J. and Colella, P., 1989, *J. Comput. Phys.* **82**, 64

Berger, M. J. and Oliger, J., 1984, *J. Comput. Phys.* **53**, 484

Biglari, H. and Diamond, P. H., 1989, *Physica D.* **37**, 206

Biskamp, D., 2003, *Magnetohydrodynamic Turbulence*, Cambridge: Cambridge University Press

Boldyrev, S., 2002, *Astrophys. J.* **569**, 841

Boldyrev, S., Nordlund, A. and Padoan, P., 2002, *Astrophys. J.* **573**, 678

Bonazzola, S., Falgarone, E., Heyvaerts, J., Perault, M. and Puget, J. L., 1987, *Astron. Astrophys.* **172**, 293

Bonazzola, S., Perault, M., Puget, J. L., Heyvaerts, J., Falgarone, E. and Panis, J. F., 1992, *J. Fluid Mech.* **245**, 1

Bonnell, I. A., Bate, M. R. and Vine, S. G., 2003, *Mon. Not. R. Astron. Soc.* **343**(2), 413

Chabrier, G., 2003, *PASP* **115**, 763

Chandrasekhar, S., 1951, *Proc. R. Soc. Lond. A* **210**, 26

Colella, P. and Woodward, P. R., 1984, *J. Comput. Phys.* **54**, 174

Commercon, B., Hennebelle, P., Audit, E., Chabrier, G. and Teyssier, R., 2007, to appear in *Proceedings of SF2A-2007: Semaine de l'Astrophysique Francaise* (J. Bouvier, A. Chalabaev and C. Charbonnel, eds.) eprint arXiv:0709.2450

Elmegreen, B. G. and Scalo, J., 2004, *Ann. Rev. Astron. Astrophys.* **42**(1), 211

Euler, L., 1757, *Principes généraux du mouvement des fluides*. Berlin: Königliche Akademie der Wissenschaften

Evans, C. and Hawley, J., 1988, *Astrophys. J.* **332**, 659

Frisch, U., 1995, *Turbulence*, Cambridge: Cambridge University Press

Fromang, S., Hennebelle, P. and Teyssier, R., 2006, *Astron. Astrophys.* **457**, 371

Glover, S. C. O. and Mac Low, M.-M., 2007, *Astrophys. J.* **659**, 1317

Heisenberg, W., 1923, *Über Stabilität und Turbulenz von Flüssigkeitsströmen*, Doctoral Dissertation, Ludwigs-Maximillian-Universität München

Hennebelle, P. and Audit, E., 2007, *Astron. Astrophys.* **65**(2), 431

Jappsen, A. K., Klessen, R. S., Larson, R. B., Li, Y. and Mac Low, M.-M., 2005, *Astron. Astrophys.* **435**(2) 611

Jeans, J. H., 1902, *Philos. Trans. R. Soc. Lond. A* **199**, 1

Klessen, R. S., 2000, *Astrophys. J.* **535**(2), 869

Klessen, R. S., 2001, *Astrophys. J.* **556**(2), 837

Klessen, R. S., Heitsch, F. and Mac Low, M.-M., 2000, *Astrophys. J.* **535**(2), 887

Kitsionas *et al.*, 2008, submitted to *Astron. Astrophys.*

Kolmogorov, A., 1941, *Dokl. Akad. Nauk SSSR* **30**, 301

Koyama, H. and Inutsuka, S.-I., 2002, *Astrophys. J.* **564**(2), L97

Kritsuk, A. G. and Norman, M. L., 2002, *Astrophys. J. Lett.* **569**, L127

Kritsuk, A. G., Norman, M. L. and Padoan, P., 2006, *Astrophys. J. Lett.* **638**, L25

Kritsuk, A. G., Norman, M. L., Padoan, P. and Wagner, R., 2007, *Astrophys. J.* **665**, 416

Landau, L. D. and Lifshitz, E. M., 1987, *Course of Theoretical Physics Vol. 6: Fluid Mechanics*, 2nd edition, Butterworth-Heinemann: Pergamon Press

Mac Low, M.-M. and Klessen, R. S., 2004, *Rev. Mod. Phys.* **76**(1), 125

Maron, J. L. and Howes, G. G., 2003, *Astrophys. J.* **595**(1), 564

Monaghan, J. J., 1992, in *Ann. Rev. Astron. Astrophys.* **30**, 543

Niemeyer, J., Schmidt, W. and Klingenberg, C., 2005, in *Ringberg Proceedings on Interdisciplinary Aspects of Turbulence*, MPA/P15, 175

Nordlund, Å. and Galsgaard, K., 1995, *A 3D MHD Code for Parallel Computers*. Technical report, Astronomical Observatory, Copenhagen University

O'Shea, B. W., Bryan, G., Bordner, J., Norman, M. L., Abel, T., Harkness, R. and Kritsuk, A., 2004, *Adaptive Mesh Refinement – Theory and Applications* (T. Plewa, T. Linde, and V. G. Weirs, eds.) Springer Lecture Notes in Computational Science and Engineering

Padoan, P. and Nordlund, Å., 2007, *Astrophys. J.* **576**, 870

Padoan, P., Nordlund, Å., Kritsuk, A. G., Norman, M. L. and Li, P. S., 2007, in *Triggered Star Formation in a Turbulent ISM* (B. G. Elmegreen and J. Palous, eds.) Cambridge: Cambridge University Press, 283

Porter, D., Pouquet, A. and Woodward, P., 2002, *Phys. Rev. E.* **66**, 026301

She, Z.-S. and Lévêque, E., 1994, *Phys. Rev. E.* **72**, 336

Schmidt, W., and Klessen, R.S., Federrath, C., 2008, submitted to *Phys. Rev. Lett.*, arXiv:0810.1397

Schmidt, W., Federrath, C., Hupp, M., Maier, A. and Niemeyer, J., 2008, submitted to *Astron. Astrophys.*, preprint arXiv:0809.1321

Schmidt, W., Hillebrandt, W. and Niemeyer, J. C., 2006, *Comp. Fluids* **35**, 353

Stone, J. M. and Norman, M. L., 1992, *Astrophys. J. Suppl.* **80**, 791

Sytine, I. V., Porter, D. H., Woodward, P. R., Hodson, S. W. and Winkler, K., 2000, *J. Comput. Phys.* **158**, 225

Teyssier, R., 2002, *Astron. Astrophys.* **385**, 337

Vàzquez-Semadeni, E., Gómez, G. C., Jappsen, A. K., Ballesteros-Paredes, J., González, R. F. and Klessen, R. S., 2007, *Astrophys. J.* **657**(2), 870

3

Numerical methods for radiation magnetohydrodynamics in astrophysics

R. Klein and J. Stone

Abstract

We describe numerical methods for solving the equations of radiation magnetohydrodynamics (MHD) for astrophysical fluid flow. Such methods are essential for the investigation of the time-dependent and multidimensional dynamics of a variety of astrophysical systems, although our particular interest is motivated by problems in star formation. Over the past few years, the authors have been members of two parallel code development efforts, and this review reflects that organization. In particular, we discuss numerical methods for MHD as implemented in the Athena code, and numerical methods for radiation hydrodynamics as implemented in the Orion code. We discuss the challenges introduced by the use of adaptive mesh refinement (AMR) in both codes, as well as the most promising directions for future developments.

3.1 Introduction

The dynamics of astrophysical systems described by the equations of radiation magnetohydrodynamics (MHD) span a tremendous range of scales and parameter regimes, from the interiors of stars (Kippenhahn & Weigert 1994), to accretion disks around compact objects (Turner *et al.* 2003), to dusty accretion flows around massive protostars (Krumholz *et al.* 2005, 2007a), to galactic-scale flows onto AGN (Thompson *et al.* 2005). All of these systems have in common that matter, radiation and magnetic fields are strongly interacting and that the energy and momentum carried by the radiation field is significant in comparison to that carried by the gas. Thus, an accurate treatment of the problem must include analysis of both the matter and the radiation, as well as the magnetic fields, and their mutual interaction.

Structure Formation in Astrophysics. ed. G. Chabrier. Published by Cambridge University Press.
© Cambridge University Press 2009.

Numerical methods are essential to generate time-dependent and multidimensional solutions to the non-linear equations of radiation MHD. In fact, numerical methods for both MHD and radiation hydrodynamics are in, and of, themselves active areas of development, let alone for the combined system of radiation MHD. Our goal in this chapter is to discuss the current status of numerical methods for radiation MHD, with emphasis on the challenges and areas where further development is needed. However, in order to clarify the discussion, we will describe methods for MHD and those for radiation hydrodynamics in separate sections, using examples drawn from two separate efforts. In particular, we will use the MHD algorithms implemented in the Athena code (developed by Stone and collaborators) to demonstrate the issues and challenges associated with higher-order Godunov methods for MHD using adaptive mesh refinement (AMR), while we will use the radiation hydrodynamics algorithms implemented in the Orion code (developed by Klein and collaborators) to demonstrate the challenges associated with higher-order Godunov methods for radiation hydrodynamics using AMR. Our focus is on a discussion of the fundamental issues for numerical algorithms in each of these areas, rather than a step-by-step description of the actual codes (for a thorough discussion of the algorithms in Athena, see Gardiner and Stone (2005, 2008) and Stone *et al.* (2008), while for the Orion code, see Truelove *et al.* (1998), Klein (1999), Fisher (2002), Crockett *et al.* (2005) and Krumholz *et al.* (2007b)).

In addition to a discussion of methods for MHD (in 3.2) and radiation hydrodynamics (in 3.3), we shall also discuss the issues associated with the implementation of AMR (in 3.4), as well as directions for future research (in 3.5).

3.2 Magnetohydrodynamic algorithms: the Athena code

3.2.1 Introduction

There are a wide range of algorithms that have been developed to solve the equations of compressible MHD. For example, the ZEUS code (Stone & Norman 1992) implements methods based on operator splitting a non-conservative formulation of the equations of motion. However, operator split methods based on the non-conservative formulation are not suitable for use with AMR, and therefore in recent years there has been a surge in the development of higher-order Godunov methods for MHD.

In this section, we describe directionally unsplit, higher-order Godunov methods for MHD as implemented in the Athena code (Gardiner & Stone 2005, 2008; hereafter GS05 and GS07, respectively). Other codes that implement similar methods include Riemann (Balsara 2001), Nirvana (Ziegler 2005), RAMSES (Fromang *et al.* 2006), PLUTO (Mignone *et al.* 2007), and AstroBEAR (Cunningham *et al.* 2007). The primary differences between the algorithms implemented in these codes are two-fold: (1) the multidimensional integration algorithm and (2) the method by which the divergence-free constraint on the magnetic field is enforced. Some of the different options that have been explored include unconstrained directionally split integrators (Dai & Woodward 1994), or directionally split and unsplit integrators that use either a Hodge projection to enforce the constraint (Zachary *et al.* 1994; Ryu *et al.* 1995; Balsara 1998; Crockett *et al.* 2005), a non-conservative formulation that allows propagation and damping of errors in the constraint (Powell 1994; Falle *et al.* 1998; Powell *et al.* 1999; Dedner *et al.* 2002), or some form of the constrained transport (CT) algorithm of Evans and Hawley (1988) to enforce the constraint (Dai & Woodward 1998; Ryu *et al.* 1998; Balsara & Spicer 1999; Tóth 2002; Pen *et al.* 2003; Londrillo & Del Zanna 2004; Ziegler 2005; Fromang *et al.* 2006; Mignone *et al.* 2007). A systematic comparison between many of these methods is provided by Tóth (2002).

Over the past several years during the development of Athena, we have explored many of the same ideas described in the above papers as the basis of explicit numerical algorithm for ideal MHD. We have focused our effort on three aspects. The first is the use of the CT algorithm to preserve the divergence-free constraint, the second is the use of a directionally unsplit integrator to update the equations of motion, and the third is the use of the piecewise parabolic method (PPM) of Colella and Woodward (1984) for the reconstruction of interface states in multidimensional MHD. Along the way, we have found a number of modifications and extensions that are required to make a robust and accurate MHD algorithm based on these three ingredients. In particular, we have found that (1) the method by which the Godunov fluxes are used to calculate the electric fields needed by CT requires a more sophisticated approach than simple arithmetic averaging, (2) using the corner transport upwind (CTU) method of Colella (1990) as a directionally unsplit integration algorithm requires the addition of 'source terms' for MHD during the interface state correction steps and (3) the extension of the dimensionally split spatial reconstruction scheme in PPM requires similar source terms for multidimensional MHD. It is beyond the scope of this chapter to describe each of these extensions in detail. However, after introducing the equations of motion for MHD and describing their discretization in the following subsections, we will provide an overview of each of the three ingredients in order to motivate the extensions implemented in Athena. Most of this section will focus on the results of a set of test problems that we have found extremely insightful for benchmarking MHD algorithms.

3.2.2 The equations of MHD

Athena solves the equations of ideal MHD in conservative form

$$\frac{\partial \rho}{\partial t} + \nabla \cdot (\rho \mathbf{v}) = 0 \tag{3.1}$$

$$\frac{\partial \rho \mathbf{v}}{\partial t} + \nabla \cdot (\rho \mathbf{v}\mathbf{v} - \mathbf{B}\mathbf{B}) + \nabla P^* = 0 \tag{3.2}$$

$$\frac{\partial E}{\partial t} + \nabla \cdot ((E + P^*)\mathbf{v} - \mathbf{B}(\mathbf{B} \cdot \mathbf{v})) = 0 \tag{3.3}$$

$$\frac{\partial \mathbf{B}}{\partial t} + \nabla \times (\mathbf{v} \times \mathbf{B}) = 0, \tag{3.4}$$

where P^* is the total pressure (gas plus magnetic), E is the total energy density and we have chosen a system of units in which the magnetic permeability $\mu = 1$. The other symbols have their usual meaning. We use an ideal gas equation of state for which $P = (\gamma - 1)\epsilon$, where γ is the ratio of specific heats, so that the internal energy density ϵ is related to the total energy E via

$$E \equiv \epsilon + \rho(\mathbf{v} \cdot \mathbf{v})/2 + (\mathbf{B} \cdot \mathbf{B})/2 . \tag{3.5}$$

Note that no explicit resistivity or viscosity is included in the above equations. We will restrict our discussion in this chapter to algorithms for ideal MHD.

3.2.3 Discretization

The above equations are discretized using a control-volume approach, with volume averages of the density, total energy and momentum stored at cell centres and area averages of the magnetic field stored at cell faces. This results in a staggered grid for the magnetic field. We have argued (GS05) that although a staggered grid introduces additional complexity into the coding of the algorithm, it is the most natural representation of the discrete form of the induction equation. That is because the integral form of the induction equation uses Stoke's law to relate the time-rate of change of area-averaged fields with line-averaged electric fields to conserve magnetic flux. This is in contrast to the first three of the equations of motion, in which Gauss' law is used to relate the time-rate of change of volume-averaged quantities to their area-averaged fluxes.

Integration of the first three equations over a grid cell and over a discrete interval of time δt gives, after application of the divergence theorem,

$$U_{i,j,k}^{n+1} = U_{i,j,k}^n - \frac{\delta t}{\delta x}\left(F_{i+1/2,j,k}^{n+1/2} - F_{i-1/2,j,k}^{n+1/2}\right)$$
$$- \frac{\delta t}{\delta y}\left(G_{i,j+1/2,k}^{n+1/2} - G_{i,j-1/2,k}^{n+1/2}\right)$$
$$- \frac{\delta t}{\delta z}\left(H_{i,j,k+1/2}^{n+1/2} - H_{i,j,k-1/2}^{n+1/2}\right), \tag{3.6}$$

where $U_{i,j,k}^n$ is a vector of volume-averaged variables, while $F_{i-1/2,j,k}^{n+1/2}$, $G_{i,j-1/2,k}^{n+1/2}$ and $H_{i,j,k-1/2}^{n+1/2}$ are their area-averaged fluxes at the x, y and z-interfaces. We have used the notation that integer subscripts denote cell centres and half-integer subscripts denote cell faces. The calculation of the time-averaged fluxes is described in Section 3.2.5. The components of $U_{i,j,k}^n$ are the mass density, each component of the momentum, and the total energy.

In contrast to the finite-volume difference formulae given above, the discrete form of the induction equation that comes from integration of Eq. (3.4) over the three orthogonal faces of the cell located at $(i - 1/2, j, k)$, $(i, j - 1/2, k)$ and $(i, j, k - 1/2)$, respectively, gives

$$B_{x,i-1/2,j,k}^{n+1} = B_{x,i-1/2,j,k}^n - \frac{\delta t}{\delta y}\left(\mathcal{E}_{z,i-1/2,j+1/2,k}^{n+1/2} - \mathcal{E}_{z,i-1/2,j-1/2,k}^{n+1/2}\right)$$
$$+ \frac{\delta t}{\delta z}\left(\mathcal{E}_{y,i-1/2,j,k+1/2}^{n+1/2} - \mathcal{E}_{y,i-1/2,j,k-1/2}^{n+1/2}\right) \tag{3.7}$$

$$B_{y,i,j-1/2,k}^{n+1} = B_{y,i,j-1/2,k}^n + \frac{\delta t}{\delta x}\left(\mathcal{E}_{z,i+1/2,j-1/2,k}^{n+1/2} - \mathcal{E}_{z,i-1/2,j-1/2,k}^{n+1/2}\right)$$
$$- \frac{\delta t}{\delta z}\left(\mathcal{E}_{x,i,j-1/2,k+1/2}^{n+1/2} - \mathcal{E}_{x,i,j-1/2,k-1/2}^{n+1/2}\right) \tag{3.8}$$

$$B_{z,i,j,k-1/2}^{n+1} = B_{z,i,j,k-1/2}^n - \frac{\delta t}{\delta x}\left(\mathcal{E}_{y,i+1/2,j,k-1/2}^{n+1/2} - \mathcal{E}_{y,i-1/2,j,k-1/2}^{n+1/2}\right)$$
$$+ \frac{\delta t}{\delta y}\left(\mathcal{E}_{x,i,j+1/2,k-1/2}^{n+1/2} - \mathcal{E}_{x,i,j-1/2,k-1/2}^{n+1/2}\right), \tag{3.9}$$

where $B_{x,i-1/2,j,k}^{n+1}$, etc. are the time- and area-averaged components of the magnetic field located at the cell faces. The time- and line-averaged electric field $\mathcal{E}_{z,i-1/2,j+1/2,k}^{n+1/2}$, etc. is located at cell edges and must be computed from the fluxes returned by the Riemann solver described in Section 3.2.5. It is easy to show that the above discretization of the induction equation preserves the divergence-free constraint on the magnetic field exactly, on a cell-by-cell basis. This is the advantage of using the staggered grid formulation of CT.

In Athena, the primary description of the magnetic field is taken to be the face-centred area averages. However, cell-centred values for the field are needed to construct the fluxes of momentum and energy. Here, we adopt the second-order accurate averages

$$B_{x,i,j,k} = \frac{1}{2}(B_{x,i+1/2,j,k} + B_{x,i-1/2,j,k}), \tag{3.10}$$

$$B_{y,i,j,k} = \frac{1}{2}(B_{y,i,j+1/2,k} + B_{y,i,j-1/2,k}), \tag{3.11}$$

$$B_{z,i,j,k} = \frac{1}{2}(B_{z,i,j,k+1/2} + B_{z,i,j,k-1/2}). \tag{3.12}$$

3.2.4 Spatial reconstruction

In a Godunov method, time-averaged fluxes of the conserved quantities at each interface are computed using a Riemann solver. This requires a spatial reconstruction step to interpolate the conserved quantities to cell faces. In Athena, we adopt the PPM reconstruction algorithm. We first transform from the primitive to conserved variables, and we perform the reconstruction using the primitive variables. Slope limiters are applied to keep the reconstruction monotonic, and these are based on the characteristic variables. The PPM algorithm also includes a time advance based on a characteristic decomposition of the primitive variables. All of these steps are too complex to describe in detail here; instead we refer the reader to Stone *et al.* (2008) and the references therein.

In GS05 and GS07, we have shown that for MHD the PPM interface reconstruction algorithm requires an important extension. The time advance of the primitive variables includes terms proportional to the transverse gradients of the magnetic field. That is, in the calculation of the interface states of B_x, it is necessary to include terms proportional to $\partial B_y/\partial y$ and $\partial B_z/\partial z$, and similarly for the calculation of the interface states of B_y and B_z. These terms arise because the divergence-free constraint relates $\partial B_x/\partial x$ with these transverse gradients. Of course, in one dimension, $\partial B_x/\partial x = 0$, and these terms are not needed. However, for multidimensional MHD, they are crucial for reconstructing complex field geometries.

3.2.5 Riemann solvers

To compute the fluxes of conserved quantities from the left and right states constructed above, a variety of Riemann solvers can be used. We have implemented Roe's linearized solver (Cargo & Gallice 1997) and various forms of the HLL (Harten–Lax–Van Leer) solver (Toro 1999), as well as exáct solvers for hydrodynamics. Roe's solver is accurate for all wave families, but often fails for strong rarefactions. Versions of the HLL solvers that do not include the contact wave are too diffusive for practical applications. We find the extension of HLL that includes the contact wave (called HLLC in Toro (1999)) for hydrodynamics, and HLLD in Miyoshi and Kusano (2005) for MHD is the most efficient, robust and accurate.

3.2.6 Constrained transport

The CT update of the magnetic field requires the line-averaged emfs at cell corners, which must be computed from the area-averaged electric fields returned by the Riemann solver. In GS05, it was shown that the relationship between the two is determined by the averaging formulae used to convert between the face-centred area-averages of the magnetic field and the cell-centred volume averages. A variety of different relationships (all consistent to second order) can be used; however, we have found the most useful to be (suppressing the index k on all the formulae below)

$$
\begin{aligned}
\mathcal{E}_{z,i-1/2,j-1/2} = \frac{1}{4} & \left(\mathcal{E}_{z,i-1/2,j} + \mathcal{E}_{z,i-1/2,j+1} + \mathcal{E}_{z,i,j-1/2} + \mathcal{E}_{z,i+1,j-1/2} \right) \\
& + \frac{\delta y}{8} \left(\left(\frac{\partial \mathcal{E}_z}{\partial y} \right)_{i-1/2,j-1/4} - \left(\frac{\partial \mathcal{E}_z}{\partial y} \right)_{i-1/2,j-3/4} \right) \\
& + \frac{\delta x}{8} \left(\left(\frac{\partial \mathcal{E}_z}{\partial x} \right)_{i-1/4,j-1/2} - \left(\frac{\partial \mathcal{E}_z}{\partial x} \right)_{i-3/4,j-1/2} \right),
\end{aligned}
\tag{3.13}
$$

where the derivative of $\mathcal{E}_z$ on each grid cell face is computed by selecting the 'upwind' direction according to the contact mode, e.g.

$$
\left(\frac{\partial \mathcal{E}_z}{\partial y} \right)_{i-1/2,j-1/4} =
\begin{cases}
\left(\frac{\partial \mathcal{E}_z}{\partial y} \right)_{i-1,j-1/4} & \text{for } v_{x,i-1/2} > 0 \\
\left(\frac{\partial \mathcal{E}_z}{\partial y} \right)_{i,j-1/4} & \text{for } v_{x,i-1/2} < 0 \\
\frac{1}{2} \left(\left(\frac{\partial \mathcal{E}_z}{\partial y} \right)_{i-1,j-1/4} + \left(\frac{\partial \mathcal{E}_z}{\partial y} \right)_{i,j-1/4} \right) & \text{otherwise}
\end{cases}
\tag{3.14}
$$

with an analogous expression for the $(\partial \mathcal{E}_z / \partial x)$. The derivatives of the electric field are computed using the face-centred electric fields (Godunov fluxes) and a cell centre 'reference' value $\mathcal{E}^r_{z,i,j}$, e.g.

$$
\left(\frac{\partial \mathcal{E}_z}{\partial y} \right)_{i,j-1/4} = 2 \left(\frac{\mathcal{E}^r_{z,i,j} - \mathcal{E}_{z,i,j-1/2}}{\delta y} \right).
\tag{3.15}
$$

Note that for the 3D CTU + CT algorithm, analogous expressions to the above are required to convert the x- and y-components of the electric field to the appropriate cell corners. GS05 discusses the derivation of the above formulae in detail and discusses a test problem based on passive advection of a field loop that demonstrates the importance of the method and potential failings of simple arithmetic averaging (this test is also introduced below).

3.2.7 Directionally unsplit integrators

In GS05, we showed that directionally split integration methods, which are extremely powerful for hydrodynamics, cannot be used in MHD without violating the divergence-free constraint. The reason for this is simple: it is impossible to enforce the divergence-free constraint between partial updates unless all three components of the magnetic field are updated together. For this reason, we adopt the unsplit CTU method of Colella (1990) to integrate the cell-centred quantities.

Even so, substantial modifications to CTU are required to make it work with MHD using CT. In particular, the transverse flux gradients used to correct the left and right states within multidimensions must be modified with 'MHD source terms', i.e. terms proportional to $\partial B_x / \partial x$, $\partial B_y / \partial y$ and $\partial B_z / \partial z$. The derivation of the exact form of the necessary terms is given in GS05 and GS07.

3.2.8 Tests

One of the most important aspects of developing new numerical methods is the adoption of a test suite that can reveal differences and shortcomings of methods. We have found that it is important to focus on multidimensional tests, since for MHD the numerical algorithms in multidimensions is much more complex than in one dimension. Even plane-parallel 1D solutions are therefore computed in multidimensions, inclined at an oblique angle in order to break any symmetries. In what follows, we present results from a series of five test problems we have found particularly useful for testing MHD codes.

3.2.8.1 Linear wave convergence

We initialize linear amplitude modes from each MHD wave family on a fully 3D grid, with a wavevector inclined at angles of α and β in the $x - y$ and $x - z$ planes respectively, where $\sin \alpha = 2/3$ and $\sin \beta = 2/\sqrt{5}$. Figure 3.1 plots the L1 error norm for linear amplitude fast, slow, Alfvén and entropy waves in a three-dimensional domain using a resolution of $2N \times N \times N$ grid points, where $N = 8, 16, 32, 64$ and 128. The HLLD Riemann solver is used. This demonstrates Athena converges at second order for all wave families in 3D. The waves propagate in a uniform medium for one crossing time of the domain. The details of the test are given in GS07 and Stone *et al.* (2008).

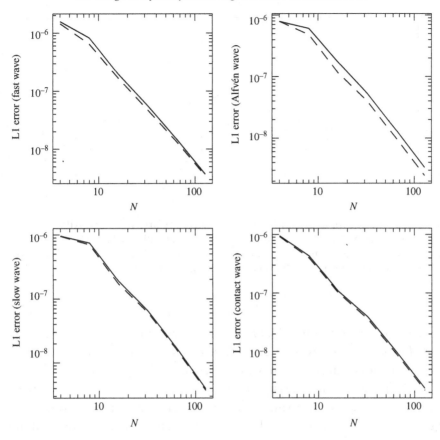

Fig. 3.1. L1 error norm for each wave family in MHD as a function of grid resolution N. The solid line uses second-order (piecewise linear) reconstruction and the MUSCL-Hancock directionally unsplit integrator. The dashed line uses second-order (piecewise linear) reconstruction and the CTU directionally unsplit integrator.

3.2.8.2 Propagation of circularly polarized Alfvén wave

This test was introduced by Tóth (2002). A non-linear amplitude circularly polarized Alfvén wave (which is an exact non-linear solution to the equations of MHD) is initialized on a fully 3D grid, with a wavevector inclined at angles of α and β in the $x - y$ and $x - z$ planes, respectively, where $\sin\alpha = 2/3$ and $\sin\beta = 2/\sqrt{5}$ (as in the linear wave test above). Figure 3.2 plots the profile of the wave after propagating a distance of five wavelengths for a variety of numerical resolutions, as well as the L1 error norm. The results show that with 32 or more grid points per wavelength, the wave profile is maintained extremely well. More importantly, the error in the solution converges at second order.

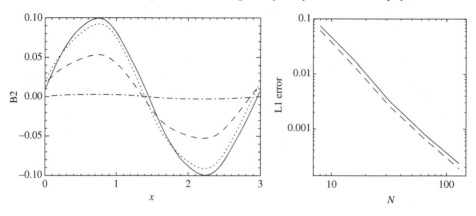

Fig. 3.2. (*Left*) Profile of non-linear amplitude circularly polarized Alfvén wave after propagating five wavelengths on a fully 3D grid with resolution $2N \times N \times N$, where $N = 64$ (solid line), 32 (dashed line), 16 (dashed line) and 8 (dash-dotted line). (*Right*) L1 errors in solution as a function of resolution N.

3.2.8.3 Advection of a field loop

This is a challenging test for Godunov schemes using CT, since the field will be distorted or the method will be unstable if the edge-centred line-averaged electric fields are not computed from the face-centred area-averaged Godunov fluxes. Figure 3.3 plots the current density from a fully 3D calculation of a field cylinder inclined at an oblique angle to the grid, after being advected around the grid twice, using a grid with a resolution of 128^3. The strength of the magnetic field in the loop is set to be very small, so the test is essentially the advection of a passive field. The current distribution should be cylindrical, with a hollow core. The distribution in the figure clearly reproduces this result; thus, the numerical algorithm has not seriously distorted or diffused the original shape. Details of the test are given in GS07.

3.2.8.4 Shocktube rotated to the grid

One-dimensional shocktubes are a standard test of both hydrodynamics and MHD codes. However, to make the test challenging, we use a fully 3D grid and rotate the initial discontinuity to the mesh at an oblique angle. Some care is required to ensure that the divergence-free constraint is maintained in the initial conditions; a complete description of the initialization of this test is given in GS07. Figure 3.4 shows profiles across the mesh of various quantities for the Riemann problem given in Ryu and Jones (1995) in their figure 2a (thus, we refer to this test as the RJ2a test). The grid resolution is $768 \times 8 \times 8$. The test is of particular interest, since the initial conditions result in discontinuities in every wave family that propagates away from the initial interface (that is both fast and slow magnetosonic shocks,

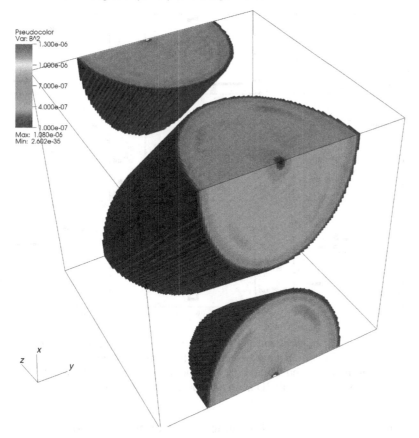

Fig. 3.3. Distribution of the amplitude of the current density $|\mathbf{J}| = |\nabla \times \mathbf{B}|$ in a cylindrical field loop inclined at an oblique angle to the grid, after being advected across the grid diagonal twice. The distribution shows little evolution from the initial conditions.

and rotational and contact discontinuities). The profiles shown in Figure 3.4 can be compared to the 1D solution shown in Ryu and Jones (1995) and elsewhere. Note that all seven waves are reproduced well, with a fidelity comparable to the 1D solution. No extraneous ringing is seen in the velocity or magnetic field components in comparison to 1D, which would be an indication of problems maintaining the divergence-free constraint.

3.2.8.5 MHD blast wave

The final test has no analytic solution, but it is a good check on how robust is the algorithm on fully dynamic problems and whether it can hold spherical symmetry. A spherical region with pressure 100 times larger than ambient is initialized at the centre of the grid, with a strong magnetic field inclined at 45° to the mesh.

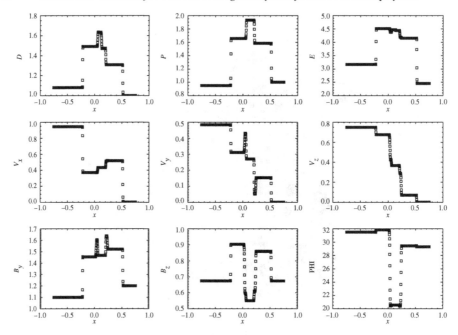

Fig. 3.4. From left to right and top to bottom, profiles of the density, pressure, total energy, three components of velocity, the two transverse components of the magnetic field B_y and B_z and the rotation angle arctan B_z/B_y for the RJ2a Riemann problem run in full 3D at an oblique angle to the grid.

This results in the production of a strong, spherical blast wave which propagates outwards, as well as a rarefaction which propagates inwards. The geometrical focusing of the rarefaction towards the centre of the region can cause numerical problems. Figure 3.5 plots contours of the density and magnetic pressure at late time during the evolution of the blast wave. Note the contours are very smooth (no oscillations) and show the appropriate symmetries. This indicates that the unsplit integrator maintains symmetries extremely well. More results and further discussion is given in GS07 and Stone *et al.* (2008).

3.3 Radiation-hydrodynamic algorithms: the Orion code

Numerical methods exist to simulate radiation-hydrodynamical systems in a variety of dimensionalities and levels of approximation. In three dimensions, treatments of the matter and radiation fields generally adopt the flux-limited diffusion approximation, first introduced by Alme and Wilson (1973), for reasons of computational cost and simplicity (Hayes *et al.* 2006). Flux-limited diffusion is optimal for treating continuum transfer in a system such as an accretion disk, stellar atmosphere or opaque interstellar gas cloud where the majority of the interesting

Density   Magnetic energy

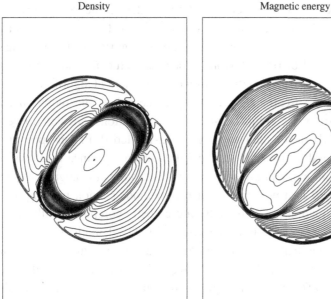

Fig. 3.5. Contours of the density and magnetic energy in a slice through the centre of a strong, spherical blast wave in a magnetized ambient medium. Thirty contours between the minimum and maximum are shown. Initially the magnetic field is inclined at 45° to the grid.

behaviour occurs in optically thick regions that are well described by pure radiation diffusion, but there is a surface of optical depth unity from which energy is radiated away. Applying pure diffusion to these problems would lead to unphysically fast radiation from this surface, so flux-limited diffusion provides a compromise that yields a computationally simple and accurate description of the interior while also giving a reasonably accurate loss rate from the surface (Castor 2004).

However, the level of accuracy provided by this approximation has been unclear because the equations of radiation hydrodynamics for flux-limited diffusion have previously only been analysed to the zeroth order in v/c. In contrast, several authors have analysed the radiation-hydrodynamic equations in the general case to beyond first order in v/c (Mihalas & Weibel-Mihalas 1999; Castor 2004). In a zeroth-order treatment, one neglects differences between quantities in the laboratory frame and the comoving frame. The problem with this approach is that, in an optically thick fluid, the radiation flux only follows Fick's law ($\mathbf{F} \propto -\nabla E$) in the comoving frame, and in other frames there is an added advective flux of radiation enthalpy, as first demonstrated by Castor (1972). In certain regimes (i.e. the dynamic diffusion limit – see below), this advective flux can dominate the diffusive flux (Mihalas & Auer 2001; Castor 2004).

Mihalas and Klein (1982) were the first to derive the mixed-frame equations of radiation hydrodynamics to order v/c in frequency-integrated and frequency-dependent forms and gave numerical algorithms for solving them. Lowrie *et al.* (1999), Lowrie and Morel (2001) and Hubeny and Burrows (2006) give alternate forms of these equations, as well as numerical algorithms for solving them. However, these treatments require that one solves the radiation momentum equation (and for the frequency-dependent equations calculated over many frequencies as well) rather than adopt the flux-limited diffusion approximation. While this is preferable from the standpoint of accuracy, since it allows explicit conservation of both momentum and energy and captures the angular dependence of the radiation field in a way that diffusion methods cannot, treating the radiation momentum equation is significantly more costly computationally than using flux-limited diffusion, making it difficult to use in 3D calculations.

In this section, we analyse the equations of radiation hydrodynamics and the algorithms implemented in the Orion code following our work described in Krumholz *et al.* (2007b) under the approximations that the radiation field obeys the flux-limited diffusion approximation and that scattering is negligible for the system. We derive an accurate set of mixed-frame equations, meaning that radiation quantities are written in the lab frame, but fluid quantities, in particular fluid opacities, are evaluated in the frame comoving with the fluid. This formulation is optimal for 3D simulations applicable to star formation regimes, because writing radiation quantities in the lab frame lets us use an Eulerian grid on which the radiative transfer problem may be solved by any number of standard methods while avoiding the need to model the direction and velocity dependence of the lab frame opacity and emissivity of a moving fluid.

We begin from the general lab frame equations of radiation hydrodynamics to first order in v/c, apply the flux-limited diffusion approximation in the frame comoving with the gas where it is applicable and transform the appropriate radiation quantities into the lab frame, thereby deriving the corresponding mixed-frame equations suitable for implementation in numerical simulations. We retain enough terms to ensure that we achieve order unity accuracy in all regimes and order v/c accuracy for static diffusion problems. We assess the significance of the higher-order terms that appear in our equations, and consider where treatments omitting them are acceptable and where they are likely to fail. We show that, in at least some regimes, the zeroth-order treatments most often used are likely to produce results that are incorrect at order unity. We also compare our equations to the comoving frame equations commonly used in other codes. We take advantage of the ordering of terms we derive for the static diffusion regime to construct a radiation hydro-dynamic simulation algorithm for static diffusion problems that is simpler and faster than those in use now, which we implement in the Orion AMR code. In

Section 3.3.6, we discuss a selection of radiation-hydrodynamic test problems. In the discussion that follows, we adopt the convention of writing quantities measured in the frame comoving with a fluid with a subscript zero. Quantities in the lab frame are written without subscripts. Also note that we follow the standard convention in radiation hydrodynamics rather than the standard in astrophysics, in that when we refer to an opacity κ we mean the total opacity, measured in units of inverse length, rather than the specific opacity, measured in units of length squared divided by mass. Since we are neglecting scattering, we may set the extinction $\chi = \kappa$.

3.3.1 Limiting regimes of radiation hydrodynamics

It is useful to first examine some characteristic dimensionless numbers for a radiation hydrodynamic system, since evaluating these quantities provides a useful guide to how we should analyze our equations. Let ℓ be the characteristic size of the system under consideration, u be the characteristic velocity in this system and $\lambda_P \sim 1/\kappa$ be the photon mean free path. Following Mihalas and Weibel-Mihalas (1999), we can define three distinct limiting cases by considering the dimensionless ratios $\tau \equiv \ell/\lambda_P$, which characterizes the optical depth of the system, and $\beta \equiv u/c$, which characterizes how relativistic it is. Since we focus on non-relativistic systems, we assume $\beta \ll 1$. We term the case $\tau \ll 1$, in which the radiation and gas are weakly coupled, the *streaming* limit. If $\tau \gg 1$, then radiation and gas are strongly coupled, and the system is in the diffusion limit. We can further subdivide the diffusion limit into the cases $\beta \gg \tau^{-1}$ and $\beta \ll \tau^{-1}$. The former is the *dynamic diffusion* limit, while the latter is the *static diffusion* limit. In summary, the limiting cases are

$$\tau \ll 1, \qquad\qquad\qquad \text{(streaming limit)} \qquad\qquad (3.16)$$

$$\tau \gg 1, \quad \beta\tau \ll 1 \quad \text{(static diffusion limit)} \qquad\qquad (3.17)$$

$$\tau \gg 1, \quad \beta\tau \gg 1 \quad \text{(dynamic diffusion limit).} \qquad (3.18)$$

Physically, the distinction between static and dynamic diffusion is that in dynamic diffusion, radiation is principally transported by advection by gas, so that terms describing the work done by radiation on gas and the advection of radiation enthalpy dominate over terms describing either diffusion or emission and absorption. In the static diffusion limit, the opposite holds. A paradigmatic example of a dynamic diffusion system is a stellar interior. The optical depth from the core to the surface of the Sun is $\tau \sim 10^{11}$, and typical convective and rotational velocities are $\gg 10^{-11}c = 0.3\,\mathrm{cm\,s^{-1}}$, so the Sun is strongly in the dynamic diffusion regime. In contrast, an example of a system in the static diffusion limit is a relatively cool, dusty, outer accretion disk around a forming massive protostar, as studied, for example, by Krumholz *et al.* (2007a). The specific opacity of gas with the standard interstellar dust abundance to infrared photons is $\kappa/\rho \sim 1\,\mathrm{cm^2\,g^{-1}}$, and at distances of more than a few AU from the central star the density is generally $\rho \lesssim 10^{-12}\,\mathrm{g\,cm^{-3}}$. For a disk of scale height $h \sim 10\,\mathrm{AU}$, the optical depth to escape is

$$\tau^{-1} \approx 6.7 \times 10^{-3} \left(\frac{\kappa/\rho}{\mathrm{cm^2\,g^{-1}}}\right)^{-1} \left(\frac{\rho}{10^{-12}\,\mathrm{g\,cm^{-3}}}\right)^{-1} \left(\frac{h}{10\,\mathrm{AU}}\right)^{-1}. \tag{3.19}$$

The velocity is roughly the Keplerian speed, so

$$\beta \approx 1.4 \times 10^{-4} \left(\frac{M_*}{10M_\odot}\right)^{1/2} \left(\frac{r}{10\,\mathrm{AU}}\right)^{-1/2}, \tag{3.20}$$

where M_* is the mass of the star and r is the distance from it. Thus, this system is in a static diffusion regime by roughly two orders of magnitude.

In the development of a self-consistent formulation of radiation hydrodynamics, our goal will be to obtain expressions that are accurate for the leading terms in all regimes. This is somewhat tricky, particularly for diffusion problems, because we are attempting to expand our equations simultaneously in the two small parameters β and $1/\tau$. The most common approach in radiation hydrodynamics is to expand expressions in powers of β alone and only analyse the equations in terms of τ after dropping terms of high order in β. However, this approach can produce significant errors, because terms in the radiation-hydrodynamic equations proportional to the opacity are multiplied by a quantity of order τ. Thus, in our derivation (Krumholz *et al.* 2007b), we will repeatedly encounter expressions proportional to $\beta^2\tau$, and in a problem that is either in the dynamic diffusion limit or close to it ($\beta\tau \gtrsim 1$), it is inconsistent to drop these terms while retaining ones that are of order β. We therefore retain all terms up to order β^2 in our derivation unless we explicitly check that they are not multiplied by terms of order τ and can therefore be dropped safely.

3.3.2 The equations of radiation hydrodynamics

We begin with the lab frame equations of radiation hydrodynamics (Mihalas & Klein 1982; Mihalas & Weibel-Mihalas 1999; Mihalas & Auer 2001)

$$\frac{\partial \rho}{\partial t} + \nabla \cdot (\rho \mathbf{v}) = 0 \tag{3.21}$$

$$\frac{\partial}{\partial t}(\rho \mathbf{v}) + \nabla \cdot (\rho \mathbf{vv}) = -\nabla P + \mathbf{G} \tag{3.22}$$

$$\frac{\partial}{\partial t}(\rho e) + \nabla \cdot [(\rho e + P)\,\mathbf{v}] = cG^0 \tag{3.23}$$

$$\frac{\partial E}{\partial t} + \nabla \cdot \mathbf{F} = -cG^0 \tag{3.24}$$

$$\frac{1}{c^2}\frac{\partial \mathbf{F}}{\partial t} + \nabla \cdot \mathcal{P} = -\mathbf{G}, \tag{3.25}$$

where ρ, $\mathbf{v}$, e and P are the density, velocity, specific energy (thermal plus kinetic) and thermal pressure of the gas, respectively, and E, $\mathbf{F}$ and $\mathcal{P}$ are the radiation energy density, flux and pressure tensor, respectively,

$$cE = \int_0^\infty dv \int d\Omega\, I(\mathbf{n}, v) \tag{3.26}$$

$$\mathbf{F} = \int_0^\infty dv \int d\Omega\, \mathbf{n}I(\mathbf{n}, v) \tag{3.27}$$

$$c\mathcal{P} = \int_0^\infty dv \int d\Omega\, \mathbf{n}\mathbf{n}I(\mathbf{n}, v), \tag{3.28}$$

$(G^0, \mathbf{G})$ is the radiation four-force density,

$$cG^0 = \int_0^\infty dv \int d\Omega\, [\kappa(\mathbf{n}, v)I(\mathbf{n}, v) - \eta(\mathbf{n}, v)], \tag{3.29}$$

$$c\mathbf{G} = \int_0^\infty dv \int d\Omega\, [\kappa(\mathbf{n}, v)I(\mathbf{n}, v) - \eta(\mathbf{n}, v)]\mathbf{n}, \tag{3.30}$$

and $I(\mathbf{n}, v)$ is the intensity of the radiation field at frequency v travelling in direction $\mathbf{n}$. Here, $\kappa(\mathbf{n}, v)$ and $\eta(\mathbf{n}, v)$ are the direction- and frequency-dependent radiation absorption and emission coefficients in the lab frame. Essentially, cG^0 is the rate of energy absorption from the radiation field minus the rate of energy emission for the fluid, and $\mathbf{G}$ is the rate of momentum absorption from the radiation field minus the rate of momentum emission. Equations (3.21)–(3.23) are accurate to first order in v/c, while Eqs (3.24)–(3.25) are exact. Since no terms involving opacity or optical depth appear explicitly in any of these equations, the fact that they are accurate to first order in β means that they include all the leading order terms.

In order to derive the mixed-frame equations, we first evaluate the radiation four-force $(G^0, \mathbf{G})$ in terms of lab frame radiation quantities and comoving frame emission and absorption coefficients. Mihalas and Auer (2001) show that, if the flux spectrum of the radiation is direction-independent, the radiation four-force on a thermally emitting material to all orders in v/c is given in terms of moments of the radiation field by

$$G^0 = \gamma[\gamma^2 \kappa_{0E} + (1 - \gamma^2)\kappa_{0F}]E - \gamma \kappa_{0P} a_R T_0^4$$

$$- \gamma \left(\frac{\mathbf{v} \cdot \mathbf{F}}{c^2}\right)[\kappa_{0F} - 2\gamma^2(\kappa_{0F} - \kappa_{0E})]$$

$$- \gamma^3(\kappa_{0F} - \kappa_{0E})(\mathbf{vv}) : \frac{\mathcal{P}}{c^2}, \tag{3.31}$$

$$\mathbf{G} = \gamma \kappa_{0F}\left(\frac{\mathbf{F}}{c}\right) - \gamma \kappa_{0P} a_R T_0^4\left(\frac{\mathbf{v}}{c}\right)$$

$$- \left[\gamma^3(\kappa_{0F} - \kappa_{0E})\left(\frac{\mathbf{v}}{c}\right)E + \gamma \kappa_{0F}\left(\frac{\mathbf{v}}{c}\right) \cdot \mathcal{P}\right]$$

$$+ \gamma^3(\kappa_{0F} - \kappa_{0E})\left[2\mathbf{v} \cdot \frac{\mathbf{F}}{c^3} - (\mathbf{vv}) : \frac{\mathcal{P}}{c^3}\right]\mathbf{v}, \tag{3.32}$$

where $\gamma = 1/\sqrt{1 - v^2/c^2}$ is the Lorentz factor and T_0 is the gas temperature. The three opacities that appear are the Planck-, energy-, and flux-mean opacities, which are defined by

$$\kappa_{0P} = \frac{\int_0^\infty d\nu_0 \, \kappa_0(\nu_0) B(\nu_0, T_0)}{B(T_0)} \tag{3.33}$$

$$\kappa_{0E} = \frac{\int_0^\infty d\nu_0 \, \kappa_0(\nu_0) E_0(\nu_0)}{E_0} \tag{3.34}$$

$$\kappa_{0F} = \frac{\int_0^\infty d\nu_0 \, \kappa_0(\nu_0) \mathbf{F}_0(\nu_0)}{\mathbf{F}_0}, \tag{3.35}$$

where $E_0(\nu_0)$ and $\mathbf{F}_0(\nu_0)$ are the comoving frame radiation energy and flux per unit frequency, E_0 and $\mathbf{F}_0$ are the corresponding frequency-integrated energy and flux, and $B(\nu, T) = (2h\nu^3/c^2)/(e^{h\nu/k_B T} - 1)$ and $B(T) = ca_R T^4/4\pi$ are the frequency-dependent and frequency-integrated Planck functions, respectively.

Note that we have implicitly assumed that the opacity and emissivity are directionally independent in the fluid rest frame, which is the case for any conventional material. We have also assumed that the flux spectrum is independent of direction, allowing us to replace the flux-mean opacity vector with a scalar. This may not be the case for an optically thin system, or one in which line transport is important, but since we are limiting our application to systems to which we can reasonably apply the diffusion approximation, this is not a major limitation.

To simplify $(G^0, \mathbf{G})$, we make several approximations (Krumholz *et al.* 2007b) such that the only two opacities remaining in our equations are κ_{0R} and κ_{0P}, both of which are independent of the spectrum of the radiation field and the direction of radiation propagation, and which may therefore be tabulated as a function of temperature for a given material once and for all.

Following our work in Krumholz *et al.* (2007b), we expand $(G^0, \mathbf{G})$ in powers of v/c, retaining terms to order v^2/c^2. The resulting expression for the radiation four-force is

$$G^0 = \kappa_{0P}\left(E - \frac{4\pi B}{c}\right) + (\kappa_{0R} - 2\kappa_{0P})\frac{\mathbf{v} \cdot \mathbf{F}}{c^2}$$
$$+ \frac{1}{2}\left(\frac{v}{c}\right)^2\left[2(\kappa_{0P} - \kappa_{0R})E + \kappa_{0P}\left(E - \frac{4\pi B}{c}\right)\right]$$
$$+ (\kappa_{0P} - \kappa_{0R})\frac{\mathbf{vv}}{c^2} : \mathcal{P} + O\left(\frac{v^3}{c^3}\right) \tag{3.36}$$

$$\mathbf{G} = \kappa_{0R}\frac{\mathbf{F}}{c} + \kappa_{0P}\left(\frac{\mathbf{v}}{c}\right)\left(E - \frac{4\pi B}{c}\right)$$
$$- \kappa_{0R}\left(\frac{\mathbf{v}}{c}E + \frac{\mathbf{v}}{c} \cdot \mathcal{P}\right) + \frac{1}{2}\left(\frac{v}{c}\right)^2\kappa_{0R}\frac{\mathbf{F}}{c}$$
$$+ 2(\kappa_{0R} - \kappa_{0P})\frac{(\mathbf{v} \cdot \mathbf{F})\mathbf{v}}{c^3} + O\left(\frac{v^3}{c^3}\right). \tag{3.37}$$

We can examine the scalings of these terms with the help of our dimensionless parameters β and τ. In the streaming limit, radiation travels freely at c and emission and absorption of radiation by matter need not balance, so $|\mathbf{F}| \sim cE$ and $4\pi B/c - E \sim E$. For static diffusion, Mihalas and Weibel-Mihalas (1999) show that $|\mathbf{F}| \sim cE/\tau$ and $4\pi B/c - E \sim E/\tau^2$. For dynamic diffusion, radiation travels primarily by advection, so $|\mathbf{F}| \sim vE$. We can show that for dynamic diffusion $4\pi B/c - E \sim \beta^2 E$. Note that the scaling $4\pi B/c - E \sim (\beta/\tau)E$ given in Mihalas and Weibel-Mihalas (1999) appears to be incorrect. Using these values, we obtain the scalings shown in Table 1 of Krumholz *et al.* (2007b) for the terms in Eqs (3.36) and (3.37).

The table shows that, despite the fact that we have kept all terms that are formally order β^2 or more, we only have leading-order accuracy in the dynamic diffusion limit, because in this limit the order unity and order β terms in G^0 vanish to order β^2. To obtain the next-order terms, we would have had to write G^0 to order β^3. A corollary of this is that treatments of the dynamic diffusion limit that do not retain order β^2 terms are likely to produce equations that are incorrect at order unity, since they will have dropped terms that are of the same order as the ones that have been retained.

At this point, we could begin dropping terms that are insignificant at the order to which we are working, but it is cumbersome to construct a table analogous to Table 1 at every step of our derivation. It is more convenient to continue our analysis retaining all the terms in Eqs (3.36) and (3.37) and to drop terms only periodically.

We now adopt the flux-limited diffusion approximation (Alme & Wilson 1973), under which we drop the radiation momentum Eq. (3.25) and set the radiation flux in the comoving frame to

$$\mathbf{F}_0 = -\frac{c\lambda}{\kappa_{0R}} \nabla E_0, \tag{3.38}$$

where λ is a dimensionless number called the flux limiter. Many functional forms for λ are possible. For the code implementation, we adopt the Levermore and Pomraning (1981) flux limiter and we adopt the corresponding approximate value for the radiation pressure tensor in the comoving frame $\mathcal{P}_0$ (Levermore 1984).

To use the approximation (3.38) and $\mathcal{P}_0$ to evaluate the radiation four-force, we must apply Lorentz transform to express the radiation quantities in the lab frame. The Lorentz transforms for the energy, flux and pressure to second order in v/c are given in Mihalas and Weibel-Mihalas (1999).

Using the same scaling arguments that are used to construct table 1 (Krumholz *et al.* 2007b), we note that $\mathcal{P}$ and $\mathcal{P}_0$ differ at order β in the streaming limit, at order β/τ for static diffusion and at order β^2 for dynamic diffusion. Since this is below our accuracy goal, we need not distinguish $\mathcal{P}$ and $\mathcal{P}_0$. The same is true of E and E_0. However, $\mathbf{F}$ is different. In the comoving frame in an optically thick system, one is in the static diffusion regime, so $\mathbf{F}_0 \sim cE_0/\tau$. Since $\mathbf{v}E_0$ and $\mathbf{v} \cdot \mathcal{P}_0$ are of order βcE_0, and in dynamic diffusion $\beta \gg 1/\tau$, this means that $\mathbf{v}E_0$ and $\mathbf{v} \cdot \mathcal{P}_0$ are the dominant components of $\mathbf{F}$ in dynamic diffusion and must therefore be retained. Thus,

$$\mathbf{F} = -\frac{c\lambda}{\kappa_{0R}} \nabla E + \mathbf{v}E + \mathbf{v} \cdot \mathcal{P}, \tag{3.39}$$

which is simply the rest frame flux plus terms describing the advection of radiation enthalpy.

Substituting $\mathcal{P}_0$ with $\mathcal{P} = \mathcal{P}_0$ and Eq. (3.39) into the four-force density Eqs (3.36) and (3.37), and continuing to retain terms to order v^2/c^2, gives

$$\begin{aligned}
G^0 = {} & \kappa_{0P}\left(E - \frac{4\pi B}{c}\right) + \left(\frac{\lambda}{c}\right)\left(2\frac{\kappa_{0P}}{\kappa_{0R}} - 1\right)\mathbf{v} \cdot \nabla E \\
& - \frac{\kappa_{0P}}{c^2}E\left[\frac{3 - R_2}{2}v^2 + \frac{3R_2 - 1}{2}(\mathbf{v} \cdot \mathbf{n})^2\right] \\
& + \frac{1}{2}\left(\frac{v}{c}\right)^2 \kappa_{0P}\left(E - \frac{4\pi B}{c}\right)
\end{aligned} \tag{3.40}$$

$$G = -\lambda \nabla E + \kappa_{0P} \frac{\mathbf{v}}{c} \left(E - \frac{4\pi B}{c} \right)$$

$$- \frac{1}{2} \left(\frac{v}{c} \right)^2 \lambda \nabla E$$

$$+ 2\lambda \left(\frac{\kappa_{0P}}{\kappa_{0R}} - 1 \right) \frac{(\mathbf{v} \cdot \nabla E)\mathbf{v}}{c^2}, \tag{3.41}$$

where R_2 is related to R in Levermore and Pomraning (1981) and $\mathbf{n}$ is the unit vector antiparallel to ∇E. Although these equations contain terms of order β^2, they are not truly accurate to order β^2 because we did not retain all the β^2 terms when applying the Lorentz transform to the flux and pressure. However, these equations include all the terms that appear at the order of accuracy to which we are working, and by retaining terms of order β^2 we guarantee that these terms will be preserved.

Inserting (G^0, G) and the lab frame flux (3.39) into the gas momentum and energy Eqs (3.22) and (3.23), and the radiation energy Eq. (3.24), and again retaining terms to order v^2/c^2 gives

$$\frac{\partial}{\partial t}(\rho \mathbf{v}) = -\nabla \cdot (\rho \mathbf{v}\mathbf{v}) - \nabla P - \lambda \nabla E$$

$$- \kappa_{0P} \frac{\mathbf{v}}{c^2} (4\pi B - cE) - \frac{1}{2} \left(\frac{v}{c} \right)^2 \lambda \nabla E$$

$$+ 2\lambda \left(\frac{\kappa_{0P}}{\kappa_{0R}} - 1 \right) \frac{(\mathbf{v} \cdot \nabla E)\mathbf{v}}{c^2}. \tag{3.42}$$

$$\frac{\partial}{\partial t}(\rho e) = -\nabla \cdot [(\rho e + P)\mathbf{v}] - \kappa_{0P}(4\pi B - cE)$$

$$+ \lambda \left(2\frac{\kappa_{0P}}{\kappa_{0R}} - 1 \right) \mathbf{v} \cdot \nabla E$$

$$- \frac{\kappa_{0P}}{c} E \left[\frac{3 - R_2}{2} v^2 + \frac{3R_2 - 1}{2} (\mathbf{v} \cdot \mathbf{n})^2 \right]$$

$$- \frac{1}{2} \left(\frac{v}{c} \right)^2 \kappa_{0P} (4\pi B - cE) \tag{3.43}$$

$$\frac{\partial}{\partial t} E = \nabla \cdot \left(\frac{c\lambda}{\kappa_{0R}} \nabla E \right) + \kappa_{0P}(4\pi B - cE)$$

$$- \lambda \left(2\frac{\kappa_{0P}}{\kappa_{0R}} - 1 \right) \mathbf{v} \cdot \nabla E$$

$$+ \frac{\kappa_{0P}}{c} E \left[\frac{3 - R_2}{2} v^2 + \frac{3R_2 - 1}{2} (\mathbf{v} \cdot \mathbf{n})^2 \right]$$

$$- \nabla \cdot \left[\frac{3 - R_2}{2} \mathbf{v} E + \frac{3R_2 - 1}{2} \mathbf{v} \cdot (\mathbf{nn}) E \right]$$

$$+ \frac{1}{2} \left(\frac{v}{c} \right)^2 \kappa_{0P} (4\pi B - cE). \tag{3.44}$$

At this point, we can construct another table (see table 2, Krumholz *et al.* 2007b) showing the scalings of the radiation terms to see which must be retained and which are superfluous. In constructing the table, we take spatial derivatives to be of characteristic scaling $1/\ell$, i.e. we assume that radiation quantities vary on a size scale of the system rather than over a size scale of the photon mean free path. In the streaming limit, $\lambda \sim \tau$ and $R_2 \sim 1 + O(\tau)$. In the diffusion limit, $\lambda \sim 1/3$ and $R_2 \sim 1/3 + O(\tau^{-2})$.

Using table 2 to drop all terms that are not significant at leading order in any regime, we arrive at our final radiation hydrodynamic equations (Krumholz *et al.* 2007b):

$$\frac{\partial}{\partial t} (\rho \mathbf{v}) = -\nabla \cdot (\rho \mathbf{vv}) - \nabla P - \lambda \nabla E \tag{3.45}$$

$$\frac{\partial}{\partial t} (\rho e) = -\nabla \cdot [(\rho e + P)\mathbf{v}] - \kappa_{0P}(4\pi B - cE)$$

$$+ \lambda \left(2 \frac{\kappa_{0P}}{\kappa_{0R}} - 1 \right) \mathbf{v} \cdot \nabla E$$

$$- \frac{3 - R_2}{2} \kappa_{0P} \frac{v^2}{c} E \tag{3.46}$$

$$\frac{\partial}{\partial t} E = \nabla \cdot \left(\frac{c\lambda}{\kappa_{0R}} \nabla E \right) + \kappa_{0P}(4\pi B - cE)$$

$$- \lambda \left(2 \frac{\kappa_{0P}}{\kappa_{0R}} - 1 \right) \mathbf{v} \cdot \nabla E$$

$$+ \frac{3 - R_2}{2} \kappa_{0P} \frac{v^2}{c} E$$

$$- \nabla \cdot \left(\frac{3 - R_2}{2} \mathbf{v} E \right). \tag{3.47}$$

These represent the equations of momentum conservation for the gas, energy conservation for the gas and energy conservation for the radiation field, which, together with the equation of mass conservation (3.21), fully describe the system under the approximations we have adopted. They are accurate and consistent to leading order in the streaming and dynamic diffusion limits. They are accurate to first order in β in the static diffusion limit, since we have had to retain all order β terms in this limit because they are of leading order in dynamic diffusion problems. Also note that if in a given problem one never encounters the dynamic diffusion regime, it is possible to drop more terms.

The equations are easy to understand intuitively. The term $-\lambda \nabla E$ in the momentum Eq. (3.45) simply represents the radiation force $\kappa_{0R} \mathbf{F}/c$, neglecting distinctions between the comoving and the laboratory frames which are smaller than leading order in this equation. Similarly, the terms $\pm \kappa_{0P}(4\pi B - cE)$ and $\pm \lambda (2\kappa_{0P}/\kappa_{0R} - 1)\mathbf{v} \cdot \nabla E$ in the two energy Eqs (3.46) and (3.47) represent radiation absorbed minus radiation emitted by the gas and the work done by the radiation field as it diffuses through the gas. The factor $(2\kappa_{0P}/\kappa_{0R} - 1)$ arises because the term contains contributions both from the Newtonian work and from a relativistically induced mismatch between emission and absorption. The term proportional to $\kappa_{0P} E/c$ represents another relativistic correction to the work, this one arising from boosting of the flux between the lab and the comoving frames. In the radiation energy Eq. (3.47), the first term on the left-hand side is the divergence of the radiation flux, i.e. the rate at which radiation diffuses, and the last term on the right-hand side represents advection of the radiation enthalpy $E + \mathcal{P}$ by the gas.

It is also worth noting that Eqs (3.43) and (3.44) are manifestly energy-conserving, since every term in one equation either has an obvious counterpart in the other with opposite sign or is clearly an advection. In contrast, the momentum Eq. (3.45) is not manifestly momentum-conserving, since there is a force term $-\lambda \nabla E$ with no equal and opposite counterpart. This non-conservation of momentum is an inevitable side effect of using the flux-limited diffusion approximation, since this approximation amounts to allowing the radiation field to transfer momentum to the gas without explicitly tracking the momentum of the radiation field and the corresponding transfer from gas to radiation.

3.3.3 The relative importance of higher-order terms

Our dynamical equations result from retaining at least some terms that are formally of order β^2. Even though our analysis shows that these terms can be the leading ones present, due to cancellations of lower-order terms, one might legitimately ask whether they are ever physically significant. In Section 3.3.3.1, we address this question by comparing our equations to those that result from lower-order treatments. In Section 3.3.3.2, we also compare our equations with those generally used in comoving frame formulations of radiation hydrodynamics.

To make our work in this section more transparent, we specialize to the diffusion regime in gray materials. Thus, we set $\lambda = R_2 = 1/3$ and $\kappa_{0P} = \kappa_{0R} = \kappa_0$. A more general analysis produces the same conclusions. We also focus on the radiation energy equation, since all the terms that appear in the gas energy equation also appear in it, and because there are no higher-order terms present in the momentum equation. Under these assumptions, our radiation energy Eq. (3.47) becomes

$$\frac{\partial}{\partial t} E = \nabla \cdot \left(\frac{c}{3\kappa_0} \nabla E \right) + \kappa_0 (4\pi B - cE)$$

$$-\frac{4}{3} \nabla \cdot (\mathbf{v} E) - \frac{1}{3} \mathbf{v} \cdot \nabla E + \frac{4}{3} \kappa_0 \frac{v^2}{c} E. \qquad (3.48)$$

3.3.3.1 Comparison to lower-order equations

A common approach in radiation-hydrodynamic problems is to expand the equations in β, rather than in both β and τ as we have done, and drop at least some terms that are of order β^2 in every regime (Mihalas & Weibel-Mihalas 1999). To determine how equations derived in this manner compare to our higher-order treatment, we compare our simplified energy Eq. (3.48) to the corresponding equation one would obtain by following this procedure with Eq. (3.47). This resulting energy equation is

$$\frac{\partial}{\partial t} E = \nabla \cdot \left(\frac{c}{3\kappa_0} \nabla E \right) + \kappa_0 (4\pi B - cE)$$

$$-\frac{4}{3} \nabla \cdot (\mathbf{v} E) - \frac{1}{3} \mathbf{v} \cdot \nabla E. \qquad (3.49)$$

We show below that Eq. (3.49) is not accurate to leading order in at least some cases and should not be used for computations unless one carefully checks that the missing terms never become important in the regime covered by the computation.

Compared to the energy Eq. (3.48) that we obtain by retaining all leading order terms in β and τ, Eq. (3.49) is missing the term $(4/3)\kappa_0 v^2 E/c$. We can describe the $\mathbf{v} \cdot \nabla E$ term as the 'diffusion work' arising from the combination of the diffusion flux and the post-Newtonian emission–absorption mismatch (as discussed in Section 3.3.2), and the $\kappa_0 v^2 E/c$ as the 'relativistic work' arising from the relativistic flux. The presence or absence of this relativistic work term is the difference between our leading order-accurate equation and the equation one would derive by dropping β^2 terms. Analysing when, if ever, this term is physically important lets us identify in which situations a lower-order treatment may be inadequate.

If we use table 2 to compare the relativistic work term to the emission–absorption term, we find that $(\kappa_0 v^2 E/c)/[\kappa_0(4\pi B - cE)]$ is of order $\beta^2 \tau^2$ for static diffusion and of order unity for dynamic diffusion. Thus, the term is never important in a static diffusion problem but is always important for a non-uniform, non-equilibrium dynamic diffusion problem system. We expect any system where variations occur on a scale for which $\beta\tau \gg 1$ to resemble a uniform, equilibrium medium, and thus we do not expect the term $(4/3)\kappa_0 v^2 E/c$ to be important in such a system.

That said, there is still clearly a problem with omitting the relativistic work term in a system where $\beta\tau \sim 1$. In this case, table 2 implies that *every* term on the right-hand side is roughly equally important regardless of whether we use the static or dynamic diffusion scalings. We have shown (Krumholz *et al.* 2007b) that one can obtain the correct structure within a radiation-dominated shock only by retaining the relativistic work term.

An interesting point to note here is that omitting the relativistic work term will not produce errors upstream or downstream of a shock, because $\beta\tau \gg 1$ in these regions. The omitted term will, however, affect radiation-gas energy exchange, not total energy conservation. The lower-order treatment will therefore only make errors within the shock. Whether this is physically important depends on whether one is concerned with structures on scales for which $\beta\tau \sim 1$. An astrophysical example of a system where one does care about structures on this scale is a radiation-dominated accretion disk subject to photon bubble instability (Turner *et al.* 2003). Such disks are in the dynamic diffusion regime over the entire disk, but photon bubbles form on small scales within them, and individual bubbles may have $\beta\tau \sim 1$ across them.

3.3.3.2 Comparison to comoving frame formulations

Many popular numerical treatments of radiation hydrodynamics (Turner & Stone 2001; Whitehouse & Bate 2004; Hayes *et al.* 2006) use a comoving formulation of the equations rather than our mixed-frame formulation. It is therefore useful to compare our equations to the standard comoving frame equations. In the comoving formulation, the evolution equation for the radiation field is usually the first law of thermodynamics for the comoving radiation field as shown by Mihalas and Klein (1982),

$$\rho\frac{D}{Dt}\left(\frac{E_0}{\rho}\right) + \mathcal{P}_0 : (\nabla\mathbf{v}) = \kappa_0(4\pi B - cE_0) - \nabla\cdot\mathbf{F}_0. \qquad (3.50)$$

This equation is accurate to first order in β in the sense that it contains all the correct leading order terms and all terms that are smaller than them by order β or less.

To compare this to our mixed-frame radiation energy Eq. (3.47), we replace the comoving frame energy E_0 in Eq. (3.50) with the lab frame energy E using the Lorentz transformation Mihalas and Weibel-Mihalas (1999) and retain all terms that are of leading order in any regime. This gives a transformed equation

$$\rho\frac{D}{Dt}\left(\frac{E}{\rho}\right) + \mathcal{P}_0 : (\nabla\mathbf{v}) = \kappa_0(4\pi B - cE) - \nabla\cdot\mathbf{F}_0$$

$$+ 2\kappa_0\frac{\mathbf{v}\cdot\mathbf{F}_0}{c} + \frac{\kappa_0}{c}\left[v^2 E + (\mathbf{vv}) : \mathcal{P}_0\right]. \qquad (3.51)$$

If we now adopt the diffusion approximation $\mathbf{F}_0 = -c/(3\kappa_0)\nabla E_0$ and $\mathcal{P}_0 = (1/3)E_0\mathcal{I}$, use the Lorentz transformation to replace E_0 with E throughout, and again only retain terms that are of leading order in some regime, then it is easy to verify that Eq. (3.51) reduces to Eq. (3.48). Thus, our evolution equation is equivalent to the comoving frame first law of thermodynamics for the radiation field, *provided that one retains all the leading order terms with respect to β and τ, including some that are of order β^2, when evaluating the Lorentz transformation.*

While the equations are equivalent, the mixed-frame formulation has two important advantages over the comoving frame formulation when it comes to practical computation. First, we are able to write the equations in a manner that allows a numerical solution algorithm to conserve total energy to machine accuracy. We present such an algorithm in Krumholz *et al.* (2007b). In contrast, it is not possible to write a conservative update algorithm using the comoving frame equations. The reason for this is that a conserved total energy only exists in an inertial frame, and for a fluid whose velocity is not a constant in space and time, the comoving frame is not inertial. The lack of a conserved energy is a serious drawback to comoving frame formulations.

A second advantage of the mixed-frame formulation is that it is far more suited to implementation in codes with dynamically modified grid structures such as AMR methods. Since the radiation energy is a conserved quantity, it is obvious how to refine or coarsen it in a conservative manner. On the contrary, there is no obviously correct method for refining or coarsening the comoving frame energy density, because it will not even be defined in the same reference frames before and after the refinement procedure.

3.3.4 Radiation hydrodynamics in the static diffusion limit

Our analysis shows that for static diffusion, the terms involving diffusion and emission minus absorption of radiation always dominate over those involving radiation work and advection. In addition, some terms are always smaller than order β. This suggests an opportunity for a significant algorithmic improvement over earlier approaches while still retaining order β accuracy in the solution. In a simulation, one must update terms for the radiation field implicitly, because otherwise stability requirements limit the update time step to values comparable to the light-crossing time of a cell. Standard approaches (Turner & Stone 2001; Whitehouse & Bate 2004; Whitehouse *et al.* 2005; Hayes *et al.* 2006) therefore update all terms involving radiation implicitly except the advection term and the radiation force term in the gas momentum equation.

However, implicit updates are computationally expensive; so the simpler the terms to be updated implicitly can be made, the simpler the algorithm will be to code and the faster it will run. Since the work and advection terms are non-dominant, we can produce a perfectly stable algorithm without treating them implicitly. Even if this treatment introduces numerically unstable modes in the work or advection terms, they will not grow because the radiation diffusion and emission/absorption terms, which are far larger, will smooth them away each time step.

For the case of static diffusion, we therefore adopt the order v/c Eqs (3.21) and (3.45) for mass and momentum conservation. For our energy equations, we adopt Eqs (3.46) and (3.47), but drop terms that are smaller than order β for static diffusion. This gives

$$\frac{\partial}{\partial t}(\rho e) = -\nabla[(\rho e + P)\mathbf{v}] - \kappa_{0P}(4\pi B - cE)$$

$$+ \lambda \left(2\frac{\kappa_{0P}}{\kappa_{0R}} - 1\right)\mathbf{v} \cdot \nabla E \qquad (3.52)$$

$$\frac{\partial}{\partial t}E = \nabla \cdot \left(\frac{c\lambda}{\kappa_{0R}}\nabla E\right) + \kappa_{0P}(4\pi B - cE)$$

$$- \lambda \left(2\frac{\kappa_{0P}}{\kappa_{0R}} - 1\right)\mathbf{v} \cdot \nabla E$$

$$- \nabla \cdot \left(\frac{3 - R_2}{2}\mathbf{v}E\right). \qquad (3.53)$$

To solve these, we operator-split the diffusion and emission/absorption terms, which we treat implicitly, from the work and advection terms, which we treat explicitly. For each update cycle, we start with the state at the old time. We first perform an implicit update to the radiation and gas energy densities using radiative terms that are handled implicitly. For our implementation of this algorithm in the Orion AMR code, we use the method of Howell and Greenough (2003), which we will not discuss in detail here. To summarize, the algorithm involves writing the equations using second-order accurate spatial discretization and a time discretization that limits to backwards Euler for large values of $\partial E/\partial t$ (to guarantee stability) and to Crank–Nicolson when $\partial E/\partial t$ is small (to achieve second-order time accuracy). This yields a matrix equation for the radiation and gas energy densities at the new time, which may be solved on both individual grids and over a hierarchy of nested grids (as is necessary for AMR) using standard multigrid techniques. The output of this procedure is an intermediate state which has been updated for the implicit terms.

Once the implicit update is done, we compute the ordinary hydrodynamic update. As with the implicit update, this may be done using the hydrodynamics method of one's choice. For our implementation, we use the Godunov method described by Truelove *et al.* (1998), Klein (1999) and Fisher (2002). This update gives us a state updated for implicit radiative terms and explicit non-radiative terms.

Finally, we explicitly compute the radiative force and advection terms and then find the new state at the advanced time.

This update is manifestly only first-order accurate in time for the explicit radiation terms, but there is no point in using a more complex update because our operator-splitting of some of the radiation terms means that we are performing our explicit update using a time-advanced radiation field, rather than the field at a half-time step. Truelove *et al.* (1998) show that one can avoid this problem for gravitational body forces because the potential is linear in the density, so it is possible to derive the half-time step potential from the whole time step states. No such fortuitous coincidence occurs for the radiation field. This necessarily limits us to first-order accuracy in time for the terms we treat explicitly. However, since these terms are always small compared to the dominant radiation terms, the overall scheme should still be closer to second order than first order in accuracy.

3.3.5 Advantages of the method

Our algorithm has two significant advantages in comparison to other approaches, in particular those based on comoving frame formulations of the equations (Turner & Stone 2001; Whitehouse *et al.* 2005; Hayes *et al.* 2006). In any of these approaches, since the radiation work terms are included in the implicit update, one must solve an implicit quartic equation arising from the combination of the terms $\kappa_{0P}(4\pi B - cE)$ and $\mathcal{P} : \nabla \mathbf{v}$. This may be done either at the same time one is iterating to update the flux divergence term $\nabla \cdot \mathbf{F}$ (Whitehouse *et al.* 2005) or in a separate iteration once the iterative solve for the flux divergence update is complete (Turner & Stone 2001; Hayes *et al.* 2006). In contrast, since our iterative update involves only $\kappa_{0P}(4\pi B - cE)$ and $\nabla \cdot \mathbf{F}$, using the Howell and Greenough (2003) algorithm we may linearize the equations and never need to solve a quartic, leading to a simpler update algorithm and a faster iteration step. Moreover, by using the Howell and Greenough (2003) time-centring, we obtain second-order accuracy in time whenever E is changing slowly, as opposed to the backwards Euler differencing of Turner and Stone (2001), Whitehouse *et al.* (2005) and Hayes *et al.* (2006), which is always first-order accurate in time. Thus, our algorithm provides a faster and simpler approach than the standard one.

A second advantage of our update scheme is that it retains the total energy-conserving character of the underlying equations. In each of the update steps involving radiation, for $\mathbf{f}_{e-rad}$ and $\mathbf{f}_i$, the non-advective update terms in the radiation and gas energy equations are equal and opposite. Thus, it is trivial to write the update scheme so that it conserves total energy to machine precision. This property is particularly important for turbulent flows with large radiation energy gradients, such as those that occur in massive star formation (Krumholz *et al.* 2007a), because numerical non-conservation is likely to be exacerbated by the presence of these features. In contrast, in comoving frame formalisms such as those of Turner and Stone (2001), Whitehouse and Bate (2004) and Hayes *et al.* (2006), the exchange terms in their gas and radiation energy equations are not symmetric. As a result, their update schemes do not conserve total energy exactly. The underlying physical reason for this asymmetry is that total energy is conserved only in inertial frames such as the lab frame; it is not conserved in the non-inertial comoving frame. For this reason, there is no easy way to write a conservative update scheme from a comoving formulation.

For dynamic diffusion problems, e.g. stellar interiors or radiation-dominated shocks, the work and advection terms can be comparable to or larger than the diffusion and heating/cooling terms and an algorithm that treats all the terms implicitly may be required.

The subtle limitation is in our treatment of the hydrodynamics. We perform the hydrodynamic update using a Riemann solver unmodified for the presence of radiation force, work, and heating and cooling terms. These terms should change the characteristic velocities of the wave families in ways that depend on the radiation-hydrodynamic regime of the system. The severity of these effects for a given problem depends on the degree of stiffness of the radiation source terms.

3.3.6 Tests in the optically thin and thick limits

As we have previously shown with MHD, strong tests of the coupling of the radiation with the hydrodynamics are crucial for gaining insight to different numerical approaches. Here we describe three tests of our radiation-hydrodynamic algorithm, done using our implementation of the algorithms in the Orion AMR code whose different components are described in detail by Truelove *et al.* (1998), Klein (1999), Fisher (2002) and Crockett *et al.* (2005). For all of these tests, we use a single fluid with no magnetic fields and no self-gravity. We describe additional tests in Krumholz *et al.* (2007b).

3.3.6.1 Radiating blast wave

We first compare to a test problem in which the gas is not at rest: a Sedov-type blast wave with radiation diffusion. Reinicke and Meyer-ter-Vehn (1991) gave the first similarity solution to the problem of a point explosion with heat conduction, and following Shestakov (1999) and Shestakov and Greenough (2001), we can adapt this solution to the case of a point explosion with radiation diffusion. This tests our code's ability to follow coupled radiation hydrodynamics in cases where radiation pressure is small.

We first summarize the semi-analytic solution. Consider an $n = 3$ dimensional space filled with an adiabatic gas with equation of state $P = (\gamma - 1)\rho e \equiv \Gamma \rho T$, where Γ is the gas constant. The Planck mean opacity κ_{OP} of the gas is very high, so the gas and radiation temperatures are always equal. The Rosseland mean opacity has a power-law form $\kappa_{OR} = \kappa_{OR,0}\rho^m T^{-n}$, and we assume that it is always high enough to place us in the diffusion regime, so $\lambda = 1/3$. Note that the choice of $-n = -3$ as the exponent of the opacity power law is a necessary condition for applying the Reinicke and Meyer-ter-Vehn (1991) conduction solution to our radiation diffusion problem. Moreover, the similarity solution does not include radiation energy density or pressure, so we consider only temperatures for which the gas energy density and pressure greatly exceed the radiation energy density and pressure, i.e. $\rho e \gg a_R T^4$.

Under the assumptions described above, we may rewrite the gas and radiation energy Eqs (3.46) and (3.47) as a single conduction-type equation for the temperature

$$\rho c_v \frac{\partial}{\partial t} T = \nabla(\chi_0 \rho^a T^b \nabla T), \tag{3.54}$$

where $c_v = \partial e/\partial T = \Gamma/(\gamma - 1)$ is the constant volume-specific heat of the gas, $\chi_0 = 4c a_R/(3\kappa_{OR,0})$, $a = -m$ and $b = n + 3$. This equation has the same form as the conduction equation considered by Reinicke and Meyer-ter-Vehn (1991).

Consider now a point explosion at the origin of a spherically symmetric region with an initial power-law density distribution $\rho(r, t = 0) = g_0 r^{-k_\rho}$. Initially, the gas temperature T and pressure P are negligible. The explosion occurs at the origin at time zero, so the initial gas energy density is $(\rho e)(r, t = 0) = E_0 \delta(\mathbf{r})$. Reinicke and Meyer-ter-Vehn (1991) show that if the initial density profile has a power-law index

$$k_\rho = \frac{(2b - 1)n + 2}{2b - 2a + 1}, \tag{3.55}$$

then one may obtain a similarity solution via the change of variables

$$\xi = \frac{r}{\zeta t^\alpha} \tag{3.56}$$

$$G(\xi) = \frac{\rho(r, t)}{g_0 r^{-k_\rho}} \tag{3.57}$$

$$U(\xi) = v(r, t)\frac{t}{\alpha r} \tag{3.58}$$

$$\Theta(\xi) = T(r, t)\Gamma\left(\frac{\alpha r}{t}\right)^2. \tag{3.59}$$

Here, ξ, $G(\xi)$, $U(\xi)$ and $\Theta(\xi)$ are the dimensionless distance, density, velocity and temperature, respectively,

$$\alpha = \frac{2b - 2a + 1}{2b - (n+2)a + n} \tag{3.60}$$

and ζ is a constant with units of [length][time]$^{-\alpha}$ whose value is determined by a procedure we discuss below.

With this similarity transformation, the equations of motion and heat conduction reduce to

$$U' - (1 - U)(\ln G)' + (n - k_\rho)U = 0 \tag{3.61}$$

$$(1 - U)U' + U(\alpha^{-1} - U) = \Theta[\ln(\xi^{2-k_\rho}G\Theta)]', \tag{3.62}$$

and

$$2[U' + nU - \mu(\alpha^{-1} - 1)] = \mu(1 - U)[\ln(\xi^2\Theta)]'$$

$$+ \beta_0\Theta^b G^{a-1}\xi^{(2b-1)/\alpha} \cdot \left((\ln\Theta)'' + [\ln(\xi^2\Theta)]'\right) \tag{3.63}$$

$$\cdot \left\{n - 2 + a[\ln(\xi^{-k_\rho}G)]' + (b+1)[\ln(\xi^2\Theta)]'\right\}, \tag{3.64}$$

where $()' \equiv d()/d\ln\xi$, $\mu = 2/(\gamma - 1)$ and

$$\beta_0 = \frac{2\chi_0(\alpha\zeta^{1/\alpha})^{2b-1}}{\Gamma^{b+1}g_0^{1-\alpha}}\,\text{sgn}(t). \tag{3.65}$$

This constitutes a fourth-order system of non-linear ordinary differential equations. All physical solutions to these equations pass through two discontinuities, a heat front and a shock front, with the heat front at larger radius. However, the jump conditions for these discontinuities are easy to determine, and one can integrate between them. For a given β_0, the solution depends only on the dimensionless parameter

$$\Omega = \frac{2\chi_0}{\Gamma^{b+1}g_0^{1-a}}\left(\frac{E_0}{g_0}\right)^{b-1/2}, \tag{3.66}$$

which measures the strength of the explosion. Large values of Ω constitute 'strong' explosions, and the ratio of heat front radius to shock front radius is a monotonically increasing function of Ω. It is important at this point to add a cautionary note: in deriving the similarity solution, we assumed that radiation energy density is negligible in comparison to gas energy density. This cannot strictly be true at early times, since at $t = 0$ the temperature diverges at the origin, and the radiation energy density varies as T to a higher power than the gas energy density. However, the true behaviour should approach the similarity solution at later times.

While we have reduced the gas dynamical equations to a system of ordinary differential equations that is trivial to integrate, solving the full problem is complex because the equations still depend on the unknown parameter β_0, which in turn depends on ζ. To solve the problem, we must determine β_0 from the given initial conditions. Reinicke and Meyer-ter-Vehn (1991) describe the iteration procedure required to do this in detail, and we only summarize it here. To find a solution, one first chooses a value $\xi_h > 1$ for the dimensionless radius of the heat front, applies the boundary conditions at the front and guesses a corresponding value of β_0. For each ξ_h, there exists a unique β_0 for which it is possible to integrate the equations back from $\xi = \xi_h$ to the location of the shock front at $\xi = \xi_s$, apply the shock jump conditions and continue integrating back to the origin at $\xi = 0$ without having the solution become double-valued and thus unphysical. One iterates to identify the allowed value of β_0 for the chosen ξ_h, and this gives the unique density, velocity and temperature profiles allowed for that ξ_h. However, the solution one finds in this way may not correspond to the desired value of Ω. Reinicke and Meyer-ter-Vehn (1991) show that

$$\Omega = \beta_0 \left[2\pi \int_0^{\xi_h} \xi^{n-k_\rho+1} G(U^2 + \mu\Theta) \, d\xi \right]^{b-1/2}. \tag{3.67}$$

Thus, each choice of ξ_h corresponds to a particular value of Ω, and one must iterate a second time to find the value of ξ_h that gives the value of Ω determined from the input physical parameters of the problem. Alternately, instead of specifying a desired value of Ω, one may specify a ratio $R = \xi_h/\xi_s$, which also determines a unique value for ξ_h.

For our comparison between the semi-analytic solution and the Orion, we adopt the parameters $\gamma = 7/5$, $c_v = 1/(\gamma - 1)$, $a = -2$, $b = 6$, $g_0 = \chi_0 = 1$ and $E_0 = 135$, which yields a strength $\Omega = 1.042 \times 10^{12}$ and a ratio $R = 2.16$. In the simulation, we turn off terms in the code involving radiation pressure and forces and we set $\lambda = 1/3$ exactly. We use 1D spherical polar coordinates rather than Cartesian coordinates; the solution procedures for this are identical to the ones outlined in Section 3.3.4, with the exception that the gradient and divergence operators have their spherical rather than Cartesian forms and the cell-centred finite differences are modified appropriately. Our computational domain falls in the range of $0 \leq r \leq 1.05$, resolved by 256, 512 or 1024 cells, and has reflecting inner and outer boundary conditions. To initialize the problem, we set initial density to the power-law profile $\rho = r^{-k_\rho}$ (with k_ρ set from Eq. (3.55)), the initial velocity to zero and the initial energy density to a small value, except in the cell adjacent to the origin, where its value is $\rho e = 135/(\gamma - 1)$.

Figures 3.6–3.8 compare the semi-analytic density, velocity and temperature profiles to the values we obtain from Orion after running to a time $t = 0.06$. As the plots show, the Orion results agree very well with the semi-analytic solution, and the agreement improves with increasing resolution. In the lowest resolution run,

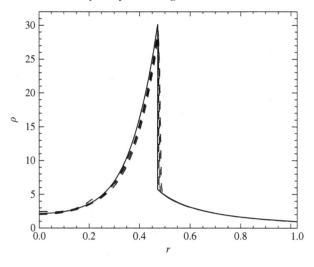

Fig. 3.6. Density ρ versus radius r for the radiating blast wave test. We show the semi-analytic solution (*solid line*) and the Orion results at resolutions of 256, 512 and 1024 cells (*dashed lines*). The 256-cell run is the dashed line furthest from the semi-analytic solution, and the 1024-cell run is the dashed line closest to it.

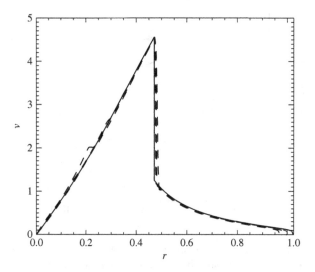

Fig. 3.7. Same as Figure 3.6, but for the velocity v.

there is a small oscillation in the density and velocity about a third of the way to the shock, which is likely due to the initial blast energy being deposited in a finite-volume region rather than as a true δ function. However, this vanishes at higher resolutions. Overall, the largest errors are in the temperature in the shocked gas.

As a metric of convergence, we plot the error of our simulation relative to the analytic solution as a function of resolution in Figure 3.9. We do this for the

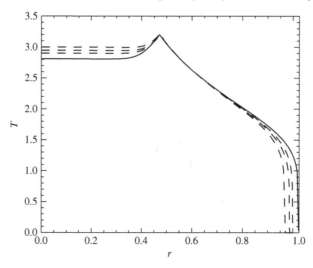

Fig. 3.8. Same as Figure 3.6, but for the temperature T.

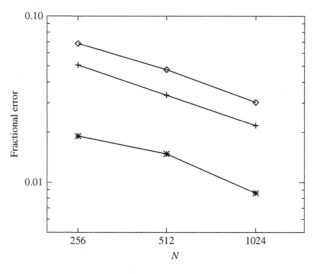

Fig. 3.9. Fractional error versus resolution N in the radiating blast wave test. The fractional error is defined as (simulation value – analytic value)/analytic value. We show error in the heat front radius r_h (*plus signs*), shock front radius r_s (*asterisks*) and their ratio $R = r_h/r_s$ (*diamonds*).

quantities r_h and r_s, the positions of the heat and shock fronts, and their ratio R. For this purpose, we define the location of the heat and shock fronts for the simulations as the positions of the cell edges where dT/dr and $d\rho/dr$ are most negative. As the plot shows, at the highest resolution the errors in all three quantities are $\lesssim 3\%$, and the calculation appears to be converging. The order of convergence is roughly

0.6 in all three quantities. It is worth noting that computing the locations of the heat and shock fronts is a particularly strong code test, because obtaining the correct propagation velocities for the two fronts requires that the code conserve total energy very well. Non-conservative codes have significant difficulties with this test (Timmes *et al.* 2006).

3.3.6.2 Radiation pressure tube

Our second test is to simulate a tube filled with radiation and gas. The gas within the tube is optically thick, so the diffusion approximation applies. The two ends of the tube are held at fixed radiation and gas temperature, and radiation diffuses through the gas from one end of the tube to the other. The radiation flowing through the tube exerts a force on the gas, and the gas density profile is such that, with radiation pressure, the gas is in pressure balance and should be stationary. For computational simplicity, we set the Rosseland- and Planck-mean opacities per unit mass of the gas to a constant value κ. A simulation of this system tests our code's ability to compute accurately the radiation pressure force in the very optically thick limit.

We first derive a semi-analytic solution for the configuration of the tube satisfying our desired conditions. Since the gas is very optically thick and we are starting the system in equilibrium, we set $T_{\text{rad}} = T_{\text{gas}} \equiv T$. The fluid is initially at rest. The condition of pressure balance amounts to setting $\partial(\rho\mathbf{v})/\partial t + \nabla \cdot (\rho\mathbf{v}\mathbf{v}) = 0$ in Eq. (3.45), so that the radiation pressure force balances the gas pressure gradient. Thus, we have

$$\frac{dP}{dx} + \lambda \frac{dE}{dx} = 0 \tag{3.68}$$

$$\left(\frac{k_B}{\mu}\rho + \frac{4}{3}a_R T^3\right) \frac{dT}{dx} + \frac{k_B}{\mu} T \frac{d\rho}{dx} = 0. \tag{3.69}$$

In the second step, we have set $E = a_R T^4$ and $P = \rho k_B T/\mu$, where μ is the mean particle mass, and we have set $\lambda = 1/3$ as is appropriate for the optically thick limit. The radiation energy Eq. (3.53) for our configuration is simply

$$\frac{d}{dx}\left(\frac{c\lambda}{\kappa\rho}\frac{dE}{dx}\right) = 0 \tag{3.70}$$

$$\frac{d^2 T}{dx^2} + 3\frac{1}{T}\left(\frac{dT}{dx}\right)^2 - \frac{1}{\rho}\left(\frac{d\rho}{dx}\right)\left(\frac{dT}{dx}\right) = 0. \tag{3.71}$$

Equations (3.69) and (3.71) are a pair of coupled non-linear ordinary differential equations for T and ρ. The combined degree of the system is three, so we need three initial conditions to solve them. Thus, let the tube run from $x = x_0$ to $x = x_1$, with temperature, density and density gradient T_0, ρ_0 and $(d\rho/dx)_0$ at x_0. For a given choice of initial conditions, it is trivial to solve Eqs (3.69) and (3.71) numerically to find the density and temperature profile. We wish to investigate both the radiation pressure- and the gas pressure-dominated regimes, so we choose parameters to ensure that our problem covers both. The choice $x_0 = 0$, $x_1 = 128\,\mathrm{cm}$, $\rho_0 = 1\,\mathrm{g\,cm^{-3}}$, $(d\rho/dx)_0 = 5 \times 10^{-3}\,\mathrm{g\,cm^{-4}}$ and $T_0 = 2.75 \times 10^7\,\mathrm{K}$ satisfies this requirement if we adopt $\mu = 2.33$, $m_P = 3.9 \times 10^{-24}\,\mathrm{g}$ and $\kappa = 100\,\mathrm{cm^2\,g^{-1}}$. Figure 3.10 shows the density, temperature and pressure as a function of position for these parameters.

We solve the equations to obtain the density and temperature as a function of position, and then set these values as initial conditions in a simulation. The simulation has 128 cells along the length of the tube on the coarsest level. We impose Dirichlet boundary conditions on the radiation field, with the radiation temperature at each end of the tube set equal to its value as determined from the analytic solution. We use symmetry boundary conditions on the hydrodynamics, so that gas can neither enter nor leave the computational domain. To ensure that our algorithm does not encounter problems at the boundaries between AMR levels, we refine the central one-fourth of the problem domain to double the resolution of the base

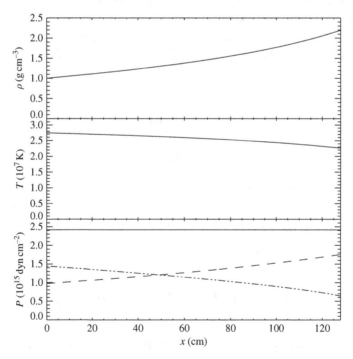

Fig. 3.10. Density, temperature and pressure versus position in the radiation tube problem. The bottom panel shows total pressure (*solid line*), gas pressure (*dashed line*) and radiation pressure (*dot-dashed line*).

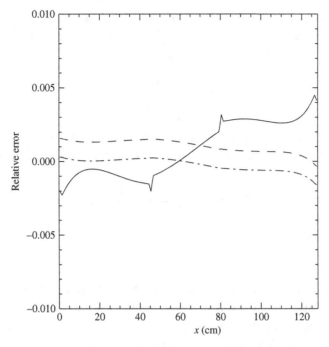

Fig. 3.11. Relative error in density (*solid line*), gas temperature (*dashed line*) and radiation temperature (*dot-dashed line*) in the radiation tube test.

grid. We evolve the system for ten sound-crossing times and measure the amount by which the density and temperature change relative to the exact solution. We plot the relative error, defined as (numerical solution − analytic solution)/analytic solution, in the density, gas temperature and radiation temperature in Figure 3.11. As the plot shows, our numerical solution agrees with the analytic result to better than 0.5% throughout the computational domain. The density error is smallest in the higher-resolution central region, as expected. There is a very small increase in error at level boundaries, but it is still at the less than 0.5% level.

3.3.6.3 Radiation-inhibited Bondi accretion

The previous test focuses on radiation pressure forces in the optically thick limit. To test the optically thin limit, we simulate accretion onto a radiating point particle. We consider a point mass M radiating with a constant luminosity L accreting from a background medium. The medium consists of gas which has zero velocity and density ρ_∞ far from the particle. We take the gas to be isothermal with constant temperature T and enforce that it is not heated or cooled radiatively by setting its Planck opacity $\kappa_{0P} = 0$. We set the Rosseland opacity of the gas to a constant non-zero value κ_{0R} and choose ρ_∞ such that the computational domain is optically thin. In this case, the radiation-free streams away from the point mass, and the radiation energy density and radiative force per unit mass on the gas are

$$E = \frac{L}{4\pi r^2 c} \tag{3.72}$$

$$\mathbf{f}_r = \frac{\kappa_{0R} L}{4\pi r^2 c} \left(\frac{\mathbf{r}}{r}\right), \tag{3.73}$$

where $\mathbf{r}$ is the radial vector from the particle and r is its magnitude. The gravitational force per unit mass is $\mathbf{f}_g = -(GM/r^2)(\mathbf{r}/r)$, so the net force per unit mass is

$$\mathbf{f} = \mathbf{f}_r + \mathbf{f}_g = -(1 - f_{Edd})\frac{GM}{r^2}\left(\frac{\mathbf{r}}{r}\right), \tag{3.74}$$

where

$$f_{Edd} = \frac{\kappa_{0R} L}{4\pi GMc} \tag{3.75}$$

is the fraction of the Eddington luminosity with which the point mass is radiating.

Since the addition of radiation does not alter the $1/r^2$ dependence of the specific force, the solution is simply the standard Bondi (1952) solution, but for an effective mass of $(1 - f_{Edd})M$. The accretion rate is the Bondi rate

$$\dot{M}_B = 4\pi \xi r_B^2 c_s \rho_\infty, \tag{3.76}$$

where

$$r_B = (1 - f_{Edd})\frac{GM}{c_s^2} \tag{3.77}$$

is the Bondi radius for the effective mass, c_s is the gas sound speed at infinity and ξ is a numerical factor of order unity that depends on the gas equation of state. For an isothermal gas, $\xi = e^{3/2}/4$, and the radial profiles of the non-dimensional density $\alpha \equiv \rho/\rho_\infty$ and velocity $u \equiv v/c_s$ are given by the solutions to the non-linear algebraic equations (Shu 1992)

$$x^2 \alpha u = \xi \tag{3.78}$$

$$\frac{u^2}{2} + \ln \alpha - \frac{1}{x} = 0, \tag{3.79}$$

where $x \equiv r/r_B$ is the dimensionless radius.

To set up this test, we make use of the Lagrangian sink particle algorithm of Krumholz *et al.* (2004), coupled with the 'star particle' algorithm of Krumholz *et al.* (2007a) which allows the sink particle to act as a source of radiation. We simulate a computational domain 5×10^{13} cm on a side, resolved by 256^3 cells, with a particle of mass $M = 10M_\odot$ and luminosity $L = 1.6 \times 10^5 L_\odot$ at its centre. We adopt fluid properties $\rho_\infty = 10^{-18}$ g cm^{-3}, $\kappa_{0R} = 0.4$ cm^2 g^{-1} and $c_s = 1.3 \times$

$10^7\,\mathrm{cm\,s^{-1}}$, corresponding to a gas of pure, ionized hydrogen with a temperature of 10^6 K. With these values, $f_{\mathrm{Edd}} = 0.5$, $r_B = 4.0 \times 10^{12}\,\mathrm{cm}$ and $\dot{M}_B = 2.9 \times 10^{17}\,\mathrm{g\,s^{-1}}$. We use inflow boundary conditions on the gas and Dirichlet boundary conditions on the radiation field, with the radiation energy density on the boundary set to the value given by Eq. (3.72).

Figure 3.12 compares the steady-state density α and velocity u computed by Orion to the analytic solution. The agreement is excellent, with differences between the analytic and the numerical solutions of $\sim 1\%$ everywhere except very near the accretion radius at $x = 0.25$. The maximum error is $\sim 10\%$ at the surface of the accretion region; this is comparable to the error in density for non-radiative Bondi accretion with similar resolution in Krumholz *et al.* (2004). In comparison, the solution is nowhere near the solution that would be obtained without

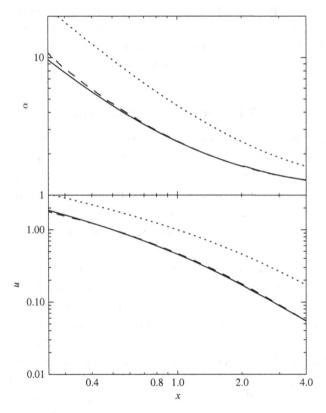

Fig. 3.12. Dimensionless density α (*upper panel*) and velocity u (*lower panel*) versus dimensionless position x for radiation-inhibited Bondi accretion. We show the analytic solution (*solid line*), the solution as computed with Orion (*dashed line*) and the analytic solution for Bondi accretion without radiation (*dotted line*). For the Orion result, the values shown are the radial averages computed in 128 logarithmically spaced bins running from the accretion radius $x = 0.25$ to the outer edge of the computational domain $x = 5$.

radiation. After running for 5 Bondi times ($=r_B/c_s$), the average accretion rate is 2.4×10^{17} g s^{-1}. While this differs from the analytic solution by 19%, the error is also not tremendously different from that obtained by Krumholz *et al.* (2004) when the Bondi radius was resolved by 4 accretion radii and is nowhere near the value of 1.2×10^{18} g s^{-1} which would occur without radiation.

We should at this point mention one limitation of our algorithm, as applied on an adaptive grid, that this test reveals. The $1/r^2$ gradient in the radiation energy density is very steep, and we compute the radiation force by computing gradients in E. We found that, in an AMR calculation, differencing this steep gradient across level boundaries introduced significant artifacts in the radiation pressure force. With such a steep gradient, we were only able to compute the radiation pressure force accurately on fixed grids, not adaptive grids. This is not a significant limitation for most applications though, since for any appreciable optical depth, the gradient will be much shallower than $1/r^2$. As the radiation pressure tube test in Section 3.3.6.2 demonstrates, in an optically thick problem, the errors that arise from differencing across level boundaries are less than 1%.

3.4 Adaptive mesh refinement

The numerical algorithm used for radiation hydrodynamics in Orion and MHD in Athena is a high-resolution conservative finite-difference method for solving the compressible Euler equations. The basic finite volume method was a higher-order extension of Godunov's method. This algorithm is second-order accurate for smooth flow problems and has a robust and accurate treatment of discontinuities. It has been used quite extensively to compute unsteady shock reflections in gases and has a demonstrated ability to resolve complex interactions of discontinuities found in astrophysical flows and star formation (Klein *et al.* 1994; Klein 1999).

The supplementary technique employed in Orion to further enhance the efficiency and resolution of our calculations is local AMR. The AMR algorithm we employ, similar to Berger and Oliger (1984), is a dynamic regridding strategy based on an underlying rectangular discretization of the spatial domain. The AMR scheme utilizes underlying rectangular grids at different levels of resolution. Linear resolution varies by integral refinement factors between levels, and a given grid is always fully contained within one at the next coarser level (excluding the coarsest grid). The AMR method dynamically resizes and repositions these grids and inserts new, finer ones within them according to adjustable refinement criteria, such as the numerical Jean's condition (Truelove *et al.* 1997). Fine grids are automatically removed as flow conditions require less resolution. During the course of the calculation, some pointwise measure of the error is computed at frequent intervals – typically every other time step. At those times, the cells that are identified are covered by a relatively small number of rectangular patches, which are refined by some even integer factor. Refinement is in both time and space, so that the calculation on the refined grids is computed at the same Courant–Freidrichs–Levy (CFL) number as that on the coarse grid. This procedure is applied recursively, i.e. the error on the refined grid monitored and the regions with large errors covered by refined rectangular patches. The overall algorithm is fully conservative: the finite difference approximations on each level are in conservation form, as is the coupling at the interface between grids at different levels of refinement.

There are several important features to this algorithm we wish to point out. The AMR uses a nested sequence of logically rectangular meshes to solve a partial differential equation (PDE). In this work, we assume the domain is a single rectangular grid although it may be decomposed into several coarse grids. It is required that the discretized solution be independent of the particular decomposition of the domain into subgrids. Grids must be properly nested such that a fine grid should be at least one cell away from the boundary of the next coarser grid unless it is touching the boundary of the physical domain. However, a fine grid can cross a coarser grid boundary and still be properly nested. In this case, the fine grid has more than one parent grid.

AMR in Orion contains five separate components (Klein 1999). The error estimation is used to estimate local truncation error. This determines where the solution is sufficiently accurate. The grid generator creates fine grid patches which cover the regions that need refinement. Data structure routines manage the grid hierarchy allowing access to the individual patches. Interpolation routines initialize a solution on a newly created fine grid and also provide the boundary conditions for integrating the fine grids. Flux correction routines ensure conservation at grid interfaces by modifying the coarse grid solution for coarse cells that are adjacent to a fine grid.

When all these components are assembled, a typical integration step proceeds as follows. The integration steps on different grids are interleaved so that before advancing a grid all the finer level grids have been integrated to the same time. One coarse grid cycle is then the basic unit of the algorithm. The mesh refinement factor in both space and time has been chosen most efficiently to be 4, although any even integer is possible. In practice, we use as many levels of refinement above the base coarse grid level as is required by the physics of the calculation. The regridding procedure is done every few time steps. The updating of the data on the locally refined grid structure is organized around the grouping of cells into rectangular grid patches, each one of which typically containing several hundred to several thousand grid cells. For example, the AMR code passes to a subroutine a rectangular grid of dependent variables and precomputed values in a set of ghost cells surrounding the grid and assumes that the subroutine updates the values in the rectangular grid by one time step, as well as passing back the fluxes at cell edges that had been used in the update. The overheads in both CPU and memory associated with the adaptive mesh structure have been kept quite small, relative to other irregular grid schemes. Typically, 80–90% of the total execution time is spent advancing cells in time using the finite difference code, while the memory required is that needed to store two copies of the solution on all of the grids. These overheads are low because they are determined by the number of rectangles into which the AMR solution has been divided; as opposed to being determined by the number of grid cells, which is the case with the irregular grid-adaptive algorithms.

In AMR, the computational volume consists of a hierarchical grid structure. A base Level 0 grid fills the computational volume, discretizing it on a rectangular grid with a resolution of Δx_0 in each direction. Multiple Level 1 grids of finer resolution $\Delta x_1 = \Delta x_0/r_1$ may be embedded within it, where $r_1 = 4$ is a typical choice. In turn, multiple Level 2 grids of resolution $\Delta x_2 = \Delta x_1/r_2$ may be embedded within Level 1 and so on. Grids at Level L always span an integral number of cells at Level $L - 1$, i.e. partial cell refinement is not permitted. Furthermore, a grid at Level L is always nested within a grid at Level $L - 1$ such that there is a buffer region of Level $L - 1$ cells surrounding it. In other words, a grid at Level L within a grid at Level $L - 1$ never shares a boundary with the Level $L - 1$ grid.

A key component of an AMR code is the procedure by which the decision is made whether or not a given portion of the flow is adequately resolved. In our code, this procedure is broken into two steps. In the first step, a specified property is measured in each cell, and the cell is flagged for refinement if a specified algorithm indicates that the measurement requires it. In the second step, the distribution of cells requiring refinement is analysed to determine the number, sizes and locations of grids to be inserted at the next finer level of resolution. These finer grids will always include every cell that was flagged for refinement, but they may also include additional cells that were not flagged. The degree to which the refinement is concentrated in the cells that require it is termed the *grid efficiency*. The grid efficiency is minimal when the smallest rectangular solid containing all flagged cells is refined. In this case, the fraction of refined volume actually containing cells that required refinement may be very small. The grid efficiency is maximal when the only cells refined are those that were flagged.

The MHD code Athena introduced in Section 3.2 also includes algorithms for local AMR. The strategy and issues for implementation are largely identical to those discussed above for Orion, with only a few additional complications introduced by MHD. For instance, the prolongation and restriction operators required to interpolate solutions between fine and coarse meshes must obey the divergence-free constraint; we use the formulation due to Tóth and Roe (2002) in Athena. Secondly, the directionally unsplit CTU integration method used in

Athena requires modification so that the divergence-free constraint is enforced during half-timestep (predict) steps in the algorithm. Several of the tests described in Section 3.2.8 have proved quite challenging for MHD AMR methods. In particular, advection of a field loop across a region of refined meshes can be most illuminating. If the divergence-free constraint is not satisfied, a variety of effects can be noted, e.g. the loop shows distortions, or slices of the field components may show oscillations, or a planar field loop advected in the $x-y$ plane in full 3D can show anomalous growth of B_z. To conclude the discussion here, in Figure 3.13 we show an image of a passive contaminant from the evolution of an MHD Rayleigh–Taylor instability computed on a 2D mesh with a base resolution of 8×16 with 5 levels of refinement (the effective resolution of the finest grid is 256×512). Initially, the magnetic field is uniform and horizontal. Further details are given in Stone *et al.* (2008). The primary point here is that with careful attention to the divergence-free constraint, AMR methods for MHD based on CT can be designed which show no perturbations or anomalies in the field components at fine/coarse boundaries.

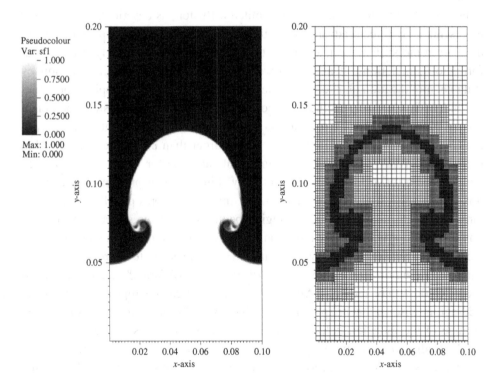

Fig. 3.13. (*Left*) Image of a passive contaminant in a single-mode MHD Rayleigh–Taylor instability computed with an AMR version of Athena. (*Right*) Grid distribution for the calculation.

3.5 Conclusions and future directions

We have described modern numerical algorithms for MHD and radiation hydrodynamics by focusing on the methods implemented in two separate codes: for MHD we have discussed Athena and for radiation hydrodynamics we have described the methods in Orion. Both codes continue to be developed in parallel directions. For example, Athena is being extended to radiation transport using variable Eddington tensors, while Orion has been extended to MHD using divergence-cleaning methods (Crockett *et al.* 2005) and a CT approach is being implemented as well.

In the future, significant advances in both algorithmic development as well as scalability and performance will be required to include all the relevant physics for star formation. For MHD, several robust and efficient algorithms seem to be available now. The focus is now turning to applications of high-order Godunov methods for MHD, including shearing-box studies of the MRI and the properties of supersonic MHD turbulence. Extensions of the methods to non-ideal MHD, including finite resistivity and viscosity, is also important for studying, e.g. protoplanetary disks.

For radiation hydrodynamics, flux-limited diffusion, as described here, already represents a significant advance over the use of the isothermal approximation or the barotropic stiffened equation of state approximation so widely used in star formation simulations. It also improves upon the Eddington approximation by suitable modification that compensates for errors made in dropping the time-dependent flux term by including a correction factor in the diffusion coefficient for the radiation flux such that the flux goes to the diffusion limit at large optical depth and correctly limits the flux to no larger than cE in the optically thin regime. Nevertheless, there are situations that can arise for which optically thick structures that arise in marginally optically thin flow can present problems for flux-limited diffusion. Radiation flow that impinges upon optically thick structures in its path tends to fill in the region behind the opaque structure eventually immersing the structure in the radiation field rather than allowing the region to correctly form a shadow (Hayes & Norman 2003). More accurate approaches such as variable Eddington tensors (Dykema *et al.* 1996), S_N (Adams & Larsen 2002) and Monte Carlo will be required. Although this situation can be improved by using an Eddington tensor moment approach, the shadow region still experiences moderate leakage of the radiation field after a few light-crossing times (Hayes & Norman 2003). Eddington tensors may also be costly with no guarantee of convergence. S_N methods and Monte Carlo methods are highly accurate and deal with the angle-dependent transport equation directly. They have not yet been developed for simulations in star formation because the cost in 3D is prohibitive. The S_N method is a short characteristic method in which a bundle of rays is created at every mesh point and are extended in the upwind direction only as far as the next spatial cell. The main problem is in finding the efficient angle set to represent the radiation field in 2D or 3D (Castor 2004). One might consider Monte Carlo methods to solve the transport equation. Although simple to implement (its great advantage), this method suffers from needing a vast number of operations per timestep to get accurate statistics in following the particles used to track the radiation field. Monte Carlo approaches, however, may be extremely well parallelized and may make the best use of massively parallel platforms scaling up to petascale machines in the future. Both of these methods will, however, avoid shadow effects and may be necessary to accurately treat optically thick inhomogeneous structures that form in accretion flows onto protostars. All of these approaches to radiation transport appear promising to us and are important to pursue in large-scale simulations in the future. To achieve the huge dynamic range in scale posed by star formation, scaling to high numbers of processors (several $\times 10^4$) will be necessary. Significant problems develop in load balancing issues (especially severe for AMR) for non-local physics. Strong coupling of non-local physics across the entire computational domain will require much more efficient parabolic and elliptic solvers for our algorithms. This too is a promising line of research for the future.

Acknowledgements

RIK thanks his collaborators Mark Krumholz, Robert Fisher and Christopher McKee for their contributions to the development of Orion. RIK is supported by NASA under NASA ATP grant NNG06GH96G, NSF grant AST-0606831 the US Department of Energy by the Lawrence Livermore National Laboratory under contract DE-AC52-07NA27344 and the hospitality of KITP at Santa Barbara. RIK is also supported by grants of high-performance computing resources from the NSF San Diego Supercomputer Center through NPACI program grant UCB267; the National Energy Research Scientific Computing Center, which is supported by the Office of Science of the US Department of Energy under Contract No. DE-AC03-76SF00098, through ERCAP grant 80325.

JMS thanks Tom Gardiner, John Hawley and Peter Teuben for their contributions to the development of Athena. In particular, the AMR version of Athena described here is the work of Tom Gardiner. JMS is supported by the DOE through DE-FG52-06NA26217. Simulations were performed on the Teragrid cluster at NCSA, the IBM Blue Gene at Princeton University and on computational facilities supported by NSF grant AST-0216105.

References

Adams, M. L. & Larsen, E. W. (2002), *Prog. Nucl. Energy*, **26**, 385

Alme, M. L. & Wilson, J. R. (1973), *ApJ.*, **186**, 1015

Balsara, D. S. (1998), *ApJ. Supp.*, **116**, 133

Balsara, D. S. (2001), *J. Comput. Phys.*, **174**, 614

Balsara, D. S. & Spicer, D. S. (1999), *J. Comput. Phys.*, **149**, 270

Berger, M. J. & Oliger, J. (1984), *J. Comput. Phys.*, **53**, 484

Bondi, H. (1952), *Mon. Not. R. Astron. Soc.*, **112**, 195

Cargo, P. & Gallice, G. (1997), *J. Comput. Phys.*, **136**, 446

Castor, J. I. (1972), *ApJ.*, **178**, 779

Castor, J. I. (2004), *Radiation Hydrodynamics* (Cambridge, UK: Cambridge University Press)

Colella, P. (1990), *J. Comput. Phys.*, **87**, 171

Colella, P. & Woodward, P. R. (1984), *J. Comput. Phys.*, **54**, 174

Crockett, R. K., Colella, P., Fisher, R. T., Klein, R. I., & McKee, C. F. (2005), *J. Comput. Phys.*, **203**, 422

Cunningham, A. J., Frank, A., Varniere, P., Mitran, S., & Jones, T. W. (2007), astro-ph0710.0424

Dai, W. & Woodward, P. R. (1994), *J. Comput. Phys.*, **115**, 485

Dai, W. & Woodward, P. R. (1998), *J. Comput. Phys.*, **142**, 331

Dedner, A., Kemm, F., Kröner, D., Munz, C.-T., Schnitzer, T., & Wesenberg, M. (2002), *J. Comput. Phys.*, **175**, 645

Dykema, P. G., Klein, R. I., & Castor, J. I. (1996), *ApJ.*, **457**, 892

Evans, C. R. & Hawley, J. F. (1988), *ApJ.*, **322**, 659

Falle, S. A. E. G., Komissarov, S. S., & Joarder, P. (1998), *MNRAS*, **297**, 265

Fisher, R. T. (2002), Single and Multiple Star Formation in Turbulent Giant Molecular Clouds, PhD thesis (Berkeley: University of California)

Fromang, S., Hennebelle, P., & Teyssier, R. (2006), *A&A*, **457**, 371

Gardiner, T. & Stone, J. M. (2005), *J. Comput. Phys.*, **205**, 509

Gardiner, T. & Stone, J. M. (2008), *J. Comput. Phys.* **227**, 4123

Hayes, J. C. & Norman, M. L. (2003), *ApJ. Suppl.*, **147**, 197

Hayes, J. C., Norman, M. L., Fiedler, R. A., Bordner, J. O., Li, P. S., Clark, S. E., ud-Doula, A., & Mac Low, M.-M. (2006), *ApJ. Suppl.*, **165**, 188

Howell, L. H. & Greenough, J. A. (2003), *J. Comput. Phys.*, **184**, 53

Hubeny, I. & Burrows, A. (2006), astro-ph/0609049

Kippenhahn, R. & Weigert, A. (1994), *Stellar Structure and Evolution* (Berlin: Springer-Verlag)

Klein, R. I. (1999), *J. Comp. App. Math.*, **109**, 123

Klein, R. I., McKee, C. F., & Colella, P. (1994), *ApJ.*, **420**, 213

Krumholz, M. R., Klein, R. I., & McKee, C. F. (2007a), *ApJ.*, **665**, 478

Krumholz, M. R., Klein, R. I., & McKee, C. F. (2007b), *ApJ.*, **667**, 626

Krumholz, M. R., McKee, C. F., & Klein, R. I. (2004), *ApJ.*, **611**, 399

Krumholz, M. R., McKee, C. F., & Klein, R. I. (2005), *ApJ. Lett.*, **618**, L33

Levermore, C. D. (1984), *J. Quant. Spectrosc. Radiat. Transf.*, **31**, 149

Levermore, C. D. & Pomraning, G. C. (1981), *ApJ.*, **248**, 321

Londrillo, P. & Del Zanna, L. (2004), *J. Comput. Phys.*, **195**, 17

Lowrie, R. B. & Morel, J. E. (2001), *J. Quant. Spectrosc. Radiat. Transf.*, **69**, 475

Lowrie, R. B., Morel, J. E., & Hittinger, J. A. (1999), *ApJ.*, **521**, 432

Mignone, A., Bodo, G., Massaglia, S., Matsakos, T., Tesileanu, O., Zanni, C., & Ferrari, A. (2007), astro-ph/070185

Mihalas, D. & Auer, L. H. (2001), *J. Quant. Spectrosc. Radiat. Transf.*, **71**, 61

Mihalas, D. & Klein, R. I. (1982), *J. Comput. Phys.*, **46**, 97

Mihalas, D. & Weibel-Mihalas, B. (1999), *Foundations of Radiation Hydrodynamics* (Mineola, NY: Dover)

Miyoshi, T. & Kusano, K. (2005), *J. Comput. Phys.*, **208**, 315

Pen, U.-L., Arras, P., & Wong, S. (2003), *ApJ. Supp.*, **149**, 447

Powell, K. G. (1994), *ICASE Report No. 94-24* (Langley, VA)

Powell, K. G., Roe, P. L., Linde, T. J., Gombosi, T. I., & de Zeeuw, D. L. (1999), *J. Comput. Phys.*, **153**, 284

Reinicke, P. & Meyer-ter-Vehn, J. (1991), *Phys. Fluids A*, **3**, 1807

Ryu, D. & Jones, T. W. (1995), *ApJ.*, **442**, 228

Ryu, D., Jones, T. W., & Frank, A. (1995), *ApJ.*, **452**, 785

Ryu, D., Miniati, F., Jones, T. W., & Frank, A. (1998), *ApJ.*, **509**, 244

Shestakov, A. I. (1999), *Phys. Fluids A*, **11**, 1091

Shestakov, A. I. & Greenough, J. A. (2001), AMRH and the High Energy Reinicke Problem, Tech. Rep. UCRL-ID-143937, Lawrence Livermore National Laboratory, Livermore, California, USA

Shu, F. H. (1992), *Physics of Astrophysics, Vol. II* (Mill Valley, CA: University Science Books)

Stone, J. M., Gardiner, T. A., Teuber, P., Hawley, J. F., & Simon, J. (2008), *ApJ. Suppl.*, **178**, 137

Stone, J. M. & Norman, M. L. (1992), *ApJ. Supp.*, **80**, 791

Thompson, T. A., Quataert, E., & Murray, N. (2005), *ApJ.*, **630**, 167

Timmes, F. X., Fryxell, B., & Hrbek, G. M. (2006), Spatial-Temporal Convergence Properties of the Tri-Lab Verification Test Suite in 1D for Code Project A, Tech. Rep. LA-UR-06-6444, Los Alamos National Laboratory

Toro, E. F. (1999), *Riemann Solvers and Numerical Methods for Fluid Dynamics* (Springer–Verlag, New York-Berlin-Heidelberg)

Tóth, G. (2002), *J. Comput. Phys.*, **161**, 605

Tóth, G., & Roe, P. (2002), *J. Comput. Phys.*, **180**, 736

Truelove, J. K., Klein, R. I., McKee, C. F., Holliman, J. H., Howell, L. H., & Greenough, J. A. (1997), *ApJ. Lett.*, **489**, L179

Truelove, J. K., Klein, R. I., McKee, C. F., Holliman, J. H., Howell, L. H., Greenough, J. A., & Woods, D. T. (1998), *ApJ.*, **495**, 821

Turner, N. J., & Stone, J. M. (2001), *ApJ. Suppl.*, **135**, 95

Turner, N. J., Stone, J. M., Krolik, J. H., & Sano, T. (2003), *ApJ.*, **593**, 992

Whitehouse, S. C., & Bate, M. R. (2004), *Mon. Not. R. Astron. Soc.*, **353**, 1078

Whitehouse, S. C., Bate, M. R., & Monaghan, J. J. (2005), *Mon. Not. R. Astron. Soc.*, **364**, 1367

Zachary, A. L., Malagoli, A., & Colella, P. (1994), *SIAM J. Sci. Comput.*, **15**, 263

Ziegler, U. (2005), *A&A*, **435**, 385

4

The role of jets in the formation of planets, stars and galaxies

R. E. Pudritz, R. Banerjee and R. Ouyed

Abstract

Astrophysical jets are associated with the formation of young stars of all masses, stellar and massive black holes, and perhaps even with the formation of massive planets. Their role in the formation of planets, stars and galaxies is increasingly appreciated and probably reflects a deep connection between the accretion flows – by which stars and black holes may be formed – and the efficiency by which magnetic torques can remove angular momentum from such flows. We compare the properties and physics of jets in both non-relativistic and relativistic systems and trace, by means of theoretical argument and numerical simulations, the physical connections between these different phenomena. We discuss the properties of jets from young stars and black holes, give some basic theoretical results that underpin the origin of jets in these systems, and then show results of recent simulations on jet production in collapsing star-forming cores as well as from jets around rotating Kerr black holes.

4.1 Introduction

The goal of this book, to explore structure formation in the cosmos and the physical linkage of astrophysical phenomena on different physical scales, is both timely and important. The emergence of multi-wavelength astronomy in the late twentieth century with its unprecedented ground- and space-based observatories, as well as the arrival of powerful new computational capabilities and numerical codes, has opened up unanticipated new vistas in understanding how planets, stars and galaxies form. Galaxy formation has turned out to be a complex problem which requires a deep understanding of star formation on galactic scales and how it feeds back on galactic gas dynamics. The discovery of significant numbers of massive black

Structure Formation in Astrophysics. ed. G. Chabrier. Published by Cambridge University Press.
© Cambridge University Press 2009.

holes in galactic nuclei now suggests that most galaxies harbour millions to billions of mass holes. The formation of these monsters and the effects that they can have on the early evolution of galaxies is still not well understood however. Star formation studies have made huge inroads, but must still take a large step forward before we can truly understand the origin of stellar masses and star formation rates in molecular clouds and galaxies. The newly discovered planetary systems bear little relation to textbook models of how our solar system is believed to have formed. It is also abundantly clear that planet and star formations are intimately coupled through the physics of protostellar disks that are their common cradles.

Astrophysical jets play an important role in the formation and evolution of stars, black holes, and perhaps massive planets. They appear to be an inescapable multiscale phenomenon that arises during structure formation. The enormous kinetic luminosity of many jets (often comparable to the bolometric luminosity of the central sources) also implies that they could have important feedback effects on structure formation – from the scale of the giant lobes of radio galaxies stretched out across many Mpc scales in the intergalactic medium (IGM), to jets from stellar mass black holes, down to the stirring up of molecular gas on sub-pc to pc scales in regions of low-mass star formation. Jets are observed to be ubiquitous during the process of star formation and associated with a very broad range of objects – from brown dwarfs to B and perhaps even O stars. It is now well established that stellar mass black holes are associated with jets and that the properties of such 'microquasars' scale very well only with the mass of the black hole (e.g., see review by Mirabel 2005) and scale naturally to the limit of massive black holes. In this regard, galactic jets are being increasingly associated with accreting massive black holes (e.g., see review by Blandford 2001).

In most cases, there is clear evidence that accretion disks are an essential part of the mechanism. The fact that outflows and jets are observed around nearly all astrophysical objects during the early stages of their existence – a time where the central objects (young stars, massive planets or black holes) undergo significant accretion from surrounding, collapsing gas structures and/or associated disks – argues for a deep link between accretion and outflow. A large body of observations and theoretical models increasingly suggests that such jets in all of these systems may be powered by the same mechanism, namely, hydromagnetic winds driven off of magnetized accretion disks (see, e.g., Livio 1999; Königl & Pudritz 2000; Pudritz *et al.* 2007 for reviews on this subject). The basic theory for hydromagnetic disk winds was worked out by Blandford and Payne (1982) in the context of models for active galactic nucleus (AGN) jets but was quickly found to be very important for understanding jets in protostellar systems (PN83). For accreting systems, the gravitational binding energy that is released as gas accretes through a disk and onto a central object that is the ultimate source of the energy for jets. Therefore, if this energy can be efficiently tapped and carried by the jet, then jet energies and other properties should simply scale with the depth of the gravitational potential well created by the central mass. The torques that can be exerted by outflows upon the underlying disks were shown to be much larger than can be produced by even strongly turbulent disks (Pudritz & Norman 1983; Pelletier & Pudritz 1992). Thus, jets can also carry off most of the angular momentum underlying disks, thereby assisting in the growth of central objects – be they stars, black holes or planets. (It should be noted that large-scale spiral waves in accretion disks can also efficiently transport angular momentum through disks.) Finally, the feedback of these powerful jets upon their surroundings can be important. In star-forming systems, outflows may help to drive turbulence within cluster-forming regions in molecular clouds, which would help to sustain such a region against global collapse.

It must be noted that isolated magnetized, spinning bodies such as magnetized A stars can also drive outflows. The extraordinary Chandra X-ray observations of Crab pulsars show that it is driving off a highly collimated, relativistic jet. The models that best describe the physics of all of these jets demonstrate that outflows can be driven and collimated from spinning bodies (be they disks, stars or compact objects). In protostellar jets, the slow spin of stars has been attributed either to star–disk coupling (Königl 1991) or to X-winds (Shu *et al.* 2000), but could also be explained as an accretion-powered stellar wind from the central star (Matt & Pudritz 2005). For black hole systems, the highly relativistic component of jets may be associated with electrodynamic processes and currents that arise from spinning holes in a magnetized environment provided by the surrounding disk (Blandford & Znajek 1977).

It is impossible to do justice to the enormous body of excellent work on the physics of jets in these different systems. In this chapter, we give specific examples of these basic themes from our own research in these fields in the context of some of the basic literature. In Section 4.2 we review aspects of unified models for non-relativistic and then relativistic jets in these systems. We follow this in Section 4.3 with a discussion of disk winds as an excellent candidate for a unified model, analysed particularly for protostellar jets. We follow this in Section 4.4 with simulations of jets during gravitational collapse and disk formation. We examine the role of feedback of jets in the specific context of star formation in Section 4.5. We then move on to the basics of relativistic jets from compact objects as well as black holes (Section 4.6) and then simulations (Section 4.7) of magnetohydrodynamic (MHD) jets in the general relativistic limit (GRMHD). We conclude with a comparison of the physics of relativistic versus non-relativistic jets.

4.2 Jets in diverse systems

4.2.1 Protostellar objects

Stars span about four decades in mass. Until recently, the study of outflows was restricted to sources that were easily detectable in millimetre surveys of molecular clouds. These studies show that outflows, over a vast range of stellar luminosities, have scalable physical properties. In particular, the observed ratio of the momentum transport rate (or thrust) carried by the CO molecular outflow to the thrust that can be provided by the bolometric luminosity of the central star (Cabrit & Bertout 1992) scales as

$$\frac{F_{\text{outflow}}}{F_{\text{rad}}} = 250 \left(\frac{L_{\text{bol}}}{10^3 L_\odot} \right)^{-0.3}, \tag{4.1}$$

This relation has been confirmed and extended by the analysis of data from over 390 outflows, ranging over six decades up to $10^6 L_\odot$ in stellar luminosity (Wu *et al.* 2004). This remarkable result shows that jets from both low- and high-mass systems, in spite of the enormous differences in the radiation fields from the central stars, are probably driven by a single, non-radiative mechanism.

Jets are observed to have a variety of structures and time-dependent behaviour – from internal shocks and moving knots to systems of bow shocks that suggest long-time episodic outbursts (e.g., see review by Bally *et al.* 2007). They show wiggles and often have corkscrew like structure, suggesting the presence either of jet precession or the operation of non-axisymmetric kink modes, or both. Numerical simulations have recently shown that, in spite of the fact that jets are predicted to have a dominant toroidal magnetic field, they are nevertheless stable against such non-linear modes (Ouyed *et al.* 2003; Nakamura & Meier 2004; Kigure & Shibata 2005). The non-linear saturation of the unstable modes is achieved by the natural regulation of the jet velocity to values near the local Alfvén speed. A second point is that jets will have some poloidal field along their 'backbone' and this too prevents a jet from falling apart.

Two types of theories have been proposed to explain protostellar jets, both of which are hydromagnetic: disk winds (e.g., see review by Pudritz *et al.* 2007) and the X-wind (e.g., see review by Shang *et al.* 2007). The latter model posits the interaction between a spinning magnetized stellar magnetosphere and the inner edge of an accretion disk as the origin of jets, while the former envisages jets as arising from large parts of magnetized disks. One of the best ways of testing such theories is by measuring jet rotation – which is a measure of how much angular momentum a jet can extract from its source. Significant observational progress has been made on this problem lately with the discovery of the rotation of jets from T-Tauri stars. The angular momentum that is observed to be carried by these rotating flows (e.g.

DG Tau) is a considerable fraction of the excess disk angular momentum – from 60 to 100% (Bacciotti 2004), which is consistent with the high extraction efficiency that is predicted by the theoretical models. Another important result is that there is too much angular momentum in the observed jet to be accounted for from the spinning star or the very innermost region of the disk as predicted by the X-wind model. The result is well explained by the disk wind model which predicts that the angular momentum derives from large reaches of the disk, typically from a region as large as 0.5 AU (Anderson *et al.* 2003).

4.2.2 *Jovian planets*

The striking similarity of the system of Galilean moons around Jupiter, with the sequence of planets in our solar system, has long suggested that these moons may have formed through a sub-disk around Jupiter (Mohanty *et al.* 2007). Recent hydrodynamical simulations of circumstellar accretion disks containing and building up an orbiting protoplanetary core have numerically proved the existence of a circumplanetary sub-disk in almost Keplerian rotation close to the planet (Kley *et al.* 2001). The accretion rate of these sub-disks is about $\dot{M}_{cp} = 6 \times 10^{-5} \, M_{\text{Jup}} \, \text{yr}^{-1}$ and is confirmed by many independent simulations. With that, the circum-planetary disk temperature may reach values up to 2000 K indicating a sufficient degree of ionization for matter–field coupling and would also allow for strong equipartition field strength (Fendt 2003). It should be possible, therefore, for lower-luminosity jets to be launched from the disks around Jovian mass planets.

The possibility of a planetary scale MHD outflow, similar to the larger-scale young stellar object (YSO) disk winds, is indeed quite likely because (i) the numerically established existence of circumplanetary disks is a natural feature of the formation of massive planets and (ii) of the feasibility of a large-scale magnetic field in the protoplanetary environment (Quillen & Trilling 1998; Fendt 2003). One may show, moreover, that the outflow velocity is of the order of the escape speed for the protoplanet, at about $60 \, \text{km s}^{-1}$ (Fendt 2003; Machida *et al.* 2006).

4.2.3 *Black holes*

Relativistic jets have been observed or postulated in various astrophysical objects, including AGNs (Urry & Padovani 1995; Ferrari 1998), microquasars in our galaxy (Mirabel & Rodríguez 1998) and gamma-ray bursts (GRBs) (Piran 2005). Table 4.1 summarizes the features or relativistic jets in different large-scale sources and demonstrates the wide range in power and Lorentz factors achieved by these systems.

Table 4.1. *The sources and features of large scale/relativistic jets*

Source	L (erg s^{-1})	Γ
AGNs/Quasars	10^{43}–10^{48}	~ 10
μ-quasars	10^{38}–10^{40}	1–10
GRBs	10^{51}–10^{52}	10^2–10^4

In the commonly accepted standard model of large scale/relativistic jets (Begelman *et al.* 1984), flow velocities as large as 99% of the speed of light (in some cases even beyond) are required to explain the apparent superluminal motion observed in many of these sources (Table 4.1). Later, considerations of stationary MHD flows have revealed that relativistic jets must be strongly magnetized (Michel 1969; Camenzind 1986; Li *et al.* 1992; Fendt & Camenzind 1996). In that case, the available magnetic energy can be transferred over a small amount of mass with high kinetic energy.

Here too, the most promising mechanisms for producing the relativistic jets involve magnetohydrodynamic centrifugal acceleration and/or magnetic pressure-driven acceleration from an accretion disk around the compact objects (Blandford & Payne 1982) or involve the extraction of rotating energy from a rotating black hole (Penrose 1969; Blandford & Znajek 1977). These models have been applied to explain jet features in the galactic microquasars GRS 1915 + 105 (Mirabel & Rodríguez 1994) and GRO J1655-40 (Tingay *et al.* 1995), or a rotating supermassive black hole in an AGN, which is fed by interstellar gas and gas from tidally disrupted stars. In general, these studies require solving special relativistic MHD (SRMHD) or GRMHD equations and often require sophisticated numerical codes. We describe the basic theory behind relativistic MHD jets and summarize recent GRMHD simulations of jets emanating from the vicinity of accreting black holes (see Sections 4.6 and 4.7) after we examine the general theory of outflows from magnetized spinning disks and objects.

4.3 Theory of disk winds

Given the highly non-linear behaviour of the force balance equation for jets (the so-called Grad–Shafranov equation), theoretical work has focused on tractable and idealized time-independent, and axisymmetric or self-similar models (BP82) of various kinds. We briefly summarize the theory in the following sections (see details in Pudritz 2004; Pudritz *et al.* 2007).

4.3.1 Conservation laws and jet kinematics

Conservation laws play a significant role in understanding astrophysical jets. This is because whatever the details, conservation laws strongly constrain the flux of mass, angular momentum and energy. What cannot be constrained by these laws will depend on the general physics of the disks such as on how matter is loaded onto field lines.

Jet dynamics can be described by the time-dependent equations of ideal MHD. The evolution of a magnetized rotating system that is threaded by a large-scale field **B** involves (i) the continuity equation for a conducting gas of density ρ moving at velocity **v** (which includes turbulence); (ii) the equation of motion for the gas which undergoes pressure (p), gravitational (with potential Φ) and Lorentz forces; (iii) the induction equation for the evolution of the magnetic field in the moving gas where the current density is $\mathbf{j} = (c/4\pi)\nabla \times \mathbf{B}$; (iv) the energy equation, where e is the internal energy per unit mass; and (v) the absence of magnetic monopoles. These are written as follows:

$$\frac{\partial \rho}{\partial t} + \nabla.(\rho \mathbf{v}) = 0 \tag{4.2}$$

$$\rho \left(\frac{\partial \mathbf{v}}{\partial t} + (\mathbf{v}.\nabla)\mathbf{v} \right) + \nabla p + \rho \nabla \Phi - \frac{\mathbf{j} \times \mathbf{B}}{c} = 0 \tag{4.3}$$

$$\frac{\partial \mathbf{B}}{\partial t} - \nabla \times (\mathbf{v} \times \mathbf{B}) = 0 \tag{4.4}$$

$$\rho \left(\frac{\partial e}{\partial t} + (\mathbf{v}.\nabla)e \right) + p(\nabla.\mathbf{v}) = 0 \tag{4.5}$$

$$\nabla.\mathbf{B} = 0 \tag{4.6}$$

Progress can be made by restricting the analysis to stationary as well as 2D (axisymmetric) flows, from which the conservation laws follow. It is useful to decompose vector quantities into poloidal and toroidal components (e.g. magnetic field $\mathbf{B} = \mathbf{B_p} + B_\phi \hat{e}_\phi$). In axisymmetric conditions, the poloidal field $\mathbf{B_p}$ can be derived from a single scalar potential $a(r, z)$ whose individual values, $a = \text{const.}$, define the surfaces of constant magnetic flux in the outflow and can be specified at the surface of the disk (Pelletier & Pudritz 1992, PP92).

4.3.1.1 Conservation of mass and magnetic flux

Conservation of mass and magnetic flux along a field line can be combined into a single function k that is called the 'mass load' of the wind which is a constant along a magnetic field line,

$$\rho \mathbf{v_p} = k \mathbf{B_p}. \tag{4.7}$$

This function represents the mass load per unit time, per unit magnetic flux of the wind. For axisymmetric flows, its value is preserved on each ring of field lines emanating from the accretion disk. Its value on each field line is determined by physical conditions – including dissipative processes – near the disk surface. It may be more revealingly recast as

$$k(a) = \frac{\rho v_p}{B_p} = \frac{d\dot{M}_w}{d\Psi}, \tag{4.8}$$

where $d\dot{M}_w$ is the mass flow rate through an annulus of cross-sectional area dA through the wind and $d\Psi$ is the amount of poloidal magnetic flux threading through this same annulus. The mass load profile is a function of the footpoint radius r_0 of the wind on the disk.

The toroidal field in rotating flows derives from the induction equation,

$$B_\phi = \frac{\rho}{k}(v_\phi - \Omega_0 r), \tag{4.9}$$

where Ω_0 is the angular velocity of the disk at the mid-plane. This result shows that toroidal fields in the jet are formed by winding up the field from the source. Their strength also depends on the mass loading as well as the jet density. Denser winds should have stronger toroidal fields. We note, however, that the density does itself depend on the value of k. Equation (4.9) also suggests that at higher mass loads, one has lower toroidal field strengths. This can be reconciled, however, since it can be shown from the conservation laws that the value of k is related to the density of the outflow at the Alfvén point on a field line, $k = (\rho_A/4\pi)^{1/2}$ (PP92). Thus, higher mass loads correspond to denser winds, and when this is substituted into Eq. (4.9), we see that this also implies stronger toroidal fields.

4.3.1.2 Conservation of angular momentum

Conservation of angular momentum along each field line leads to the conserved angular momentum per unit mass,

$$l(a) = r v_\phi - \frac{r B_\phi}{4\pi k} = \text{const.} \tag{4.10}$$

The form for l reveals that the total angular momentum is carried by both the rotating gas (first term) and the twisted field (second term), the relative proportion being determined by the mass load.

The value of $l(a)$ that is transported along each field line is fixed by the position of the Alfvén point in the flow, where the poloidal flow speed reaches the Alfvén speed for the first time ($m_A = 1$). It is easy to show that the value of the specific angular momentum is

$$l(a) = \Omega_0 r_A^2 = \left(\frac{r_A}{r_0}\right)^2 l_0. \tag{4.11}$$

where $l_0 = v_{K,0} r_0 = \Omega_0 r_0^2$ is the specific angular momentum of a Keplerian disk. For a field line starting at a point r_0 on the rotor (disk in our case), the Alfvén radius is $r_A(r_0)$ and constitutes a lever arm for the flow. The result shows that the angular momentum per unit mass that is being extracted from the disk by the outflow is a factor of $(r_A/r_0)^2$ greater than it is for gas in the disk. For typical lever arms, one particle in the outflow can carry the angular momentum of ten of its fellows left behind in the disk.

4.3.1.3 Conservation of energy

Conservation of energy along a field line is expressed as a generalized version of Bernoulli's theorem (this may be derived by taking the dot product of the equation of motion with $\mathbf{B_p}$). Thus, there is a specific energy $e(a)$ that is a constant along field lines, which may be found in many papers (BP82 and PP92). Since the terminal speed $v_p = v_\infty$ of the disk wind is much greater than its rotational speed, and for cold flows, the pressure may also be ignored, one finds the result:

$$v_\infty \simeq 2^{1/2} \Omega_0 r_A = \left(\frac{r_A}{r_0}\right) v_{esc,0}. \tag{4.12}$$

There are three important consequences for jet kinematics here: (i) the terminal speed exceeds the *local* escape speed from its launch point on the disk by the lever arm ratio; (ii) the terminal speed scales with the Kepler speed as a function of radius, so that the flow will have an onion-like layering of velocities, the largest inside and the smallest on the larger scales, as seen in the observations and (iii) the terminal speed depends on the depth of the local gravitational well at the footpoint of the flow – implying that it is essentially scalable to flows from disks around YSOs of any mass and therefore universal.

Another useful form of the conservation laws is the combination of energy and angular momentum conservation to produce a new constant along a field line (PP92): $j(a) \equiv e(a) - \Omega_0 l(a)$. This expression has been used (Anderson *et al.* 2003) to deduce the rotation rate of the launch region on the Kepler disk, where the observed jet rotation speed is $v_{\phi,\infty}$ at a radius r_∞ and which is moving in the poloidal direction with a jet speed of $v_{p,\infty}$. Evaluating j for a cold jet at infinity and noting that its value (calculated at the foot point) is $j(a_0) = -(3/2)v_{K,0}^2$, one solves for the Kepler rotation at the point on the disk where this flow was launched:

$$\Omega_0 \simeq \frac{v_{p,\infty}^2}{2v_{\phi,\infty} r_\infty}. \tag{4.13}$$

When applied to the observed rotation of the large velocity component (LVC) of the jet DG Tau (Bacciotti *et al.* 2002), this yields a range of disk radii for the observed rotating material in the range of disk radii 0.3–4 AU, and the magnetic lever arm is $r_A/r_0 \simeq 1.8$–2.6.

4.3.2 Angular momentum extraction

How much angular momentum can such a wind extract from the disk? The angular momentum equation for the accretion disk undergoing an external magnetic torque may be written as follows:

$$\dot{M}_a \frac{d(r_0 v_0)}{dr_0} = -r_0^2 B_\phi B_z|_{r_0, H},$$ (4.14)

where we have ignored transport by MRI turbulence or spiral waves. By using the relation between poloidal field and outflow on the one hand, as well as the link between the toroidal field and rotation of the disk on the other, the angular momentum equation for the disk yields one of the most profound scaling relations in disk wind theory, namely, the link between disk accretion and mass outflow rate (see KP00, PP92 for details):

$$\dot{M}_a \simeq \left(\frac{r_A}{r_0}\right)^2 \dot{M}_w.$$ (4.15)

The observationally well-known result that in many systems $\dot{M}_w/\dot{M}_a \simeq 0.1$ is a consequence of the fact that lever arms are often found in numerical and theoretical work to be $r_A/r_0 \simeq 3$ – the observations of DG Tau being a perfect example.

4.3.3 Jet power and universality

These results can be directly connected to the observations of momentum and energy transport in the molecular outflows. Consider the total mechanical power that is carried by the jet, which may be written as (Pudritz 2003)

$$L_{jet} = \frac{1}{2} \int_{r_i}^{r_j} d\dot{M}_w v_\infty^2 \simeq \frac{GM_* \dot{M}_a}{2r_i}\left(1 - \frac{r_i^2}{r_j^2}\right) \simeq \frac{1}{2} L_{acc}.$$ (4.16)

This explains the observations of Class 0 outflows wherein $L_w/L_{bol} \simeq 1/2$, since the main luminosity of the central source at this time is due to accretion and not nuclear reactions. (The factor of $1/2$ arises from the dissipation of some accretion energy as heat at the inner boundary.) The ratio of wind to stellar luminosity decreases at later stages because the accretion luminosity becomes relatively small compared to the bolometric luminosity of the star as it nears the zero age main sequence (ZAMS).

This result states that the wind luminosity taps the gravitational energy release through accretion in the gravitational potential of the central object – and is a direct consequence of Bernoulli's theorem. This, and the previous results, imply that jets may be produced in any accreting system. The lowest mass outflow that has yet been observed corresponds to a protobrown dwarf of luminosity $\simeq 0.09 L_\odot$, a stellar mass of only 20–45M_{Jup}, and a very low mass disk $<10^{-4} M_\odot$ (Bourke *et al.* 2005).

On very general grounds, disk winds are also likely to be active during massive star formation (Königl 1999). Such outflows may already start during the early collapse phase when the central YSO still has only a fraction of a solar mass (Banerjee & Pudritz 2006, 2007). Such early outflows may actually enhance the formation of massive stars via disk accretion by punching a hole in the infalling envelope and releasing the building radiation pressure (Krumholz *et al.* 2005).

4.3.4 Jet collimation

In the standard picture of hydromagnetic winds, collimation of an outflow occurs because of the increasing toroidal magnetic field in the flow resulting from the inertia of the gas. Beyond the Alfvén surface, Eq. (4.8) shows that the ratio of the toroidal field to the poloidal field in the jet is of the order $B_\phi / B_p \simeq r/r_{\mathrm{A}} \gg 1$, so that the field becomes highly toroidal. In this situation, collimation is achieved by the tension force associated with the toroidal field which leads to a radially inward-directed component of the Lorentz force (or 'z-pinch'): $F_{\mathrm{Lorentz,r}} \simeq j_z B_\phi$ (see Ouyed, Clarke and Pudritz 2003 for the stability of such systems).

In Heyvaerts and Norman (1989), it was shown that two types of solution are possible depending upon the asymptotic behaviour of the total current intensity in the jet,

$$I = 2\pi \int_0^r j_z(r', z')\mathrm{d}r' = \left(\frac{c}{2}\right) r B_\phi. \tag{4.17}$$

In the limit that $I \to 0$ as $r \to \infty$, the field lines are paraboloids which fill space. On the contrary, if the current is finite in this limit, then the flow is collimated to cylinders. The collimation of a jet therefore depends upon its current distribution – and hence on the radial distribution of its toroidal field – which, as we saw earlier, depends on the mass load. Mass loading therefore must play a very important role in controlling jet collimation.

It can be shown (PRO Pudritz *et al.* 2006) from this that jets should show different degrees of collimation depending on how they are mass loaded. As an example, neither the highly centrally concentrated magnetic field lines associated with the initial split-monopole magnetic configuration used in simulations by Romanova *et al.* (1997) nor the similar field structure invoked in the X-wind (see review by Shu *et al.* 2000) should become collimated in this picture. On the contrary, less centrally (radially) concentrated magnetic configurations such as the potential configuration of Ouyed *et al.* (1997) and BP82 should collimate to cylinders.

This result also explains the range of collimation that is observed for molecular outflows. Models for observed outflows fall into two general categories: the jet-driven bow shock picture and a wind-driven shell picture in which the molecular gas is driven by an underlying wide-angle wind component such as given by the X-wind (see review by Cabrit *et al.* 1997). A survey of molecular outflows by Lee *et al.* (2000) found that both mechanisms are needed in order to explain the full set of systems observed.

4.4 Gravitational collapse, disks and the origin of outflows

Jets are expected to be associated with gravitational collapse because disks are the result of the collapse of rotating molecular cloud cores. One of the first simulations to show how jets arise during gravitational collapse is the work of Tomisaka (1998, 2002). The author used as initial conditions the collapse of a magnetized rotating filament (cylinder) of gas and showed that this gave rise to the formation of a disk from which a centrifugally driven disk wind was produced.

Banerjee and Pudritz revisited the problem of collapsing magnetized molecular cloud cores in Banerjee and Pudritz (2006) (low-mass cores; BP06) and Banerjee and Pudritz (2007) (high-mass cores; BP07). Here, the initial cores are modelled as supercritical Bonnor–Ebert (B–E) spheres with a slight spin. B–E spheres are well-controlled initial setups for numerical experiments, and there are plenty of observed cores which can fit by a B–E profile (see Lada *et al.* 2007 on a compilation of hydrostatic cores). Recent efforts have focused on understanding the evolution of magnetized B–E spheres (Matsumoto & Hanawa 2003; Machida *et al.* 2005a,b; Hennebelle & Fromang 2008; Hennebelle & Teyssier 2008). Whereas purely hydrodynamic collapses of such objects never show outflows (Foster & Chevalier 1993; Banerjee *et al.* 2004), the addition of a magnetic field produces them.

Outflows are also implicated in playing a fundamental role in controlling the overall amount of mass that is incorporated into a new star from the original core. Observations of the so-called core mass function (CMF) that describes the spectrum of core masses, show that it has nearly the identical shape as the initial mass function (IMF) that describes the spectrum of initial stellar masses – with one caveat. The CMF needs to be multiplied by a factor of 1/3 (in one well-studied case – see Alves *et al.* 2007) in order to align them. The suggestion is that prostellar outflows are responsible for removing as much as 2/3 of a core's mass in the process of collapse and star formation.

In BP06, the FLASH adaptive mesh refinement (AMR) code (Fryxell *et al.* 2000) was used to follow the gravitational collapse of magnetized molecular cloud cores. This code allowed the Jeans length to be resolved with at least eight grid points

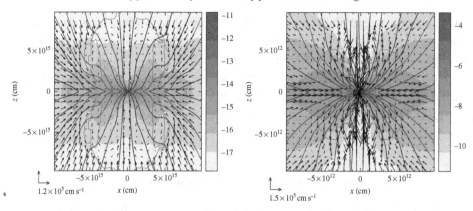

Fig. 4.1. Large-scale outflow (*left panel*, scale of hundreds of AU) and small-scale disk wind and jet formed (*right panel*, scale of a fraction of an AU) during the gravitational collapse of a magnetized B–E rotating cloud core. Cross-sections through the disk and outflows are shown – the contour lines marks the Alfvén surface. Snapshots taken of an adaptive mesh calculation at about 70 000 years into the collapse. (Adapted from Banerjee and Pudritz (2006).)

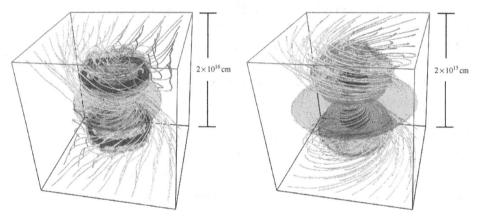

Fig. 4.2. Magnetic field line structure, outflow and disk. The two 3D images show the magnetic field lines, isosurfaces of the outflow velocities and isosurfaces of the disk structure at the end of our simulation ($t \simeq 7 \times 10^4$ years) at the two different scales as shown in Figure 4.1. The isosurfaces of the upper panel refer to velocities $0.18 \, \mathrm{km \, s^{-1}}$ and $0.34 \, \mathrm{km \, s^{-1}}$ and a density of $2 \times 10^{-16} \, \mathrm{g \, cm^{-3}}$ whereas the lower panel shows the isosurfaces with velocities $0.6 \, \mathrm{km \, s^{-1}}$ and $2 \, \mathrm{km \, s^{-1}}$ and the density at $5.4 \times 10^{-9} \, \mathrm{g \, cm^{-3}}$. (Adapted from Banerjee and Pudritz (2006).)

throughout the collapse calculations. This ensures compliance with the Truelove criterion (Truelove *et al.* 1997) to prevent artificial fragmentation.

The results of the low-mass core simulation are shown in Figures 4.1 and 4.2. They show the end state (at about 70 000 years) of the collapse of a B–E sphere that is chosen to be precisely the Bok globule observed by Alves *et al.* (2001) – whose mass is $2.1 M_\odot$ and radius $R = 1.25 \times 10^4$ AU at an initial temperature of 16 K. Two types of outflow can be seen: (i) an outflow that originates at scale of $\simeq 130$ AU on the forming disk that consists of a wound-up column of toroidal magnetic field whose pressure gradient pushes out a slow outflow and (ii) a disk-wind that collimates into a jet on scale of 0.07 AU. A tight protobinary system has formed in this simulation, whose masses are still very small $\leq 10^{-2} M_\odot$, which is much less than the mass of the disk at this time $\simeq 10^{-1} M_\odot$. The outer flow bears the hallmark of a magnetic tower, first observed by Uchida and Shibata (1985) and studied analytically by Lynden-Bell (2003). Both flow components are unbound, with the disk wind reaching 3 km s^{-1} at 0.4 AU which is quite super-Alfvénic and above the local escape speed. The outflow and jet speeds will increase as the central mass grows.

In Banerjee and Pudritz (2007), a similar setup to the one described above is used to study the formation of massive stars. In a unified picture of star formation, one would also assume that outflows and jets will also be launched around young massive stars. Due to the deeply embedded nature of massive star formation, observations of jets and outflows are much more difficult than for the more isolated low-mass stars. Nevertheless, there is increasing observational evidence for outflows around young massive stars (Arce *et al.* 2007; Zhang *et al.* 2007).

The simulations of collapsing magnetized massive cores ($\sim 170 M_\odot$) show clear signs of outflows during the early stages of massive star formation. Figure 4.3 shows two 2D snapshots from this simulations where the onset of an outflow is visible. Early outflows of this kind might have a large impact on the accretion history of the young massive star. Krumholz *et al.* (2005) showed that cavities blown

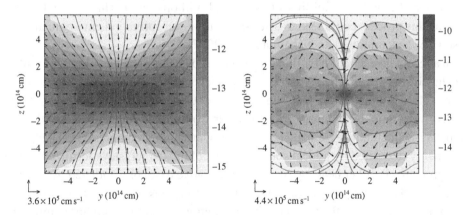

Fig. 4.3. Snapshots of the central region of a collapsing magnetized massive cloud core ($M_{\text{core}} \sim 170 M_\odot$). The *left panel* shows the situation at $t = 1.45 \times 10^4$ years ($1.08\, t_{\text{ff}}$) into the collapse and before the flow reversal and *right panel* shows the configuration 188 years later when the outflow is clearly visible. (Adapted from Banerjee and Pudritz (2007).)

by outflows help to release the radiation pressure from the newly born massive star which in turn relaxes radiation pressure-limited accretion onto the central object.

From these simulations, one can conclude that the theory and computation of jets and outflows is in excellent agreement with new observations of many kinds. Disk winds are triggered during magnetized collapse and persist throughout the evolution of the disk. They efficiently tap accretion power, transport a significant part portion of the disk's angular momentum and can achieve different degrees of collimation depending on their mass loading.

4.5 Feedback from collimated protostellar jets?

The interstellar medium (ISM) and star-forming molecular clouds are permeated by turbulent, supersonic gas motions (e.g., see recent reviews by Elmegreen & Scalo 2004; Mac Low & Klessen 2004; Ballesteros-Paredes *et al.* 2007 and references therein). Supersonic turbulence in molecular clouds is known to play at least two important roles: it can provide pressure support to help support molecular clouds against rapid collapse and it can also produce the system of shocks and compressions throughout such clouds which fragment the cloud into the dense cores that are the actual sites of gravitational collapse and star formation. One of the major debates in the literature is on the question of just how long this turbulence can be sustained – and with it, star formation. Despite the importance of supersonic turbulence for the process of star formation, its origin is still unclear. Norman and Silk (1980) proposed that Herbig-Haro (HH) outflows could provide the energetics to drive the turbulence in molecular clouds that could keep star formation active for a number of cloud free-fall times (i.e. several million years). This is an attractive idea as this process could lead to a self-regulating star formation environment (see also Li & Nakamura 2006; Nakamura & Li 2007).

In a recent investigation by Banerjee *et al.* (2007), this idea was addressed within a detailed study of feedback from collimated jets on their supra-core scale environment. These studies show that supersonic turbulence excited by collimated jets decays very quickly and does not spread far from the driving source, i.e. the jet. Supersonic fluctuations, unlike subsonic ones, are highly compressive. Therefore, the energy deposited by a local source, like a collimated jet, stays localized as the compressed gas either heats up (in the non-radiative case) or is radiated by cooling processes. The re-expansion of the compressed regions excites only subsonic or marginally supersonic fluctuations. This is contrary to almost incompressible subsonic fluctuations. These fluctuations travel like linear waves with little damping into the ambient medium and make up most of the overall velocity excitations.

In the study of Banerjee *et al.* (2007), a series of numerical experiments were performed with the AMR code FLASH. Here, the jet is modelled as a kinetic energy injection from the box boundary. The energy injection could be switched on and off after a certain amount of time, i.e. the jets are either continuously driven or transient. These jets are interacting either with a homogeneous or a clumpy environment, where the latter is modelled as a spherical overdensity. Additionally, the influence of magnetic fields on jet-excited fluctuations were considered in this study.

The authors quantify the jet-excited motions of the gas using velocity probability distribution functions (PDFs) which can be regarded as volume-weighted histograms. From these PDFs, the distinction between subsonic and supersonic fluctuations comes about naturally: the supersonic regime is strongly suppressed compared to the subsonic regime. Additionally, separate decay laws of the kinetic energy for the subsonic and supersonic regimes were derived to quantify their different behaviours with time.

Figure 4.4 shows that the kinetic energy of the supersonic fluctuations decays much faster than subsonic contributions. Furthermore, one can see from Figure 4.5 that the jet-excited fluctuations are mainly subsonic which is due to the fact that the supersonic excitations do not travel far from the edge of the jet (see *right panel* of Figure 4.5).

This study shows that supersonic fluctuations are damped quickly because they excite mainly compressive modes. The re-expansion of the compressed overdensities drives mainly subsonic velocity fluctuations that then propagate further into

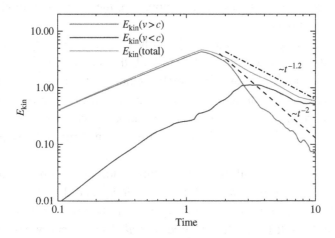

Fig. 4.4. Time evolution of the kinetic energies, E_{kin}, for a transient jet with a speed of Mach 5. The quantities are divided into a supersonic regime, $v > c$, and a subsonic regime, $v < c$. The decay of supersonic energy contributions is much faster (faster than $\propto t^{-2}$) than the subsonic one. The time shift between the peaks in the kinetic energies shows that the subsonic fluctuations are powered by the decay of the supersonic motions. (Adapted from Banerjee *et al.* 2007.)

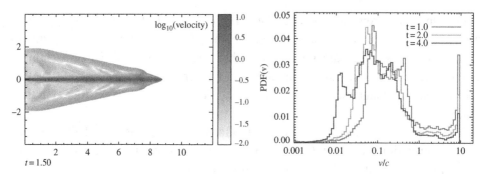

Fig. 4.5. Velocity structure and velocity PDFs at different times of a continuously driven Mach 10 jet. Essentially no supersonic fluctuations in the ambient gas get excited by the jet (the peak at $v/c = 10$ is the jet itself). (Adapted from Banerjee *et al.* 2007.)

the ambient medium. This is in spite of the appearance of bow shocks and instabilities. In particular, instabilities, such as Kelvin–Helmholtz modes at the edge of the jet, develop most efficiently for transonic or slower velocities. High-velocity jets, on the contrary, are bullet-like and stay very collimated, transiting the surrounding cloud without entraining much of its gas. From the point of view of jet-driven supersonic turbulence in molecular clouds, this is a dilemma which is difficult to circumvent. Even in the case of overdense jets which affect more gas of the surrounding media and have higher momenta, the supersonic motions do not propagate far from their source.

Simulations of magnetized jets in this study show that jets stay naturally more collimated if the magnetic field is aligned with the jet axis and therefore entrain less gas. Furthermore, perpendicular motions are damped by magnetic tension preventing a large spread of high-amplitude fluctuations. On the contrary, perpendicular field configurations support the propagation of such modes which are able to spread into a large volume. Nevertheless, the vast majority of these motions are still subsonic.

This study shows that collimated jets from young stellar objects are unlikely drivers of large-scale *supersonic* turbulence in molecular clouds. Alternatively, it could be powered by large-scale flows which might be responsible for the formation of the cloud itself (Ballesteros-Paredes *et al.* 1999). Energy cascading down from the driving scale to the dissipation scale will then produce turbulent density and velocity structure in the inertial range in between (Lesieur 1997). If the large-scale dynamics of the ISM is driven by gravity (as suggested, e.g., by Li *et al.* 2005, 2006), gravitational contraction would also determine to a large extent the internal velocity structure of the cloud. Otherwise, blast waves and expanding shells from supernovae are also viable candidates to power supersonic turbulence in molecular clouds (see, e.g., Mac Low & Klessen 2004).

4.6 Relativistic jets: theory

The theory of relativistic MHD jets is best described in the SRMHD regime. The essential parameter is the *magnetization* parameter (Michel 1969),

$$\sigma = \frac{\Psi^2 \Omega_F^2}{4 \dot{M} c^3}.$$ (4.18)

The iso-rotation parameter Ω_F is frequently interpreted as the angular velocity of the magnetic field lines. The function $\Psi = B_p r^2$ is a measure of the magnetic field distribution (see Li 1993) and $\dot{M} \equiv \pi \rho v_p R^2$ is the mass flow rate within the flux surface. Equation (4.18) demonstrates that the launch of a highly relativistic (i.e. highly magnetized) jet essentially requires at least one of three conditions: rapid rotation, strong magnetic field and/or a comparatively low mass load.

In the case of a spherical outflow ($\Psi = $ const) with negligible gas pressure, one may derive the Michel scaling between the asymptotic Lorentz factor and the flow magnetization (Michel 1969),

$$\Gamma_\infty = \sigma^{1/3}.$$ (4.19)

Depending on the exact magnetic field distribution $\Psi(r, z)$ (which describes the *opening* of the magnetic flux surfaces), in a *collimating jet* the matter can be substantially accelerated beyond the fast point magnetosonic point (Begelman & Li 1994; Fendt & Camenzind 1996), as it is moved from infinity to a finite radius of several Alfvén radii. As a result, the power-law index in Eq. (4.19) can be different from the Michel scaling (see Fendt & Camenzind 1996; Vlahakis & Königl 2001, 2003).

The *light cylinder* (hereafter l.c.) is located at the cylindrical radius $r_l = c/\Omega_F$. At the l.c., the velocity of the magnetic field lines 'rotating' with angular velocity Ω_F coincides with the speed of light.[1] The l.c. has to be interpreted as the Alfvén surface in the limit of vanishing matter density (force-free limit). The location of the l.c. determines the relativistic character of the magnetosphere. If the *l.c.* is comparable to the dimensions of the object investigated, a relativistic treatment of MHD is required.

While SRMHD can be invoked to study jets at scales larger than the gravitational radius of the black hole, close to the horizon, *general relativity* becomes relevant and the MHD equations need to be coupled to general relativity. This often requires

---

[1] Outside the l.c., the magnetic field lines 'rotate' faster than the speed of light. As the field line is not a physical object, the laws of physics are not violated.

the use of sophisticated numerical codes in order to capture the complexity of jet physics in GRMHD. In the (non-rotating) black hole's Schwarzschild spacetime, the GRMHD equations are identical to the SRMHD equations in general coordinates, except for the gravitational force terms and the geometric factors of the lapse function.

4.7 SRMHD and GRMHD simulations

Full GRMHD numerical simulations on the formation of jets near a black hole were first performed by Kudoh *et al.* (1998). This was followed by a plethora of GRMHD codes with fixed spacetimes used to investigate jets from accreting black holes (De Villiers & Hawley 2003; Gammie *et al.* 2003; Komissarov 2004; Anninos *et al.* 2005; Antón *et al.* 2006). Here, we focus on the most recent simulations that have managed to reach the highest Lorentz factors for the study of GRMHD jets (Figure 4.6).

Recent state-of-the-art GRMHD simulations have shown that the accretion flow launches energetic jets in the axial funnel region (i.e. low-density region) of the disk/jet system, as well as a substantial coronal wind (De Villiers *et al.* 2005). The jets feature knot-like structures of extremely hot, ultra-relativistic gas; the gas in these knots begins at moderate velocities near the inner engine and is accelerated to ultra-relativistic velocities (Lorentz factors of 50 and higher) by the Lorentz force in the axial funnel. The increase in jet velocity takes place in an acceleration zone extending to at least a few hundred gravitational radii from the inner engine. The overall energetics of the jets are strongly spin-dependent, with high-spin black holes producing the highest energy and mass fluxes. In addition, with high-spin black holes, the ultra-relativistic outflow is cylindrically collimated within a few hundred gravitational radii of the black hole, whereas in the zero-spin case the jet retains a constant opening angle of approximately 20°.

Figure 4.7 shows the Lorentz factor, Γ, in the funnel outflow from the GRMHD simulations discussed in De Villiers *et al.* (2005). Elevated values of the Lorentz factor are found in compact, hot, evacuated knots that ascend the funnel radially; the combination of low density and high temperature in the knots.[2] The highest values of Lorentz factor reach the maximum allowed by the code ($\sim$50, and test runs show that much higher values could be reached with a higher ceiling), and these are only found at large radii, suggesting the presence of an extensive region where the knots are gradually accelerated to higher Lorentz factors.

While the details of the mechanism by which GRMHD jets are launched are yet to be fully understood, the following summary addresses the salient points of the

---

[2] Ouyed *et al.* (1997) discuss the appearance of knots in MHD jet simulations; the knots seen in those simulations consisted of high-density material, in contrast to what is observed in the present simulations.

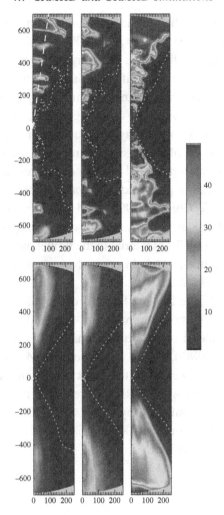

Fig. 4.6. Lorentz factor from GRMHD simulations. In all panels, the axes are in units of M (here $1M \approx 4$ km). The black hole is located at the origin. The *top three panels* show, left to right, plots of Lorentz factor for decreasing black hole rotation $a/M = 0.995, 0.9, 0.0$, respectively. The dotted contour marks the boundary of the jets. The only region where elevated values of Γ are found is in the jets (and also in the bound plunging inflow near the black hole, which is not resolved at the scale of this figure). Maximum values of Γ are found in knots that appear episodically in the upper and lower parts of the funnel. The *bottom three panels* show, from left to right, the corresponding time-averaged value of the Lorentz factor. The plots show evidence of spin-dependent collimation: cylindrical collimation is seen in the high-spin models, while the zero-spin model shows no such collimation. These plots also show evidence of an extended acceleration zone: large Lorentz factors are built up over the full radial range. (Adapted from De Villiers *et al.* (2005).)

complex dynamics by which the plasma is unbound from the vicinity of the compact star (see discussion in De Villiers *et al.* 2005; McKinney & Narayan 2007). In the region of the accretion flow near the marginally stable orbit, both pressure gradients and the Lorentz force act to lift material away from the equatorial plane. Some of this material is launched magneto-centrifugally in a manner reminiscent of the scenario of Blandford and Payne (1982), generating the coronal wind; some of this material, which has too much angular momentum to penetrate the centrifugal barrier, also becomes part of the massive funnel-wall jet. There is also evidence that the low-angular momentum funnel outflow originates deeper in the accretion flow. Some of this material is produced in a gravitohydromagnetic interaction in the ergosphere (Punsly & Coroniti 1990), and possibly in a process similar to that proposed by Blandford and Znajek (1977) where conditions in the ergosphere approach the force-free limit. The material in the funnel outflow is accelerated by a relatively strong, predominantly radial Lorentz force; gas pressure gradients in the funnel do not contribute significantly.

In the context of GRBs, the Lorentz factor Γ of the relativistic wind must reach high values ($\Gamma \sim 10^2$–10^4) both to produce γ-rays and to avoid photon–photon annihilation along the line of sight, whose signature is not observed in the spectra of GRBs (Goodman 1986). As we have seen, GRMHD simulations indicate Lorentz factors of up to $\sim$50 close to the black hole with indication of higher Γ far beyond the central object where SRMHD treatment is sufficient. The extreme Lorentz factors have been confirmed by recent SRMHD simulations which have found solutions with γ as high as $\simeq$5000 as shown in Figure 4.7 (see Fendt and Ouyed (2004) for more discussion).

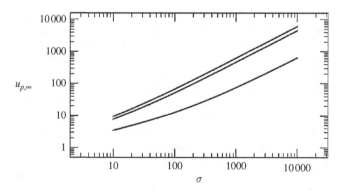

Fig. 4.7. Asymptotic Lorentz factor ($u_{p,\infty}$) from SRMHD simulations for different magnetization parameter σ. (Adapted from Fendt and Ouyed (2004)). Lorentz values as high as 5000 can be reached for strongly magnetized sources.

4.8 Non-relativistic versus relativistic MHD jets

While SRMHD and GRMHD jet simulations show similarities with non-relativistic regime, there are nevertheless some important differences in the underlying physics.

- The existence of the l.c. as a natural length scale in relativistic MHD is not consistent with the assumption of a *self-similar* jet structure (as is often assumed in non-relativistic MHD jet models). The latter holds even more when general relativistic effects are considered.
- Contrary to Newtonian MHD in the relativistic case, *electric fields* cannot be neglected. The poloidal electric field component is directed perpendicular to the magnetic flux surface. Its strength scales with the l.c. radius, $E_p = E_\perp = (r/r_l)B_p$. As a consequence of $E_p \simeq B_p$, the effective magnetic pressure can be lowered by a substantial amount (Begelman & Li 1994).
- In relativistic MHD, the poloidal Alfvén speed u_A becomes complex for $r > r_l$, $u_A^2 \sim B_p^2 (1 - (r/r_l)^2) = b_p^2 - E_\perp^2$. Therefore, Alfvén waves cannot propagate beyond the l.c. and only fast magnetosonic waves are able to exchange information across the jet.
- In the relativistic case, break-up of the MHD approximation is an issue. The problem is hidden in the fact that one may find arbitrarily high velocities for an arbitrarily high-flow magnetization. However, an arbitrarily high magnetization may be in conflict with the intrinsic *MHD condition* which requires a sufficient density of charged particles in order to be able to drive the electric current system (Michel 1969). Below the Goldreich–Julian particle density n_G (Goldreich & Julian 1969), the concept of MHD as applied to relativistic jet breaks down.

The points above imply that with due regard to the breakdown of the MHD approximation, it is feasible to scale the physics of jets from non-relativistic to ultra-relativistic MHD regime. GRMHD simulations suggest that a continuous scaling is more likely for slowly rotating black holes (where the radius of marginal stability is rather large and in a comparatively low-gravity region of spacetime) but show that a full general relativistic electrodynamic treatment is required to robustly treat the case of rapidly rotating black holes. GRMHD simulations also show that the situation is even more complex in the 'plunge region'. This is the region of the disk within the radius of marginal stability in which the accretion flow is undergoing rapid inwards acceleration (ultimately crossing the event horizon at the velocity of light as seen by a locally non-rotating observer). Unless the magnetic field is extremely strong, this is a region where inertial forces will dominate and the commonly employed force-free approximation breaks down. As a particle gets closer to the central object, it crosses three regions defining the MHD, force-free and inertial regimes, respectively. Another crucial aspect of the physics inherent to ultra-relativistic jets emanating from the near vicinity of rapidly rotating central objects is the role played by the dynamically important electric field

(Lyutikov & Ouyed 2007; Ouyed *et al.* 2007) in driving plasma instabilities generally leading to pair creation. Pair-creation and subsequent annihilation into radiation has yet to be taken into account, at least consistently, in GRMHD or/and GREMD codes which are mostly based on mass conservation schemes.

In summary, it appears that a universal aspect of all jets is their ability to tap their energy from the accretion of gas into the gravitational potential of the underlying central object. Jets of all stripes may also share common morphological features far beyond the source. However, complete unification of non-relativistic with relativistic regimes for jets probably breaks down close to the central black holes, where jet physics will also depend on the rotation of the hole and the production of relativistic plasmas.

Acknowledgements

We thank the organizers for the opportunity to present this work in such a stimulating conference in such a spectacular setting. Some material in this chapter (see Section 1.3) has appeared in different form in another recent review. REP thanks the KITP in Santa Barbara for a stimulating environment enjoyed during the composition of this chapter. RB is supported by the Deutsche Forschungsgemeinschaft under grant KL 1358/4-1. REP and RO are supported by grants from the National Science and Engineering Research Council of Canada.

References

Alves J. F., Lada C. J., Lada E. A., 2001, *Nature*, **409**, 159
Alves J. F., Lombardi M., Lada C. J., 2007, *A&A*, **462**, L17
Anderson J. M., Li Z.-Y., Krasnopolsky R., Blandford R. D., 2003, *ApJ.*, **590**, L107
Anninos P., Fragile P. C., Salmonson J. D., 2005, *ApJ.*, **635**, 723
Antón L., Zanotti O., Miralles J. A., Martí J. M., Ibáñez J. M., Font J. A., Pons J. A., 2006, *ApJ.*, **637**, 296
Arce H. G., Shepherd D., Gueth F., Lee C.-F., Bachiller R., Rosen A., Beuther H., 2007, in Reipurth B., Jewitt D., Keil K., eds, *Protostars and Planets V, Molecular Outflows in Low- and High-Mass Star-forming Regions.* University of Arizona Press, Tucson, pp. 245–260
Bacciotti F., 2004, *Ap&SS*, **293**, 37
Bacciotti F., Ray T. P., Mundt R., Eislöffel J., Solf J., 2002, *ApJ.*, **576**, 222
Ballesteros-Paredes J., Hartmann L., Vázquez-Semadeni E., 1999, *ApJ.*, **527**, 285
Ballesteros-Paredes J., Klessen R. S., Mac Low M.-M., Vázquez-Semadeni E., 2007, in Reipurth B., Jewitt D., Keil K., eds, *Protostars and Planets V, Molecular Cloud Turbulence and Star Formation.* University of Arizona Press, Tucson, pp. 63–80
Bally J., Reipurth B., Davis C. J., 2007, in Reipurth B., Jewitt D., Keil K., eds, *Protostars and Planets V, Observations of Jets and Outflows from Young Stars.* University of Arizona Press, Tucson, pp. 215–230

Banerjee R., Klessen R. S., Fendt C., 2007, *ApJ.*, **668**, 1028
Banerjee R., Pudritz R. E., 2006, *ApJ.*, **641**, 949
Banerjee R., Pudritz R. E., 2007, *ApJ.*, **660**, 479
Banerjee R., Pudritz R. E., Holmes L., 2004, *MNRAS*, **355**, 248
Begelman M. C., Blandford R. D., Rees M. J., 1984, *Rev. Mod. Phys.*, **56**, 255
Begelman M. C., Li Z.-Y., 1994, *ApJ.*, **426**, 269
Blandford R. D., 2001, *Prog. Theor. Phys. Suppl.*, **143**, 182
Blandford R. D., Payne D. G., 1982, *MNRAS*, **199**, 883
Blandford R. D., Znajek R. L., 1977, *MNRAS*, **179**, 433
Bourke T. L., Crapsi A., Myers P. C., Evans II N. J., Wilner D. J., Huard T. L., Jørgensen
    J. K., Young C. H., 2005, *ApJ.*, **633**, L129
Cabrit S., Bertout C., 1992, *A&A*, **261**, 274
Cabrit S., Raga A., Gueth F., 1997, in Reipurth B., Bertout C., eds, Herbig-Haro Flows
    and the Birth of Stars Vol. 182 of IAU Symposium, *Models of Bipolar Molecular
    Outflows*. Kluwer Academic Publishers, pp. 163–180
Camenzind M., 1986, *A&A*, **162**, 32
De Villiers J.-P., Hawley J. F., 2003, *ApJ.*, **589**, 458
De Villiers J.-P., Staff J., Ouyed R., 2005, e-prints arXiv:astro-ph/0502225
Elmegreen B. G., Scalo J., 2004, *ARA&A*, **42**, 211
Fendt C., 2003, *A&A*, **411**, 623
Fendt C., Camenzind M., 1996, *Astrophys. Lett. Comm.*, **34**, 289
Fendt C., Ouyed R., 2004, *ApJ.*, **608**, 378
Ferrari A., 1998, *ARA&A*, **36**, 539
Foster P. N., Chevalier R. A., 1993, *ApJ.*, **416**, 303
Fryxell B., Olson K., Ricker P., Timmes F. X., Zingale M., Lamb D. Q., MacNeice P.,
    Rosner R., Truran J. W., Tufo H., 2000, *ApJS*, **131**, 273
Gammie C. F., McKinney J. C., Tóth G., 2003, *ApJ.*, **589**, 444
Goldreich P., Julian W. H., 1969, *ApJ.*, **157**, 869
Goodman J., 1986, *ApJ.*, **308**, L47
Hennebelle P., Fromang S., January 2008, *A&A*, **477**(1), 9–24
Hennebelle P., Teyssier R., January 2008, *A&A*, **477**(1), 25–34
Heyvaerts J., Norman C., 1989, *ApJ.*, **347**, 1055
Kigure H., Shibata K., 2005, *ApJ.*, **634**, 879
Kley W., D'Angelo G., Henning T., 2001, *ApJ.*, **547**, 457
Komissarov S. S., 2004, *MNRAS*, **350**, 1431
Königl A., 1991, *ApJ.*, **370**, L39
Königl A., 1999, *New Astron. Rev.*, **43**, 67
Königl A., Pudritz R. E., 2000, Mannings V., Boss A. P., Russell S. S., eds, *Protostars
    and Planets IV*. University of Arizona Press, Tucson, p. 759
Krumholz M. R., McKee C. F., Klein R. I., 2005, *ApJ.*, **618**, L33
Kudoh T., Matsumoto R., Shibata K., 1998, *ApJ.*, **508**, 186
Lada C. J., Alves J. F., Lombardi M., 2007, in Reipurth B., Jewitt D., Keil K., eds,
    Protostars and Planets V, Near-Infrared Extinction and Molecular Cloud Structure.
    University of Arizona Press, Tucson, pp. 3–15
Lee C.-F., Mundy L. G., Reipurth B., Ostriker E. C., Stone J. M., 2000, *ApJ.*, **542**, 925
Lesieur M., 1997, *Turbulence in Fluids*. Kluwer Academic Publishers, Dordrecht
Li Y., Mac Low M.-M., Klessen R. S., 2005, *ApJ.*, **626**, 823
Li Y., Mac Low M.-M., Klessen R. S., 2006, *ApJ.*, **639**, 879
Li Z.-Y., 1993, *ApJ.*, **415**, 118
Li Z.-Y., Chiueh T., Begelman M. C., 1992, *ApJ.*, **394**, 459

Li Z.-Y., Nakamura F., 2006, *ApJ.*, **640**, L187

Livio M., 1999, *Phys. Rep.*, **311**, 225

Lynden-Bell D., 2003, *MNRAS*, **341**, 1360

Lyutikov M., Ouyed R., 2007, *Astropart. Phys.*, **27**, 473

Mac Low M.-M., Klessen R. S., 2004, *Rev. Mod. Phys.*, **76**, 125

Machida M. N., Inutsuka S.-I., Matsumoto T., 2006, *ApJ.*, **649**, L129

Machida M. N., Matsumoto T., Hanawa T., Tomisaka K., 2005a, *MNRAS*, **362**, 382

Machida M. N., Matsumoto T., Tomisaka K., Hanawa T., 2005b, *MNRAS*, **362**, 369

Matsumoto T., Hanawa T., 2003, *ApJ.*, **595**, 913

Matt S., Pudritz R. E., 2005, *ApJ.*, **632**, L135

McKinney J. C., Narayan R., 2007, *MNRAS*, **375**, 513

Michel F. C., 1969, *ApJ.*, **158**, 727

Mirabel F., 2005, in Novello M., Perez Bergliaffa S., Ruffini R., eds, *The Tenth Marcel Grossmann Meeting. On recent developments in theoretical and experimental general relativity, gravitation and relativistic field theories Black Hole Jet Sources.* pp. 606

Mirabel I. F., Rodríguez L. F., 1994, *Nature*, **371**, 46

Mirabel I. F., Rodríguez L. F., 1998, *Nature*, **392**, 673

Mohanty S., Jayawardhana R., Huélamo N., Mamajek E., 2007, *ApJ.*, **657**, 1064

Nakamura F., Li Z.-Y., 2007, *ApJ.*, **662**, 395

Nakamura M., Meier D. L., 2004, *ApJ.*, **617**, 123

Norman C., Silk J., 1980, *ApJ.*, **238**, 158

Ouyed R., Clarke D. A., Pudritz R. E., 2003, *ApJ.*, **582**, 292

Ouyed R., Pudritz R. E., Stone J. M., 1997, *Nature*, **385**, 409

Ouyed R., Sigl G., Lyutikov M., 2007, arXiv:0706.2812, 706

Pelletier G., Pudritz R. E., 1992, *ApJ.*, **394**, 117

Penrose R., 1969, *Nuovo Cimento*, **1**, 252

Piran T., 2005, *Rev. Mod. Phys.*, **76**, 1143

Pudritz R. E., 2003, NATO ASI, Les Houches, Session LXXVIII, *Accretion Discs, Jets and High Energy Phenomena in Astrophysics*, pp. 187–230

Pudritz R. E., 2004, *Ap&SS*, **292**, 471

Pudritz R. E., Norman C. A., 1983, *ApJ.*, **274**, 677

Pudritz R. E., Ouyed R., Fendt C., Brandenburg A., 2007, in Reipurth B., Jewitt D., Keil K., eds, *Protostars and Planets V, Disk Winds, Jets, and Outflows: Theoretical and Computational Foundations.* University of Arizona Press, Tucson, pp. 277–294

Pudritz R. E., Rogers C. S., Ouyed R., 2006, *MNRAS*, **365**, 1131

Punsly B., Coroniti F. V., 1990, *ApJ.*, **354**, 583

Quillen A. C., Trilling D. E., 1998, *ApJ.*, **508**, 707

Romanova M. M., Ustyugova G. V., Koldoba A. V., Chechetkin V. M., Lovelace R. V. E., 1997, *ApJ.*, **482**, 708

Shang H., Li Z.-Y., Hirano N., 2007, in Reipurth B., Jewitt D., Keil K., eds, *Protostars and Planets V, Jets and Bipolar Outflows from Young Stars: Theory and Observational Tests.* University of Arizona Press, Tucson, pp. 261–276

Shu F. H., Najita J. R., Shang H., Li Z.-Y., 2000, in Mannings V., Boss A. P., Russell S. S., eds, *Protostars and Planets IV.* University of Arezona Press, Tucson, p. 789

Tingay S. J., Jauncey D. L., Preston R. A., Reynolds J. E., Meier D. L., Murphy D. W., Tzioumis A. K., McKay D. J., Kesteven M. J., Lovell J. E. J., Campbell-Wilson D., Ellingsen S. P., Gough R., Hunstead R. W., Jones D. L., McCulloch P. M., Migenes V., Quick J., 1995, *Nature*, **374**, 141

Tomisaka K., 1998, *ApJ.*, **502**, L163

Tomisaka K., 2002, *ApJ.*, **575**, 306

Truelove J. K., Klein R. I., McKee C. F., Holliman J. H., Howell L. H., Greenough J. A., 1997, *ApJ.*, **489**, L179

Uchida Y., Shibata K., 1985, *PASJ*, **37**, 515

Urry C. M., Padovani P., 1995, *PASP*, **107**, 803

Vlahakis N., Königl A., 2001, *ApJ.*, **563**, L129

Vlahakis N., Königl A., 2003, *ApJ.*, **596**, 1104

Wu Y., Wei Y., Zhao M., Shi Y., Yu W., Qin S., Huang M., 2004, *A&A*, **426**, 503

Zhang Q., Sridharan T. K., Hunter T. R., Chen Y., Beuther H., Wyrowski F., 2007, *A&A*, **470**, 269

5

Advanced numerical methods in astrophysical fluid dynamics

A. Hujeirat and F. Heitsch

Abstract

Computational gas dynamics has become a prominent research field in both astrophysics and cosmology. In the first part of this chapter, we intend to briefly describe several of the numerical methods used in this field, discuss their range of application and present strategies for converting conditionally stable numerical methods into unconditionally stable solution procedures. The underlying aim of the conversion is to enhance the robustness and unification of numerical methods and subsequently enlarge their range of applications considerably. In the second part, Heitsch presents and discusses the implementation of a time-explicit magneto hydrodynamic (MHD) Boltzmann solver.

PART I

5.1 Numerical methods in AFD

Astrophysical fluid dynamics (AFD) deals with the properties of gaseous matter under a wide variety of circumstances. Most astrophysical fluid flows evolve over a large variety of different time and length scales, henceforth making their analytical treatment unfeasible.

On the contrary, numerical treatments by means of computer codes have witnessed an exponential growth during the last two decades due to the rapid development of hardware technology. Nowadays, the vast majority of numerical codes are capable of treating large and sophisticated multi-scale fluid problems with high resolutions and even in 3D.

The numerical methods employed in AFD can be classified into two categories (see Figure 5.1):

Structure Formation in Astrophysics. ed. G. Chabrier. Published by Cambridge University Press.
© Cambridge University Press 2009.

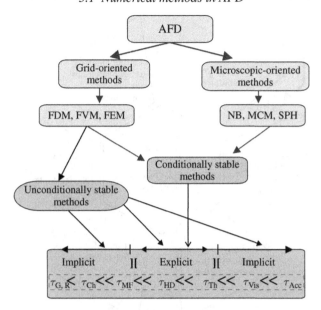

Fig. 5.1. Numerical methods (FDM, FVM, FEM, NB, MC and the SPH) employed in AFD and their possible regime of application from the timescale point of view. The timescales read as follows: the radiative, τ_R; gravitative, τ_G; chemical, τ_{Ch}; magnetic, τ_{MF}; hydrodynamic, τ_{HD}; thermal, τ_{Th}; viscous, τ_{Vis} and the accretion timescale, τ_{Acc}.

1. Microscopic-oriented methods: These are mostly based on N-body (NB), Monte Carlo (MC) and on the Smoothed Particle Hydrodynamics (SPH).
2. Grid-oriented methods: To this category belong the finite difference (FDM), finite volume (FVM) and finite element methods (FEM).

Most numerical methods used in AFD are conditionally stable. Hence, they may converge if the Courant–Friedrichs–Levy (CFL) condition for stability is fulfilled. As long as efficiency is concerned, these methods are unrivalled candidates for flows that are strongly time-dependent and compressible. They may stagnate, however, if important physical effects are to be considered or even if the flow is weakly incompressible. On the contrary, only a small number of the numerical methods employed in AFD are unconditionally stable (Figure 5.2). These are implicit methods, but they are effort-demanding from the programming point of view.

It has been shown that strongly implicit (henceforth IM) and explicit (henceforth EM) methods are different variants of the same algebraic problem (Hujeirat 2005). Hence, both methods can be unified within the context of the hierarchical solution scenario (henceforth HSS, see Figure 5.3).

In Table 5.1, we have summarized the relevant properties of several numerical methods available.

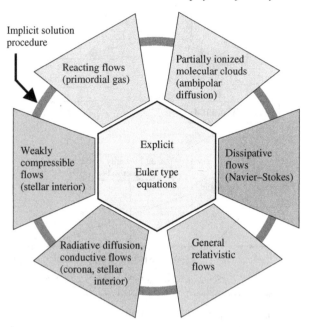

Fig. 5.2. The regime of application of explicit method is severely limited to Euler-type flows, whereas sophisticated treatment of most flow problems in AFD require the employment of much more robust methods.

5.2 Timescales in AFD

As will be shown later, the time scales characterizing typical astronomical phenomena can be so different to span 12 orders of magnitude (Table 5.3). The gas is said to be radiating, magnetized, chemical reacting, partially ionized and under the influence of its own/external gravitational field. Let the initial state of the gas be characterized by a constant velocity, density, temperature and a constant magnetic field. The timescales associated with the flow can be obtained directly from the radiative magnetohydrodynamic (MHD) equations as follows (see Hujeirat (2005) for detailed description of the set of equations).

- Continuity equation:

$$\frac{\partial \rho}{\partial t} + \nabla \cdot \rho V = 0, \tag{5.1}$$

where ρ and V stand for the density and the velocity field, respectively. Using scaling variables (Table 5.2), we may approximate the terms of this equation as follows:

$\partial \rho / \partial t \sim \rho / \tau$ and $\nabla \cdot \rho V \sim \rho V / L$. This yields the hydrodynamical timescale $\tau_{HD} = L / V$.

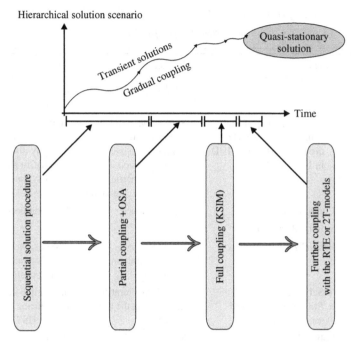

Fig. 5.3. A schematic description of the HSS. The HSS is based on dynamically varying the efficiency and robustness of the numerical method to leapfrog the transient phase. The method is most suitable for searching quasi-stationary flow configurations that depend weakly on the initial conditions. Here, the coupling between the equations can be enhanced gradually, by solving them sequentially, then partial coupling in combination with the operator-splitting approach (OSA), full coupling using the Krylov subspace iterative method (KSIM) and finally extending the coupling to include the radiative transfer equation (RTE) and energy equation of multi-temperature plasmas.

The so-called accretion timescale can be obtained by integrating the continuity equation over the whole fluid volume. Specifically,

$$\int_{Vol} \frac{\partial \rho}{\partial t}\, \mathrm{d}Vol = \frac{\partial M}{\partial t} \sim \frac{M}{\tau}, \quad \int_{Vol} (\nabla \cdot \rho V)\, \mathrm{d}Vol = \int_S \rho V \cdot n \cdot \mathrm{d}S = \Delta \dot{M} \sim \dot{M},$$

where *Vol* denotes the total volume of the gas and '*S*' corresponds to its surface. Equating the latter two terms, we obtain

$$\frac{M}{\tau} \sim \dot{M} \Rightarrow \tau_{acc} \sim \frac{M}{\dot{M}}.$$

In general, τ_{acc} is one of the longest timescales characterizing astrophysical flows connected to the accretion phenomena.

- The momentum equations:

$$\frac{\partial V}{\partial t} + \nabla V \otimes V = -\frac{1}{\rho}\nabla P + f_{cent} + \frac{f_{rad}}{\rho} + \nabla \psi + \frac{\nabla \times B \times B}{4\pi\rho} + Q^r_{vis}, \qquad (5.2)$$

Table 5.1. *A list of only a part of the grid-oriented codes in AFD and their algorithmic properties*

	Explicit	Implicit	HSS
Solution method	$q^{n+1} = q^n + \delta t\, d^n$	$q^{n+1} = q^n + \delta t\, \tilde{A}^{-1} d^*$	$q^{n+1} = \alpha q^n + (1-\alpha)\delta t\, \tilde{A}_d^{-1} d^*$
Type of flows	Strongly time-dependent, compressible, weakly dissipative HD and MHD in 1, 2 and 3D	Stationary, quasi-stationary, highly dissipative, radiative and axi-symmetric MHD flows in 1, 2 and 3D	Stationary, quasi-stationary, weakly compressible, highly dissipative, radiative and axi-symmetric MHD-flows in 1, 2 and 3D
Stability	Conditioned	Unconditioned	Unconditioned
Efficiency	1 (normalized/2D)	$\sim m^2$	$\sim m_d^2$
Efficiency: enhancement strategies	Parallelization	Parallelization, preconditioning, multigrid	HSS, parallelization, preconditioning, prolongation
Robustness: enhancement strategies	i. Subtime-stepping ii. Stiff terms are solved semi-implicitly	i. Multiple iteration ii. Reducing the time step size	i. Multiple iteration ii. Reducing the time step size, HSS
Newtonian numerical codes	Solvers1[a] ZEUS & ATHENA[b], FLASH[c], NIRVANA[d], PLUTO[e], VAC[f]	Solver2[g]	IRMHD[h]
Relativistic numerical codes	Solvers3[i] GRMHD[j], ENZO[k], PLUTO[l], HARM[m], RAISHIN[n], RAM[o], GENESIS[p], WHISKY[q]	Solver4[r]	GR-I-RMHD[s]

[a] Bodenheimer *et al.* (1978), Clarke (1996).
[b] Stone, Norman (1992), Gardiner, Stone (2006).
[c] Fryxell *et al.* (2000).
[d] Ziegler (1998).
[e] Mignone, Bodo (2003); Mignone *et al.* (2007).
[f] Tóth *et al.* (1998).
[g] Wuchterl (1990), Swesty (1995).
[h] Hujeirat (1995, 2005), Hujeirat, Rannacher (2001), Falle (2003).
[i] Koide *et al.* (1999), Komissarov (2004).
[j] De Villiers, Hawley (2003).
[k] O'Shea *et al.* (2004).
[l] Mignone *et al.* (2007).
[m] Gammie *et al.* (2003).
[n] Mizuno *et al.* (2006).
[o] Zhang, MacFadyen (2006).
[p] Aloy *et al.* (1999).
[q] Baiotti *et al.* (2003).
[r] Liebendörfer *et al.* (2002).
[s] Hujeirat *et al.* (2007).

In these equations, $q^{n,n+1}$, δt, $\tilde{A}$, α and d^* denote the vector of variables from the old and new time levels, time step size, a preconditioning matrix, a switch on/off parameter and a time-modified defect vector, respectively. 'm' in row 4 denotes the bandwidth of the corresponding matrix.

Table 5.2. *A list of possible scaling variables typical for three different astrophysical phenomena: giant molecular clouds, accretion onto supermassive black holes (SMBHs) and accretion onto ultra-compact objects (UCO)*

Scaling	Variables	Molecular cloud	Accretion (onto SMBH)	Accretion (onto UCO)
$\tilde{L}$	Length	$\mathcal{O}(pc)$	$\mathcal{O}(AU)$	$\mathcal{O}(10^6 \, \text{cm})$
$\tilde{\rho}$	Density	$10^{-22} \, \text{g cm}^{-3}$	$10^{-6} \, \text{g cm}^{-3}$	$10^{-8} \, \text{g cm}^{-3}$
$\tilde{T}$	Temperature	$10 \, \text{K}$	$10^6 \, \text{K}$	$10^7 \, \text{K}$
$\tilde{V}$	Velocity	$0.3 \, \text{km s}^{-1}$	$10^2 \, \text{km s}^{-1}$	$10^{2-3} \, \text{km s}^{-1}$
$\tilde{B}$	Magnetic fields	$30 \, \mu\text{G}$	$10^2 \, \text{G}$	$10^4 \, \text{G}$
$\tilde{M}$	Mass	$10^3 M_\odot$	$10^6 M_\odot$	$M_\odot$
$\tilde{\dot{M}}$	Accretion rate		$10^{-2} M_\odot \, \text{yr}^{-1}$	$10^{-10} M_\odot \, \text{yr}^{-1}$

These variables may be used for reformulating the radiative MHD equations in non-dimensional form.

Table 5.3. *A list of the timescales relative to the hydrodynamical timescale for three different astrophysical phenomena*

Timescales	Molecular cloud	Accretion (onto SMBH)	Accretion (onto UCO)
τ_{HD}	$\sim 10^6 \, \text{yr}$	$\sim$months	$\sim 1 \, \text{s}$
τ_{rad}/τ_{HD}	$\sim 10^{-6}$	$\sim 10^{-3}$	$\sim 10^{-3}$
τ_{grav}/τ_{HD}	$\sim 10^{-2}$	$\sim 10^{-3}$	$\sim 10^{-3}$
τ_{ch}/τ_{HD}	$\sim 10^{-1}$	$\sim 10^{-5}$	$\sim 10^{-4}$
τ_{mag}/τ_{HD}	$\sim 10^{-2}$	$\sim 10^0$	$\sim 10^{-1}$
τ_{vis}/τ_{HD}	$\sim 10^1$	$\sim 10^2$	$\sim 10^2$
τ_{acc}/τ_{HD}		$\sim 10^4$	$\sim 10^{12}$

where P, f_{cent}, f_{rad}, ψ, B and Q^r_{vis} denote gas pressure, centrifugal force, radiative force, gravitational potential, magnetic field and viscous operators, respectively. From this equation, we may obtain the following timescales:

1. The sound speed crossing time can be obtained by comparing the following two terms:

$$\frac{\partial V}{\partial t} \approx \frac{\nabla P}{\rho}, \quad \text{which yields} \quad \tau_s \approx \tau_{HD} \left(\frac{V}{V_s} \right)^2,$$

where V_s is the sound speed.

2. The gravitational timescale:

$$\frac{\partial V}{\partial t} \approx \nabla \psi \Rightarrow \tau_G = \tau_{HD} \left(\frac{V}{V_g}\right)^2,$$

where $V_g^2 = GM/L$ and G is the gravitational constant.

3. Similarly, the Alfvén wave crossing time:

$$\frac{\partial V}{\partial t} \approx \frac{\nabla \times B \times B}{4\pi\rho} \Rightarrow \tau_{mag} = \tau_{HD} \left(\frac{V}{V_A}\right)^2,$$

where $V_A^2 = (B^2/4\pi\rho)$ denotes the Alfvén speed squared.

4. Radiative effects in moving flows propagate on the radiative scale, which is obtained from

$$\frac{\partial V}{\partial t} \approx \frac{f_{rad}}{\rho} \Rightarrow \tau_{rad} = \tau_{HD} \left(\frac{V}{c}\right)^2,$$

where c is the speed of light.

5. The viscous timescale:

$$\frac{\partial V}{\partial t} \approx Q_{vis}^r \sim \frac{\nu V}{L^2} \Rightarrow \tau_{vis} = \frac{L^2}{\nu}$$

where ν is a viscosity coefficient.

- The induction equation, taking into account the effects of α_{dyn}–dynamo, magnetic diffusivity ν_{diff} and of ambipolar diffusion, reads as follows:

$$\frac{\partial B}{\partial t} = \nabla \times \langle V \times B + \alpha_{dyn} B - \nu_{mag} \nabla \times B\rangle + \nabla \times \left\{\frac{B}{4\pi \gamma \rho_i \rho_n} \times [B \times (\nabla \times B)]\right\}, \quad (5.3)$$

where $\rho_{i,n}$ denote the ion and neutral densities.

Thus, the induction equation contains several important timescales:

1. The dynamo amplification timescale, which results from the equality:

$$\frac{\partial B}{\partial t} = \nabla \times \alpha_{dyn} B \Rightarrow \tau_{dyn} = \frac{L}{\alpha_{dyn}}$$

2. The magnetic diffusion timescale:

$$\frac{\partial B}{\partial t} = \nabla \times (\nu_{mag} \nabla \times B) \Rightarrow \tau_{diff} = \frac{L^2}{\nu_{mag}}$$

3. The ambipolar diffusion timescale:

$$\frac{\partial B}{\partial t} = \nabla \times \left\{\frac{B}{4\pi \gamma \rho_i \rho_n} \times [B \times (\nabla \times B)]\right\}$$

$$\Leftrightarrow \frac{B}{\tau} \sim \frac{1}{L}\left(\frac{B^2}{4\pi\rho_n}\right)\left(\frac{1}{\gamma\rho_i}\right)\left(\frac{B}{L}\right) \sim \frac{V_A^2}{\gamma\rho_i}\frac{B}{L^2} = \mathcal{D}_{amb}\frac{B}{L^2} \Rightarrow \tau_{amb} = \frac{L^2}{\mathcal{D}_{amb}},$$

where $\mathcal{D}_{amb} = (V_A^2/(\gamma\rho_i))$ is the ambipolar diffusion coefficient.

- The chemical reaction equations:

  The equation describing the chemical evolution of species 'i' is

  $$\frac{\partial \rho_i}{\partial t} = \sum_m \sum_n k_{mn} \rho_m \rho_n + \sum_m I_m \rho_m, \tag{5.4}$$

  where k_{mn} denotes the reaction rate between the species m and n. I_m stands for other external sources. For example, the reaction equation of atomic hydrogen in a primordial gas reads as follows:

  $$\frac{\partial \rho_H}{\partial t} = \frac{k_2}{m_H} \rho_{H^+} \rho_e - \frac{k_1}{m_H} \rho_H \rho_e \Leftrightarrow \frac{\rho_H}{\tau} \sim \frac{k_2}{m_H} \rho_H \rho_e \Rightarrow \tau_{ch} \sim \frac{m_H}{k_2 \rho_e},$$

  where ρ_e, $k_2 (10^{-10} \, \text{cm}^3 \, \text{s}^{-1})$ correspond to the electron density and to the generation rate of atomic hydrogen through the capture of electrons by ionized atomic hydrogen. m_H corresponds to the mass of atomic hydrogen.
- Relativistic MHD equations:

  The velocities in relativistic flows are comparable to the speed of light. This implies that the hydrodynamical τ_{HD} and radiative τ_{rad} timescales are comparable and that both are much shorter than in Newtonian flows.

We note that although the dynamical timescale in relativistically moving flows is relatively short, there are still several reasons that justify the use of implicit numerical procedures, in particular, the following:

1. The relativistic MHD equations are strongly non-linear, giving rise to fast growing non-linear perturbations, imposing thereby a further restriction on the size of the time step.
2. The deformation of the geometry grows non-linearly when approaching the black hole. Thus, in order to capture flow configurations in the vicinity of a black hole accurately, a non-linear distribution of the grid points is necessary, which, again, may destabilize explicit schemes.
3. Initially, non-relativistic flows may become ultra-relativistic or vice versa. However, almost all non-relativistic astrophysical flows known to date are considered to be dissipative and diffusive. Therefore, in order to track their time evolution reliably, the employed numerical solver should be capable of treating the corresponding second-order viscous terms properly.
4. The accumulated round off errors resulting from performing a large number of time extrapolations for time-advancing a numerical hydrodynamical solution may easily cause divergence. The constraining effects of boundary conditions may fail to configure the final numerical solution.

5.3 Numerical methods: a unification approach

In this section, we show that explicit and implicit methods are special cases of a more general solution method in higher dimensions.

Assume we are given the following evolution equation of a vector variable q:

$$\frac{\partial q}{\partial t} + L(q) = f, \tag{5.5}$$

where L and f correspond to an advection operator and to external forces, respectively.

Adopting a time-forward discretization procedure, the unknown vector q at the new time level can be extrapolated as follows:

$$q^{n+1} = q^n + \delta t \cdot RHS^n, \tag{5.6}$$

where $RHS = f - L(q)$.

Depending on the time step size and on the number of grid points, the numerical procedure can be made sufficiently accurate in space and time.

On the Eq. (5.6) can be viewed as an equality of two 1D contrary, vectors:

$$[\text{vector of unknowns}] = [\text{vector of knowns}] \iff q^{n+1} = \bar{b}, \tag{5.7}$$

where $\bar{b} = q^n + \delta t \cdot RHS^n$.

In higher dimensions, however, Eq. (5.7) is a special case of the matrix equation:

$$Aq^{n+1} = \bar{b}, \tag{5.8}$$

in which it is projected along the diagonal elements. It is obvious that the matrix $I/\delta t$ is a further simplification of the matrix that contains just the diagonal elements of A.

Therefore, we may adopt the higher dimension formulation to gain a better understanding of the stability of the solution procedure. According to matrix algebra, a necessary condition for the matrix A to have a stable inversion procedure is that A must be strictly diagonally dominant. Equivalently, the entries in each row of the matrix A must fulfil the following condition: the module of the diagonal element $d_{i,i}$ is larger than the sum of all off-diagonal elements $\sum_{j \neq i} |a_{i,j}|$, where i and j denote the row and column numbers of the matrix . Applying a conservative and monotonicity preserving scheme, the latter inequality may be rewritten in the following form:

$$\left| \frac{1}{\delta t} + \text{positive contributions} \right| > \sum_{j \neq i} |a_{i,j}|. \tag{5.9}$$

We note that since δt is a free parameter, it can be chosen sufficiently small, so that $1/\delta t$ largely dominates all other off-diagonal elements, or so large that $1/\delta t$ becomes negligibly small.

We may further simplify this inequality by choosing the time step size even smaller, such that

$$\frac{1}{\delta t} > \sum_j |a_{i,j}|, \quad \text{for } \forall \; j \neq i, \tag{5.10}$$

can be safely fulfilled. We may decompose the matrix A as follows:

$$A = D + R = D(I + D^{-1}R),$$

where D is the matrix consisting of the diagonal entries of A and R ($=A - D$) consists of the off-diagonals. Thus, the elements of D are proportional to $1/\delta t$, whereas those of R are proportional to δt. This implies that A can be expanded around $I/\delta t$ in the form

$$A = A^{(0)} + A^{(1)} + A^{(2)} + \cdots, \tag{5.11}$$

where the leading matrix $A^{(0)} \approx \frac{1}{\delta t} I$ and $A^{(1)} \sim \delta t \, I$. In this case, the inversion of the matrix A is no more necessary, and the resulting numerical procedure would correspond to a classical time-explicit method.

On the other hand, this restriction can be relaxed by incorporating the off-diagonal entries in the coefficient matrix (see Eq. (5.13)), so that purely Eulerian, Navier-Stokes, strongly or weakly compressible flows can be treated stably (Figures 5.5 and 5.6).

5.3.1 Example

The time evolution of density in 1D is described by the continuity equation:

$$L_\rho = \frac{\partial \rho}{\partial t} + \frac{\partial \rho U}{\partial x} = 0. \tag{5.12}$$

The corresponding Jacobian matrix is $A = \partial L_\rho / \partial \rho$. The non-zero entries of A read as follows:

$$a_{ii} = \frac{1}{\delta t} + \frac{|U_i|}{\Delta x} \quad \text{and the off-diagonal} \quad a_{ij} = -\frac{|U_{j+1}|}{\Delta x} \quad \text{for } i \neq j, \tag{5.13}$$

where Δx and i, j denote the grid spacing and grid point numbering, respectively. Applying a first-order upwind discretization, then the condition of diagonal dominance demands

$$\left| \frac{1}{\delta t} + \frac{U_i}{\Delta x} \right| > \frac{|U_{j+1}|}{\Delta x}. \tag{5.14}$$

This condition can be further simplified by choosing the time step size so small, such that

$$\frac{1}{\delta t} > \frac{2 \max(|U_j|, |U_{j+1}|)}{\Delta x} \quad \Leftrightarrow \quad \frac{\delta t \max(|U_j|, |U_{j+1}|)}{\Delta x} < \frac{1}{2}. \tag{5.15}$$

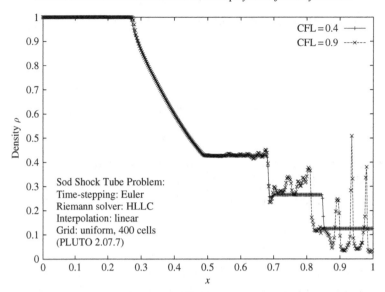

Fig. 5.4. The profile of the shock tube problem obtained with Courant–Friedrichs–Levy numbers, CFL = 0.4 and 0.9 using the PLUTO code. Although both CFL numbers are smaller than unity, the numerical solution procedure does not appear to be stable even with CFL = 0.9.

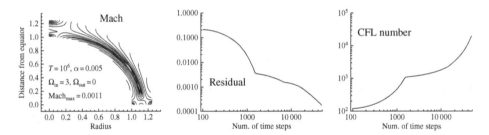

Fig. 5.5. Weakly incompressible flow between two concentric rotating spheres. *Left panel*: the 2D-distribution of Mach number ($=V/V_s$) is displayed (25 isolines) for extreme weakly incompressible flows (Max (Mach) $\sim 10^{-3}$). The maximum residual (*middle panel*) and the CFL number (*right*) versus the number of time steps are shown.

Thus, the condition of diagonal dominance is more restrictive than the normal CFL condition. This may explain, why most explicit methods fail to converge for Courant–Friedrichs–Levy number, CFL $= 1 - \epsilon$ (see Figure 5.4).

5.4 Converting time-explicit into implicit solution methods

In a series of publications, we have shown that the robustness of explicit methods can be enhanced gradually to recover full-implicit solution procedures (Hujeirat

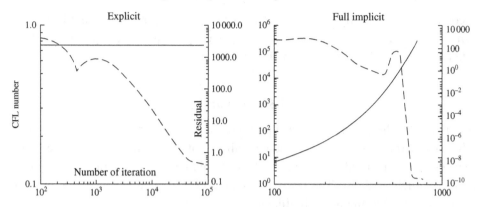

Fig. 5.6. The profiles of the CFL number (*solid line*) versus the number of iteration for both explicit and implicit solution procedures (*dashed line*). The profiles correspond to the free-fall of spherical plasma onto a non-magnetized Schwarzschild black hole, in which the final solution is time-independent.

2005). In the following, we outline the main algorithmic steps towards extending classical explicit methods into implicit:

1. Use the same mathematical form of RHS^n of Eq. (5.6) to compute $RHS^{n+1} = RHS(q^{n+1})$ and subsequently the mean $\overline{RHS} = \alpha \cdot RHS^n + (1-\alpha) \cdot RHS^{n+1}$, where $0 \le \alpha \le 1$ is a parameter that may depend also on the time step size.
2. Define the defect

$$d = -\left(\frac{q^{n+1} - q^n}{\delta t}\right) + \overline{RHS}. \tag{5.16}$$

3. Compute the Jacobian $J^{\text{real}} = \partial L_q/\partial q$, where L_q denotes the set of equations in operator form.
4. Construct a simplified matrix $\tilde{A}$ (preconditioner), which is easy to invert, but still share the spectral properties of J^{real} (Hackbusch 1994).
5. Solve the system of equation:

$$\tilde{A}\mu = d, \tag{5.17}$$

where μ is a vector of small correction, so that $q^{l+1} = q^l + \mu$.
In general, $\tilde{A} \neq J^{\text{real}}$, which implies that Eq. (5.17) should be solved iteratively to assure that the maximum norm of the defect, $||d||_\infty$, is sufficiently small.

We note that for sufficiently small δt, the matrix $I/\delta t$ can be made similar to J^{real}, hence they share the same spectral space. As a consequence, a variety of solution procedures can be constructed that range from purely explicit up to strongly implicit, depending on how similar the preconditioner $\tilde{A}$ is to the real

Jacobian. This naturally suggests the HSS as a highly powerful numerical algorithm for enhancing the robustness of explicit schemes and optimizing their efficiency (Figure 5.3; see also Hujeirat 2005)

5.5 Summary

In this part of the chapter, we have presented a method for converting conditionally stable explicit methods into numerically stable implicit solution procedures. The conversion method allows a considerable enlargement of the range of application of explicit methods. The HSS is best suited for gradual enhancement of their robustness and optimizing their efficiency.

PART II

5.6 (Magneto-)hydrodynamic Boltzmann solvers

In this part, I will discuss the implementation of a time-explicit gas-kinetic grid-based integrator for non-relativistic hydrodynamics introduced by Prendergast and Xu (1993), Xu (1999) and Tang and Xu (2000), and its extension to non-ideal MHD (Heitsch *et al.* 2004, 2007). Some properties of Boltzmann solvers are discussed in Section 5.6, the equations and the implementation are described in Section 5.7, followed by a selection of test cases and applications (Section 5.8) and a summary (Section 5.9).

5.7 Why Boltzmann Solvers?

It is the physical model for the fluid equations which distinguishes gas-kinetic schemes from the widely popular Godunov methods. The latter are formulated on the basis of the Vlasov equation, i.e. assuming that any dynamical timescale is larger than the collision time between particles, setting the collision term in the Boltzmann equation to zero. The distribution function is then given by a Maxwellian at all times. In contrast, gas-kinetic schemes keep the collision term in the Boltzmann equation, but because of the impractibility to compute all the collisions between particles, they need to come up with a model for the collision term.

One such model has been introduced by Bhatnagar *et al.* (1954), formulating the collision term as the difference between the equilibrium distribution function g (the Maxwellian) and the initial distribution function f, resulting in a Boltzmann equation of the form

$$\partial_t f + u \partial_x f + \dot{u} \partial_u f = \frac{g - f}{\tau}, \tag{5.18}$$

where τ is the collision time. Integrating Eq. (5.18) over a time t gives (at position x)

$$f(x, t, u) = \frac{1}{\tau} \int_0^t g(x - u(t - t'), t', u) \, e^{-(t-t')/\tau} dt' + e^{-t/\tau} \, f_0(x - ut, 0, u),$$
(5.19)

where τ is the collision time and f_0 the initial distribution function. For a complete description, see Xu (2001). Thus, the distribution function f at time t gets two contributions: one from the decaying initial conditions $f(t = 0)$ and one from the growing equilibrium distribution g.

The zeroth-, first- and second-order velocity moments of the distribution function (here for a monatomic gas),

$$g \equiv \rho \left(\frac{\lambda}{\pi} \right)^{3/2} \exp(\lambda(\mathbf{u} - \mathbf{U})^2),$$
(5.20)

result in the (macroscopic) conserved quantities density ρ, momentum density $\rho \mathbf{U}$ and total energy density ρE. The quantity $\lambda \equiv m/(2kT)$. The corresponding moments of the Boltzmann Eq. (5.18) give the conservation equations. The Bhatnagar–Gross–Krook (BGK) collision term in Eq. (5.18) then gives rise to a viscous flux, depending on the ratio of the CFL time step and a specified collision time. Thus, the Reynolds number of the flow can be controlled. The Prandtl number is 1 by construction. The scheme is upwind and it satisfies the entropy condition (Prendergast & Xu 1993, Xu 2001). The fully controlled dissipative terms come at (close to) no extra computational cost. Fragmentation of hydrodynamically unstable systems due to numerical noise can thus be suppressed. Specifically, gas-kinetic schemes can provide a viscosity independent of grid geometry, thus allowing the modelling of disks on a cartesian grid (see Slyz *et al.* 2002).

In the following, I will discuss a specific implementation of a gas-kinetic solver, namely Proteus (Heitsch *et al.* 2007).

5.8 Equations and implementation: Proteus

Proteus solves the equations of non-ideal-MHD, with an Ohmic resistivity λ_Ω and a shear viscosity ν.

$$\partial_t \rho + \nabla \cdot (\rho \mathbf{v}) = 0$$
(5.21)

$$\partial_t \rho \mathbf{v} + \nabla \cdot \left[\rho \mathbf{v} \mathbf{v} - \frac{\mathbf{B} \mathbf{B}}{4\pi} + p + \frac{\mathbf{B}^2}{8\pi} \right] = \nabla \cdot \bar{\Pi}$$
(5.22)

$$\partial_t \rho E + \nabla \cdot \left[\rho E \mathbf{v} + \left(p + \frac{\mathbf{B}^2}{8\pi} \right) \mathbf{v} - \frac{(\mathbf{v} \cdot \mathbf{B})\mathbf{B}}{4\pi} \right] = \mathbf{v} \cdot \left(\nabla \cdot \bar{\bar{\Pi}} \right) + \lambda_\Omega \mathbf{J}^2 \quad (5.23)$$

$$\partial_t \mathbf{B} + \nabla \cdot (\mathbf{v}\mathbf{B} - \mathbf{B}\mathbf{v}) = \lambda_\Omega \nabla^2 \mathbf{B}. \quad (5.24)$$

The mechanism of how to split the fluxes at the cell walls is described in detail by Xu (1999) and will not be repeated here. Viscosity and resistivity are implemented as dissipative fluxes. They require spatially constant coefficients λ_Ω and ν. Ambipolar drift is implemented in the two-fluid description, though currently only for an isothermal equation of state.

Higher-order time accuracy is achieved by a Time–Variation–Diminishing (TVD) Runge-Kutta time stepping method (Shu & Osher 1988). For second-order spatial accuracy, a choice of reconstruction prescriptions is available.

Proteus offers two gas-kinetic solvers, the one just described and a one-step integrator at second order in time and space for hydrodynamics. The latter has been discussed in detail by Slyz and Prendergast (1999) and Slyz *et al.* (2005), so we refer the interested reader to those papers.

5.9 Test cases and applications

5.9.1 *1D: resistively damped linear Alfvén wave*

This 1D test checks the resistive flux implementation as well as the accuracy of the overall scheme. A linear Alfvén wave under weak Ohmic dissipation is damped at a rate of

$$\omega_i = \frac{1}{2}\lambda_\Omega k^2, \quad (5.25)$$

where λ_Ω is the Ohmic resistivity and $k = 2\pi\kappa/L$ is the wave number of the Alfvén wave, with κ a natural number. The strongly damped case, where the decay dominates the time evolution, is uninteresting for our application, since the Ohmic resistivity is mainly used to control numerical dissipation. Figure 5.7 shows the damping rate against Ohmic resistivity λ_Ω for $\kappa = 1, 2, 4$ at a grid resolution of $N = 64$. The damping rate is derived by measuring the amplitude of the wave at each full wave period.

From Figure 5.7, it is clear that, as one diminishes the value of λ_Ω, there comes a point when the numerical resistivity of the scheme becomes comparable to the physical one, causing the measured damping rate to flatten out and depart from the analytical solution. For $\kappa = 4$ and $\lambda_\Omega = 0.1$, the wave decays too quickly to allow a reliable measurement, and the system enters the strongly damped branch of the dispersion relation. However, we emphasize that even at 16 cells per wave length, the resistivity range available to Proteus spans nearly two orders of magnitude.

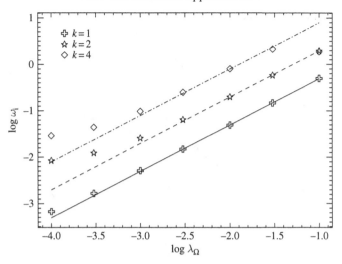

Fig. 5.7. Logarithm of the damping rate (Eq. (5.25)) of a linear Alfvén wave against logarithm of the Ohmic resistivity for $k = 1, 2, 4$. The resolution is $N = 64$. Lines denote the analytical solution.

5.9.2 1D: linear Alfvén waves in weakly ionized plasmas

The dispersion relation for a linear Alfvén wave in a weakly ionized plasma splits into two branches (Kulsrud & Pearce 1969): a strongly coupled branch, for which the ion Alfvén frequency $\omega_k \equiv kB/\sqrt{4\pi\rho_i} \ll \nu_{\mathrm{in}} \equiv \gamma\rho_n$, the ion-neutral collision frequency, and a weakly coupled branch, for which $\omega_k \gg \nu_{\mathrm{in}}\sqrt{\rho_i/\rho_n}$. The strongly coupled case leads to a dispersion relation of

$$\omega = \pm\left(\omega_k^2\epsilon - \frac{\omega_k^4}{4\nu_{\mathrm{in}}}\right)^{1/2} - \imath\frac{\omega_k^2}{2\nu_{\mathrm{in}}}, \tag{5.26}$$

with $\epsilon \equiv \rho_i/\rho_n$. Thus, the strongly coupled Alfvén wave travels at the neutral Alfvén speed $c_{\mathrm{An}} \equiv B/\sqrt{4\pi\rho_n}$ and is increasingly damped with decreasing collision frequency. The weakly coupled branch leads to

$$\omega = \pm\left(\omega_k^2 - \frac{\nu_{\mathrm{in}}^2}{4}\right)^{1/2} - \imath\frac{\nu_{\mathrm{in}}}{2}. \tag{5.27}$$

Now, the wave travels at the ion Alfvén speed, and damping is proportional to ν_{in}. Since $c_{\mathrm{An}}\sqrt{\rho_n/\rho_i}$, the speeds can be widely disparate.

Figure 5.8 shows the real and imaginary part of the Alfvén wave frequency in a weakly ionized plasma. For simplicity, we vary the collision coefficient γ_{AD} and keep the densities constant. Wave speed (*upper panel*) and damping term (*lower panel*) are well reproduced.

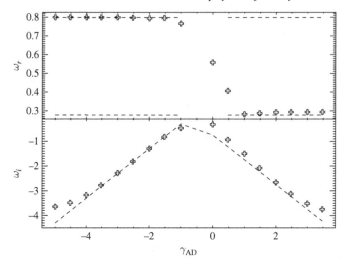

Fig. 5.8. Logarithm of the frequency (*upper panel*) and damping rate (*lower panel*) for the linear Alfvén wave in a partially ionized plasma. For simplicity, we vary the collision coefficient γ_{AD} instead of the density.

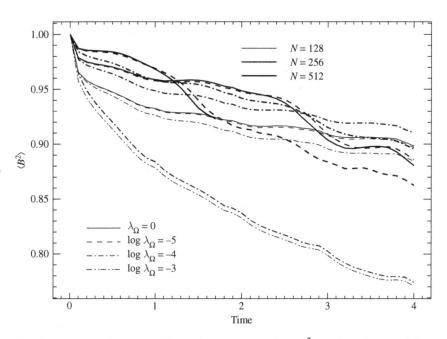

Fig. 5.9. Current sheet test. Magnetic energy density $\langle B^2 \rangle$ against time. A finite resistivity λ_Ω helps stabilize the code. Line thickness stands for resolution, line style for resistivity.

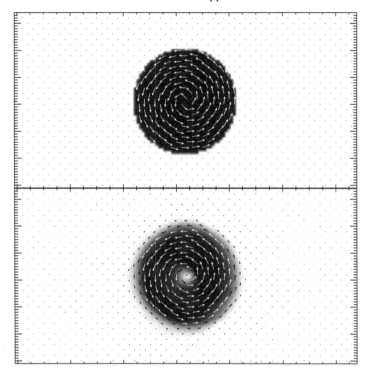

Fig. 5.10. Field loop advection test: magnetic energy density B^2 at $t = 0$ (*top*) and at $t = 2$ corresponding to two horizontal crossing times (*bottom*), with over-plotted field vectors. The grid resolution is $N_x \times N_y = 128 \times 64$.

5.9.3 2D: current sheet

This test is taken from Gardiner and Stone (2005). A square domain of extent $0 \le x, y \le 2$ and of constant density $\rho_0 = 1$ and pressure $p_0 = 0.1$ is permeated by a magnetic field along the y direction such that $B_y(0.5 < x < 1.5) = -1$ and $B_y = 1$ elsewhere. The ratio of thermal over magnetic pressure is $\beta = 0.2$. This setup results in two magnetic null lines, which then are perturbed by velocities $v_x = v_0 \sin(\pi y)$. Here, we use an adiabatic exponent of $\gamma = 5/3$ and employ the conservative formulation of the scheme. Figure 5.9 summarizes the test results in the form of the magnetic energy density $\langle B^2 \rangle$ against time. Different line styles stand for resistivities, and the line thickness denotes the model resolution. We ran tests at $N = 128^2$, 256^2 and 512^2. All models ran up to $t = 4$ and farther except for the 512^2-model at $\lambda_\Omega = 0$. A finite resistivity helps stabilizing the code.

The evolution of the system follows that described by Gardiner and Stone (2005), including the merging of magnetic islands until there are two islands per magnetic null line left, located approximately at the velocity anti-nodes. For zero

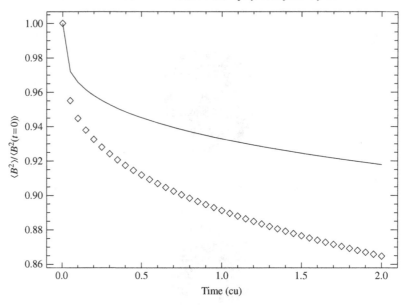

Fig. 5.11. Normalized magnetic energy density against time (in units of horizontal crossing time) with the same parameters as in Figure 5.10. Diamonds stand for
Proteus results, and the energy evolution as observed in ATHENA is shown by
the *solid line*.

resistivity (*solid lines*), the magnetic energy decay depends strongly on the resolution. This effect is reduced by increasing λ_Ω. For $\log \lambda_\Omega = -5$ (*dashed lines*),
the energy evolution follows pretty much the curves for $\lambda_\Omega = 0$ (*solid lines*), indicating insufficient resolution. For $\log \lambda_\Omega = -4$, the two higher resolutions start
to separate from the lower-resolution run, while at $\log \lambda_\Omega = -3$, the two higher
resolutions lead to indistinguishable curves (*dash-3dot lines*).

5.9.4 2D: advection of a field loop

A cylindrical current distribution (i.e. a field loop) is advected diagonally across the
simulation domain. Again, we follow the implementation presented by Gardiner
and Stone (2005). Density and pressure are both initially uniform at $\rho_0 = 1$ and
$p_0 = 1$, and the fluid is described as an ideal gas with an adiabatic exponent of
$\gamma = 5/3$. The computational grid at a resolution of $N_x \times N_y = 128 \times 64$ extends
over $-1.0 \leq x \leq 1.0$ and $-0.5 \leq y \leq 0.5$. The field loop is initialized via the
z-component of the vector potential $A_z = a_0(R - r)$, where $a_0 = 10^{-3}$, $R = 0.3$
and $r \equiv (x^2 + y^2)^{1/2}$. The loop is advected at an angle of $30°$ with respect to the
x-axis. Thus, two round trips in x correspond to one crossing in y. Figure 5.10

shows the initial magnetic energy density B^2 with the magnetic field vectors over-plotted (*top*) and the B^2 distribution after two time units measured in horizontal crossing times (*bottom*). The overall shape is preserved although some artifacts are visible. These results concerning the shape are similar to those of Gardiner and Stone (2005); specifically, Proteus preserves the circular field lines. This test uses $\lambda_\Omega \equiv 0$.

The time evolution of the magnetic energy density corresponding to Figure 5.10 is shown in Figure 5.11. Diamonds stand for Proteus results, the energy decay observed by Gardiner and Stone (2005) with ATHENA is indicated by the *solid line*, following their analytical fit. The energies are normalized to 1. Clearly, Proteus is somewhat more diffusive.

In summary, these numerical test cases demonstrate that Proteus models dissipate MHD effects accurately. Furthermore, it can advect geometrically complex magnetic field patterns properly.

5.10 Summary

Gas-kinetic schemes provide a robust and physical mechanism to solve the equations of MHD. Dissipative effects can be fully controlled. I discussed a specific implementation of a gas-kinetic solver – Proteus – including resistivity and (two-fluid) ambipolar diffusion. Details of the implementation have been presented elsewhere (Tang & Xu 2000; Heitsch *et al.* 2004, 2007), and an application to shear flows in magnetized fluids will be discussed by Palotti *et al.* (2008).

References

Aloy, M.-A., Ibanez, J. M., Mart, J. M., Müller, E., 1999, *ApJS*, **122**, 151

Baiotti, L., Hawke, I., Montero, P. J., Rezzolla, L., 2003, *MSAIS*, 1, 210

Bhatnagar, P. L., Gross, E. P., Krook, M., 1954, *Phys. Rev.*, **94**, 511–525

Bodenheimer, P., Tohline, J. E., Black, D. C., 1978, *BAAS*, **10**, 655

Clarke, D. A., 1996, *ApJ.*, **457**, 291

De Villiers, J.-P., Hawley, J. F., 2003, *ApJ.*, **589**, 458

Falle, S. A. E. G., 2003, astro-ph/0308396

Fryxell, B., *et al.*, 2000, *ApJS*, **131**, 273–334

Gammie, C. F., McKinney, J. C., Tóth, G., 2003, *ApJ.*, **589**, 444–457

Gardiner, T. A., Stone, J. M., 2005, *J. Comput. Phys.* **205**, 509–539

Gardiner, T. A., Stone, J. M., 2006, *ASPC*, 359, 143

Hackbusch, W., 1994, *Iterative Solution of Large Sparse Systems of Equations*,
    Springer-Verlag, New York-Berlin-Heidelberg

Heitsch, F., Slyz, A. D., Devriendt, J. E. G., Hartmann, L. W., Burkert, A., 2007, *ApJ.*,
    665, 445–456

Heitsch, F., Zweibel, E. G., Slyz, A. D., Devriendt, J. E. G., 2004, *ApJ.*, **603**, 165–179
Hujeirat, A., 1995, *A&A*, **295**, 268
Hujeirat, A., 2005, *CoPhC*, **168**, 1
Hujeirat, A., Rannacher, R., 2001, *New Astr. Rev.*, **45**, 425
Hujeirat, A., Camenzind Keil, B., 2007, arXiv, 0705.125
Koide, S., Shibata, K., Kudoh, T., 1999, *ApJ.*, **522**, 727
Komissarov, S. S., 2004, *MNRAS*, **350**, 1431
Kulsrud, R., Pearce, W. P., 1969, *ApJ.*, **156**, 445–469
Liebendörfer, M., Rosswog, S., Thielemann, F.-K., 2002, *ApJS*, **141**, 229L
Mignone, A., Bodo, G., 2003, *NewAR*, **47**, 581
Mignone, A., Bodo, G., Massaglia, S., Matsakos, T., Tesileanu, O., Zanni, C., Ferrari, A., 2007, *ApJS*, **170**, 228–242 (PLUTO)
Mizuno, Y., Nishikawa, J.-I., *et al.*, 2006, arXiv: astro-ph/0609004
O'Shea, B. W., Bryan, G., Bordner, J., Norman, M. L., Abel, T., Harkness, R., Kritsuk, A., 2004, arXiv: astro-ph/0403044 (ENZO)
Palotti, M. L., Heitsch, F., Zweibel, E. G., Huang, Y.-M., 2008, *ApJ.*, **678**, 234
Prendergast, K.-H., Xu, K., 1993, *J. Comput. Phys.*, **109**, 53–66
Shu, C.-W. and Osher, S., 1988, *J. Comp. Phys.*, **77**, 439–471
Slyz, A. D., Devriendt, J. E. G., Bryan, G., Silk, J., 2005, *MNRAS* **356**, 737–752
Slyz, A. D., Prendergast, K. H., 1999. *Astro. & Astrophys. Supp.*, **139**, 199–217
Slyz, A. D., Devriendt, J. E. G., Silk, J., Burkert, A., 2002, *MNRAS*, **333**, 894
Stone, J. M., Norman, M. L., 1992, *ApJS*, **80**, 753
Swesty, F. D., 1995, *ApJ.*, **445**, 811
Tang, H.-Z., Xu, K., 2000, *J. Comput. Phys.*, **165**, 69–88
Tóth, G., Keppens, R., Botchev, M. A., 1998, *A&A*, **332**, 1159
Wuchterl, G., 1990, *A&A*, **238**, 83
Xu, K., 1999, *J. Comput. Phys.*, **153**, 334–352.
Xu, K., 2001, *J. Comput. Phys.*, **171**, 289–335.
Zhang, W., MacFadyen, A. I., 2006, *ApJS*, **164**, 255–279
Ziegler, U., 1998, *Comp. Phys. Comm.*, **109**, 111

Part II

Structure and Star Formation in the
Primordial Universe

6

New frontiers in cosmology and galaxy formation: challenges for the future

R. Ellis and J. Silk

Abstract

Cosmology faces three distinct challenges in the next decade. (i) The dark sector, both dark matter and dark energy, dominates the universe. Key questions include determining the nature of the dark matter and whether dark energy can be identified with, or if dynamical, replace, the cosmological constant. Nor, given the heated level of current debates about the nature of gravity and string theory, can one yet unreservedly accept that dark matter or the cosmological constant/dark energy actually exists. Improved observational probes are crucial in this regard. (ii) Galaxy formation was initiated at around the epoch of reionization: we need to understand how and when the universe was reionized, as well as to develop probes of what happened at earlier epochs. (iii) Our simple dark matter–driven picture of galaxy assembly is seemingly at odds with several observational results, including the presence of ultraluminous infrared galaxies (ULIRGS) at high redshift, the 'downsizing' signature whereby massive objects terminate their star formation prior to those of lower masses, chemical signatures of α-element ratios in early-type galaxies and suggestions that merging may not be important in defining the Hubble sequence. Any conclusions, however, are premature, given current uncertainties about possible hierarchy-inverting processes involved with feedback. Understanding the physical implications of these observational results in terms of a model of star formation in galaxies is a major challenge for theorists and refining the observational uncertainties is a major goal for observers.

6.1 Introduction

We live in interesting times where we can measure, with good and improving precision, the constituents of our universe. However, we should not confuse

Structure Formation in Astrophysics. ed. G. Chabrier. Published by Cambridge University Press.

measurement with understanding: a major puzzle is that only 4% of the present energy density of the cosmos is in familiar baryonic form. The first challenge we will discuss in this chapter is how to move from measurement to understanding in the most basic question of all – what constitutes our universe?

The key, we will argue, is improved observations. Promising tools are available to tackle the distribution of dark matter on various scales, as well as to trace the history of the dark energy that causes the recently discovered cosmic acceleration. On a more fundamental level, the theoretical challenge of the cosmological constant and dark energy is so demanding in terms of fundamental physics that significant numbers of theorists are searching for alternative explanations of the observed acceleration. The existence of dark matter also is being questioned in schemes of alternative theories of gravitation. Refinement of the observational probes is essential in order to consolidate both the nature of the acceleration that motivates dark energy and the theoretical infrastructure of Einstein/Newton gravity that provides the framework for dark matter. Much investment will be needed to calibrate and understand the limitations of these various approaches, and dedicated missions will likely be needed, but the returns to science will be considerable. Whatever its origin, dark energy promises to fundamentally change our views of the physical world. And dark matter already has brought fundamental change to our understanding of structure formation, although more questions have been raised than answered.

Excellent observational progress is also being made in pushing the redshift frontiers. The most distant known galaxy, at the time of writing, lies at $z \simeq 7$, and promising candidates are now being claimed to $z \simeq 10$. Searches for the earliest stellar systems are motivated by more than redshift records. Poorly understood processes govern the distribution of luminosities and masses of the first stellar systems, yet these objects act as the basis from which galaxies later assemble.

Detailed agreement between the standard model and the spectrum of fluctuations in the microwave background gives us confidence that our basic picture of structure formation is correct. It predicts that the first galaxy-size halos will accrete cooling baryons to form stars by a redshift of around 20 or so. Many expect that intergalactic hydrogen was reionized by the first substantial generation of star-forming low-mass galaxies. However, the detailed predictions are very uncertain. Indeed, intermediate mass black holes (IMBHs), visible as miniquasars, are a theoretically compelling, although hitherto unobserved, component of the high z universe that could also provide an important ionization source. Observations of the earliest galaxies, with the James Webb Space Telescope (JWST) and the new generation of extremely large telescopes (ELTs), and the nature of the intergalactic medium at high redshift, as revealed by upcoming radio surveys, will be crucial in advancing our understanding of the astrophysical nature of reionization.

Finally, although the interface between theory and observation is richest in the area of low redshift ($z < 2$) galaxy formation, it is clear that the hierarchical dark matter–driven assembly picture promoted in the 1980s and 1990s is too simplistic. Although new phenomenological ingredients are being added to the semi-analytical models in order to patch agreement between the theory and observations of new multi-wavelength surveys, it is clear we seek a breakthrough in our understanding. All current galaxy formation theories are refined by 'interplay' with the data: this is satisfactory for generating mock catalogues for future surveys but falls short if one wants to arrive at a fundamental understanding of the physical processes governing galaxy formation.

The missing ingredient seems to be an understanding of how star formation is regulated in galaxies. This is perhaps not really surprising. We cannot readily simulate how stars form in nearby clouds, from Orion to Taurus. Inevitably this leads to new challenges when we try to cope with conditions that almost certainly were different at high redshift in the early universe.

A good example is the Kennicutt (1998) law, which relates the star formation rate to the gas content. Local data suggest that this may be universal. However, whether phenomenological modelling of the star formation rate, efficiency and initial mass function derived from studies of the nearby universe is relevant to the high redshift universe is far from certain. Some have argued that two modes of star formation are likely required: one to explain young, gas-rich, star-forming disks that are currently forming stars with low efficiency and another to explain the presence of old, gas-poor massive spheroids that formed stars with high efficiency long ago.

Another example is the initial stellar mass function (IMF). There are periodic attempts to argue that the IMF featured either more massive stars or fewer low-mass stars at high redshift or in extreme environments. The advantages of turning up the relative massive star frequency include accounting for the far infrared galaxy counts, rapid chemical evolution and boosting the number of ionizing photons for early reionization. Systematically lowering the number of low-mass stars may help with the colour–magnitude relation.

Finally, we cannot avoid remarking on the role of supermassive black holes in spheroid formation. There is little doubt that active galactic nuclei (AGN) are intimately connected to star formation. Unfortunately, neither theorists nor observers agree on the sign of the effect. Did black holes quench or trigger star formation? Most theorists favour quenching, but there are well-known counter examples. Did black hole growth precede or follow spheroid assembly? Most but not all data suggest that black hole masses were supercritical (relative to the Magorrian relation) in the past. Making progress in this complex area at the interface of star and galaxy formation represents our third major challenge.

6.2 Constituents of the universe

6.2.1 Dark matter

Dark matter studies go back to the pioneering efforts based on galaxy radial velocities by Zwicky for clusters and Rubin for galaxies. This technique has to some extent been superceded by a more fundamental approach first predicted by Einstein, namely gravitational lensing by dark matter. Here, we differentiate between 'strong lensing', where the foreground lensing mass density is often sufficient to create multiple images of the background source, and 'weak lensing', the statistical distortion of the background population by large-scale structure. Although lensing is more often used to promote studies of dark energy (see Section 6.2.2), it is important to realize its unique potential to track the properties of dark matter.

Weak gravitational lensing was only detected in 2000, but is now a highly-developed field. Dedicated facilities are being constructed to exploit the effect, and significant progress is expected from high-quality imaging surveys. In the short term, on the ground, these include PanSTARRS, the Dark Energy Survey, the VLT Survey Telescope, VISTA and the Subaru HyperSuPrime Camera. Ultimately, in space, we may realize the NASA/DoE Joint Dark Energy Mission (JDEM) and the Dark UNiverse Explorer (DUNE) proposed for ESA's Cosmic Visions programme around 2015.

Existing facilities such as Subaru's SuPrimeCam imager, the CFHT Megacam and the ACS onboard the Hubble Space Telescope have demonstrated the ability of weak lensing to map the dark matter (DM) distribution, both in 2-D projection on the sky (Figure 6.1) and, via the use of photometric redshift data, in reconstruction of the 3-D distribution. Comparing the distribution of dark matter, revealed by lensing, and the baryonic matter, revealed by broad-band imaging, has verified the important assumption that dark matter provides the basic 'scaffolding' within which baryons cool to form stars and assemble (Massey *et al.* 2007a).

Strong lensing is likewise effective in providing important constraints on the distribution of dark matter on small scales ($<100\,\mathrm{kpc}$). A key question is how can extragalactic data provide constraints on the *nature* of the dark matter? Numerical simulations based on the assumption that the dark matter is 'cold' and non-interacting suggest that the radial profile of the dark matter should be sharply peaked to small scales with a central density cusp $\rho \propto r^{-1}$ (Navarro *et al.* 1997).

By combining measures of strong lensing in clusters with stellar dynamics for the central cluster galaxy and its halo, Sand *et al.* (2004, 2008) have shown how it is possible to isolate the distribution of dark matter on small scales for direct comparison with the numerical simulations. Preliminary data, based on a few clusters, suggest that the dark matter is not as sharply peaked as predicted by CDM theory. Conceivably, this could arise via prolonged gravitational interactions between

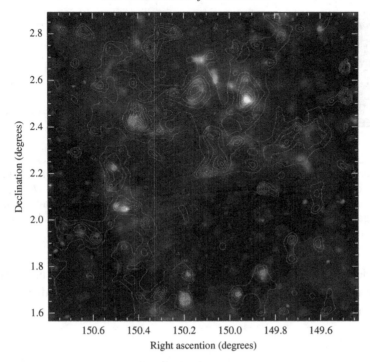

Fig. 6.1. The distribution of dark matter derived by weak lensing in the Hubble ACS COSMOS Field (Massey *et al.* 2007a). Such studies are influential in correlating the dark matter distribution (contours) with that of baryons.

the dark matter distribution and the baryons, the latter, for example, receiving dynamical input by non-gravitational forces.

Although few galaxy clusters have so far been studied with the detail required to make a convincing attack on this problem, the list of possible systems to explore is quite large and, with sufficient effort, the outcome of such a programme could be very interesting. As an extreme option, the presence of cores could imply that the dark matter is not entirely cold. For example, cluster dark matter cores might be the result of dwarf galaxy aggregation. The apparent ubiquity and universality of cores in nearby dark matter–dominated dwarf galaxies has revived interest in the possibility of warm dark matter consisting of sterile neutrinos in the few keV mass range.

The combination of strong and weak lensing data offers a powerful dynamical probe. The Sloan lens advanced camera survey (SLACS) project (Koopmans *et al.* 2006; Gavazzi *et al.* 2007) has demonstrated that dynamical masses for elliptical galaxy lenses span a wide range in redshift. The sum of the luminous and DM components 'conspire' to form an isothermal distribution, and the uniformity of the mass profile with redshift is striking. Absence of any significant evolution in

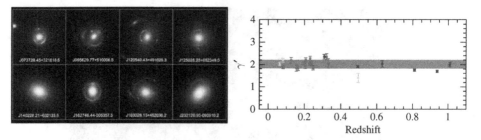

Fig. 6.2. (*Left*) The Sloan Lens ACS Survey led to a spectacular increase in the number of known Einstein rings via ACS imaging of ellipticals whose spectra revealed additional line emission from background lensed sources. (*Right*) The combination of strong lensing and stellar dynamics constrains the inner slope of the logarithmic mass profile, $\gamma = \mathrm{d} \log \rho / \mathrm{d} \log r$ which is remarkably close to isothermal at all redshifts (Gavazzi *et al.* 2007).

the inner density slopes suggest a collisional scenario in which gas and dark matter strongly couple during galaxy formation, leading to a total mass distribution that rapidly converges to dynamical isothermality (Figure 6.2).

6.2.2 Dark energy

The past 2 years have seen unprecedented interest in developing new facilities to address the perplexing question of the existence and nature of dark energy. Dedicated ground-based telescopes and future space missions are being developed or proposed with a variety of observational probes. But, given that there is no consensus on the physical basis of dark energy, one must exercise care in deciding what is the optimal observational target and which of the various observational probes is likely to be the most effective. More to the point, is a costly space observatory really required to make the necessary progress?

Recent studies in the United States and Europe (Albrecht *et al.* 2006; Peacock *et al.* 2007) have emphasized a step-wise approach where the most immediate target is to determine, or otherwise, whether dark energy is consistent with the cosmological constant (a vacuum with an equation of state such that $p = w \rho$ with $w = -1$). In the following, we explore the hurdles that we must overcome with the four most promising techniques in order to realize the necessary precision.

6.2.2.1 Type Ia SNe

This is the most well-developed method for probing dark energy. SNe Ia are events rich in detail, and the one parameter light curve 'stretch' correction yields individual luminosity distances accurate to about 7%. The latest results (Astier *et al.* 2006), when combined with local baryonic oscillation data for an assumed flat universe, yield consistency with $w = 1$ to about 9% accuracy.

Despite this progress, it is unclear whether SNe Ia are suitable for the required next step in precision cosmology. Host-galaxy correlations (Sullivan *et al.* 2006) demonstrate that SNIa explosions can occur with a wide range of delay times. This arouses the spectre of environmental and evolutionary biases. Could SNIa dimming with redshift, generally accepted as the unique reliable measure of acceleration currently available, be predominantly or even partly due to evolution? A significant UV dispersion has been seen in detailed spectra of $z \simeq 0.5$ events (Ellis *et al.* 2008) which seems not to correlate with the light curves. Conceivably, this effect arises from metallicity variations which may, ultimately, create a *systematic floor* particularly for the distant events that are essential for tracking the time dependence of dark energy (Figure 6.3). Great efforts are needed to understand these possible limitations, both via local high-quality surveys and via improved modelling of the spectra.

6.2.2.2 *Weak lensing*

Weak lensing complements SNe as a measure of dark energy since, unlike SNe which track the distance–redshift relation, lensing tracks the growth rate of density fluctuations. However, lensing is less well developed as a precision tool. Key issues include the accuracy with which the weak shear signal can be calibrated and whether it is linearly recovered, and the reliability of photometric redshifts, essential for 'tomography', i.e. slicing the data in depth to recover the growth rate. The current state of the art, following pioneering surveys at the Canada–France–Hawaii telescope, is consistent with the SN results but with larger error bars (e.g. $w < 0.8$, Semboloni *et al.* 2006).

Ground- and space-based observations both offer advantages. Space offers increased depth, a higher surface density of useable background galaxies and a more stable point-spread function (Kasliwal *et al.* 2007). Infrared coverage from space enables more precise photometric redshifts, particularly for $z > 0.8$ (Abdalla *et al.* 2008). Ground-based surveys offer a more cost-effective areal coverage, the potential of associated wide-field spectroscopy and associated non-cosmology science.

Although problems remain, the progress since weak lensing was detected in 2000 is impressive. The Shear Testing Evaluation Program (STEP), a collaborative effort using simulations of known shear, is being used to test the various image-processing algorithms (Heymans *et al.* 2006, Massey *et al.* 2007b). Although most

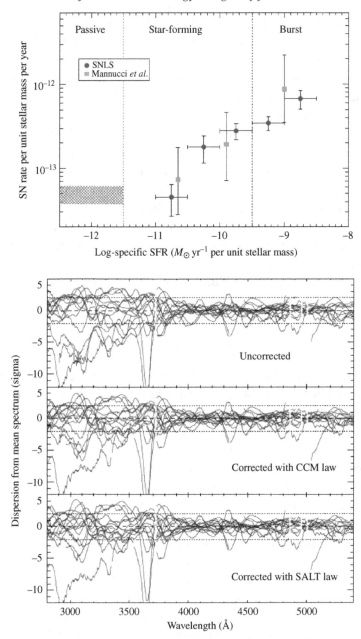

Fig. 6.3. Challenges to the precision use of SNe Ia as distance indicators. (*Top*) The Ia rate depends on the star formation history of the host galaxies (Sullivan *et al.* 2006). This suggests a range of delay times from birth to the explosion and possibly more than one progenitor mechanism. (*Bottom*) Spectral dispersion at maximum light as a function of wavelength from the Keck survey of Ellis *et al.* (2008). The significantly increased UV dispersion is not removed by standard calibration procedures and may indicate a new parameter is at play (e.g. metallicity), which evolves with redshift.

of the algorithms in regular use do not yet recover shear at the necessary precision, one can hope to learn from these and related developments.

6.2.2.3 Baryon Acoustic oscillations

Baryon wiggles are an imprint on the dark matter of the gravitational potential variations due to acoustic oscillations of baryons from the epoch of matter-radiation decoupling. This relic of the horizon scale in the galaxy power spectrum may be the perfect 'cosmic ruler'. As the signal is weak and on very large scales ($\gtrsim 120$ Mpc), well-sampled redshift surveys in huge volumes are needed to achieve the necessary precision. A major hurdle to be overcome will be the biasing of galaxies relative to DM. This can, in principle, be modelled via numerical simulations.

The most substantial project currently underway is WiggleZ at the AAT (Glazebrook *et al.* 2007) which plans to obtain more than 200 000 emission line redshifts to $z \sim 1$, thereby constraining w to 10% (Figure 6.4). A similar Sloan Digital Sky Survey (SDSS)-based extension is likely to exceed this precision by a factor of 2 or so. Such accuracy ($\sim 5\%$) cannot easily be superceded with imaging surveys that utilize photometric redshifts, which have intrinsic limitations especially at $z \gtrsim 1$. Thus, the next-generation instruments for baryonic acoustic oscillations

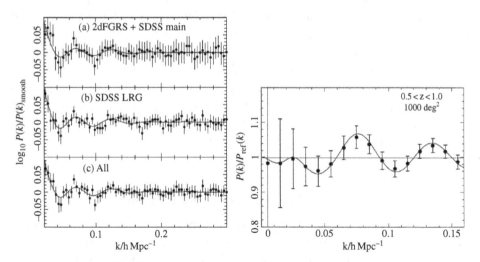

Fig. 6.4. How baryonic oscillations work: (*Left*) Detection of baryonic oscillations in the local 2dF and SDSS redshift surveys (Percival *et al.* 2007). The galaxy power spectrum is divided by a smooth function to reveal oscillations on large scales which represent the expanded relic of the horizon scale at recombination. (*Right*) With sufficient data, the wavelength of the oscillations acts as a standard ruler enabling the expansion history of the universe to be directly detected (simulation of AAT WiggleZ project, Glazebrook *et al.* 2007).

(BAO) studies will necessarily have to be quite ambitious. Gemini/Subaru are funding conceptual designs for a 1.5 square degree wide field multi-object spectrograph (WFMOS). Although motivated in part as a dark energy experiment, such an instrument would represent a logical 8 m successor to SDSS and 2dF and offer much valuable science, particularly in studies of galactic archeology.

A longer-term development involves a space-based spectroscopic survey such as DESTINY (a grism-type instrument proposed for NASA JDEM) or SPACE (a microlenslet instrument proposed for ESA Visions), which will obtain, in the latter case, up to a billion galaxy spectra. Whether ground- or space-based, one is looking at huge investments, $\sim$\$50 million or $\sim$\$500 million, respectively, to obtain a factor of order 2 improvement in precision for w. The possible pay-off, which cannot be overstated, is a signature of new physics such as $w \neq -1$. The downside is that one may just end up confirming Einstein's cosmological constant.

6.2.2.4 Galaxy clusters

Large samples of clusters are planned from X-ray and Sunyaev–Zeldovich (SZ) surveys. The path to dark energy again arises from the growth of structure. The most direct route is via the cluster number dependence on mass and redshift, $N(M, z)$, which constrains w via the perturbation growth factor. A key requirement is thus the mass of each cluster. Calibrating masses to the necessary precision, and with adequate survey uniformity, via the X-ray luminosity or temperature or the SZ decrement is the challenge. Ideally, one combines both approaches once the systematics are understood.

Current planning includes a survey of 10^5 X-ray clusters on eROSITA (projected launch 2010). The claimed precision on w is 5% (Haiman *et al.* 2005), assuming a mass calibration that is precise to 1%. Much progress on calibration accuracy will be necessary to realize this goal given that the current mass uncertainties are of order 10% for simulated data (Nagai *et al.* 2007).

6.3 First light and cosmic reionization

We now turn to our second frontier topic: the location of the first star-forming galaxies and evaluation of their role in bringing the so-called 'Dark Ages' to an end and reionizing the intergalactic medium. We have very few indicators of when cosmic reionization occurred but evidence is accumulating that star-forming galaxies were the responsible agents.

(i) The polarization–temperature cross-correlation derived from the Wilkinson microwave anisotropy probe (WMAP) 3-year data set suggests a redshift window of $7 < z < 15$ for the location of the bulk of the scattering electrons (Dunkley 2008).

(ii) The assembled stellar mass at $z \simeq$ 5-6 (Eyles *et al.* 2006; Stark *et al.* 2007a) indicates a substantial amount of star formation at earlier epochs, perhaps sufficient to explain reionization. Many of the more massive galaxies which can be studied in detail at this epoch show 'Balmer breaks' – signatures of old (>100 Myr) stars in galaxies (Eyles *et al.* 2005, Figure 6.5).

(iii) Carbon, produced only in stellar cores, is present in absorption in the highest redshift quasi-stellar object (QSO) spectra and seems ubiquitous in the intergalactic medium out to $z \simeq$ 6 (Ryan-Weber *et al.* 2006). This suggests an early period of enrichment from supernovae.

The negatives are that one has no idea of the escape fraction for ionizing photons in the first galaxies and that there is at least one plausible alternative source of ionizing photons. This consists of IMBHs, which act as miniquasars and are prolific sources of ionizing photons at very early epochs. They must be present in considerable numbers in the early universe if one is to understand how supermassive black holes were in place by $z \sim$ 6 as evidenced by the presence of ultraluminous quasars. Theoretical arguments suggest that the first generation of dissipating gas clouds at $z \sim$ 10 could as easily form IMBHs as population III stars, and indeed probably form both. Confirmation of such a high redshift population of non-thermal ionizing sources could eventually come from a combination of X-ray background, high-ℓ CMB (Cosmic microwave background) and LOFAR (low frequency array) observations.

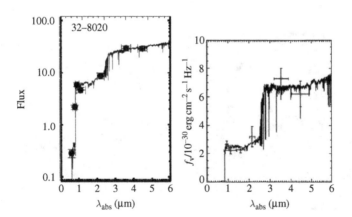

Fig. 6.5. Spectral energy distribution of two spectroscopically confirmed high redshift galaxies with established (>100 Myr) stellar populations as revealed via their significant 'Balmer break' at 2–3 μm with Spitzer. (*Left*) A $z = 5.55$ galaxy with a stellar mass of $1.1 \times 10^{11} M_{\odot}$ from the survey of Stark *et al.* (2007a). (*Right*) A $z = 5.83$ galaxy with a stellar mass of $2–4 \times 10^{10} M_{\odot}$ from the survey of Eyles *et al.* (2006).

Given the evidence for early activity, several groups are now probing the era $z > 7$ for the first glimpse of the sources responsible for reionization (for a review, see Ellis 2008). Unfortunately, prior to the launch of JWST and the commissioning of the next generation of 30 to 40-m telescopes, progress will inevitably be very slow. A key question is whether the bulk of the sources responsible for reionization are luminous and rare, or abundant and feeble. As there is growing evidence over $3 < z < 7$ that the luminosity function of both continuum 'dropouts' and Lyman-α emitters is steepening at the faint end with redshift (Bouwens *et al.* 2007, Kashikawa *et al.* 2006), it seems prudent to search for an abundant population of sources whose star formation rates are less than $1 M_\odot \, \mathrm{yr}^{-1}$.

One possible route to discovering such faint sources, if they are very numerous, is via strong gravitational lensing. A rich cluster of galaxies can magnify selectively distant sources by factors of $\times 25$ along their 'critical lines'. As several distant sources have been successfully recovered using this technique (Ellis *et al.* 2001, Kneib *et al.* 2004), a number of groups have begun searching for faint $z > 7$ galaxies in these regions (Figure 6.6). Through this approach, Stark *et al.* (2007b) have published six candidate Lyman-α emitters, some of which they claim lie in the redshift range $8.6 < z < 10.2$ with (unlensed) star formation rates as low as $0.2 M_\odot \, \mathrm{yr}^{-1}$. Since the lensed searches cover only a small area (0.3 arcmin^2 in Stark *et al.*'s case), if even a fraction of these sources are truly at these redshifts, the abundance of low-luminosity galaxies is within striking distance of that required for cosmic reionization.

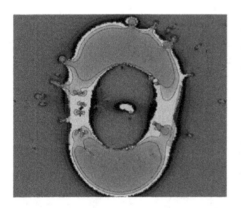 

Fig. 6.6. Probing to high redshift with gravitational lensing. (*Left*) Magnification map for the rich lensing cluster Abell 370. Two curves of maximum magnification ('critical lines') encircle the cluster. By searching in a blind manner spectroscopically within these narrow regions, highly magnified line emitters from $z > 7$ can be located. (*Right*) 2-D spectral image of one of six candidate Lyman-α emitters at $z = 10.0$ found in the Keck infrared survey of Stark *et al.* (2007b).

6.4 Galaxy formation

Our final challenge is the search for a full understanding of how the star-forming galaxies that we see at $z \simeq$ 2–3 assemble into the varied forms which we see locally.

6.4.1 Disk galaxies

The disk mode of galaxy formation is motivated by the gravitational instability of gas-rich disks. Disk galaxies form late, slowly and inefficiently. The inefficiency is plausibly interpreted in terms of momentum input into the interstellar medium from expanding supernova shells. Supernovae provide the quantitative amount of feedback required to account for the star formation rate observed in disk galaxies. Star formation is sustained by a combination of disk self-gravity and supernova feedback. The self-gravity drives large-scale instability provided the disk is cold. Continuing accretion of cold gas guarantees the gaseous disk continues to fragment into large molecular cloud complexes. These accrete smaller clouds, cool and themselves fragment into stars. Calculation of the star formation rate is complicated by the role of magnetic fields, whose pressure also plays a role in supporting clouds against collapse and decelerating star formation. The magnetic fields themselves are ionization-coupled to the cold gas, and the ionization fraction is itself due to radiation from massive stars. The situation is further complicated by the fact that many clouds are supported by turbulent pressure, and the turbulence driver is not well understood.

Despite these non-linear complications, the global star formation rate in galactic disks is described by a remarkably simple formula:

$$SFR = 0.02 \frac{\Sigma_{\text{gas}}}{t_{\text{dyn}}}.$$

This simple model fits a wide range of data, including quiescent and star-bursting galaxies, and even the individual star-forming complexes in M51 (Kennicutt *et al.* 2007) as well as submillimetre galaxies and ULIRGs at $z \sim$ 2 (Bouche *et al.* 2007). Evidence for local cold gas feeding is found in nearby disk galaxies that have extensive HI envelopes.

While one can phenomenologically understand star formation in disks, there is no successful model for disk formation. All attempts to date lead to overly massive or concentrated bulges. Detailed studies of self-gravitating hydrodynamic collapse with realistic initial conditions find that while the initial specific angular momentum matches that of observed disks, some 90% is lost to the halo during collapse. Late infall implies gas-rich halos, which in turn require large SN-heating efficiencies (Governato *et al.* 2007) and result in predicted large X-ray luminosities (Sommer-Larsen 2006). The implied large amounts of hot gas are at best only rarely seen for massive disk galaxies, where coronal absorption lines set strong constraints (Yao *et al.* 2008).

In fact, Fraternali and Binney (2008) argue that the HI 'beards' found in nearby spiral galaxies such as NGC 891 and NGC 2403 are analogous to galactic high-velocity clouds and are signposts of a substantial and otherwise mostly hidden halo gas accretion rate that interacts with supernova-driven galactic fountains. While the required accretion rate is consistent with lambda cold dark matter (LCDM) expectations and is comparable to the star formation rate, the infalling gas is required to have low angular momentum in contrast with expectations from simulations (Gottloeber & Ypes 2007).

6.4.2 Early-type galaxies

There is a long-standing debate about how early-type galaxies formed. Was the process primarily monolithic or hierarchical? The response seems to centre on understanding star formation as opposed more simply to mass assembly.

The following observations argue for a monolithic origin.

- $[\alpha/Fe]$: The ratio of type II supernova (SNII)-generated alpha elements to iron, mostly produced by SNIa, is a powerful clock that demonstrates star formation in massive galaxies proceeded more rapidly than in lower-mass systems (Thomas *et al.* 2005, Figure 6.7).
- *Star formation time:* SED (spectral energy distribution) analysis of distant galaxies favours systematically lower specific star formation rates for increasing stellar mass. This supports downsizing – massive galaxy formation preceded low-mass galaxy formation. Star formation in massive galaxies was complete before the peak in cosmic star formation history at $z \sim 1.5$ (Papovich *et al.* 2006).
- *Fundamental plane:* The low dispersion in the fundamental plane for elliptical galaxies indicates a narrow spread in ages at a given stellar mass.
- *Colour–magnitude relation:* The low dispersion argues for a conspiracy between age and metallicity.

A hierarchical origin is supported by the following arguments.

- *Environment:* The dependence of morphological type on environment, especially mean density and cluster membership, argues for formation via a merging history.
- *Major mergers:* Ultraluminous infrared and submillimetre galaxies invariably show evidence of triggering by major mergers.
- *Morphological evolution:* Deep surveys show that the Hubble sequence is essentially unchanged to $z \sim 1$, but rapid evolution in both size and morphology occurs at higher redshift.

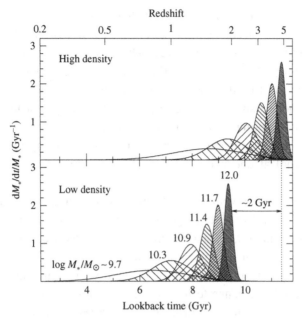

Fig. 6.7. The evidence from the abundance ratios of alpha elements to iron demonstrates downsizing in star formation history (Thomas *et al.* 2005). Massive early-type galaxies systematically form earlier (from SED fitting) and have shorter star formation timescales (from $[\alpha/\text{Fe}]$).

- *Star formation history:* The colour–magnitude relation conspiracy may be broken with GALEX data, which, however, leads to new issues concerning the gas supply needed to account for the residual star formation found in 30% of nearby early-type galaxies. Significant minor mergers provide a plausible explanation.

The reality most likely is a combination of both hierarchical and monolithic scenarios. The former controls disk formation, the latter seems appropriate to massive spheroid formation. However, even monolithic star formation is reconcilable with hierarchical assembly of dark matter and gas, although stars could only track hierarchical aggregation to a much lesser extent. Most stars in massive early-type galaxies must be formed over a period of order of the dynamical timescale and hence monolithically. However, a pure monolithic scenario is incompatible with the observed cosmic star formation and galaxy assembly history at high redshift. It also results in far too low a rate of ionizing photon production at $z \gtrsim 6$.

6.4.3 Feedback

Semi-analytical modelling gives an excess of both small and massive galaxies relative to the observed luminosity function. Resolving the galaxy luminosity function problems is possible with astrophysical feedback. There is an active debate as to the optimal feedback mechanism. A majority view is that AGN-driven outflows provide substantial feedback in the early phases of massive spheroid formation (Bower *et al.* 2006, Croton *et al.* 2006). This provides a means of quenching star formation by heating the gas reservoir and dispersing it. AGN outflows provide a heating mode that is effective on the intracluster gas at low redshift. It is essential to avoid continued star formation via the accreting gas. Indeed, there is evidence for entropy injection, from the observed entropy profiles, the cooling cores that cool at unexpectedly low rates and the presence of AGN-driven bubbles.

A clue that favours blowout comes from the correlation between central black hole mass and spheroid velocity dispersion. If gas accretion fuels both black hole growth and star formation, exhaustion of the gas reservoir occurs when the black hole mass is high enough for blowout to drive out the residual gas. Collimation could affect this argument, although jet instability in an inhomogeneous protogalaxy would rapidly degenerate into a cocoon. Blowout occurs, star formation terminates and the supermassive black holes (SMBHs)–σ relation saturates. The resulting momentum balance condition between self-gravity of the protogalactic gas and Eddington-limited outflow yields a relation between supermassive black hole mass and spheroid velocity dispersion that fits the observed normalization and slope of the correlation (Silk & Rees 1998). Correlations between ultraluminous starbursts, winds and AGN may be a manifestation of this process at high redshift. Observations of X-ray-selected quasars show that AGN feeding and SMBH growth coevolve and peak along with the cosmic star formation history at $z \sim 2$ (Silverman *et al.* 2007).

6.4.4 *Downsizing*

A major enigma has come from the phenomenon of galaxy downsizing. In one study, the Palomar-DEEP2 galaxy stellar mass functions over $0.4 < z < 1.4$ have been obtained using rest-frame U-B colours as a discriminant (Bundy *et al.* 2006). A threshold galactic stellar mass is apparent, above which there is no star formation. This mass threshold increases from $10^{11}M_\odot$ at $z \sim 0.3$ to $10^{12}M_\odot$ at $z \sim 1$ (Figure 6.8). The most massive galaxies are already in place by $z \sim 3$. The evidence from NIR stellar mass estimates demonstrates downsizing in mass assembly over a broad range in redshift (Perez-Gonzalez *et al.* 2008). Parallel studies focusing on X-ray-selected AGN similarly find that AGN luminosities have a luminosity threshold that increases to beyond $z \sim 2$ from the weakest to the strongest AGN (Hasinger *et al.* 2005).

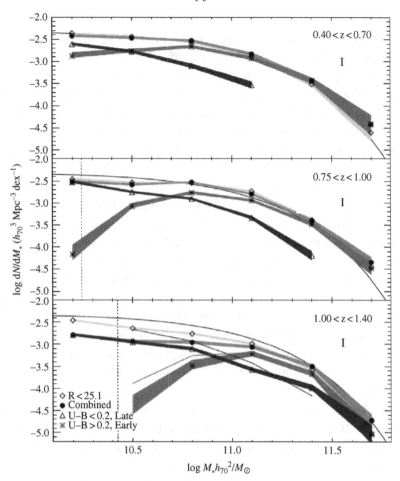

Fig. 6.8. A detailed analysis of downsizing in the Palomar-DEEP2 survey (Bundy *et al.* 2006). The panels show the stellar mass function in three redshift intervals partitioned according to a measure of star formation: – quiescent and – active. An evolving mass threshold is apparent above which star formation is quenched.

The conclusion is that both early-type galaxies and SMBH masses are anti-hierarchical, that is to say, the most massive objects are present early, whereas the least massive are absent at early epochs. This phenomenon was *not* predicted by the pioneering models on galaxy formation in a CDM-dominated universe. Since low-mass halos predominated early, star formation must have been more efficient in the most massive galaxies. Why is there downsizing of both massive spheroids and AGN, with a similar rise between redshift 0 and ~2?

There are alternative schools of thought on the resolution of this paradox. Gravitational heating is environmentally induced and mostly affects massive galaxies which are found in the most massive halos. This can terminate late star formation, while the environment provides intense gas feeding and fuelling provided by minor mergers (Khochfar & Ostriker 2008). This scenario leads to downsizing, at the price of prolonged star formation. Cooling primarily affects the lower mass halos where there is late infall of cold gas along filaments and star formation is correspondingly more efficient (Dekel & Birnboim 2008). This helps explain the scaling relations (colour, magnitude, metallicity), where the relations flatten above a characteristic galactic stellar mass of $\sim 3 \times 10^{10} M_\odot$ (Tremonti *et al.* 2004).

The major, and prevalent, rival view appeals to AGN heating of the ICM to suppress cooling flows at late epochs (Bower *et al.* 2006; Croton *et al.* 2006). Early AGN activity is Eddington-limited, and the cooling times are shorter. Massive galaxies form early as a consequence. AGN feedback at late times primarily suppresses gas cooling and star formation in massive halos. This terminates both massive black hole growth and spheroid star formation and results in downsizing as the late cold gas supply is quenched. One obtains simultaneous downsizing both for AGN and for the massive host galaxies. However, the star formation timescale in massive galaxies again is long. Neither model gives the short timescales seen in SED modelling of massive galaxies and especially in the alpha element ratios. Nor do the models account for the low dispersion in scaling relations such as the fundamental plane.

What could be the missing link? One hope is that better simulations at higher resolution may help resolve this discrepancy. But additional physics most likely is needed. AGN triggering could provide a solution. If indeed the Eddington luminosities are high enough to drive the gas out, the AGN-driven outflows could interact with the gas strongly enough to trigger star formation by jet-driven cocoon overpressuring of massive interstellar clouds. The jet drives a series of weakly supersonic shocks into the cocoon, which fills with disrupted cloud gas. The cores of the massive clouds survive and are overpressured. Moreover, the turbulent backflow overpressures the protospheroid and protodisk. The enhanced pressure induces gravitational instability and star formation (Antonuccio & Silk 2008). The jet/cocoon timescales associated with compact radio sources are short, of order 10^6 years. Moreover, the interstellar gas kinematics in high redshift radio galaxies are seen in some cases to be controlled by cocoon-driven turbulence (Nesvada *et al.* 2007). The timescale of triggering, associated with the jet/cocoon propagation speed and necessarily short compared to the dynamical time-scale, results in an enhanced rate of star formation. Downsizing is a natural corollary if the the SMBH feeding, jet power and outflow strength are a strong function of galaxy and

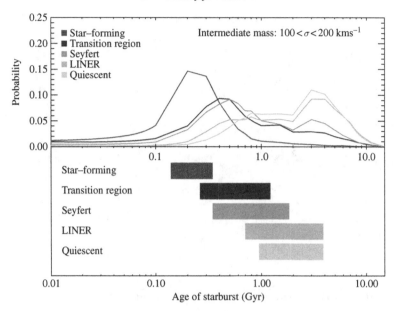

Fig. 6.9. The link between starbursts and AGN is demonstrated by this time sequence for low- and intermediate-mass galaxies, with time delay indicated from onset of star formation for different types of activity in early-type galaxies (Schawinski *et al.* 2007)

SMBH mass, as inferred observationally. For example, the radio power is found to correlate with the Bondi accretion rate for an X-ray sample of massive radio galaxies (Allen *et al.* 2006). The next phase would be quenching, aided and abetted by the boosted star formation and SN rate. It is only at this later stage that the AGN may be observable as an X-ray source. In the earliest stage, there is correlated AGN and star formation activity. This phenomenon may be visible in the mid-IR for massive galaxy precursors such as SMGs and ULIRGs at $z \sim 2$ (cf. Sajina *et al.* 2007; Pope *et al.* 2008). This is a chicken-and-egg problem of course: it remains to be shown which is the driver. Possibly, it may be neither: theory suggests that a major merger may both feed the AGN and drive the starburst, as discussed earlier.

Some direct evidence for jet-induced star formation at $z > 3$ comes from high redshift radio galaxy studies (Klamer *et al.* 2004; Feain 2007), where one sees the molecular gas reservoir as well as the young star distribution aligned with the radio lobes. Minkowski's Object is a classical nearby example of jet-induced star formation (Croft *et al.* 2006). More circumstantially, there is expected to be a connection between AGN and ultraluminous starbursts. ULIRGs are usually associated

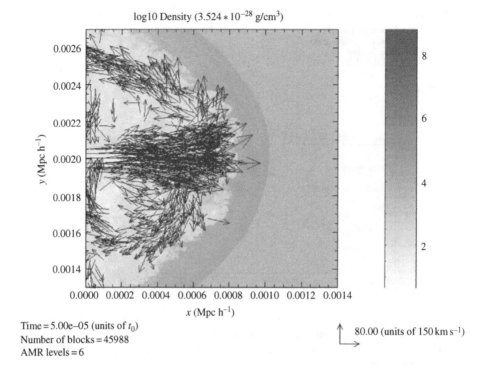

Fig. 6.10. Simulation of radio jet injection of energy into the ISM. The jet is supersonic and up to half the injected energy is dissipated by driving turbulence into an expanding overpressured cocoon. Density and velocity distributions are shown. The high-density enhancement within the cocoon is the stripped material from a cloud which has been shocked by the jet. The large-scale circulation in the cocoon induces a backwards compression flow onto the protogalaxy (Antonuccio & Silk 2008).

with a massive galaxy merger, also favoured as supporting the case for the monolithic gas conversion phase evidenced by the trend in the abundance ratios of alpha elements to iron with spheroid mass. Simulations suggest that mergers both fuel supermassive black hole growth and account for formation of early-type galaxies. Simulations suggest that the massive merger rate at $z \sim 4$ may suffice, up to one per massive halo, to account for massive spheroid formation (Fakhouri & Ma 2008). Observational data on merger rates are consistent with the expectation from the simulations that the merger rate per halo per unit time increases approximately as $(1 + z)^2$.

However, it has been argued that major mergers do not seem to be sufficiently frequent at $z \sim 2$ to account for the observed frequency of ULIRGs (Daddi *et al.* 2007). This critique is partly based on the result from the Millennium simulations that major mergers are predominantly dry by this epoch, as a consequence of the

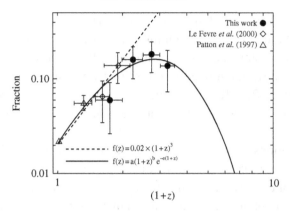

Fig. 6.11. Observed major merger fraction for massive galaxies ($\gtrsim 10^{10} M_\odot$) as function of redshift. Data are from Patton *et al.* (1997) and Le Fevre *et al.* (2000), with added points (*filled circles*) from Ryan *et al.* (2008). The major merger fraction is found to peak at $z = 1.8 \pm 0.5$. Major merger activity coincides with the observed peaks in star formation history and AGN feeding.

AGN feedback modelling recipe that is adopted (Springel *et al.* 2005). One solution is appeal to minor mergers which are sufficiently common, to fuel generic spheroid formation, along with providing a more extended star formation history. Another option occurs if AGN feedback is only effective at late epochs ($z \lesssim 2$) as inferred from observations and modelling of cooling cores and cavities in the intracluster gas (Nusser *et al.* 2006; Best *et al.* 2007).

Early major mergers would be gas-rich (wet), but late major mergers would then be gas-poor (dry). Semi-analytical modelling without AGN feedback indeed predicts that gas-rich (wet) mergers dominate in the past, and dry mergers are more abundant at low redshift (Khochfar & Burkert 2003). The early mergers could both feed the AGN and drive starbursts. Of course, late epoch feedback, presumably by AGN, is still needed to account for the large current epoch population of red galaxies.

6.5 Where next?

The prospects are bright for advances in cosmology. Although all current observations are consistent with the hypothesis that $w = -1$ and constant, new physics may be hiding within the error bars. The basic aim is to empirically track luminosity distance and perturbation growth factor over redshift to $z \sim 3$. More than one technique will be essential, with the aim of achieving better than 5% accuracy in w. In addition to supernovae, there are potentially three methods that appeal to different techniques and instrumentation: weak lensing, baryon oscillations and galaxy cluster surveys.

Another goal will be to find the first galaxies. Theory is not sufficiently precise to define the optimum search programmes: for example, it is not clear whether AGN precede, are contemporaneous with, or follow galaxy formation. Nor is dust formation and evolution understood, so that we cannot tell whether to focus on X-ray, optical, NIR, FIR or submillimetre frequencies. At present, observers are undertaking or proposing exploratory searches over all of these frequency regimes.

Current motivations centre on developing a census of the earliest galaxies at $z = 6$, or an age of 0.95 Gyr. With this in hand, one should be able to evaluate the contribution of early star formation to cosmic reionization, provided we understand the ionizing photon escape fraction. An important outcome will be improved constraints on mass assembly and feedback.

6.5.1 ELTs and JWST

In terms of probing the early universe, we are near the limits of all current facilities. The planned restoration of HST in 2008 with installation of a new camera and spectrograph (WFC3, COS) and the repair of the existing camera and spectrograph (ACS, STIS) will provide a welcome boost to ultradeep survey cosmology. However, the limitations of a 2.4 m telescope are self-evident and now provide a major motivator for the ELTs and JWST.

ELT design studies are focusing on adaptive optics (AO)-optimized 20 to 40-m telescopes. The TMT (30 m) is leading the pack at present with good progress on a $80-million design study (2004–2009) and a construction proposal now being considered by various sponsors. ESO has launched a 57-million Euro design study for a 42-m telescope (2007–2010). These projects envisage first light in 2016. AO will play a crucial role in ELT science and in particular multiconjugate AO systems will be essential to give the fields essential for cosmology. Exploring the territory with laser-guide star systems on our existing telescopes is critical.

6.5.2 21-cm facilities

The intergalactic medium at early epochs is the primary reservoir for galaxy formation, and its structure will be probed by new generations of radio interferometers. An entirely new domain will be opened with LOFAR and MWA via 21-cm surveys for tomography of primordial HI. The Mileura Wide-Field Array will have a collecting area of 8000 m^2 over 1.5 km with arc-min resolution at 150 MHz. In the era of TMT+JWST+LOFAR/MWA, we probably will not be interested in the simple issue of when reionization occurred but rather how it occurred. Complex physical processes involving varied astrophysical inputs will be studied and tracked by the topology and structure of ionization bubbles. JWST can find luminous sources, and TMT will scan their vicinities to determine topology of ionized shells via the distribution of fainter emitters. This will be done in conjunction with HI surveys at the reionization epoch.

6.5.3 Theory

Theory trails behind observation. It is reactive but lacks enough depth to be really predictive. Our current understanding of recent star formation is hierarchical, as viewed in disk galaxies. The star formation history is extended, and there is evidence for discrete gas feeding events at $z \sim 1$. Most likely, these are minor mergers. Cold gas feeding is required by theory in order to account for the continued instability of self-gravitating galactic disks to cloud and star formation. However, the instability timescale is a dynamical time, and one risks achieving far too high a star formation rate. This well-known problem is customarily resolved via feedback by supernova heating and momentum injection into the interstellar medium. This accounts for the low fraction of gas converted into stars per dynamical time. However, the physics is poorly modelled at current numerical resolutions. New approaches need to be developed to model feedback in multiphase interstellar media.

For massive spheroids, the situation is at least superficially different. It is common to refer to star formation efficiency. Parenthetically, we recall that efficiency is an imprecise concept that combines timescale and gas mass fraction converted into stars. To be more precise, we can specify star fraction formed per local dynamical time. Star formation was undoubtedly more efficient at high redshift than found today in disks. Downsizing occurred, with the most massive spheroids forming at even higher efficiency. This seems to be a distinct starburst mode of star formation in which the negative feedback by supernovae is unimportant. The starbursts may be merger-triggered, but mergers also drive AGN feeding. Circumstantial evidence for AGN triggering, mostly in the past, suggests AGN outflows may also play a role in triggering the quasi-monolithic mode. What seems more certain is that AGN outflows quench star formation by truncating the gas supply. Here, not only is current numerical resolution insufficient, but the relevant feedback physics is uncertain. From jets propagating in an inhomogeneous interstellar medium to star formation is a major step that will require implementation of skills taken from two distinct communities of numerical simulators.

6.6 Summary

We require a healthy balance between general purpose facilities, projects targeting key science questions and theory. Panoramic imaging with high resolution will revolutionize studies of dark matter on various scales. Multi-object and integral field spectroscopy will enormously boost our understanding of galaxy evolution and provide far more reliable probes of dark energy.

A step-wise plan for constraining dark energy is emerging. Baryonic oscillation surveys are the newest ingredient and will generate much additional science. Natural synergy is envisaged in studies of $z \sim 10$ galaxies between 30-m telescopes with adaptive optics and mid-IR studies with JWST. Laser-guide star-adaptive optics will revolutionize studies of $z \sim 1$–3 emission line galaxies; ELTs will extend this to absorption line systems.

Massively parallel computing focusing on memory per node is essential to incorporate refined subgrid physics into the large-scale simulations. Armed with improved physics on star formation and AGN, we may eventually be able to develop robust predictions. From these and the new generation of observations, there will emerge answers to such key questions for the future as whether hierarchical assembly of dark matter halos governs growth of galaxies. We need to ascertain what physical processes curtail galaxy growth in a time-dependent manner and what processes stimulate star formation. There is no doubt that exciting times are ahead with immense challenges for observers and theorists alike.

References

Abdalla, F. *et al.*, 2008, *MNRAS*, **387**, 969
Albrecht, A. *et al.*, 2006, Dark Energy Task Force report (astro-ph/0609951)
Allen, S. *et al.*, 2006, *MNRAS*, **372**, 21
Antonuccio, V. and Silk, J., 2008, *MNRAS*, **389**, 1750
Astier, P. *et al.*, 2006, *A&A*, **447**, 31
Best, P. *et al.*, 2007, *MNRAS*, **379**, 894
Bouche, N. *et al.*, 2007, *ApJ.*, **671**, 303
Bouwens, R. *et al.*, 2007, *ApJ.*, **670**, 928
Bower, R. *et al.*, 2006, *MNRAS*, **370**, 645
Bundy, K. *et al.*, 2006, *ApJ.*, **651**, 120
Croft, S. *et al.*, 2006, *ApJ.*, **647**, 1040
Croton, D. *et al.*, 2006, *MNRAS*, **365**, 11

Daddi, E. *et al.*, 2007, *ApJ.*, **670**, 156

Dekel, A. and Birnboim, Y., 2008, *MNRAS*, **383**, 119

Dunkley, J. *et al.*, 2008, *ApJ.*, in press (astro-ph/0803.0586)

Ellis, R. S. 2008, *First Light in the Universe*, Saas Fee Lectures Series, Springer-Verlag, p. 259 (arXiv:astro-ph/0701024)

Ellis, R. S. *et al.*, 2001, *ApJ.*, **560**, L119

Ellis, R. S. *et al.*, 2008, *ApJ.*, **674**, 51

Eyles, L. *et al.*, 2005, *MNRAS*, **364**, 443

Eyles, L. *et al.*, 2006, *MNRAS*, **374**, 910

Fakhouri, O. and Ma, C. 2008, *MNRAS*, **386**, 577

Feain, I. 2007, *ApJ.*, **662**, 872

Fraternali, P. and Binney, J. 2008, *MNRAS*, **386**, 935

Gavazzi, R. *et al.*, 2007, *ApJ.*, **667**, 176

Glazebrook, K. *et al.*, 2007, in *Cosmic Frontiers*, eds. Metcalfe, N. and Shanks, T., ASP Conference Series, Vol. 379, p. 72

Gottloeber, S. and Ypes, G., 2007, *ApJ.*, **664**, 117

Governato, F. *et al.*, 2007, *MNRAS*, **374**, 1479

Haiman, Z. *et al.*, 2005, arXiv:astro-ph/0507013

Hasinger, G., Miyaji, T., and Schmidt, M., 2005, *A&A*, **441**, 417

Heymans, C. *et al.*, 2006, *MNRAS*, **368**, 1323

Kashikawa, N. *et al.*, 2006, *ApJ.*, **648**, L7

Kasliwal, M. *et al.*, 2007, *ApJ.*, **684**, 34

Kennicutt, R. 1998, *ApJ.*, **498**, 541

Kennicutt, R. *et al.*, 2007, *ApJ.*, **671**, 333

Khochfar, S. and Burkert, A. 2003, *ApJ.*, **597**, L117

Khochfar, S. and Ostriker, J. 2008, *ApJ.*, **680**, 54

Klamer, I. *et al.*, 2004, *ApJ.*, **612**, L97

Kneib, J.-P. *et al.*, 2004, *ApJ.*, **607**, 697

Koopmans, L. *et al.*, 2006, *ApJ.*, **649**, 599

Le Fevre, O. *et al.*, 2000, *MNRAS*, **311**, 565

Massey, R. *et al.*, 2007a, *Nature*, **445**, 286

Massey, R. *et al.*, 2007b, *MNRAS*, **376**, 13

Nagai, D. *et al.*, 2007, *ApJ.*, **655**, 98

Navarro, J. *et al.*, 1997, *ApJ.*, **490**, 493

Nesvada, N. *et al.*, 2007, *A&A*, **475**, 145

Nusser, A., Silk, J. and Babul, A., 2006, *MNRAS*, **373**, 739

Papovich, C. *et al.*, 2006, *ApJ.*, **640**, 92

Patton, *et al.*, 1997, *ApJ.*, **475**, 29

Peacock, J.A. *et al.*, 2007, ESA-ESO Report on Fundamental Cosmology (arXiv:astro-ph/0610906)

Percival, W. *et al.*, 2007, *MNRAS*, **381**, 1053

Perez-Gonzalez, P. *et al.*, 2008, *ApJ.*, **675**, 234

Pope, A. *et al.*, 2008, *ApJ.*, **675**, 1171

Ryan-Weber, E. *et al.*, 2006, *MNRAS*, **371**, L78

Ryan, R. *et al.*, 2008, *ApJ.*, **678**, 751

Sajina, A. *et al.*, 2007, *ApJ.*, **664**, 713

Sand, D. J. *et al.*, 2004, *ApJ.*, **604**, 88

Sand, D. J. *et al.*, 2008, *ApJ.*, **674**, 711

Schawinski, K. *et al.*, 2007, *MNRAS*, **382**, 1415

Semboloni, E. *et al.*, 2006, *A & A*, **452**, 51

Silk, J. and Rees, M., 1998, *A&A*, **331**, 1
Silverman, J. *et al.*, 2007, *ApJ.*, **679**, 118
Sommer-Larsen, J., 2006, *ApJ.*, **644**, L1
Springel, V. *et al.*, 2005, *Nature*, **435**, 629
Stark, D. P. *et al.*, 2007a, *ApJ.*, **659**, 84
Stark, D. P. *et al.*, 2007b, *ApJ.*, **663**, 10
Sullivan, M. *et al.*, 2006, *ApJ.*, **648**, 868
Thomas, D. *et al.*, 2005, *ApJ.*, **621**, 673
Tremonti, C. *et al.*, 2004, *ApJ.*, **613**, 898
Yao, Y. *et al.*, 2008, *ApJ.*, **672**, 21

7

Galaxy formation physics

T. Abel, G. Bryan and R. Teyssier

7.1 Introduction

A significant fraction of all efforts in astronomy is expended on studying the properties of galaxies and their spatial distributions over a very large fraction of all of cosmic time. Much of this is motivated by the mysteries of understanding dark matter and dark energy. The accuracy at which we can use galaxy clustering and other properties to determine the more fundamental parameters describing the universe is determined by the sophistication of our understanding of galaxies themselves.

As is by now well known, hierarchical cosmological models like ΛCDM predict that the first objects to collapse out of the expanding homogeneous universe are also the smallest. Since the early universe was also generally a simpler place (fewer heavy elements, etc.), this produces the anti-intuitive result that it is often easier to model the early universe. For the present-day universe, the significant complexities of the star formation process and the enormous range of relevant spatial and temporal scales forces us to employ phenomenological sub-grid models. In this contribution, we explore some aspects of our current understanding of galaxy formation with a distinctly numerical bias. We cannot hope to comprehensively review this topic, so instead we begin with a general overview of some of the important physical processes which are operating and then delve into a few current issues in more detail.

7.2 Fast versus slow structure formation

The non-relativistic mass content of the universe is dominated by dark matter, which is thought to generally have a low-interaction cross-section with other forms of matter, including itself. The result of this is that dark matter does not, by itself, add any scales to the problem of structure formation (or more precisely, the scale dictated by the dark matter cross-section is generally too low to be important

Structure Formation in Astrophysics. ed. G. Chabrier. Published by Cambridge University Press.
© Cambridge University Press 2009.

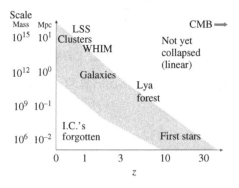

Fig. 7.1. The grey band shows a region surrounding the non-linear mass and length scale as a function of redshift. This band indicates objects which have recently gone non-linear (i.e. collapsed) and so have not yet forgotten their initial conditions. Also labelled are the positions in this diagram of a number of observational probes, ranging from large-scale structure (LSS) to the formation of present-day galaxies at $z \sim 3$ to the formation of the first stars at high redshift.

for most structure formation problems). The result is that in structure formation (excluding baryons), there is only one non-linear scale: the scale at which cosmological structures are currently collapsing. We show this scale as a function of redshift in Figure 7.1 (the details of cosmological structure growth are reviewed elsewhere, such as in Ciardi and Ferrara (2005)).

In the absence of baryons, and their rich physical processes, all of these structures would have a nearly universal structure (Navarro *et al.* 1997); however, additional physicals such as radiative cooling, star formation, magnetic fields and radiation transfer combine to make important differences along this sequence. The single most important of these processes is radiative cooling because this is how gas loses its binding energy and continues collapsing. Without cooling, the gas would end up with a density distribution similar to that observed by the dark matter. To obtain a rough gauge of the relative importance of cooling for a particular halo, we can treat the gas in a spherical symmetric manor with a density obtained from the assumption of hydrostatic equilibrium with the dominant dark matter and a temperature given by the virial temperature of the halo (White & Frenk 1991). Using this model, we can then find the radius at which the cooling time of the gas equals the dynamical time and convert that into a fraction of baryons which can cool.

The result of performing this calculation is shown in Figure 7.2, for a wide range of halo masses. Also shown are the results from numerical simulations for the low- and high-mass ends, indicating that the model has some rough correspondence to more detailed predictions (although it is clear that other processes will become important for the middle range of masses). As can be seen from this figure, there

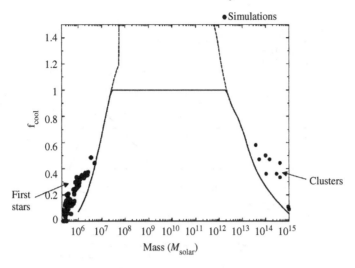

Fig. 7.2. The fraction of a halo$ baryonic mass which can cool radiatively in a dynamical time, as a function of halo mass. This is based on a simple spherically-symmetric model described in the text; however, we also show results from numerical simulations for low-mass (Machacek *et al.* 2001) and high-mass (Cien & Bryan 2008) halos.

are two distinct kinds of objects which can cool. The first class are those for which the predicted cool fraction is substantially less than one and includes both low-mass objects (forming at high redshift), which cool primarily through molecular hydrogen, and high-mass objects (forming at late times), which are mostly galaxy clusters.[1] For these types of objects, the cooling time is longer than the dynamical time and so cooling progresses more slowly than collapse, and therefore we denote this as a *slow* mode of collapse. In the second class are objects with intermediate masses, for which cooling via hydrogen line emission (and metal lines at higher temperatures) is important. These collapse in a *fast* mode, as their cooling time is shorter than their dynamical time. There is an additional complexity in that the accretion rate of the halo, i.e. its growth in mass naturally leads to heat input which needs to be radiated as well. Consequently, fast accretion can delay cooling and collapse. This has been demonstrated in numerical simulations of first structure formation (Yoshida *et al.* 2003) and by more straightforward analytic arguments (Wang & Abel 2008).

This division into fast and slow modes of collapse is similar to the hot and cold modes of accretion discussed in Section 7.7 and in Birnboim and Dekel (2003). As that reference points out, the simple spherical model described above assumes the halo gas is able to establish itself at the virial temperature before cooling, an

[1] In fact, due to the hierarchical nature of our cosmology, the gas in high-mass objects must have progressed via a phase of lower-mass objects, and so one can question how clean this division is for the high-mass end.

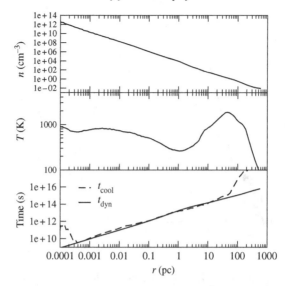

Fig. 7.3. The radial profile of the collapse of a $10^6 M_\odot$ halo from Abel *et al.* (2002), showing the density (*top*), temperature (*middle*) and two timescales (*bottom*) over more than six orders of magnitude on radius. The equality between the cooling and the dynamical times over the vast majority of this range is a product of the slow, quasi-static inflow seen in slow-mode collapse.

assumption which is probably false for many objects which collapse in the fast mode. Nevertheless, this distinction is important and may play a key role in galaxy properties at late times (see Section 7.7). Here we also want to stress that it plays an important role for low-mass objects.

In slow-mode structure formation, the collapse is quasi-static and results in largely spherical objects. For example, as discussed in Abel *et al.* (2002) and Bromm *et al.* (2002), the collapse of a $10^6 M_\odot$ halo proceeds in a largely spherical way and does not result in fragmentation. The reason for this can be seen in Figure 7.3, which demonstrates the equality between the dynamical and the cooling time scales over many orders of magnitude in radius. This arises because collapse is limited by the rate at which the binding energy can be radiated. A side effect of this is that fragmentation does not occur (coupled with the fact that the cooling rate of molecular hydrogen is a steep function of temperature and so thermal instabilities do not occur, see Abel *et al.* (2002) and Ripamonti and Abel (2004).

7.3 The importance of H_2 cooling in early galaxy formation

The relevance of H_2 for early structure formation has been recognized since the late 1960s (Peebles & Dicke 1968). However, because its formation relies on non-equilibrium chemistry and some claims that photo-dissociation may be effective

(Dekel & Silk 1986), early models of galaxy formation (Rees & Ostriker 1977; White & Rees 1978; White & Frenk 1991 chose to neglect H_2 altogether. With the implementation of fast ordinary differential equation solvers in three-dimensional cosmological codes (Anninos *et al.* 1997; Gnedin & Ostriker 1997) in the last decade, this short-coming can be avoided. Very quickly it became clear that even substantial radiation backgrounds have only modest effects on the formation of early objects (Machacek *et al.* 2001). This conclusion has recently been strengthened (Wise & Abel 2007) as it has been shown that central shocks can lead to small electron fractions that catalyse molecular hydrogen formation in halos whose formal virial temperature would seem too low ($<10^4$ K) to allow for that. In even the most extreme cases studied by Wise and Abel (2007), cooling and collapse occurred faster than in the identical setup that did not allow H_2 formation. This clearly demonstrates that H_2 as a coolant will always be dominant in the smallest of galaxies.

Even for postulated unphysically large radiation backgrounds, the halo masses that cool and collapse are up to two orders of magnitude smaller than the halos that cool via atomic hydrogen line cooling alone. This abundance of cooling halos and the cosmic mass fraction contained within them depends exponentially on this critical mass scale. Consequently, the majority of current models of cosmological reionization, chemical evolution, supermassive black hole formation and galaxy formation underestimate the number of star-forming progenitors of any given system by at least one order of magnitude.

Interestingly, these first objects are prone to make very massive stars early on (Abel *et al.* 2002; Bromm *et al.* 2002), which have a dramatic impact on the gas surrounding them (Abel *et al.* 2007). Most notably, their photo-heating expels all baryons from their host halo at ten times the escape velocity. This changes the initial conditions for the subsequent generations of cosmic objects all of which have such early small-scale structure as progenitors. Without the inclusion of H_2, Population III (Pop III) star formation and their feedback, one would have found the first objects to form in $\sim 10^8 M_\odot$ halos in which $10^5 M_\odot$ of gas would have undergone runaway collapse and most likely have formed an accreting supermassive black hole as the first luminous object in the universe (Wise *et al.* 2008). Instead isolated very massive Pop III stars shape the physics of the assembly of the first true galaxies that host a large number of stars.

7.4 Simulating a $10^8 M_\odot$ galaxy one star at a time

Since it has recently become possible to carry out the *ab initio* calculation of the formation of the first stars and their impact on the surrounding medium, it is now feasible to follow the formation of galaxies *one star at a time*. The first example

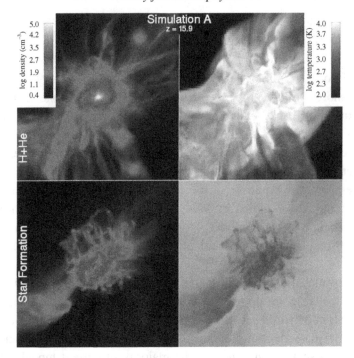

Fig. 7.4. Density-squared weighted projections of gas density (*left*) and temperature (*right*) of the most massive halo in simulation A of Wise and Abel (2008). The field of view is 1.2 proper kpc. The *top* row shows the model without star formation and only atomic hydrogen and helium cooling. The *bottom* row shows the same halo affected by primordial star formation. Note the filamentary density structures, clumpy interstellar medium and the counter-intuitive effect that feedback leads to lower temperatures.

of these calculations have been presented by Wise and Abel (2008). These models follow the formation of stars, their HII regions and the supernovae at unprecedented resolution. The density projections of one of the galaxies of Wise and Abel (2008) is shown in Figure 7.4. It is remarkable how clearly a turbulent interstellar medium with a markedly clumpy structure and filamentary extensions out of the halo become apparent even when only radiative feedback from the first stars is included. The stark contrast to the same model neglecting feedback from Pop III stars in the upper panels of the figure visually demonstrates the importance of these early stars and the build-up of cosmological objects.

The phase diagrams of the same galaxy is given in the *top panel* of Figure 7.5 showing the wide spread of temperatures at a given density and vice versa caused by the feedback. This figure also shows the phase diagrams for another galaxy of similar mass with the added assumptions that the first stars end their lives without

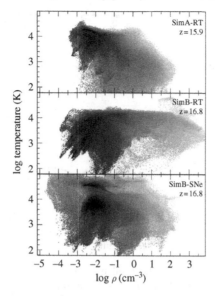

Fig. 7.5. Mass-weighted ρ–T phase diagrams of a sphere with radius 1 kpc, centred on the most massive halo in SimA-RT (*top*), SimB-RT (*middle*) and SimB-SNe (*bottom*). At $T > 10^4$ K, one can see the HII regions created by current star formation. The warm, low-density ($\rho < 10^{-3}$ cm^{-3}) gas in SimB-SNe is contained in SNe shells.

explosion (*middle panel*) and that all of them explode as 170 solar mass pair-instability supernovae (*bottom panel*). The latter models show a much more distinct hot phase and larger outflow velocities.

For the topic of this chapter, perhaps the most relevant findings from these simulations are twofold. First, the earliest galaxies may be enriched to one thousandth of solar metallicity by Pop III stars alone. Second, the spin parameters and the baryon fractions of high-redshift galaxies is determined by feedback and is independent of the large-scale gravitational torques that set the spin of the underlying dark matter component. This latter effect is illustrated in Figure 7.6 which shows the evolution of the spin parameter and the baryon fraction of the most massive progenitor of a dark matter halo that has a total mass of $4 \times 10^7 M_\odot$ by $z = 17$.

These simulations of early dwarf galaxies at high redshift require a minimum of one-tenth of a parsec resolution to accurately capture the transition from D- to R-type ionization fronts of the first HII regions. Simulations with poorer spatial resolution fail to attain the correct outflow velocities from these earliest objects and would result in unrealistically high baryon fraction of the subsequent objects they become part of. Current computational resources will allow us to simulate galaxies at this level of detail perhaps up to a mass of $10^{10} M_\odot$ for up to a billion years of evolution, i.e. down to redshifts of ∼5. To simulate galaxies like our own

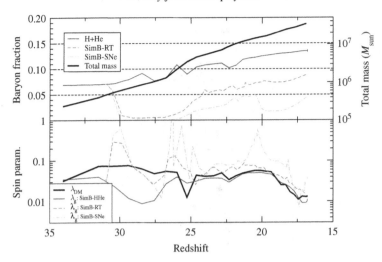

Fig. 7.6. The *top graph* shows the total mass (*thick line*) of the most massive progenitor in simulation B. The *thin lines* depict the baryon fraction, M_{gas}/M_{tot}, of simulation B without star formation (*thin*), with star formation (*dashed*) and with SNe (*dot-dashed*). The *bottom graph* shows the spin parameter of the same halo in the dark matter (*thick line*) and gas (*thin lines*, same legend as top graph).

with $10^{12} M_{\odot}$ existing 13.7 billions years after the Big Bang requires more phenomenological descriptions for the effects of star formation and their feedback at tens of parsec scales. Such models have been developed and give very promising results of which we will give a number of examples in the rest of this chapter.

7.5 A minimum mass for galaxies?

One important concept for hierarchical structure formation relates to the extrema. What sets the minimum and maximum masses of galaxies? Let us discuss the question of the smallest possible galaxies here. The role of active galactic nucleus (AGN) feedback on the most massive galaxies will be a topic later on.

The relevant processes have been touched on above and in these proceedings already: radiative cooling and heating. Inelastic collisions between atomic species (H, He and metals) and electrons as well as neutral hydrogen atoms and hydrogen molecules in the gas are radiating away its potential energy which was converted to thermal energy by shock- and PdV-heating. For large enough dark matter halos, this loss of pressure support triggers the formation of centrifugally supported disks of warm (10^4 K) neutral hydrogen. These disks can further fragment and form stars. The presence of an existing UV background, due to quasars and OB stars in distant galaxies, will also heat the cosmological gas before it collapses into dark matter halos. Also, all the feedback from the smallest progenitors of a given object will strongly affect the properties of the baryons as we have seen in Section 7.4. This pre-heating, at the level of 10^4–10^5 K, prevents the collapse into small-mass dark matter halos, before pressure gradients overcome the gravitational acceleration. Gnedin (2000), Hoeft *et al.* (2004) and Rasera and Teyssier (2006) have shown that this translates into a critical mass above which, on average, dark matter halos host more than 50% of the maximum baryons fraction. Below this critical mass, dark matter halos have a deficit of baryons by more than 50%. Moreover, these gas-poor dark matter halos cannot cool fast enough to host a star-forming disk. Clearly, the star formation efficiency in the smallest objects will be significantly smaller than in the more massive halos.

What the exact minimal mass for efficiently star-forming dark matter halos may be has been computed both analytically and numerically and its exact value is still under debate (Gnedin 2000; Hoeft *et al.* 2004; Rasera & Teyssier 2006), but qualitatively, it is of great importance in our hierarchical picture of galaxy formation, since it sets a cutoff in the quasi-infinite mass hierarchy of dark matter halos. This mass scale is only related to atomic physics and to the underlying model for the UV background. Whether in fact the complex interplay between feedback physics and the net effects of all progenitors will conspire to be expressed by a dark matter halo mass alone is clearly questionable and may explain why it remains elusive. It is quite likely that such a mass scale would depend on environment and the details of the accretion history of the object of interest.

7.6 The cosmic star formation rate

From a cosmological point of view, the fragmentation of galactic disks into cold neutral medium (CNM), molecular clouds and ultimately into stars is totally out of reach. Although recent studies have tried to approach the problem using idealized disk simulations Tasker & Bryan (2006), cosmological simulations still rely on sub-grid physical models to describe the interstellar medium (ISM). The model of Yepes *et al.* (1997) and Springel and Hernquist (2003), inspired by the model of McKee and Ostriker (1977), captures the essential physics of multiphase ISM and the associated star formation. This rather complex model can be simplified into a single polytrophic equation of state for the galactic gas of the form $P \propto \rho^\Gamma$, with Γ in the range 4/3–2, depending on the density range. This stiffer equation of state captures the thermal heating due to exploding supernovae and helps stabilizing disks in cosmological simulations. The resulting star-formation rate is calibrated on local observations of star-forming regions (the so-called Kennicutt law (Kennicutt 1998). Star formation is finally implemented by spawning collisionless particles at a rate consistent with the chosen star formation law.

One last and very important ingredient is related to the kinetic feedback from supernovae. Indeed, one missing ingredient in the previous multiphase model is the turbulence induced by supernovae and cloud formation and fragmentation. Supernovae bubbles coalesce and form giant superbubbles of typical size around 100–200 pc that eventually break out of the disk and can trigger galactic winds or galactic fountains (see Dubois and Teyssier (2008) and reference therein). The physics of these galactic winds is very complex and depends critically on the internal structure of the disk. Current studies indicate that such winds are very inefficient in removing gas from the galactic disks. The wind efficiency, usually defined as the ratio of the wind ejection rate to the disk star formation rate, is found to be below 10%. Observations of distant starbursts and Lyman Break galaxies, on the contrary, indicate wind efficiencies above 100%.

We now have in hand all the necessary ingredients to compute the global star formation rate in the universe, also called the Madau plot. This theoretical calculation was performed by Rasera and Teyssier (2006), using both analytical and numerical approaches, and compared with observations. The outcome of this work is that strong supernovae-driven winds are needed (around 100% efficiency) in order to reproduce the star formation history in the universe (Figure 7.7). Using extended Press–Schechter theory, Rasera and Teyssier (2006) have also computed

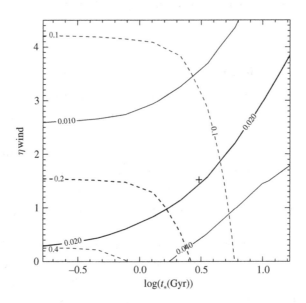

Fig. 7.7. Parameter space allowed by current observations of the cosmic star formation history. η is the wind efficiency and t_* is the star formation timescale. The cross is our best model. It corresponds roughly to the Kennicutt efficiency for star formation and to the observed efficiency in star burst galaxies for supernovae-driven winds (Reprinted from Rasera & Teyssier 2006).

the star formation history of halos of a given mass range. Massive ellipticals or even galaxy clusters tend to form stars earlier than dwarf galaxies, the reason being that the mass fraction of baryons in halos exceeding the minimal mass defined above is less for smaller objects. This leads to an apparent *anti-hierarchical* behaviour, as observed in the Sloan galaxy sample. On the contrary, it is clear from observation that massive objects do not form stars at late time, leading to the so-called 'galaxy bimodality' (Kauffmann *et al.* 2003). The galaxy population can be divided into two groups: the blue sequence with small galaxies still forming stars actively and the red sequence composed of large galaxies devoid of gas and young stars. The previous model can explain this behaviour on a qualitative level, but not quantitatively. The origin of this star formation shutdown in massive galaxies is therefore still a puzzle.

7.7 The role of diffuse gas accretion

Due to the ever 5 increasing size of present-day supercomputers, it has now become realistic to design large-scale, 3D simulations with the galaxy formation physics we have just described. These ambitious projects require not only the mobilization of astrophysicists, but also computer scientists of various expertise. They are often referred to as 'Grand Challenge Projects'. Such a project has been prepared and selected at the Barcelona Supercomputing Centre and is one of the most exciting and important projects in computational cosmology in the last decade. The goal is to perform the largest cosmological simulations so far, with two different state-of-the-art numerical methods (smoothed particle hydrodynamics (SPH) and adaptive mesh refinement (AMR)) and a wealth of physical ingredients to study galaxy formation with unprecedented accuracy. This project is a collaboration between the Horizon team, a French Science Foundation funded project, and a European team led by Gustavo Yepes from the University of Madrid. Each team has access to a different code (RAMSES versus GADGET) based on different numerical techniques (AMR versus SPH). Each code has its own advantages and drawbacks. Since both teams are using the same set of initial conditions, we are in the unique position of being able to directly compare the outcomes of both simulations.

After the first run, we were able to visualize the data, discovering an impressive level of detail. This reminded us of the first very large N-body simulations, such as the Hubble Volume Simulation performed in 2000 by the Virgo Consortium, except that we have now a precise description of gas dynamics, star formation and other complex physical processes. The large-scale density distribution is typical of self-gravitating Gaussian random fields, with a filamentary structure often referred to as the 'Cosmic Web'. It is worth mentioning that these filaments are in quasi-pressure equilibrium, so that, in this respect, the Mare Nostrum simulation can be

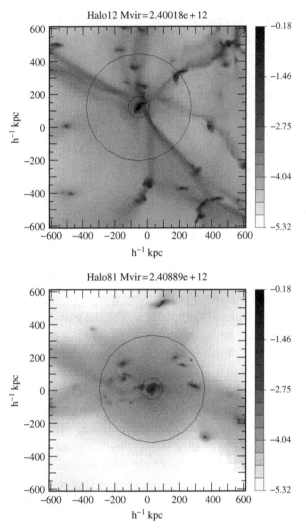

Fig. 7.8. Projection maps of the gas density for a typical $M = 2 * 10^{12} M_\odot$. *Top*, $z = 4$; *Bottom*, $z = 2.5$. The outer circle is the virial radius of the halo, while the inner circle defines the central galaxy vicinity.

considered as 'converged'. These filaments constitute the 'Lyman alpha' forest that appears in the distant quasars' spectral energy distribution. The current epoch of the simulation (redshift 1.5) is optimal to compare directly with observed quasar spectra.

At smaller scales, it is possible to discover in our simulation one very important aspect of galaxy formation, namely, the accretion flows around forming galaxies. Figure 7.8 shows two halos of the same mass (roughly $2 \times 10^{12} M_\odot$) containing a large disk in the centre with several satellites orbiting around, but one lives at redshift 4, while the other lives at $z = 2$. Clearly visible around the $z = 4$ halo are elongated filaments, often referred to as 'cold streams', which are feeding the central galaxies with fresh gas. This gas will be processed into stars by our sub-grid model describing star formation. In between these filaments exist hot shocks, typical of large halos such as the one shown in the image. These cold streams have completely disappeared by redshift 2. What we are witnessing here is the transition from the 'cold mode' to the 'hot mode' of accretion. It is very tempting to relate this bimodality in the accretion mode of smooth gas to the 'blue' versus 'red' dichotomy in the present-day galaxy population (Dekel & Birnboim 2006). Following the pioneering work of Kravtsov (2003) and of Keres *et al.* (2005), we have been able to observe such a phenomenon in the Mare Nostrum simulation, because of the large box size, allowing for rare events to occur in the simulated volume and the good shock-capturing properties of our AMR scheme (Godunov method) that allows for an accurate treatment of shock-heating.

The origin of this bimodality in the accretion flow is related to gas cooling of warm filaments feeding the galactic disk. As long as these filaments can survive disruptions from the halo environment (shocks, satellites, tidal forces, etc.), cold gas can be channelled into the disk and star formation can proceed. On the contrary, if these filaments are destroyed, fresh infalling gas will be spread over the entire halo and will be shock-heated to the virial temperature. The halo gas can then cool down and fall onto the disk, through the hot mode of accretion. If, for some reason, cooling is prevented from occurring in the halo, star formation might be shut down.

7.8 The role of AGN feedback

One of the preferred mechanisms to prevent gas from cooling in galaxy cluster cores is AGN feedback. X-ray observations of galaxy clusters show cavities related to the presence of AGN-driven jets seen in radio maps (Arnaud *et al.* 1984; Carilli *et al.* 1994; Fabian *et al.* 2000, 2002; McNamara *et al.* 2000, 2001; David *et al.* 2001). In the last few years, a number of groups have made an intense effort to simulate the impact of radio jets and bubbles on the structure of the ICM (Churazov *et al.* 2001; Quilis *et al.* 2001; Reynolds *et al.* 2001, 2002; Basson & Alexander 2003; Omma & Binney 2004; Omma *et al.* 2004; Ruszkowski *et al.* 2004; Sijacki & Springel 2006; Vernaleo & Reynolds 2006). Most of these studies usually considered a jet with a prescribed power, fine-tuned to have a maximum impact on the cooling flow in the cluster core. It turned out that powerful jets escape the cluster core as a laminar flow, while weak jets trigger a turbulent flow, leading potentially to gas-heating through shocklets dissipation or viscous damping of sound waves. Cattaneo and Teyssier (2007) showed that if one computes the jet power

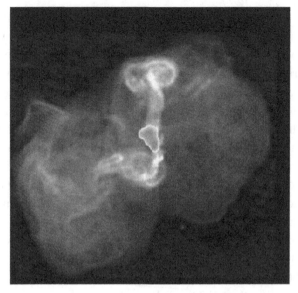

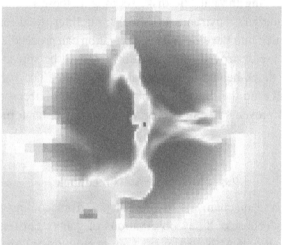

Fig. 7.9. *Top*: observation of M87 in radio from Owen *et al.* (2000). *Bottom*: column density of the cluster core with AGN feedback obtained by Cattaneo and Teyssier (2007).

as a fraction of the Bondi rate of accretion, one can self-consistently relate the jet power to the cluster core thermal state. This leads naturally to a self-regulated state, for which turbulence plays a central role in destabilizing the jet and maximizing the energy feedback. The main assumption in this model is that black hole accretion is controlled by the large-scale Bondi accretion rate. Small-scale processes, especially around the black hole itself, are assumed to be much faster. Despite many simplifying assumptions, AGN feedback together with filamentary accretion of diffused gas might be key ingredients in the physics of galaxy formation, as the comparison between our simulation and the radio map of M87 suggests (Figure 7.9).

7.9 Star formation in quiescent disk galaxies

Given the difficulty of studying star formation and feedback in cosmological simulations of galaxy formation, one alternative approach is to perform high-resolution simulations of idealized disk galaxies. This has been done by a number of groups, as summarized in Bryan (2007). Here, we explore some of the lessons of this approach, described in more detail in Tasker and Bryan (2006, 2008).

To model the galaxy disk, we used the structured AMR grid code, *Enzo*, described in Bryan (1999) and O'Shea *et al.* (2004). The disk is modelled in a 3D periodic box of side $1 h^{-1}$ Mpc. The size of the parent grid is 128^3 and we proceed down to an additional eight sub-grids of refinement which gives us a maximum resolution (i.e. minimum cell size) of about 50 pc (and include some runs with resolution down to 25 pc). The simulations start with an isothermal gas disk with a temperature of 10^4 K and a density profile given by $\rho_0 \exp(-r/r_0) sech^2(z/2z_0)$. The disk sits in a $10^{12} M_\odot$ dark matter profile which takes the form described by Navarro *et al.* (1997). Once set up, the disk was allowed to evolve over a period of $\Delta z = 0.1 \approx 1.4$ Gyr. Radiative gas cooling was allowed using rates given in Rosen and Bregman (1995).

Two different star formation prescriptions were used, the first, inspired by cosmological simulations, was termed C-type and forms stars at a rate equal to $\epsilon \rho_{gas}/t_{dyn}$, where $\epsilon \approx 0.01$ is an efficiency parameter and t_{dyn} is the dynamical time. This form basically has the Schmidt law built into it. We also experimented with a variant (density, or D-type), which assumes that stars will only form in the giant (molecular) clouds resolved in the simulation, and so we adopt a high efficiency and a high threshold (corresponding to a number density of 10^3 cm^{-3}).

The code also allows the inclusion of stellar feedback from type II supernovae explosions. This form of feedback has often been suggested as the main driving force for self-regulated star formation. If used, then 10^{-5} of the rest-mass energy of generated stars is added to the thermal energy of the gas over a time period equal to t_{dyn}. This is equivalent to a supernova of 10^{51} erg for every $55 M_\odot$ of stars formed.

In Figure 7.10, we show the gas and stellar densities of the simulations with and without feedback, including a non-feedback star formation run at high resolution. It is clear that the typical clump size is strongly affected by resolution, with smaller fragments appearing at higher resolution. This is consistent with the fact that we do not always resolve the Jeans length in the centre of the disk, particularly for the

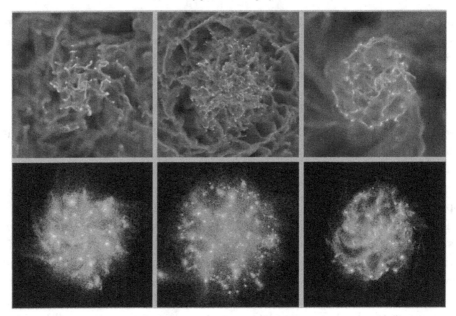

Fig. 7.10. This figure shows the gas density (*top*) and stellar density (*bottom*) in a region 22 kpc on a side after 330 Myr of evolution from a simulation of a quiescent disk (see Tasker and Bryan (2006) for more details). The *left-most images* are from the simulation with star formation but no feedback; the *central images* are from the high-resolution version of this simulation, while the *right-most images* are from the feedback simulation.

standard D run. On the contrary, there is a well-defined radius beyond which star formation does not occur, and this does appear to be well resolved.

This star formation threshold can be understood with the Toomre stability parameter (Toomre 1964), defined as $Q = \kappa c_s / \pi G \Sigma_g$, where κ is the usual epicyclic frequency, c_s is the thermal sound speed as measured in the disk and Σ_g is the gas surface density. This parameter is plotted as a function of radius in Figure 7.11, along with the current star formation rate (averaged over the last 20 Myr) at that radius for a variety of simulations. In each case, there is a sharp cutoff at a particular value of the Q parameter. From the runs represented in the *top panel* of Figure 7.11, we see that this cutoff is unaffected by feedback. As we describe below, feedback acts to reduce the star formation rate, but it has no bearing on this stability cutoff point. The critical Q parameter derived this way is about 0.6. A linear analysis for a 2D dimensional disk predicts 1 (Toomre 1964).

For comparison, we also calculated the star formation rate and Q for a disk with four times the gas mass, which is shown in the *bottom panel* of Figure 7.11. The heavier disk draws out the star formation cutoff to a greater radius, in close

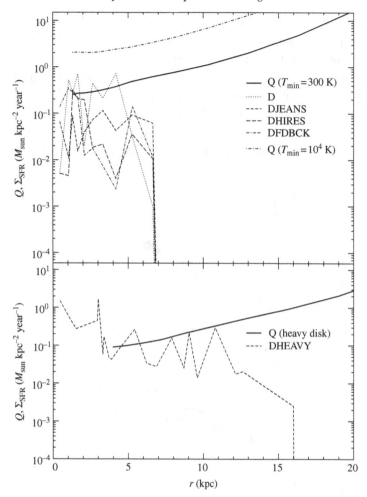

Fig. 7.11. In the *top panel*, the *solid line* shows the Q parameter computed shortly after the start of the quiescent disk simulation discussed in the text, when the gas has reached its minimum value (300 K) but before non-linear instability formation. The instantaneous rate of star formation (averaged over 20 Myr) is shown for four variants of the standard simulation (see text). Note that gravitational instability (and hence star formation) only occurs below a critical Toomre Q parameter. The *dot-dashed curve* shows the Q parameter for a simulation with $T_{min} = 10^4$ K, which never falls below this critical value and so never forms stars. The *bottom panel* shows the same plot for a disk with a gas mass four times greater than the other simulations. The cut-off this time is at a larger radius and corresponds to a Q value which is similar (but slightly larger).

agreement with the Milky Way's own stellar radius of ~15 kpc. The value for Q at the absolute cutoff for the star formation, $r = 16$ kpc, is 1, slightly higher than for the original disk. However, the star formation rate starts to decrease at a smaller radius than this, at around $r = 13$ kpc. At this point, $Q \sim 0.5$, in closer agreement to the lighter disk. The Q scaling therefore works well with the changing weight of the disk, especially since the variation of Q over the disk is of order 100.

Next, we examine how star formation depends on density, as observations indicate it does. In Figure 7.12, we show the global star formation rate averaged over the star-forming part of the disk (corresponding to the radius containing 95% of the star formation). The *top panel* shows that for a given set of parameters, the two types of star formation (C and D types) produce relations which parallel the Kennicutt relation, although at a higher star formation rate. The *middle*

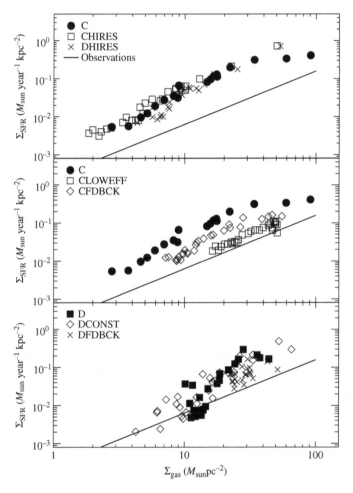

Fig. 7.12. The global Schmidt law: variation over time of the relationship between total star formation rate and gas surface density for a variety of simulations. Star formation rate is plotted as a function of density averaged over the disk for a variety of models. Each point with the same symbol represents the same simulation at a different time, equally spaced along the 1.4 Gyr simulation run. The *solid line* is a best fit from observations (Kennicutt 1989).

panel demonstrates that this overproduction can be improved either by adjusting the ϵ parameter in the star formation law or by introducing feedback. Finally, the *last panel* shows the results for a simulation (DCONST) in which the star formation was directly proportional to the gas density. Remarkably, this still generates a steep Kennicutt-like relation between the average gas surface density and the star formation rate.

These results demonstrate that the star formation prescription used appears not to be a crucial factor in at least some disk properties, such as the cutoff radius or the disk-averaged star formation rate. On the contrary, other disk properties (such as gas outflows, see Tasker and Bryan (2006) and Section 7.8) do depend crucially on the feedback method chosen.

7.10 Conclusions

In this contribution, we have attempted to introduce the important physical processes which dictate the properties of high-redshift galaxies, as well as some of the outstanding problems. We have shown that the number of physical processes thought relevant to understanding galaxy formation is increasing rapidly in recent years. Note that we have not even touched on magnetic fields or cosmic ray pressure despite the observational fact that they dominate the force balance in the cold and warm ISM in the Milky Way. That this would be coincidental and the Milky Way just recently acquired such large fields and them having been unimportant in the past is very unlikely. It will not be surprising if the effects of magnetic fields and cosmic ray pressure would become an integral part of galaxy formation physics in the next decade since numerical simulations are just starting to be able to include these processes (Fromang *et al.* 2006; Enßlin *et al.* 2007; Wang & Abel 2008).

While some of the physical conditions are different from those in other fields, it is clear that many of the issues and physical processes are remarkably similar between planetary, stellar and galactic scales. The equations of radiation magnetohydrodynamics underlie the understanding of the most astronomical objects we seek to understand and are addressed in this volume. The rapid advances in computational astrophysics over the past decades and in particular the advent of spatially and temporally adaptive techniques allow calculations in unprecedented detail. We are now in a situation where it is conceivable that we can develop models, sufficiently accurate and valid at all cosmic epochs, for the micro-physics of atomic, molecular kinetics and radiation losses coupled to the latest 3D radiation transport approaches to arrive at an essentially complete set of equations governing galaxy formation and evolution. At that point, the continued adaptation and optimization of our numerical codes to the peta- and exo-scale supercomputers being planned today will lead to models of galaxies as large as the Milky Way while capturing all the stars it and its progenitors ever hosted. The importance of developing an understanding of galaxy formation to this level of detail cannot be underestimated. Not only will it aid the enormous observational programmes executed today and planned for the future, but it is only at that point that the direct connections between planet, star and galaxy formation will be understood.

Acknowledgements

GB acknowledges the support from AST-05-07161, AST-05-47823 and AST-06-06959, as well as computational resources from the National Centre for Supercomputing Applications. The Horizon Project is supported by the 'Agence National pour la Recherche' under project number NT05-4-41478. The Mare Nostrum simulation was run at Barcelona Supercomputing Center under project ID AECT-2006-3-0011, AECT-2006-4-0009, AECT-2007-2-0009 and AECT-2007-3-0009. TA acknowledges the support by the NSF CAREER award AST-0239709 from the National Science Foundation.

References

Abel, T., Bryan, G. L., & Norman, M. L. 2002, *Science*, **295**, 93
Abel, T., Wise, J. H., & Bryan, G. L. 2007, *ApJL*, **659**, L87
Anninos, P., Zhang, Y., Abel, T., & Norman, M. L. 1997, *New Astron.*, **2**, 209
Arnaud, K. A., Fabian, A. C., Eales, S. A., Jones, C., & Forman, W. 1984, *MNRAS*, **211**, 981
Basson, J. F. & Alexander, P. 2003, *MNRAS*, **339**, 353
Birnboim, Y. & Dekel, A. 2003, *MNRAS*, **345**, 349
Bromm, V., Coppi, P. S., & Larson, R. B. 2002, *ApJ.*, **564**, 23
Bryan, G. L. 1999, *Comp. Phys. Eng.*, **1:2**, 46
Bryan, G. L. 2007, in *Proceedings of Chemodynamics: From First Stars to Local Galaxies*, ed. E. Emsellem, H. Wazniak, G. Massacrier, J.-F. Gonzalez, J. Devriendt and N. Champavert, EAS Publications Series, **24**, 77 (arXiv:0707.1856)
Carilli, C. L., Perley, R. A., & Harris D. E. 1994, *MNRAS*, **270**, 173
Cattaneo, A. & Teyssier, R. 2007, *MNRAS*, **376**, 1547
Churazov, E., Bruggen, M., Kaiser, C. R., Bohringer, H., & Forman, W. 2001, *ApJ.*, **554**, 261
Ciardi, B. & Ferrara, A. 2005, *Space Sci. Rev.*, **116**, 625
Cien, S. & Bryan, G. L. 2008, in preparation
David, L. P., Nulsen, P. E. J., McNamara, B. R., Forman, W., Jones, C., Ponman, T., Robertson, B., & Wise, M. 2001, *ApJ.*, **557**, 546
Dekel, A. & Birnboim, Y. 2006, *MNRAS*, **368**, 2
Dekel, A. & Silk, J. 1986, *ApJ.*, **303**, 39
Dubois, Y. & Teyssier, R. 2008, *A&A*, **477**, 79
Enßlin, T. A., Pfrommer, C., Springel, V., & Jubelgas, M. 2007, *A&A*, **473**, 41
Fabian, A. C., Celotti, A., Blundell, K. M., Kassim, N. E., & Perley, R. A. 2002, *MNRAS*, **331**, 369

Fabian, A. C. *et al.* 2000, *MNRAS*, **318**, L65
Fromang, S., Hennebelle, P., & Teyssier, R. 2006, *A&A*, **457**, 371
Gnedin, N. 2000, *ApJ.*, **542**, 535
Gnedin, N. Y. & Ostriker, J. P. 1997, *ApJ.*, **486**, 581
Hoeft, M., Yepes, G., Gottloeber, G., & Springel, V. 2004, *ApJ.*, **371**, 401
Kauffmann, G., *et al.* 2003, *MNRAS*, **341**, 54
Kennicutt, R. C. 1989, *ApJ.*, **344**, 685
Kennicutt, R. C. 1998, *ApJ.*, **498**, 541
Keres, D., Katz, N., Weinberg, D. H., & Dave, R. 2005, *MNRAS*, **363**, 2
Kravtsov, A. 2003, *ApJ.*, **590**, L1
Machacek, M. E., Bryan, G. L., & Abel, T. 2001, *ApJ.*, **548**, 509
McKee, C. & Ostriker, J. P. 1977, *ApJ.*, **218**, 148
McNamara, B. R. *et al.* 2000, *ApJ.*, **534**, L135
McNamara, B. R. *et al.* 2001, *ApJ.*, **562**, L149
Navarro, J. F., Frenk, C. S., & White, S. D. M. 1997, *ApJ.*, **490**, 493
Omma, H. & Binney, J. 2004, *MNRAS*, **350**, L13
Omma, H., Binney, J., Bryan, G., & Slyz, A. 2004, *MNRAS*, **348**, 1105
O'Shea, B. W., Bryan, G., Bordner, J., Norman, M. L., Abel, T., Harkness, R.,
     & Kritsuk, A. 2004, arXiv Astrophysics e-prints, astro-ph/0403044
Owen, F. N., Eilek, J. A., & Kassim, N. E. 2000, *ApJ.*, **543**, 611
Peebles, P. J. E. & Dicke, R. H. 1968, *ApJ.*, **154**, 891
Quilis, V., Bower, R. G., & Balogh, M. L. 2001, *MNRAS*, **328**, 1091
Rasera, Y. & Teyssier, R. 2006, *A&A*, **445**, 1
Rees, M. J. & Ostriker, J. P. 1977, *MNRAS*, **179**, 541
Reynolds, C. S., Heinz, S., & Begelman, M. C. 2001, *ApJ.*, **549**, L179
Reynolds, C. S., Heinz, S., & Begelman, M. C. 2002, *MNRAS*, **332**, 271
Ripamonti, E. & Abel, T. 2004, *MNRAS*, **348**, 1019
Rosen, A. & Bregman, J. N. 1995, *ApJ.*, **440**, 634
Ruszkowski, M., Bruggen, M., & Begelman, M. C. 2004, *ApJ.*, **615**, 675
Sijacki, D. & Springel, V. 2006, *MNRAS*, **366**, 397
Springel, V. & Hernquist, L. 2003, *MNRAS*, **339**, 289
Tasker, E. J. & Bryan, G. L. 2006, *ApJ.*, **641**, 878
Tasker, E. J. & Bryan, G. L. 2008, *ApJ.*, **673**, 810
Toomre, A. 1964, *ApJ.*, **139**, 1217
Vernaleo, J. C. & Reynolds, C. S. 2006, *ApJ.*, **645**, 83
Wang, P. & Abel, T. 2008, *ApJ.*, **672**, 752
Wang, P. & Abel, T. 2008, *ApJ.*, in press, arXiv:astro-ph/0712.0872
Wise, J. H. & Abel, T. 2007, *ApJ.*, **671**, 1559
Wise, J. H. & Abel, T. 2008, *ApJ.*, **685**, 40
Wise, J. H., Turk, M. J., & Abel, T. 2008, *ApJ.*, **682**, 745
White, S. D. M. & Frenk, C. S. 1991, *ApJ.*, **379**, 52
White, S. D. M. & Rees, M. J. 1978, *MNRAS*, **183**, 341
Yepes, G., Kates, R., Khokhlov, A., & Klypin, A. 1997, *MNRAS*, **284**, 235
Yoshida, N., Abel, T., Hernquist, L., & Sugiyama, N. 2003, *ApJ.*, **592**, 645

8

First stars: formation, evolution and feedback effects

V. Bromm, A. Ferrara and A. Heger

Abstract

The formation of the first stars at redshifts $z \sim 20$–30 marked the transition from the simple initial state of the universe to one of ever increasing complexity. We here review recent progress in understanding their formation process with numerical simulations. We discuss the physics behind the prediction of a top-heavy primordial initial mass function (IMF) and focus on protostellar accretion as the key unsolved problem. We continue by describing their evolution and their death as energetic supernovae (SNe) or massive black holes. Finally, we address feedback processes from the first stars that are now realized to hold the key to our understanding of structure formation in the early universe. We discuss three broad feedback classes (radiative, chemical and mechanical) and explore the enrichment history of the intergalactic medium (IGM).

8.1 Introduction

How did the first stars in the universe form, how did they evolve and die and what was their impact on cosmic history (Woosley *et al.* 2002; Bromm & Larson 2004; Ciardi & Ferrara 2005)? The first stars formed at the end of the cosmic dark ages beyond the current horizon of observability (Couchman & Rees 1986; Haiman *et al.* 1996; Tegmark *et al.* 1997). These so-called Population III (Pop III) stars ionized (Kitayama *et al.* 2004; Whalen *et al.* 2004; Alvarez *et al.* 2006; Johnson *et al.* 2007) and metal-enriched (Furlanetto & Loeb 2003; Tornatore *et al.* 2007) the intergalactic medium (IGM) and consequently had important effects on subsequent galaxy formation (Barkana & Loeb 2001; Mackey *et al.* 2003). They imprinted their signature on the large-scale polarization anisotropies of the cosmic microwave background (CMB; see Kaplinghat *et al.* 2003).

Structure Formation in Astrophysics. ed. G. Chabrier. Published by Cambridge University Press.
© Cambridge University Press 2009.

8.1.1 When did the cosmic dark ages end?

In the context of popular cold dark matter (CDM) models of hierarchical structure formation, the first stars are predicted to have formed in dark matter halos of mass $\sim 10^6 M_\odot$ that collapsed at redshifts $z \simeq 20$–30 (Barkana & Loeb 2001; Yoshida *et al.* 2003). Understanding the properties of the first sources of light, in particular their expected luminosities and spectral energy distributions, is important for the design of upcoming instruments, such as the *James Webb Space Telescope* (JWST) (see http://ngst.gsfc.nasa.gov) or the next generation of large (>10 m) ground-based telescopes. The hope is that over the next decade, it will become possible to confront current theoretical predictions about the properties of the first stars with direct observational data.

Results from recent numerical simulations of the collapse and fragmentation of primordial clouds suggest that the first stars were predominantly very massive, with typical masses $M_* \geq 100 M_\odot$ (Bromm *et al.* 1999, 2002; Abel *et al.* 2000, 2002; Nakamura & Umemura 2001; Yoshida *et al.* 2006; Gao *et al.* 2007; O'Shea & Norman 2007). Despite the progress already made, many important questions remain unanswered: How does the primordial initial mass function (IMF) look like? We have constrained the characteristic mass scale, indicating the typical outcome of the Pop III star formation process, in the sense of an average mass. This, however, leaves still undetermined the overall range of stellar masses and the precise functional shape of the IMF (e.g. log-normal or broken power-law?). In addition, it is presently unknown whether binaries or, more generally, clusters of zero-metallicity stars can form (Saigo *et al.* 2004; Clark *et al.* 2008). How did Pop III protostars grow by accretion (Omukai & Palla 2001, 2003)? Did radiation pressure from the protostar shut off accretion in the absence of dust grains, thus setting the final mass of a Pop III star (Omukai & Inutsuka 2002; McKee & Tan 2008)?

8.1.2 What is the nature of the feedback exerted by the first stars on their surroundings?

Once the first stars have formed, their mass deposition, energy injection and emitted radiation can deeply affect the subsequent galaxy formation process and influence the evolution of the IGM via a number of so-called feedback effects. The word 'feedback' is by far one of the most used in modern cosmology, where it is applied to a vast range of situations and astrophysical objects. However, for the same reason, its meaning is often unclear or fuzzy. Generally speaking, the concept of feedback invokes a *back reaction of a process on itself or on the causes that have produced it*. This feedback can be either negative or positive, and it is difficult to predict the overall sign (Greif *et al.* 2007).

Although a rigorous classification of the various effects is not feasible, they can be divided into three broad classes: *radiative, mechanical* and *chemical* feedback. In the first class fall all those effects associated, in particular, with ionization/dissociation of hydrogen atoms/molecules; the second class is produced by the mechanical energy injection of massive stars in the form of winds or supernova (SN) explosions; and chemical feedback is instead related to the postulated existence of a critical metallicity, governing the cosmic transition from very massive stars (VMSs) to 'normal' stars.

8.2 Population III star formation

8.2.1 First-generation stars

The metal-rich chemistry, magnetohydrodynamics and radiative transfer involved in present-day star formation is complex, and we still lack a comprehensive theoretical framework that predicts the IMF from first principles (see Larson 2003 for a recent review). Star formation in the high redshift universe, on the other hand, poses a theoretically more tractable problem due to a number of simplifying features, such as (i) the initial absence of heavy elements and therefore of dust and (ii) the absence of dynamically significant magnetic fields in the pristine gas left over from the big bang. The cooling of the primordial gas then only depends on hydrogen in its atomic and molecular form. Whereas the initial state of the star-forming cloud is poorly constrained in the present-day interstellar medium (ISM), the corresponding initial conditions for primordial star formation are simple, given by the popular ΛCDM model of cosmological structure formation (Spergel *et al.* 2007).

8.2.1.1 How did the first stars form?

A complete answer to this question would entail a theoretical prediction for the precise Pop III IMF, which is rather challenging. Let us begin by addressing the simpler problem of estimating the characteristic mass scale of the first stars. This mass scale is observed to be $\sim 1 M_\odot$ in the present-day universe. To investigate the collapse and fragmentation of primordial gas, a number of groups have carried out 3D numerical simulations, using the smoothed particle hydrodynamics (SPH) method (Bromm *et al.* 1999, 2002; Yoshida *et al.* 2003, 2006; Gao *et al.* 2007) or the adaptive mesh refinement (AMR) technique (Abel *et al.* 2000, 2002; O'Shea & Norman 2007). These simulations have included the chemistry and cooling physics relevant for the evolution of metal-free gas. At the low temperatures ($<10^4$ K) found in the so-called minihalos of total mass $\sim 10^6 M_\odot$ that were the sites for Pop III star formation, the only viable coolant is molecular hydrogen. The microphysics of H_2 cooling imprints characteristic values for temperature, $T_{\text{char}} \sim 200$ K, and hydrogen number density, $n_{char} \sim 10^4 \, cm^{-3}$. The former derives from the large energy spacing ($k_B 512 \, K$) between the lowest-lying rotational levels of H_2. The primordial gas can cool slightly below 512 K due to hydrogen atom–molecule collisions with velocities in the tail of the Maxwell–Boltzmann distribution, but not much below $\sim 200 \, K$. The latter reflects the transition from NLTE to LTE H_2 rotational-level populations at such low temperatures. This transition leads to a 'saturation' in the cooling, such that the H_2 cooling rate scales as $\Lambda_{H_2} \propto n^2$ below n_{char} and as $\Lambda_{H_2} \propto n$ above. When the primordial gas collapses into a minihalo, it reaches a state with T_{char} and n_{char}, sometimes called the 'loitering regime' (Bromm *et al.* 2002), at which point the collapse is temporarily slowed. This slow, quasi-hydrostatic phase ends when sufficient mass has accumulated to trigger gravitational runaway collapse: $M > M_J$, where

$$M_J \simeq 700 M_\odot \left(\frac{T_{char}}{200 \, K} \right)^{3/2} \left(\frac{n_{char}}{10^4 \, cm^{-3}} \right)^{-1/2}$$

is the Jeans mass, which is closely related to the Bonnor–Ebert mass (for details, see Bromm & Larson 2004; Glover 2005). The cloud at the 'loitering state' is the immediate progenitor for the formation of a Pop III star and thus plays an analogous role to the dense cores in present-day star formation (Larson 2003). This characteristic behaviour is rather robust and is found in most simulations, both SPH- and AMR-based, as long as one considers variants of the CDM model of cosmological structure formation. To move away from it, one needs to consider rather drastic changes in the cosmology, such as warm dark matter (WDM) scenarios (Gao & Theuns 2007).

8.2.1.2 How massive were the first stars?

Star formation typically proceeds from the 'inside-out', through the accretion of gas onto a central hydrostatic core. Whereas the initial mass of the hydrostatic core is very similar for primordial and present-day star formation (Omukai & Nishi 1998), the accretion process – ultimately responsible for setting the final stellar mass – is expected to be rather different. On dimensional grounds, the accretion rate is simply related to the cube of the sound speed over Newton's constant (or equivalently given by the ratio of the Jeans mass and the free-fall time): $\dot{M}_{acc} \sim c_s^3 / G \propto T^{3/2}$. A simple comparison of the temperatures in present-day star-forming regions ($T \sim 10 \, K$) with those in primordial ones ($T \sim 200–300 \, K$) already indicates a difference in the accretion rate of more than two orders of magnitude.

Studying the 3D accretion flow around the protostar marks the current frontier of the subject (Omukai & Palla 2001, 2003; Ripamonti *et al.* 2002; Tan & McKee 2004). Starting from constrained cosmological initial conditions, Bromm and Loeb

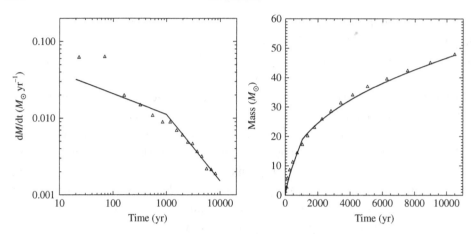

Fig. 8.1. Accretion onto a primordial protostar. *Left:* Accretion rate (in $M_\odot$ yr^{-1}) versus time (in yr) since molecular core formation. *Right:* Mass of the central core (in $M_\odot$) versus time. *Solid line:* Accretion history approximated as $M_* \propto t^{0.75}$ at $t < 10^3$ yr, and $M_* \propto t^{0.4}$ afterwards. Using this analytical approximation, we extrapolate that the protostellar mass has grown to $\sim 120 M_\odot$ after $\sim 10^5$ yr, and to $\sim 500 M_\odot$ after $\sim 3 \times 10^6$ yr, the total lifetime of a VMS (Adapted from Bromm & Loeb 2004).

(2004) have studied the accretion problem with very high resolution, employing SPH with particle splitting (Kitsionas & Whitworth 2002; Bromm & Loeb 2003b). Making the idealized assumption of negligible protostellar feedback, their simulation allows the gas to reach densities of 10^{12} cm^{-3} before being incorporated into a central sink particle. At these high densities, three-body reactions (Palla *et al.* 1983) have converted the gas into a fully molecular form. In Figure 8.1, we show how the molecular core grows in mass over the first $\sim 10^4$ yr after its formation. The accretion rate (*left panel*) is initially very high, $\dot{M}_{acc} \sim 0.1 M_\odot$ yr^{-1} and subsequently declines with time. The mass of the molecular core (*right panel*), taken as an estimator of the protostellar mass, grows approximately as $M_* \sim \int \dot{M}_{acc} dt \propto t^{0.75}$ at $t < 10^3$ yr and $M_* \propto t^{0.4}$ afterwards. A rough upper limit for the final mass of the star is then $M_*(t = 3 \times 10^6 \, \text{yr}) \sim 500 M_\odot$. In deriving this upper bound, we have conservatively assumed that accretion cannot go on for longer than the total lifetime of a VMS. Similar upper mass limits have been obtained with AMR simulations (Abel *et al.* 2002; O'Shea & Norman 2007).

8.2.1.3 Can a Pop III star ever reach this asymptotic mass limit?

The answer to this question is not yet known with any certainty, and it depends on whether the accretion from a dust-free envelope is eventually terminated by feedback from the star (Omukai & Palla 2001, 2003; Omukai & Inutsuka 2002; Ripamonti *et al.* 2002; Tan & McKee 2004). The standard mechanism by which accretion may be terminated in metal-rich gas, namely radiation pressure on dust grains (Wolfire & Cassinelli 1987), is evidently not effective for gas with a primordial composition. Recently, it has been speculated that accretion could instead be turned off through the formation of an H II region (Omukai & Inutsuka 2002) or through the radiation pressure exerted by trapped Lyα photons (McKee & Tan 2008). The latter authors argue that the combination of a declining accretion rate and photoevaporation of material from the protostellar accretion disk imposes a mass limit of $\sim 140 M_\odot$. The termination of the accretion process defines the current unsolved frontier in studies of Pop III star formation.

8.2.2 Second-generation stars

How and when did the transition take place from the early formation of massive stars to that of low-mass stars at later times? In contrast to the formation mode of massive stars (Pop III) at high redshifts, fragmentation is observed to favour stars below a solar mass (Pop I and II) in the present-day universe. Understanding the physics driving the transition between these fundamental modes is another frontier of current research. It has often been argued that the progressive enrichment of the cosmic gas with heavy elements was the key factor, enabling the gas to cool to lower temperatures. The concept of a 'critical metallicity', Z_{crit}, has been used to characterize the transition between Pop III and Pop II formation modes, where Z denotes the mass fraction contributed by all heavy elements (Omukai 2000; Bromm *et al.* 2001; Schneider *et al.* 2002; Mackey *et al.* 2003; Schneider *et al.* 2003; Yoshida *et al.* 2004; Venkatesan 2006; Smith & Sigurdsson 2007). These studies have constrained this important parameter to only within a few orders of magnitude, $Z_{crit} \sim 10^{-6} - 10^{-3} Z_\odot$, under the implicit assumption of solar relative abundances of metals. This assumption is likely to be violated by the metal yields of the first SNe at high redshifts, for which strong deviations from solar abundance ratios are predicted (Heger & Woosley 2002; Qian & Wasserburg 2002; Umeda & Nomoto 2002, 2003).

The current debate is centred on the question of whether dust or atomic fine-structure cooling dominates the transition to low-mass star formation. The argument in favour of fine-structure cooling suggests singly ionized carbon or neutral atomic oxygen as the key species (Bromm & Loeb 2003a). Earlier estimates of Z_{crit} which did not explicitly distinguish between different coolants are refined by introducing separate critical abundances for carbon and oxygen, $[C/H]_{crit}$ and $[O/H]_{crit}$, respectively, where $[A/H] = \log_{10}(N_A/N_H) - \log_{10}(N_A/N_H)_\odot$. Since C and O are also the most important coolants throughout most of the cool atomic ISM in present-day galaxies, it is not implausible that these species might be responsible for the global shift in the star formation mode. Numerically, the critical C and O abundances are estimated to be $[C/H]_{crit} \simeq -3.5 \pm 0.1$ and $[O/H]_{crit} \simeq -3.1 \pm 0.2$. Other atoms, in particular Si and Fe, have also been proposed to be important (Santoro & Shull 2006). The competing theory argues in favour of dust grains to enable efficient cooling up to high densities (Schneider *et al.* 2003, 2006; Tsuribe & Omukai 2006). The dust theory has the attractive feature of allowing the low-metallicity gas to remain cold even at densities where fine-structure lines become unimportant. Recently, Jappsen *et al.* (2007, 2008) have argued that fine-structure cooling cannot compete with H_2 cooling, provided the latter is not photodissociated by soft UV radiation (see Section 8.4.1), so that there would be no critical fine-structure threshold. There is some empirical support for the C- and O-based fine-structure theory, however, in the observed abundance patterns of extremely low-metallicity galactic halo stars (Frebel *et al.* 2007). All currently known stars exceed the critical C and O abundances calculated by Bromm and Loeb (2003a), and it is not obvious how to account for this if fine-structure cooling were not relevant for enabling low-mass star formation.

Even if sufficient coolants are present to cool the gas further, there will be a minimum attainable temperature that is set by the interaction with the thermal CMB: $T_{CMB} = 2.7\,K(1 + z)$ (Larson 1998; Clarke & Bromm 2003). At $z \simeq 15$, this results in a characteristic stellar mass of $M_* \sim 20 M_\odot (n_f/10^4\,\mathrm{cm}^{-3})^{-1/2}$, where $n_f > 10^4\,\mathrm{cm}^{-3}$ is the density at which opacity prevents further fragmentation. It is possible that the transition from the high-mass to the low-mass star formation mode was modulated by the CMB temperature and was therefore gradual, involving intermediate-mass ('Pop III.2') stars at intermediate redshifts (Mackey *et al.* 2003; Johnson & Bromm 2006; see McKee & Tan 2008 for an exposition of the terminology). This transitional population could give rise to the faint SNe that have been proposed to explain the observed abundance patterns in metal-poor stars (Umeda & Nomoto 2002, 2003; Christlieb *et al.* 2002; Frebel *et al.* 2005). When and how uniformly the transition in the cosmic star formation mode did take place was governed by the detailed enrichment history of the IGM. This in turn was determined by the hydrodynamical transport and mixing of metals from the first SN explosions (Mori *et al.* 2002; Bromm *et al.* 2003; Scannapieco *et al.* 2003; Wada & Venkatesan 2003; Greif *et al.* 2007).

Recently, the additional boost to the cooling of still metal-free gas provided by HD has been investigated (Johnson & Bromm 2006; Yoshida *et al.* 2007). If the primordial gas goes through a strongly shocked, fully ionized phase prior to the onset of protostellar collapse, cooling is possible down to the temperature of the CMB which sets the minimum floor accessible via radiative cooling (Figure 8.2). The lower temperatures in turn could allow the fragmentation into intermediate-mass stars, with masses of order a few tens of $M_\odot$, giving rise to a possible 'Pop III.2' (Figure 8.3).

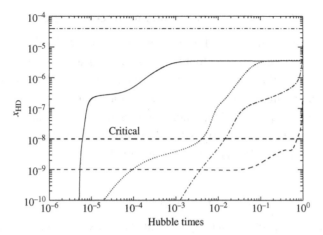

Fig. 8.2. Evolution of hydrogen deuteride (HD) abundance in four distinct cases. *Solid line:* Gas compressed and heated by a SN explosion with $u_{sh} = 100\,\mathrm{km\,s^{-1}}$. *Dotted line:* Gas shocked in the build-up of a 3σ fluctuation dark matter halo collapsing at $z \simeq 15$. *Dash-dotted line:* Gas collapsing inside a relic H II region, which is left behind after the death of a very massive Pop III star. *Dashed line:* Gas collapsing inside a minihalo at $z \simeq 20$. In this case, the gas does not experience a strong shock and is never ionized. Contrary to the other three cases, where the gas went through a fully ionized phase, HD cooling is not important here. The critical HD abundance, shown by the *bold dashed line*, is defined such that primordial gas is able to cool to the CMB temperature within the fraction of a Hubble time. The CMB sets a minimum floor to the gas temperature, because radiative cooling below this floor is thermodynamically not possible. The HD abundance exceeds the critical value in a time which is short compared to the Hubble time for all fully ionized, strongly shocked cases (Adapted from Johnson & Bromm 2006).

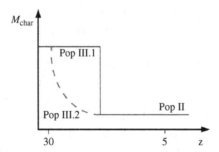

Fig. 8.3. Characteristic mass of stars as a function of redshift. Pop III.1 stars, formed out of metal-free gas and not going through a fully ionized phase prior to the onset of collapse, have typical masses of $M_* \sim 100 M_\odot$. Pop II stars, formed out of already metal-enriched gas, are formed at lower redshifts and have typical masses of $M_* \sim 1 M_\odot$. Pop III.2 stars, formed out of strongly shocked, but still virtually metal-free gas, are hypothesized to have typical masses that are intermediate between Pop III.1 and Pop II with $M_* \sim 10 M_\odot$ (Adapted from Johnson & Bromm 2006).

8.3 Structure, evolution and nucleosynthesis

The first stars are very unique in their properties. Not only that they may have been typically more massive than stars that form today but also that these stars may have kept most of their initial mass till the end of their evolution while stars today may lose most of their initial mass due to stellar winds (Heger *et al.* 2003). Second, these stars have a unique initial chemical composition – just what was made by the big bang. This unique initial composition has very marked effects on the evolution of these stars, their final deaths, remnants and nucleosynthesis.

8.3.1 Structure and evolution

The main features of primordial (Pop III) stars as far as their evolution is concerned can be summarized as follows (Table 8.1):

8.3.1.1 Hydrogen burning

- Massive stars[1] burn hydrogen dominantly by the CNO (carbon–nitrogen–oxygen) cycle rather than the pp (proton–proton) chains (see, e.g., Clayton 1968; Kippenhahn & Weigert 1990). Primordial stars, however, are born without initial metals; i.e. they also lack the CNO isotopes that are needed for the CNO cycle. What these stars do instead is to continue contracting until a high enough central temperature and density is reached to start making ^{12}C by the triple alpha reaction. This seeds the CNO cycle. Due to the high temperature dependence of the CNO cycle (about T^{14} at the temperatures reached when the triple alpha sets in in massive primordial stars, around $1 - 1.2 \times 10^8$ K), only a small amount of CNO isotopes is needed to efficiently run the CNO cycle. Typically, we find the contraction is halted when a CNO mass fraction of about 10^{-9} is reached. After that, the star slightly re-expands and the structure adjusts to settle on the primordial zero-age main sequence. This unique transition phase takes about 10,000–20,000 years.
- The centre of these primordial stars, however, remains at a high temperature beyond 10^8 K (whereas $2 - 3 \times 10^7$ K are typical for present-day, metal-rich massive stars) and hence continues to experience some continuing triple alpha process, in particular as the core gets hotter and contracts more while the star uses up its central hydrogen. As a result, towards the end of central hydrogen burning, e.g. when a central hydrogen mass fraction of about 1 % is reached, the CNO isotope mass fraction has reached about 10^{-7}. This value of 10^{-7} is also a typical mass fraction in the hydrogen-burning shell of primordial stars. At the same time, a breakout and weak p/αp-process occurs, building up traces of elements up to ^{40}Ca already during hydrogen burning.

8.3.1.2 Mass loss

- While the first generation of stars was likely born massive, current theory also favours that these stars 'died massive'. That is, mass loss from these stars may have been much

---

[1] As massive stars we denote stars that are massive enough to make SNe; for the limiting lower mass see, e.g., Poelarends *et al.* (2008); typical values in the literature are $8\ldots10M_\odot$.

Table 8.1. *Burning stages, central temperatures and timescales in* $20M_\odot$
and $200M_\odot$ *primordial stars*

Burning stages		$20M_\odot$ star		$200M_\odot$ star	
Fuel	Main Product	$T(10^9\,\text{K})$	Time (yr)	$T(10^9\,\text{K})$	Time (yr)
H	He	0.02	10^7	0.1	2×10^6
He	O, C	0.2	10^6	0.3	2×10^5
C	Ne, Mg	0.8	10^3	1.2	10
Ne	O, Mg	1.5	3	2.5	$3 \times 10-6$
O	Si, S	2.0	0.8	3.0	$2 \times 10-6$
Si, S	Fe	3.5	0.02	4.5	3×10^{-7}

less than what we find in stars today. For example, it has been shown by Kudritzki (2002) that line-driven winds, the key mechanism for mass loss in present-day stars, would die off as the metallicity goes to zero. To counter a common misconception that people may take out of Section 8.3.1.1, the metals made by the star itself during hydrogen burning are created in the centre of the star, and even if they were to get to the surface, say, due to rotationally induced mixing, it would be far from sufficient to drive a powerful wind.

- While it is well known for some time that for stars above some $60M_\odot$ a nuclear-powered vibrational instability sets in that could lead to significant mass loss (Schwarzschild & Härm 1959), it has now been shown that *primordial* massive stars avoid this fate – due mostly to their higher central temperature for CNO hydrogen burning which results in a lower temperature-sensitivity than in present-day stars, but also because these stars are more compact (Baraffe *et al.* 2001). More importantly, the absence of metals makes these stars avoid opacity-driven pulsations that may become quite prominent in metal-rich VMSs (see also Baraffe *et al.* 2001).

- Recently, it has been suggested that continuum-driven and super-Eddington mass loss may operate in VMSs, and this mechanism would work in the same way in stars of primordial composition (Smith & Owocki 2006). The magnitude and efficiency of this mechanism, however, is not clear at the present.

- It has been argued that rotationally induced mass loss may lead to significant mass loss allowing the most massive primordial stars to shrink down in mass almost similar to their modern counterparts (Ekström *et al.* 2008). We feel, however, that predicting the mass loss rates of critically rotating primordial stars, without any observational bounds, is still very uncertain at best.

- Finally, during helium burning, at least some of the primordial stars may encounter a mixing of the helium-burning products, mainly carbon, with the CNO-deficient envelope. This has been found to occur due to rotationally induced mixing (Hirschi 2007) or even without it (Heger *et al.* 2000). Once the stellar envelope is enriched, the star may expand to become a red super-giant and encounter significant mass loss. The magnitude of this mass loss is hard to estimate based on theory and, again, there are no observational bounds.

8.3.2 The fates of the first stars

A schematic of the fate of the first stars is given in Figure 8.4. In this figure, no mass loss is assumed. This implies that stars die essentially as massive as they were born. At low masses, we anticipate stellar fates very similar to those of modern stars. But while initially massive stars may eventually lose their hydrogen envelopes, uncover their helium cores and further shrink down, Pop III stars might be able to avoid this fate. Instead of a birth mass above which the final mass starts to decrease again, thus limiting the maximum mass of remnants, Pop III stars may continue to be increasingly more massive at the time of their final collapse or explosion.

One of the first unique instabilities for these big stars is nuclear-powered vibrational instability during core oxygen burning for stars above around $80 M_\odot$. Since it is rather weak and this occurs very briefly before the death of the star, it has little further effect. At higher masses, instabilities due to pair production set in (see Sec. 8.3.3). Above about $260 M_\odot$ instabilities due to photodisintegration of nuclei, in particular the onset of photodisintegration of helium nuclei into free nucleons, cause a very strong instability that probably leads to direct collapse of much of the star into an 'intermediate mass' black hole (tens to hundreds of solar masses; Fryer *et al.* 2001).

There is also a regime of 'supermassive' stars, usually attributed to stars above some $100,000 M_\odot$. These very highly radiation-dominated stars can collapse due to a general relativistic instability even before hydrostatic hydrogen burning is achieved. This is because the radiation-dominated equation of state has an adiabatic index very close to $\gamma_{ad} = 4/3$, whereas due to the increased strength of gravity in general relativity the critical adiabatic index for stability of the star rises above $4/3$ until eventually a crossover occurs (more accurately, the average adiabatic index of the star needs to be above $4/3$, i.e. $\int (\gamma_{ad} - 4/3) P/\rho \, dm > 0$ is needed for stability even for non-general relativistic stars). For Pop III stars, Fuller *et al.* (1986) have found that these objects collapse to black holes.

8.3.3 Very massive stars and pair instability supernovae

Pop III stars above some $100 M_\odot$ have helium cores big enough to encounter the electron–positron pair instability. For the lowest masses at which it sets in, the energy released during the explosive burning is not enough to unbind and disrupt the star. Instead, only the outer layers of the star are ejected. During the first pulse, this usually comprises just the hydrogen envelope. Later, pulses will successively

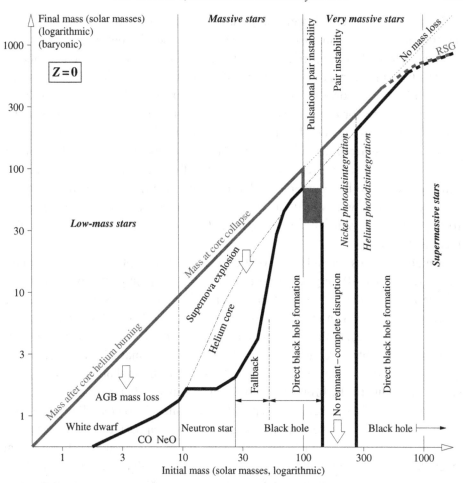

Fig. 8.4. Initial-final mass function of non-rotating primordial stars. The x-axis gives the initial stellar mass. The y-axis gives both the final mass of the col-lapsed remnant (*thick black curve*) and the mass of the star when the event begins that produces that remnant (e.g., mass loss in asymptotic giant-branch (AGB) stars, SN explosion for those stars that make a neutron star, etc.; *thick grey curve*). We distinguish four regimes of initial mass: *low-mass stars* below ∼$10M_\odot$ that form white dwarfs; *massive stars* between ∼10 and ∼$100M_\odot$; *very massive stars* between ∼100 and ∼$1000M_\odot$ and *supermassive stars* (arbitrar-ily) above ∼$1000M_\odot$. Assuming no wind mass loss for primordial stars, the *grey curve* coincides with the line of no mass loss (*dotted*), except for ∼100–$140M_\odot$, where the pulsational pair instability ejects the outer layers of the star before it collapses, and above ∼$500M_\odot$, where pulsational instabilities in red supergiants may lead to significant mass loss. Since the magnitude of the latter is uncertain, lines are *dashed*. The defining characteristic of VMSs is the electron-positron *pair instability* after carbon burning. Starting at about $100M_\odot$ the pulsational pair instability sets in, for 140 . . . $260M_\odot$ the pair instability entirely disrupts the star. Above $260M_\odot$, complete collapse to a black hole results.

erode the helium core and deeper layers. Eventually, the star either drops below the critical mass for pair instability or has progressed its burning in the centre far enough to build up an iron core and the star ends its life as a core collapse SN. The time between subsequent outbursts can vary from days to 10,000 years – depending on how powerful the burst was. If it expands the star enough that it leaves the regime for efficient neutrino cooling with central temperatures above $\sim 10^9$ K, it will have to cool down and contract on the Kelvin–Helmholtz timescale for radiation loss from the surface of the star; otherwise, it can cool down much faster, resulting in shorter recurrence times. The first case means that the star has come very close to being unbound, that is having had a very powerful outburst. On the contrary, how powerful an explosion is depends on the detailed chemical structure inside the star – a very complicated function of the preceding evolution stages and pulses.

At higher masses, complete disruption occurs. The explosion energies range from a few times, 10^{51} erg, to almost 100×10^{51} erg (compared to about 1.2×10^{51} erg for SN 1987A). While at the lower end of the pair instability range, the ejecta are mostly oxygen with essentially no radioactive ^{56}Ni being produced (that later decays to ^{56}Fe), at the high-mass end more than $50M_\odot$ of ^{56}Ni can be made and power the light curve tail of the SN. Due to the large mass, however, radiation may take some time to come out and in fact a lot is lost in adiabatic expansion. Yet, these events can still be bright if the progenitor was a red supergiant star, due to self-enrichment of the envelope (see p. 189 and Scannapieco *et al.* 2005).

While the most massive stars in our galaxy today appear to exhibit vast outbursts of mass loss, e.g. Eta Carinae (Humphreys & Davidson 1979), the recent SN 2006gy, one of the brightest SNe ever observed (Smith *et al.* 2007), may have been a pulsational pair instability SN (Woosley *et al.* 2007). In this scenario, a first pulsational pair instability pulse ejects the hydrogen layers as a shell. Then, a few years later, a second pulse with just a 'typical' core collapse SN energy of around 10^{51} erg ejects a shell of helium that collides with the previously ejected hydrogen envelope. When they collide, the shock is optically thin and the kinetic energy of the ejecta is converted into radiation that can escape at close to 100 % efficiency. This is in contrast to normal core collapse SNe and the (non-pulsational) pair instability SNe where only about 1 % or less of the total energy can escape as radiation during the initial explosion.

If this observation can be taken as an indication that pair instability SNe from VMSs can still occur in the present universe even from stars with present-day composition, such SNe should also occur for Pop III stars. Moreover, the pulsational pair instability, in contrast to iron core collapse SNe, is very robust as it only involves well-understood physics and therefore seems to be an unavoidable event for such massive stars. As a consequence, it is the pulsational pair instability SNe that are likely to be those bright explosions which we will observe as the first SNe from Pop III stars.

8.3.4 Nucleosynthesis of the first stars

The main distinguishing feature of nucleosynthesis by the first stars is the absence of initial metals. In metal-rich stars, initial abundances of carbon, oxygen and nitrogen are converted into ^{14}N during CNO cycle hydrogen burning. Later, during helium burning, this ^{14}N is eventually converted into ^{22}Ne, which, in turn, is a source of neutrons due to the ^{22}Ne(α,n)^{25}Mg reaction. These neutrons allow the weak component of the s-process (on iron seeds) as well as the production of elements with odd mass numbers which only have isotopes with neutron excess. Since these initial seeds are not present in primordial stars, a rather strong 'odd–even' effect of nucleosynthesis production, an underproduction of the elements with odd charge numbers results when comparing the yields of these stars with the abundance ratios in the sun (Heger & Woosley 2002).

Since, however, some neutron excess also can result in hydrostatic burning of carbon, this effect is somewhat lessened in the lower mass ranges of massive stars (Heger & Woosley 2008). For pair instability SNe, in particular, however, carbon burning is very weak and sweeps through the star quickly. Typical for these stars hence is a very strong odd–even effect, by an order of magnitude or more, for elements with charge numbers 10–30. The prediction then is that if a second generation of stars formed from the ejecta of a pair instability SN with sufficiently low mass to still be around today, it should show this abundance pattern at its surface. This very clear theoretical nucleosynthetic signature of pair instability SNe has, however, not been observed to date in any star (but see Karlsson *et al.* 2008).

8.3.5 The remnants of the first stars

Several effects distinguish the remnant mass function of primordial stars from later generations. First, the IMF of these stars may have been different, with a possible bias towards higher initial masses (see p. 182), resulting in more remnants from more massive stars. If a large fraction of stars falls into the pair instability regime, however, that fraction of stars does not leave any remnants. Second, Pop III stars may retain most of their initial mass and hence leave behind more massive remnants. But lastly, these stars may also end their lives as blue stars with tightly bound hydrogen envelopes (Heger & Woosley 2008, in prep.). For a given explosion energy, these stars have a significantly larger fallback mass – mass that initially is driven outwards by the SN explosion but due to a strong reverse shock resulting from this tightly bound hydrogen envelope turns around and is accreted by the central remnant within the first minutes to days (Zhang *et al.* 2008). All three factors contribute to produce more massive remnants for Pop III stars compared to the present-day case.

8.4 Feedback effects

Pop III stars transformed the early universe in a number of fundamental ways via radiative, mechanical and chemical feedback effects (Ciardi & Ferrara 2005). The idea of feedback is intimately linked to the possibility that *a system can become self-regulated.* Although some types of feedback processes are disruptive, the most important ones in astrophysics are probably those that are able to drive the systems towards a steady state of some sort. To exemplify this, think of a galaxy which is witnessing a burst of star formation. The occurrence of the first SNe will evacuate/heat the gas, thus suppressing the subsequent star formation process. This feedback is then acting back on the energy source (star formation); it is of a negative type and it could drive the star formation activity in such a way that only a sustainable amount of stars is formed. However, feedback can fail to produce such regulation either in small galaxies, where the gas can be ejected by a handful of SNe, or in cases when the star formation timescale is too short compared to the feedback timescale. In the spirit of the present review, we will only discuss feedback occurring at high redshifts and hence shaping the first structures.

8.4.1 Radiative feedback

Radiative feedback is related to the ionizing/dissociating radiation produced by massive stars or quasars. This radiation can have local effects (i.e. on the same galaxy that produces it) or long-range effects, either affecting the formation and evolution of nearby objects or joining the radiation produced by other galaxies to form a background. In spite of the different scenarios implied, the physical processes are very similar.

8.4.1.1 Photoionization/evaporation

The collapse and formation of primordial objects exposed to a UV radiation field can be inhibited or halted for two main reasons: (i) cooling is considerably suppressed by the decreased fraction of neutral hydrogen and (ii) gas can be photo-evaporated out of the host halo. In fact, the gas incorporated into small mass objects that were unable to cool efficiently can be boiled out of the gravitational potential well of the host halo if it is heated by UV radiation above the virial temperature. Such effects are produced by the same radiation field and act simultaneously. For this reason, it is hard to separate their individual impact on the final outcome. The problem has been extensively studied by several authors (see Ciardi and Ferrara (2005) and references therein). In particular, by assuming spherical symmetry, one can solve self-consistently radiative transfer of photons, non-equilibrium H_2 chemistry and gas hydrodynamics of a collapsing halo. Several studies have found that at weak UV intensities ($J < 10^{-23} \, \mathrm{erg \, s^{-1} \, cm^{-2} \, sr^{-1} \, Hz^{-1}}$), objects as small as $v_c \sim 15 \, \mathrm{km \, s^{-1}}$ are able to collapse, owing to both self-shielding of the gas and H_2 cooling. At stronger intensities, though, objects as large as $v_c \sim 40 \, \mathrm{km \, s^{-1}}$ can be photoevaporated and prohibited from collapsing, in agreement with previous investigations based on the optically thin approximation. More refined 3D hydrodynamic simulations confirm the result that the presence of UV radiation delays or suppresses the formation of low-mass objects. In contrast with these studies, Dijkstra *et al.* (2004), applying a 1D code to $z > 10$ objects, found that objects as small as $v_c \sim 10 \, \mathrm{km \, s^{-1}}$ can self-shield and collapse because the collisional cooling processes at high redshift are more efficient and the amplitude of the ionizing background is lower. The second condition might not always apply, as the ionizing flux at high redshift is dominated by the direct radiation from neighbouring halos rather than the background (Ciardi *et al.* 2000).

The photoevaporation effect might be particularly important for Pop III objects, as their virial temperatures are below the typical temperatures achieved by a primordial photoionized gas ($T \approx 10^4 \, \mathrm{K}$). For such an object, the ionization front gradually burns its way through the collapsed gas, producing a wind that blows backwards into the IGM and that eventually evaporates all the gas content. According to recent studies, these sub-kpc galactic units were so common as to dominate the absorption of ionizing photons. This means that estimates of the number of ionizing photons per H atom required to complete reionization should not neglect their contribution to absorption. The number of ionizing photons absorbed per initial minihalo atom, ξ, increases gradually with time (Shapiro *et al.* 2004); in addition, it depends on the ionizing spectrum assumed. For the hard quasi-stellar object (QSO) spectrum, the ionization front is thicker and penetrates deeper into the denser and colder parts of the halo, increasing the rate of recombinations per atom, compared to stellar-type sources. However, this same pre-heating effect shortens the evaporation time, ultimately leading to a rough cancellation of the two effects and the same total ξ as for a black-body spectrum with $T = 5 \times 10^4 \, \mathrm{K}$ (mimicking a low-metallicity Pop II stellar emission). An even lower ξ is needed for a black-body spectrum with $T = 10^5 \, \mathrm{K}$ (more typical of a Pop III stellar emission; e.g., Bromm *et al.* (2001)), because of an increased evaporation versus penetration ability. Thus, overall, Pop III stellar sources appear significantly more efficient than Pop II or QSO sources in terms of the total number of ionizing photons needed to complete the photoevaporation process.

8.4.1.2 H_2 photodissociation

As intergalactic H_2 is easily photodissociated, a soft-UV background in the
Lyman–Werner bands could quickly build up and have a negative feedback on the
gas cooling and star formation inside small halos. In addition to an external background, the evolution of structures can also be affected by internal dissociating
radiation. In fact, once the first generation of stars has formed in an object, it can
affect the subsequent star formation process by photodissociating molecular hydrogen in star-forming clouds; for example, if the molecular cloud has a metallicity
smaller than about $10^{-2.5} Z_\odot$, a single O star can seriously deplete the H_2 content
so that subsequent star formation is almost quenched. Thus, it seems plausible that
stars do not form efficiently before the metallicity becomes larger than about 10^{-2}
solar.

Machacek *et al.* (2001) find that the fraction of gas available for star formation
in Pop III objects of mass M exposed to a flux with intensity J_{LW} in the Lyman–
Werner band is $\sim 0.06 \ln(M/M_{th})$, where the mass threshold, M_{th}, is given by:

$$\frac{M_{th}}{M_\odot} = 1.25 \times 10^5 + 8.7 \times 10^5 \left(\frac{J_{LW}}{10^{-21} \, \mathrm{erg^{-1} \, s^{-1} \, cm^{-2} \, Hz^{-1}}} \right). \qquad (8.1)$$

The same problem has been analysed by Susa and Umemura (2004) by means of
3D SPH calculations, where radiative transfer is solved by a direct method and
the non-equilibrium chemistry of primordial gas is included. They find that star
formation is suppressed appreciably by UVB, but baryons at high-density peaks
are self-shielded, eventually forming some amount of stars.

The negative feedback described above could be counterbalanced by the *positive
feedback* of H_2 re-formation, e.g. in front of H II regions, inside relic H II regions,
once star formation is suppressed in a halo and ionized gas starts to recombine, in
cooling gas behind shocks produced during the ejection of gas from these objects.
Thus, a second burst of star formation might take place also in the small objects
where it has been suppressed by H_2 dissociation. H_2 production could also be promoted by an X-ray background, which would increase the fractional ionization of
protogalactic gas. Such a positive feedback, though, is not able to balance UV pho-
todissociation in protogalaxies with $T_{vir} < 2000 \, \mathrm{K}$. Similar arguments apply to
high-energy cosmic rays (Jasche *et al.* 2007; Stacy & Bromm 2007).

8.4.1.3 Photoheating filtering

Cosmic reionization might have a strong impact on subsequent galaxy formation,
particularly affecting low-mass objects (Greif & Bromm 2006). In fact, the heating
associated with photoionization causes an increase in the temperature of the IGM
gas which will suppress the formation of galaxies with masses below the Jeans

mass. One can expect that the effect of reionization depends on the reionization history and thus is not universal at a given redshift. More precisely, one should introduce a 'filtering' scale, k_F, (or, equivalently, filtering mass M_F) over which the baryonic perturbations are smoothed as compared to the dark matter, yielding the approximate relation $\delta_b = \delta_{dm} e^{-k^2/k_F^2}$. The filtering mass as a function of time is related to the Jeans mass by

$$M_F^{2/3} = \frac{3}{a} \int_0^a da' M_J^{2/3}(a') \left[1 - \left(\frac{a'}{a} \right)^{1/2} \right]. \tag{8.2}$$

Note that at a given moment in time, the two scales can be very different. Also, in contrast to the Jeans mass, the filtering mass depends on the full thermal history of the gas instead of the instantaneous value of the sound speed, so it accounts for the finite time required for pressure to influence the gas distribution in the expanding universe. The filtering mass increases from roughly $10^7 M_\odot$ at $z \approx 10$ to about $10^9 M_\odot$ at redshift $z \approx 6$, thus efficiently suppressing the formation of objects below that mass threshold. Of course such a result is somewhat dependent on the assumed reionization history.

An analogous effect is found inside individual H II regions around the first luminous sources. Once an ionizing source turns off, its surrounding H II region Compton cools and recombines. Nonetheless, the 'fossil' H II regions left behind remain at high adiabats, prohibiting gas accretion and cooling in subsequent generations of Pop III objects.

8.4.2 Chemical feedback

The concept of chemical feedback is relatively recent, having been first explored by Schneider *et al.* (2002). According to the most popular scenario, the first stars forming out of gas of primordial composition might be very massive, with masses $\approx 10^2 - 10^3 M_\odot$ (see p. 182). The ashes of these first SN explosions pollute with metals the gas out of which subsequent generations of low-mass Pop II/I stars form, driving a transition from a top-heavy IMF to a 'Salpeter-like' IMF when locally the metallicity approaches the critical value $Z_{crit} = 10^{-5 \pm 1} Z_\odot$ (Schneider *et al.* 2002). Thus, the cosmic relevance of Pop III stars and the transition to a Pop II/I star formation epoch depends on the efficiency of metal enrichment from the first stellar explosions, the so-called chemical feedback, which is strictly linked to the number of Pop III stars that explode as pair instability supernova (PISN), the metal ejection efficiency, transport and mixing in the IGM. It is very likely that the transition occurred rather smoothly because the cosmic metal distribution is observed to be highly inhomogeneous: even at moderate redshifts, $z \approx 3$, the clustering properties of C IV and Si IV QSO absorption systems are consistent with a metal filling factor $<10\%$, showing that metal enrichment is incomplete and inhomogeneous.

As a consequence, the use of the critical metallicity as a global criterion is somewhat misleading because chemical feedback is a *local process*, with regions close to star formation sites rapidly becoming metal-polluted and overshooting Z_{crit}, and others remaining essentially metal-free. Thus, Pop III and Pop II star formation modes could have been coeval, and detectable signatures from Pop III stars could be found well after the volume-averaged metallicity has become larger than critical.

Scannapieco *et al.* (2003) have studied, using an analytical model of inhomogeneous structure formation, the separate evolution of Pop III/Pop II stars as a function of star formation and wind efficiencies. They parametrized the chemical feedback through a single quantity, E_g, which represents the kinetic energy input from SNe per unit gas mass into stars. This quantity governs the transport of metals in regions away from their production site and therefore the metallicity distribution; it is related to the number of exploding Pop III stars and encodes the dependence on the assumed IMF. For all values of the feedback parameter, E_g, Scannapieco *et al.* (2003) found that while the peak of Pop III star formation occurs at $z \approx 10$, such stars continue to contribute appreciably to the star formation rate density at much lower redshifts, even though the mean IGM metallicity has moved well past the critical transition metallicity. This finding has important implications for the development of efficient strategies for the detection of Pop III stars in primeval galaxies, as discussed previously (Scannapieco *et al.* 2005).

8.4.3 Mechanical feedback and IGM metal enrichment

Mechanical feedback results from the energy deposited in the surrounding medium by winds from massive stars and SN explosions (quasar energy deposition might have similar effects, although their physics is less understood). The powerful shocks originating from such events heat and accelerate the gas, among other effects. Under such circumstances, the gas outflows from the parent galaxy might easily take place; even more importantly, the nucleosynthetic products of stellar evolution are carried away from their production sites (i.e. the star-forming regions) and possibly injected into the IGM. The first question we need to address is then what is the fate of such gas.

8.4.3.1 Are metals in galaxies or in the IGM?

Ferrara *et al.* (2000, FPS) investigated the metal escape fraction from galaxies. They recognized that SNe in galaxies do not occur in isolation as massive stars tend to cluster into OB associations. Thus, the energy deposition from such associations is large and spread in the disk, so that each explosion site acts incoherently. In nearby galaxies, it is found that the luminosity function of OB associations is well approximated by a power law

$$\phi(N) = \frac{d\mathcal{N}_{OB}}{dN} = AN^{-\beta} \tag{8.3}$$

with $\beta \approx 2$. Here $\mathcal{N}_{OB}$ is the number of associations containing N OB stars; normalization of $\phi(N)$ to unity requires $A = 1$. Thus, the probability for a cluster of OB stars to host N SNe is $\propto N^{-2}$, where $N = L_{OB}t_{OB}/\epsilon_0$ and $t_{OB} = 40\,\text{Myr}$ is the time at which the lowest mass ($\approx 8M_\odot$) SN progenitors expire. The total mechanical luminosity is then found to be

$$L_t(z) = \int_{N_m}^{N_M} L_{OB}(N)\phi dN, \tag{8.4}$$

where $N_m = 1$ (N_M) is the minimum (maximum) possible number of SNe in a cluster. This gives

$$L_t(z) = \text{const.} \frac{\epsilon_0}{t_{OB}} \ln\frac{N_M}{N_m}. \tag{8.5}$$

The contribution to the total luminosity from clusters powerful enough to lead to blowout is

$$L_B(z, > L_c) = \text{const.} \frac{\epsilon_0}{t_{OB}} \ln\frac{N_M}{N_c}, \tag{8.6}$$

where N_c is the number of SNe in a cluster with mechanical luminosity equal to L_c, i.e.

$$N_c = \frac{L_c t_{OB}}{\epsilon_0}. \tag{8.7}$$

Thus, the fraction of the mechanical energy which can be blown out is

$$\delta_B = \frac{\ln(N_M/N_c)}{\ln(N_M/N_m)} < 1. \tag{8.8}$$

Clearly, N_M (and therefore δ_B) is an intrinsically stochastic number. To determine its dependence on the total number of SNe, $N_t = L_t(z)t_{OB}/\epsilon_0$, produced by a galaxy during the lifetime of an OB association, we have used a Monte Carlo procedure applied to the distribution function in Eq. (8.3). For low values of N_t, the quantity N_M is larger than N_c, implying that in every galaxy at least some star bursts are able to blow out. However, near $N_t = 10^4$, N_M flattens and eventually becomes equal to N_c at $N_t \simeq 45,000$. Above this limit (corresponding to a galaxy with $\dot{M}_\star \approx 0.35M_\odot\,\text{year}^{-1}$ or $M_h \approx 10^{12}(1+z)^{-3/2}M_\odot$), blowout is inhibited. The fraction δ_B can be shown (FPS) to be a decreasing function of N_t; an approximate analytical form is

$$\delta_B(N_t) = 1 \ \text{ for } N_t < 100$$
$$\delta_B(N_t) = a + b\ln(N_t^{-1}) \ \text{ for } N_t > 100, \tag{8.9}$$

with $a = 1.76$ and $b = 0.165$. Clearly, in small galaxies, even the smallest associations are capable of producing blowouts, so that the issue of coherence discussed above is irrelevant.

The cosmic ejection fraction, $f_{ej}(z)$, (i.e. the value of δ_B averaged over the entire population of halos as predicted by the concordance cosmology) is very close to unity at high redshift where predominantly small galaxies are present, but it steadily decreases to about 50% at $z = 0$, as the number of more massive galaxies able to retain their metals increases. Stated differently, today about 50% of the cosmic metals are not in galaxies but reside in the IGM. Are these the metals that we detect in the Lyα forest?

Acknowledgements

We are grateful to Gilles Chabrier for organizing a unique and stimulating conference. VB acknowledges support from NSF grant AST-0708795 and NASA *Swift* grant NNX07AJ636. AH performed this work under the auspices of the US Department of Energy at the University of California Los Alamos National Laboratory under contract no. W-7405-ENG-36 and the DOE Program for Scientific Discovery through Advanced Computing (SciDAC; DE-FC02-01ER41176).

References

Abel, T., Bryan, G., & Norman, M. L. 2000, *ApJ.*, **540**, 39
Abel, T., Bryan, G., & Norman, M. L. 2002, *Science*, **295**, 93
Alvarez, M. A., Bromm, V., & Shapiro, P. R. 2006, *ApJ.*, **639**, 621
Baraffe, I., Heger, A., & Woosley, S. E. 2001, *ApJ.*, **550**, 890
Barkana, R., & Loeb, A. 2001, *Phys. Rep.*, **349**, 125
Bromm, V., Coppi, P. S., & Larson, R. B. 1999, *ApJ.*, **527**, L5
Bromm, V., Coppi, P. S., & Larson, R. B. 2002, *ApJ.*, **564**, 23
Bromm, V., Ferrara, A., Coppi, P. S., & Larson, R. B. 2001, *MNRAS*, **328**, 969
Bromm, V., Kudritzki, R. P., & Loeb, A. 2001, *ApJ.*, **552**, 464
Bromm, V., & Larson, R. B. 2004, *ARA&A*, **42**, 79
Bromm, V., & Loeb, A. 2003a, *Nature*, **425**, 812
Bromm, V., & Loeb, A. 2003b, *ApJ.*, **596**, 34
Bromm, V., & Loeb, A. 2004, *New Astron.*, **9**, 353
Bromm, V., Yoshida, N., & Hernquist, L. 2003, *ApJ.*, **596**, L135

Christlieb, N., *et al.* 2002, *Nature*, **419**, 904

Ciardi, B., & Ferrara, A. 2005, *Space Sci. Rev.*, **116**, 625

Ciardi, B., Ferrara, A., Governato, F., & Jenkins, A. 2000, *MNRAS*, **314**, 611

Clark, P. C., Glover, S. C. O., & Klessen, R. S. 2008, *ApJ.*, **672**, 757

Clarke, C. J., & Bromm, V. 2003, *MNRAS*, **343**, 1224

Clayton, D. D. 1968, *Principles of Stellar Evolution and Nucleosynthesis* (New York: McGraw-Hill)

Couchman, H. M. P., & Rees, M. J. 1986, *MNRAS*, **221**, 53

Dijkstra, M., Haiman, Z., Rees, M. J., & Weinberg, D. H. 2004, *ApJ.*, **601**, 666

Ekström, S., Meynet, G., & Maeder, A. 2008, in *AIP conf. proc. 990, First Stars III*, ed. B. O'Shea, A. Heger & T. Abel (Mebuille: AIP), 220

Ferrara, A., Pettini, M., & Shchekinov, Yu. 2000, *MNRAS*, **319**, 539 (FPS)

Frebel, A., Johnson, J. L., & Bromm, V. 2007, *MNRAS*, **380**, L40

Frebel, A., *et al.* 2005, *Nature*, **434**, 871

Fryer, C. L., Woosley, S. E., & Heger, A. 2001, *ApJ.*, **550**, 372

Fuller, G. M., Woosley, S. E., & Weaver, T. A. 1986, *ApJ.*, **307**, 675

Furlanetto, S. R., & Loeb, A. 2003, *ApJ.*, **588**, 18

Gao, L., & Theuns, T. 2007, *Science*, **317**, 1527

Gao, L., Yoshida, N., Abel, T., Frenk, C. S., Jenkins, A., & Springel, V. 2007, *MNRAS*, **378**, 449

Glover, S. C. O. 2005, *Space Sci. Rev.*, **117**, 445

Greif, T. H., & Bromm, V. 2006, *MNRAS*, **373**, 128

Greif, T. H., Johnson, J. L., Bromm, V., & Klessen, R. S. 2007, *ApJ.*, **670**, 1

Haiman, Z., Thoul, A. A., & Loeb, A. 1996, *ApJ.*, **464**, 523

Heger, A., Fryer, C. L., Woosley, S. E., Langer, N., & Hartmann, D. H. 2003, *ApJ.*, **591**, 288

Heger, A., & Woosley, S. E. 2002, *ApJ.*, **567**, 532

Heger, A., & Woosley, S. E. 2008, *ApJ.*, submitted (arXiv:0803.3161)

Heger, A., Woosley, S. E., & Waters, R. 2000, in *Proc. of the MPA/ESO Workshop: The First Stars*, ed. A. Weiss, T. Abel & V. Hill (Berlin: Springer), 121

Hirschi, R. 2007, *A&A*, **461**, 571

Humphreys, R. M., & Davidson, K. 1979, *ApJ.*, **232**, 409

Jappsen, A.-K., Glover, S. C. O., Klessen, R. S., & Mac Low, M.-M. 2007, *ApJ.*, **660**, 1332

Jappsen, A.-K., Klessen, R. S., Glover, S. C. O., & Mac Low, M.-M. 2008, *ApJ.*, submitted (arXiv:0709.3530)

Jasche, J., Ciardi, B., & Ensslin, T. A. 2007, *MNRAS*, **380**, 417

Johnson, J. L., & Bromm, V. 2006, *MNRAS*, **366**, 247

Johnson, J. L., Greif, T. H., & Bromm, V. 2007, *ApJ.*, **665**, 85

Kaplinghat, M., Chu, M., Haiman, Z., Holder, G. P., Knox, L., & Skordis, C. 2003, *ApJ.*, **583**, 24

Karlsson, T., Johnson, J. L., & Bromm, V. 2008, *ApJ.*, **679**, 6

Kippenhahn, R., & Weigert, A. 1990, *Stellar Structure and Evolution* (Heidelberg: Springer)

Kitayama, T., Yoshida, N., Susa, H., & Umemura, M. 2004, *ApJ.*, **613**, 631

Kitsionas, S., & Whitworth, A. P. 2002, *MNRAS*, **330**, 129

Kudritzki, R. P. 2002, *ApJ.*, **577**, 389

Larson, R. B. 1998, *MNRAS*, **301**, 569

Larson, R. B. 2003, *Rep. Prog. Phys.*, **66**, 1651

Machacek, M. M., Bryan, G. L., & Abel, T. 2001, *ApJ.*, **548**, 509

Mackey, J., Bromm, V., & Hernquist, L. 2003, *ApJ.*, **586**, 1
McKee, C. F., & Tan, J. C. 2008, *ApJ.*, **681**, 771
Mori, M., Ferrara, A., & Madau, P. 2002, *ApJ.*, **571**, 40
Nakamura, F., & Umemura, M. 2001, *ApJ.*, **548**, 19
Omukai, K. 2000, *ApJ.*, **534**, 809
Omukai, K., & Inutsuka, S. 2002, *MNRAS*, **332**, 59
Omukai, K., & Nishi, R. 1998, *ApJ.*, **508**, 141
Omukai, K., & Palla, F. 2001, *ApJ.*, **561**, L55
Omukai, K., & Palla, F. 2003, *ApJ.*, **589**, 677
O'Shea, B. W., & Norman, M. L. 2007, *ApJ.*, **654**, 66
Palla, F., Salpeter, E. E., & Stahler, S. W. 1983, *ApJ.*, **271**, 632
Poelarends, A. J. T., Herwig, F., Langer, N., & Heger, A. 2008, *ApJ.*, **675**, 614
Qian, Y.-Z., & Wasserburg, G. J. 2002, *ApJ.*, **567**, 515
Ripamonti, E., Haardt, F., Ferrara, A., & Colpi, M. 2002, *MNRAS*, **334**, 401
Saigo, K., Matsumoto, T., & Umemura, M., 2004, *ApJ.*, **615**, L65
Santoro, F., & Shull, J. M. 2006, *ApJ.*, **643**, 26
Scannapieco, E., Madau, P., Woosley, S., Heger, A. & Ferrara, A. 2005, *ApJ.*, **633**, 1031
Scannapieco, E., Schneider, R., & Ferrara, A. 2003, *ApJ.*, **589**, 35
Schneider, R., Ferrara, A., Natarajan, P., & Omukai, K. 2002, *ApJ.*, **571**, 30
Schneider, R., Ferrara, A., Salvaterra, R., Omukai, K., & Bromm, V. 2003, *Nature*, **422**, 869
Schneider, R., Salvaterra, R., Ferrara, A., & Ciardi, B. 2006, *MNRAS*, **369**, 825
Schwarzschild, M., & Härm, R. 1959, *ApJ.*, **129**, 637
Shapiro, P. R., Iliev, I. T., & Raga, A. C. 2004, *MNRAS*, **348**, 753
Smith, B. D., & Sigurdsson, S. 2007, *ApJ.*, **661**, L5
Smith, N., & Owocki, S. P. 2006, *ApJ.*, **645**, L45
Smith, N., *et al.* 2007, *ApJ.*, **666**, 1116
Spergel, D. N., *et al.* 2007, *ApJS*, **170**, 377
Stacy, A., & Bromm, V. 2007, *MNRAS*, **382**, 229
Susa, H., & Umemura, M. 2004, *ApJ.*, **600**, 1
Tan, J. C., & McKee, C. F. 2004, *ApJ.*, **603**, 383
Tegmark, M., Silk, J., Rees, M. J., Blanchard, A., Abel, T., & Palla, F. 1997, *ApJ.*, **474**, 1
Tornatore, L., Ferrara, A., & Schneider, R. 2007, *MNRAS*, **382**, 945
Tsuribe, T., & Omukai, K. 2006, *ApJ.*, **642**, L61
Umeda, H., & Nomoto, K. 2002, *ApJ.*, **565**, 385
Umeda, H., & Nomoto, K. 2003, *Nature*, **422**, 871
Venkatesan, A. 2006, *ApJ.*, **641**, L81
Wada, K., & Venkatesan, A. 2003, *ApJ.*, **591**, 38
Whalen, D., Abel, T., & Norman, M. L. 2004, *ApJ.*, **610**, 14
Wolfire, M. G., & Cassinelli, J. P. 1987, *ApJ.*, **319**, 850
Woosley, S. E., Blinnikov, S., & Heger, A. 2007, *Nature*, **450**, 390
Woosley, S. E., Heger, A., & Weaver, T. A. 2002, *Rev. Mod. Phys.*, **74**, 1015
Yoshida, N., Abel, T., Hernquist, L., & Sugiyama, N. 2003, *ApJ.*, **592**, 645
Yoshida, N., Bromm, V., & Hernquist, L. 2004, *ApJ.*, **605**, 579
Yoshida, N., Omukai, K., & Hernquist, L. 2007, *ApJ.*, **667**, L117; 2006, *ApJ.*, **652**, 6
Yoshida, N., Omukai, K., Hernquist, L., & Abel, T. 2006, *ApJ.*, **652**, 6
Zhang, W., Woosley, S. E., & Heger, A. 2008, *ApJ.*, **679**, 639

Part III

Contemporary Star and Brown Dwarf Formation

9

Diffuse interstellar medium and the formation of molecular clouds

P. Hennebelle, M.-M. Mac Low and E. Vazquez-Semadeni

9.1 Introduction

The formation of molecular clouds (MCs) from the diffuse interstellar gas is a necessary step for star formation, as young stars invariably occur within them. However, the mechanisms controlling the formation of MCs remain controversial. In this contribution, we focus on their formation in compressive flows driven by interstellar turbulence and large-scale gravitational instability.

Turbulent compression driven by supernovae appears insufficient to explain the bulk of cloud and star formation. Rather, gravity must be important at all scales, driving the compressive flows that form both clouds and cores. Cooling and thermal instability allow the formation of dense gas out of moderate, transonic compressions in the warm diffuse gas and drive turbulence into the dense clouds. MCs may be produced by an overshoot beyond the thermal-pressure equilibrium between the cold and the warm phases of atomic gas, caused by some combination of the ram pressure of compression and the self-gravity of the compressed gas.

In this case, properties of the clouds such as their mass, mass-to-magnetic flux ratio, and total kinetic and gravitational energies are in general time-variable quantities. MCs may never enter a quasi-equilibrium or virial equilibrium state but rather continuously collapse to stars. Gravitationally collapsing clouds exhibit a pseudo-virial energy balance $|E_{\mathrm{grav}}| \sim 2E_{\mathrm{kin}}$, which, however, is representative of contraction rather than of virial equilibrium in this case. However, compression-driven cloud and core formation still involves significant delays as additional material accretes, leading to lifetimes longer than the free-fall time. In this case, the star formation efficiency (SFE) may be determined by the combined effect of the dispersive action of the early stellar products formed in the density fluctuations produced by the initial turbulence and of the magnetic support of large fractions of the volume of the MCs.

Structure Formation in Astrophysics. ed. G. Chabrier. Published by Cambridge University Press.
© Cambridge University Press 2009.

9.2 Large-scale interstellar medium

At the scale of galactic disks, gravitational instability occurs not just in the gas alone (Goldreich & Lynden-Bell 1965) but in the combined medium of collisionless stars and collisional gas (Gammie 1992; Rafikov 2001). The combination is always more unstable than either component in isolation (Gammie 1992). In relatively gas-poor galaxies like the modern Milky Way, the instability of the stars and gas is often analysed independently, in terms of gas flowing into stellar spiral arms (Roberts 1969). However, this is just an approximation to the general gravitational instability. Analysis of the instability must include both mass distributions (Yang *et al.* 2007).

Numerical experiments on the behaviour of the gravitational instability in disks have been done by Li *et al.* (2005), using isothermal gas, collisionless stars and live dark matter halos computed with GADGET (Springel *et al.* 2001). They controlled the initial gravitational instability of the disk and then computed its subsequent behaviour, as shown in Figure 9.1. Using sink particles, they measured the amount of gas that collapsed as a function of time and related it to the initial instability, as expressed by the initial minimum Toomre parameter for stars and gas combined $Q_{sg,\mathrm{min}}$.

Li *et al.* (2005) measured the collapse timescale τ_{sf} by fitting curves of the form $M_* = M_0(1 - \exp(-t/\tau_{sf}))$, where the amount of collapsed mass is M_*, the initial gas mass M_0 and the elapsed time t. Figure 9.2 shows that the collapse timescale depends exponentially on the initial strength of the instability

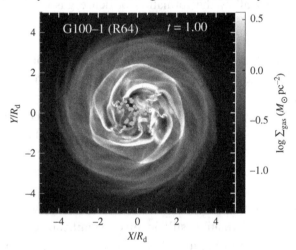

Fig. 9.1. Gas surface density map from the models of Li *et al.* (2005, Figure 5c), showing a model with 6.4 million particles distributed evenly between gas and collisionless populations (stars and dark matter). Filled circles represent sink particles (for colour image, see astro-ph:2007arXiv0711.2417). The model galaxy has rotation velocity at the virial radius of $100\,\mathrm{km\,s^{-1}}$ and sound speed $c_s = 6\,\mathrm{km\,s^{-1}}$.

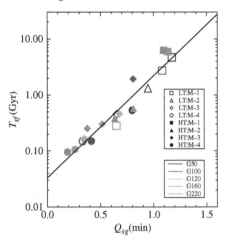

Fig. 9.2. Star formation timescale τ_{sf} correlates exponentially with the initial disk instability $Q_{sg,\min}$ for both low-temperature (sound speed $c_s = 6\,\mathrm{km\,s}^{-1}$; *open symbols*) and high-temperature ($c_s = 15\,\mathrm{km\,s}^{-1}$; *filled symbols*) models. The *solid line* is a least-squares fit to the data (Li *et al.* 2005).

$$\tau_{sf} = (34.7 \pm 7\,\mathrm{Myr}) \exp\left(\frac{Q_{sg,\min}}{0.24}\right). \qquad (9.1)$$

Kravtsov (2003) and Li *et al.* (2006) demonstrate that gravitational instability can explain the Schmidt law (Kennicutt 1998) as a natural outcome of the evolution of galactic disks. Kravtsov (2003) computed a cosmological volume and followed the star formation in individual disks, using a star formation law $\dot{M}_* \propto \rho_g$ deliberately chosen to not automatically reproduce the Schmidt law, as compared to the frequently chosen $\dot{M}_* \propto \rho_g^{1.5}$. The sink particles used by Li *et al.* (2006) effectively give a similar star formation law, as they measure collapsed gas above a fixed threshold. Kravtsov (2003) found that including feedback made little difference so long as cooling was prevented below 10^4 K, roughly the temperature chosen for their isothermal equation of state by Li *et al.* (2006). The resulting Schmidt law is shown in Figure 9.3.

Gravitational instability drives compressive flows at the largest scales. Another candidate to drive compressive flows is turbulence driven by the expansion of H II regions or supernovae. Substantial observational evidence exists for multiple generations of massive star formation in molecular clouds (MCs), apparently triggered by H II regions and supernova explosions (see, e.g., Elmegreen & Palouš 2007). However, supersonic-driven turbulence inhibits collapse rather than enhancing it (Mac Low Klessen 2004 and references therein).

An examination of the triggering effect of H II regions in the Milky Way by Mizuno *et al.* (2007) led to the conclusion that, although triggering does occur,

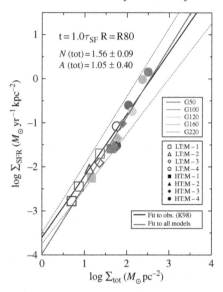

Fig. 9.3. Comparison of the global Schmidt laws between the simulations of Li *et al.* (2006, Figure 5) and the observations. The *grey line* is the least-squares fit to the total gas of the simulated models, the *black solid line* is the best fit of observations from Kennicutt (1998) and the *black dotted lines* indicate the observational uncertainty (for colour image, see astro-ph:2007arXiv0711.2417).

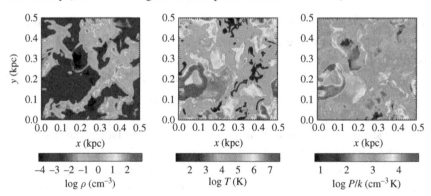

Fig. 9.4. Cuts through the midplane ($z = 0$) of the numerical model of Joung and Mac Low (2006), showing distributions of the density (*left*), temperature (*middle*) and pressure (*right*) at $t = 79.3$ Myr. Density and temperature both vary by about seven orders of magnitude. Note, however, that high density regions do not correlate well with high pressure regions in the absence of self-gravity (for colour image, see astro-ph:2007arXiv0711.2417).

only 10–30% of star formation occurs in triggered regions. A numerical model, shown in Figure 9.4, of large-scale supernova-driven turbulence by Joung and Mac Low (2006) allowed them to reach a similar conclusion. They measured the mass of gas that was Jeans-unstable in their flow, allowing them to estimate the star

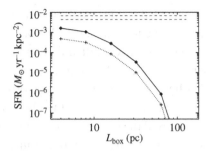

Fig. 9.5. Predicted star formation rate from the model plotted against the sub-box sizes used to measure the Jeans stability of the gas. The smallest boxes most accurately measure the amount of Jeans-unstable gas available for star formation. (Note that, absent self-gravity, collapse does not actually occur.) The *dotted line* is drawn assuming that 30% of the mass in Jeans-unstable regions turns into stars. The *dashed lines* show the star formation rates consistent with the assumed galactic supernova rate, assuming 130 or 200$M_\odot$ of stellar mass is required per supernova.

formation rate that would be induced by the turbulent motions alone. They found that the corresponding star formation rate was an order of magnitude lower than the star formation rate required to maintain the assumed supernova rate (Figure 9.5). This again suggests that triggering has a 10% effect.

Gravitational instability thus appears capable of explaining observed star formation rates, while turbulent compression alone seems to fail by roughly an order of magnitude. Considering galactic scale magnetic fields does not alter this conclusion. Parker instabilities alone appear insufficient to form giant molecular clouds (GMCs) (Kim *et al.* 2000; Santillán *et al.* 2000), but magneto-Jeans instabilities are effective (Kim & Ostriker 2006).

The first consequence of gravitational instability at large scales, though, is large-scale compressive flows, first in the spiral arms that form the first manifestation of gravitational instability in disks and then in smaller collapsing regions (Field *et al.* 2008). Therefore, a general physical understanding of MC formation in compressive flows is a vital link in understanding star formation.

9.3 Neutral interstellar medium

The interplay between compressible turbulence and heating and cooling processes in the neutral interstellar gas leads to the production of density fluctuations, including tiny-scale atomic structures, cold atomic clouds and ultimately MCs.

Observationally, the neutral atomic hydrogen is mostly studied through the HI 21-cm line, both in emission and in absorption (Dickey & Lockman 1990; Heiles & Troland 2005). These observations show that the interstellar atomic

hydrogen spans a wide range of densities and temperatures, from what has traditionally been called the warm neutral medium (WNM), with $n \simeq 0.3$–0.5 cm^{-3}
and $T \simeq 5000$–8000 K, to what has been called the cold neutral medium (CNM),
roughly 100 times denser and colder, although a significant fraction of the gas
mass (possibly up to 50%) lies at intermediate values of n and T (Heiles 2001).
This picture is significantly more complex than the classical two-phase model of
Field *et al.* (1969), which proposed the existence of discrete phases (the WNM and
the CNM) in pressure equilibrium. In the remainder of this section, we discuss our
present understanding of the extent to which this picture is modified by the presence of compressive turbulence in the atomic gas. This turbulence turns out to be
transonic with respect to WNM sound speed but is supersonic with respect to CNM
sound speed.

9.3.1 Thermal balance and thermal instability

The detailed thermal balance of the interstellar medium (ISM) was first investigated by Field *et al.* (1969) and Dalgarno and McCray (1972) and more recently
by Wolfire *et al.* (1995, 2003). The interstellar atomic hydrogen is thought to be
mainly heated by ultraviolet and soft X-ray radiation through the photoelectric
effect on small dust grains and polycyclic aromatic hydrocarbons. The heating
term is therefore proportional to the gas number density, n, and has a weak dependence on temperature. On the contrary, for temperatures larger than a few thousand
kelvins, the most important cooling mechanism is due to the Lyman-α HI line, and
at lower temperatures, due to the [O$_\mathrm{I}$] and [C$_\mathrm{II}$] fine-structure lines. Since these
lines are excited by collisions, the cooling term is proportional to n^2. Thus, the gas
is thermally *unstable* when the cooling function varies slowly with the temperature,
as is the case at a few thousand kelvins, when the [O$_\mathrm{I}$] and [C$_\mathrm{II}$] lines are saturated,
and thermally *stable* when the cooling function varies rapidly with the temperature,
as happens in the wings of the atomic lines responsible for the cooling. This is the
physical origin of the famous two-phase model for interstellar atomic hydrogen
(Field *et al.* 1969).

Field (1965) showed that the criterion applicable to the atomic interstellar gas is
the isobaric stability condition (see, e.g., the review by Vázquez-Semadeni *et al.*
2003), which can be written as

$$\left(\frac{\partial P}{\partial \rho}\right)_{\mathcal{L}} \leq 0, \tag{9.2}$$

where $\mathcal{L}$ is the loss function equal to the cooling minus heating terms. When the
pressure along the cooling curve decreases with increasing density, the density

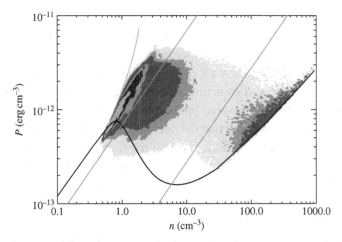

Fig. 9.6. Thermal equilibrium curve for the ISM (*black line*), and isothermal curves with $T = 5000\,$K and 200 K (*grey lines*). The color represents the gas mass fraction in a simulation of a turbulent two-phase medium. (Adapted from Audit & Hennebelle 2005.) In arbitrary units, they correspond to 1, 5, 10, 50, 100, 200 and 1000, respectively (for colour image, see astro-ph:2007arXiv0711.2417).

fluctuation tends to be amplified because the surrounding gas has a larger thermal pressure and compresses it further.

The cooling curve computed by Wolfire *et al.* (1995, 2003) presents in a pressure–density diagram two branches of positive slope joined by one branch of negative slope. The WNM branch cannot exist at pressures higher than a certain critical value, P_{max}, and the CNM branch cannot exist at pressures lower than another critical value, P_{min} (see Figure 9.6). The density and temperature of the stable branches are in good agreement with the density and temperature ranges of the WNM and CNM inferred from observations.

9.3.2 Dynamical formation of cold atomic clouds

The influence of turbulence on the interstellar gas and its role in cloud formation was first investigated by Vázquez-Semadeni *et al.* (1995), who used a cooling curve that implied thermal stability at all temperatures, although still causing denser gas to be colder. They found that converging flows form strong density fluctuations even in relatively weak HII-region-driven turbulence, because of the strongly compressible nature of the flow. Such converging flows have been proposed to rapidly form MCs (Ballesteros-Paredes *et al.* 1999a; Hartmann *et al.* 2001).

The influence of dynamical motions on the warm stable phase of a thermally bistable flow (i.e. the WNM) was considered in 1D by Hennebelle and Pérault

(1999), Koyama and Inutsuka (2000) and Vázquez-Semadeni *et al.* (2006). These studies showed that either a shock or a converging flow can trigger the formation of a long-lived cold structure. The unperturbed incoming WNM flow undergoes a shock that heats the gas and throws it out of thermal equilibrium. Behind the shock, the gas continues to flow and cool, until finally, roughly one cooling length behind, it undergoes a transition to the cold phase, forming a thin, cold, dense layer. The phase transition occurs provided that the fluctuation amplitude is sufficiently strong to reach the pressure threshold P_{max} and that the fluctuation lasts long enough for the gas to cool. Perturbations that fail to satisfy either of these conditions produce weak fluctuations of WNM instead. The response of the flow is therefore very non-linear and depends sensitively on whether the perturbation is able to push the gas into the thermally unstable area.

The detailed evolution of density and velocity perturbations in a thermally unstable medium as a function of their associated crossing times was investigated further by Sánchez-Salcedo *et al.* (2002) and Vázquez-Semadeni *et al.* (2003) starting with initially thermally unstable gas. For *density* perturbations whose crossing time is shorter than the cooling time, the gas condenses more or less isobarically, while if the crossing time exceeds the cooling time, then the gas follows the thermal equilibrium curve while condensing. On the contrary, *velocity* perturbations with crossing times shorter than the cooling time tend to behave adiabatically, at least before the gas has time to cool.

In the 1D studies, once the cooling gas reaches the CNM branch, it cools and contracts along it, until the thermal pressure equals the ram pressure of the shocked warm incoming gas. When the perturbation has relaxed, the density within the structure decreases until its internal pressure is equal to the thermal pressure of the surrounding WNM. The cold structure is then pressure confined, and therefore stable, unlike what happens in an isothermal medium. In 2D and 3D, however, the evolution is more complicated, as we describe in p. 213.

9.3.3 Front stability and thermal fragmentation

By performing a linear stability analysis, Inoue *et al.* (2006) show that evaporation fronts between cold and warm phases suffer instabilities similar to the Darrius–Landau instability that occurs in combustion fronts, rendering them unstable under corrugational deformations. On the contrary, condensation fronts are stable. The fastest growth rate of the evaporation front instability corresponds to wavelengths slightly larger than the Field length, while larger wavelengths grow proportionally to the wavenumber k, and smaller wavelengths are stable. Numerical simulations performed by Kritsuk and Norman (2002) and Koyama and Inutsuka (2006) show that the non-linear development of the instability can sustain weak turbulence. The source of energy is the heating term, which is not fully compensated by the cooling term within the thermal front.

Koyama and Inutsuka (2002) have also investigated the propagation of a shock through the WNM in 2D. They found that the post-shock gas is very unstable and fragments into many small CNM structures. Pittard *et al.* (2005) have shown that radiative shocks become more prone to overstability at a given upstream Mach number as the final post-shock temperature is lower with respect to the upstream one. All of these results suggest that weak turbulence can easily be driven in the cold gas. Koyama and Inutsuka (2002) and Heitsch *et al.* (2005) have further shown that the CNM structures have a velocity dispersion that is a fraction of the sound speed of the warm phase, but still supersonic with respect to the internal sound speed of the CNM clumps.

Similar results for a shock-bounded layer in a radiatively cooling gas have been obtained by Walder and Folini (1998, 2000). Indeed, shock-bounded layers were shown to be non-linearly unstable even in the isothermal case by Vishniac (1994), through the non-linear thin-shell instability.

9.3.4 Colliding flows and thermally bistable turbulence

The formation of CNM structures induced by dynamical motions within the WNM has been further studied in 2D by Audit and Hennebelle (2005) and Heitsch *et al.* (2005, 2006) and in 3D by Vázquez-Semadeni *et al.* (2006) (see also Kritsuk and Norman (2002) for the decaying thermally bistable turbulence and Gazol *et al.* (2005) for the driven case). Typically, these studies consider a computational box of a few tens of parsecs and a resolution of about 1000^2 cells in 2D and up to 400^3 cells in 3D. They all consider a converging flow of WNM that produces a turbulent shocked layer, which fragments into CNM structures, producing a turbulent, clumpy medium.

When the incoming flow is initially nearly laminar, the turbulence in the dense layer is believed to be driven by the combined action of the thermal Kelvin–Helmholtz and non-linear thin-shell instabilities (Heitsch *et al.* 2005, 2006). However, when non-linear fluctuations are initially present, the importance of these instabilities remains to be clarified (astrophysical flows are not expected to be initially laminar). Indeed, Audit and Hennebelle (2005) found that adding non-linear perturbations to the converging flow strongly changes the level of turbulence within the computational box, including in the CNM phase.

The overall structure of the resulting flow appears to be complex (see Figure 9.7). The two phases are strongly mixed and a substantial fraction of thermally unstable gas is produced by the turbulent motions (Gazol *et al.* 2001) as shown in Figures 9.6–9.8, in agreement with observational determinations (Heiles 2001). Note,

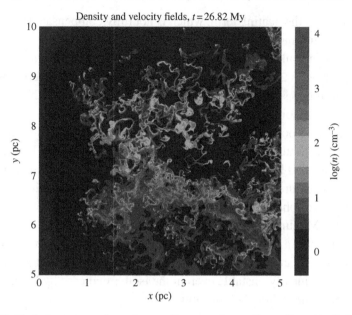

Fig. 9.7. Turbulent, two-phase interstellar atomic medium. Density field of a high-resolution 2D simulation. (Adapted from Hennebelle & Audit (2007); for colour image, see astro-ph: 2007arXiv0711.2417).

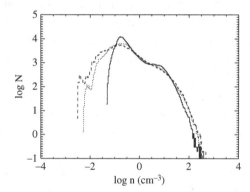

Fig. 9.8. Density probability distribution function as a function of rms Mach number M (with respect to the warm unstable gas). *Solid line, $M = 0.5$; dotted line, $M = 1$; dashed line, $M = 1.25$.* (Adapted from Gazol *et al.* (2005).)

however, that even when the turbulence is strongly driven, producing transonic velocity dispersion within WNM, the tendency towards two-phase behaviour persists, although the fraction of thermally unstable gas increases for more strongly turbulent regimes. In particular, in the simulations that have been performed, a large fraction of the CNM structures remain bounded by contact discontinuities and pressure confined by the surrounding WNM. Note that it is certainly the case that for high-Mach number conditions, this structure may change. The question of how and at which Mach number remains an open issue.

As in the numerical experiment of Koyama and Inutsuka (2002), the CNM structures present velocity dispersion somewhat smaller than the WNM sound speed. Since these motions are supersonic with respect to the CNM internal sound speed, the CNM structures undergo high-Mach number collisions, which create strong density fluctuations that may be reminiscent of the small atomic structures observed in HI (Heiles 1997). Note that interestingly enough, this relatively high-velocity dispersion within CNM structures is sustained self-consistently by the WNM turbulence, which in the study of colliding flows is triggered from the turbulence in the incoming WNM as well as by the turbulence spontaneously generated as part of the cloud formation process.

9.3.5 *Dense structure statistics in thermally bistable turbulent flows*

Studying the statistical physical properties of the dense structures produced by interstellar turbulence is of great interest since such structures constitute both atomic and MCs in the ISM (Sasao 1973, Elmegreen 1993; Ballesteros-Paredes *et al.* 1999b). The statistics from the numerical models can be compared with the observational properties of dense clouds such as, e.g., mean densities, pressures and temperatures, typical sizes, scaling relations among variables, etc., allowing tests of the various theoretical models.

The simulations by Gazol *et al.* (2005) and Audit and Hennebelle (2005) have shown that even at Mach numbers ~ 1, local values of the density and pressure can be reached that exceed 1000 cm^{-3} and $\sim 10^5 \text{ K cm}^{-3}$, respectively. Also, the bimodal nature of the density histogram, which in the extreme case of a static two-phase medium would be two Dirac delta functions at the densities of the WNM and CNM, is seen to gradually lose its bimodal character as the rms Mach number is increased (Figure 9.8), signalling the increasing amounts of thermally unstable gas in the flow. It is also interesting that sometimes the highest pressures can be obtained in the transient unstable warm gas rather than in the densest gas (see also Vázquez-Semadeni *et al.* 2006).

More recently, much higher resolution simulations have been performed, using up to 10000^2 cells in 2D (Hennebelle & Audit 2007) and 1200^3 cells in 3D (Audit & Hennebelle 2008). Figures 9.9 and 9.10 show the mass spectrum, the mass–size relations and the velocity dispersion as a function of size of the CNM structures extracted from the 3D simulations. Corresponding results for the 2D case can be found in Hennebelle and Audit (2007) and Hennebelle *et al.* (2007). The CNM structures are simply determined by a clipping algorithm that uses a

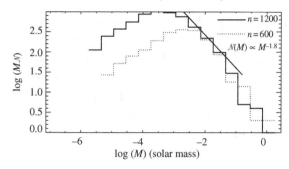

Fig. 9.9. Mass distribution of the structure identified in the simulations (Audit & Hennebelle 2008).

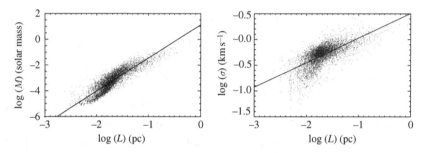

Fig. 9.10. Mass versus size (*left*) and velocity dispersion versus size (*right*) relations for the CNM structures extracted from the 1200^3 cells simulations (Audit & Hennebelle 2008). The slope of the *solid lines* are 2.5 (*left panel*) and 0.5 (*right panel*).

density threshold lying in the thermally unstable domain. The mass spectrum follows $dN/dM = \mathcal{N} \propto M^{-1.8}$. Numerical convergence seems to have been reached at a resolution of 1200 zones for masses between $\simeq 3 \times 10^{-3}$ and $\simeq 3 \times 10^{-2} M_{\odot}$. The mass–size relation is given approximately by $M \propto L^{2.3-2.5}$, the smallest value of the index being obtained for the largest structures. Finally, the internal velocity follows about $\sigma \propto L^{0.5}$. These indices are very similar to the values inferred from observations of CO clumps. In particular, Heithausen *et al.* (1998) (see also Kramer *et al.* 1998), probing clumps of mass as small as one Jupiter mass, therefore directly comparable with the masses of the clouds produced in the simulation, report a very similar mass spectrum and mass–size relation.

It is worth comparing these results with results obtained in supersonic isothermal turbulent flows. Ballesteros-Paredes and Mac Low (2002) (see also Vázquez-Semadeni *et al.* 1997) have extracted the clumps formed in their simulations both in physical space and by a procedure mimicking the observational one, using a clump finding algorithm. In both cases, they found a log-normal mass spectrum and an internal velocity dispersion $\sigma \propto L^{0.5}$ compatible with the observations. The mass–size relation appears to be more complex. In physical space, they found no correlation between the mean density and the size of the clumps. However, structures extracted from the observational procedure followed approximately $M \propto L^2$. They concluded that this is a projection effect that artificially connects structures along the line of sight.

It appears therefore that the two-phase model and the supersonic isothermal turbulent model lead to statistically different structure distributions. In principle, this could constitute a nice test. One should, however, keep in mind that, at this point, the formation of molecular hydrogen has not been included in the two-phase model and that structures have been extracted in the physical space. Further work is needed before definite conclusions can be reached.

9.3.6 Influence of the magnetic field

Hennebelle and Pérault (2000) consider in 1D the formation of a single structure when the incoming flow makes an angle with the magnetic field. They find that the thermal condensation is possible provided that the angle between the flow and the field is sufficiently small. In this process, the magnetic tension plays an important role in unbending the field lines and therefore reducing the magnetic pressure. As a result, the condensation occurs mainly along the field lines. Thus, the magnetic intensity does not increase with the gas density – this seems to be in good agreement with the observations (Troland & Heiles 1986). The role of the magnetic waves has been further investigated by Hennebelle and Passot (2006). They conclude that over a large range of parameters, the waves tend to trigger the formation of CNM structures rather than preventing the condensation.

Recently, the formation of cold structures in 1D via ambipolar diffusion has been investigated by Inoue *et al.* (2007). They show that the value of the final magnetic intensity within the structure is relatively independent of the value of the magnetic field within the WNM, implying a weak correlation between the density and the magnetic intensity (see also Heitsch *et al.* 2004). Such a weak correlation between the magnetic field strength and the density is a general feature of magnetohydrodynamic (MHD) turbulence simulations, both isothermal (Padoan & Nordlund 1999; Ostriker *et al.* 2001) and multi-temperature (Passot *et al.* 1995; de Avillez & Breitschwerdt 2005). In the case of an isothermal gas, the weakness of the correlation has been interpreted by Passot and Vázquez-Semadeni (2003) in terms of the fact that the various types of (nonlinear, or simple) MHD waves are characterized by different scalings of the field strength with the density. Thus, in a turbulent flow in which all kinds of waves pass through one given point, the field strength there is a function of the history of wave passages rather than of the local density.

Thus, it can be seen that, in general, a significant correlation between magnetic field strength and density is *not* expected in the atomic gas as a consequence of the ability of the gas to flow freely along field lines, except perhaps at the highest densities, which probably require focused external compressions and which in turn increase the magnetic pressure as well (Gazol *et al.* 2007). This seems to be in good agreement with the observations (Troland & Heiles 1986; Heiles & Troland 2005).

Concerning the interaction of the magnetic field with the non-linear thin-shell instability, Heitsch *et al.* (2007) have begun to investigate the role of the field in possibly suppressing the instability in the isothermal case, finding that under a variety of circumstances it may still grow. Studies in the presence of thermal bistability are still pending.

9.4 Formation of molecular clouds

9.4.1 Context

MCs are the densest regions in the ISM. Their distribution in external galaxies shows that GMCs are in general the tip of the iceberg of the gas distribution, appearing at column densities above roughly $8M_\odot$ cm^{-2} (Blitz *et al.* 2007). As the latter authors conclude, MCs seem to form out of the HI gas. Their formation thus requires significant compressions of the diffuse gas. As discussed in Section 9.2, the ultimate drivers of these compressions may be several large-scale instabilities, such as the global gravitational instability of the combined stars and gas, the magneto-Jeans instability, the magneto-rotational instability or else local disturbances such as passing supernova shocks or the general motions of the transonically turbulent warm diffuse ISM (Kulkarni & Heiles 1987; Heiles & Troland 2003).

Whatever the source of the compressions, the results described in the previous sections suggest that transonic compressions in the WNM can induce a transition to the CNM, followed by an overshoot to physical conditions typical of GMCs, which are colder and denser than the CNM clouds. This occurs because the compressed gas is in pressure balance with the total pressure of the inflowing WNM, including its ram pressure, implying that the gas at MC densities must be systematically overpressured with respect to the mean thermal pressure in the ISM (Vázquez-Semadeni *et al.* 2006), as is indeed the case (Blitz & Williams 1999). Moreover, as described in the previous sections, the formation of dense clouds by this process naturally produces at least some of the turbulence observed within them.

However, transonic compressions in the WNM alone are not sufficient to produce the extreme densities and pressures found in the interiors of MCs. For example, Vázquez-Semadeni *et al.* (2006) found that the pressure of the dense gas (there defined as gas with $n > 100$ cm^{-3}) formed by Mach 2.5 compressions lies in the range 1.5–4 times the mean ISM pressure, while mean observed MC pressures are actually closer to ten times the mean ISM pressure (Blitz & Williams 1999). This suggests that self-gravity must be crucially involved in the formation and evolution of MCs to produce the observed additional pressure enhancement.

9.4.2 Numerical results

The process of dense cloud formation out of compressions in the WNM in the presence of thermal bistability and self-gravity has been first studied by Vázquez-Semadeni *et al.* (2007) by means of numerical simulations using GADGET (Springel *et al.* 2001), which, being Lagrangian, allows very high effective resolution at very dense points. Sink particles (Jappsen *et al.* 2005) are used to avoid the need for prohibitively high resolutions while allowing for the simulation of individual or clustered star formation.

The simulations start with a converging flow immersed in a much larger box, so that the cloud can later interact freely with its environment, without artificial confinement. The box is initially filled with WNM gas at $n = 1$ cm^{-3} and $T = 5000$ K. The inflows have finite durations, in order to be able to follow the subsequent evolution of the cloud. Non-equilibrium chemistry was not implemented, but densities and temperatures are reached that allow quick formation of molecular hydrogen (see Glover & Mac Low 2007b).

The clouds formed in these simulations exhibit a secular evolutionary process that proceeds along the following lines:

1. The turbulence in the dense clouds continues to be driven for as long as the inflows last (see also Folini & Walder 2006). The turbulent velocity dispersion is maintained at a roughly constant level that depends on the inflow Mach number, as can be seen for $0 \lesssim t \lesssim 10$ Myr in the *bottom panel* of Figure 9.11.
2. The clouds *do not ever* reach an equilibrium state. As the gas transits from the warm to the cold phase, it becomes denser and colder than its surroundings, and the cloud's mass is constantly varying. During the early stages, the mass increases continuously through accretion from the inflows ($0 \leq t \lesssim 20$ Myr in the *top panel* of Figure 9.11). Later, the dense gas mass begins to decrease because of its conversion to stars.
3. At some point during the evolution, the cloud's gravitational energy becomes larger than the sum of its internal and turbulent energies; the cloud becomes gravitationally unstable and begins to contract ($t \sim 10$ Myr in Figure 9.12). This contraction begins roughly at the time when the inflows have begun to weaken and the turbulence produced by the cloud formation mechanism has begun to decay. This decay is, however, very slow, because the flows weaken slowly, and so does the rate of turbulent energy

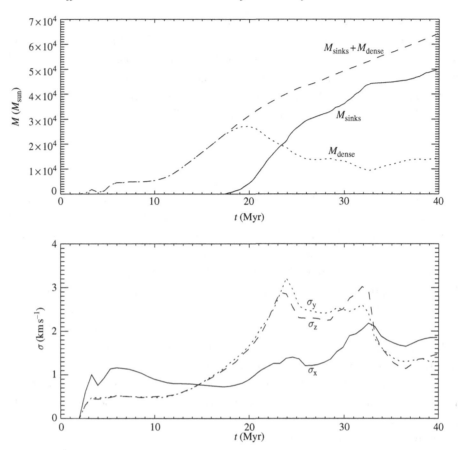

Fig. 9.11. *Top panel:* Evolution of the dense ($n > 50$ cm^{-3}, *dotted line*) gas mass, the mass in sink particles (*solid line*) and the sum of the two (*dashed line*) for a simulation of dense cloud formation with self-gravity in a 256-pc cubic box. The colliding inflows had a diameter of 64 pc and a length of 112 pc each. *Bottom panel:* Evolution of the velocity dispersion along the direction of the colliding inflows (x) and the two directions perpendicular to it (y and z). (Adapted from Vázquez-Semadeni *et al.* 2007.)

injection. The turbulence in the clouds is thus in a state intermediate between being continuously driven and absolutely decaying.

4. While the clouds are contracting, they exhibit a near-equipartition energy balance satisfying $|E_g| \approx 2E_k$, which *appears as* virial equilibrium. However, in this case, the energy balance is a signature of gravitational contraction rather than of virial equilibrium, contrary to standard notions about MCs (Blitz & Williams 1999), and the velocity dispersion of the cloud contains a dominant fraction of infall motion rather than random turbulence.

5. Stars begin forming after the cloud has already been contracting for a long time (presumably still in atomic form). Specifically, the cloud begins contracting at $t \sim 10$ Myr,

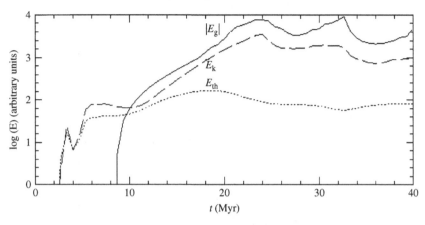

Fig. 9.12. Evolution of the thermal (*dotted line*), kinetic (*dashed line*) and absolute value of the gravitational (*solid line*) for the cloud formed in the same
simulation as in Figure 9.11. (Adapted from Vázquez-Semadeni *et al.* 2007.)

while the first stars begin to appear at $t \sim 17$ Myr. In this simulation, however, they form
at local density fluctuations produced by the initial turbulence (in turn produced by the
flow collision) long before the global collapse is completed.

6. The SFE, i.e. the mass converted from gas into stars after some characteristic time,
appears to be too large, with 15% of the mass having already been converted into stars
3 Myr after star formation starts ($t \sim 20$ Myr) and roughly three times more mass in stars
than in dense gas by $t \sim 40$ Myr. Instead, observational estimates of the SFE range from
a few percent for whole MC complexes (Myers *et al.* 1986) up to 30–50% for cluster-
forming cores (Lada & Lada 2003). Thus, the control of the SFE in the scenario of
these simulations must probably rely on magnetic support of the clouds (Mouschovias
1978; Shu *et al.* 1987, Elmegreen 2007) or energy feedback from the stellar products,
as suggested by various authors (Norman & Silk 1980; Franco *et al.* 1994; Matzner &
McKee 2000; Hartmann *et al.* 2001; Krumholz *et al.* 2006; Nakamura & Li 2007).

9.4.3 Formation of molecular hydrogen

One argument often advanced in favour of cloud lifetimes longer than 10 Myr is the
apparent difficulty involved in producing sufficient H_2 in only 1–2 Myr to explain
observed clouds, given the relatively slow rate at which H_2 forms in the ISM. The
H_2 formation timescale in the ISM is approximately (Hollenbach *et al.* 1971)

$$t_{form} \simeq \frac{10^9 \text{ yr}}{n}, \tag{9.3}$$

where n is the number density in cm^{-3}, which suggests that in gas with a mean
number density $\bar{n} \sim 100$ cm^{-3}, characteristic of most GMCs (Blitz & Shu 1980),

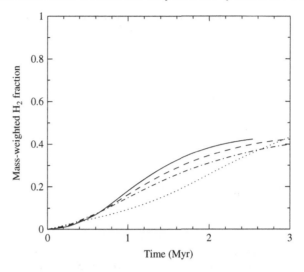

Fig. 9.13. Time evolution of the mass-weighted average H_2 fraction in several runs with supersonically turbulent initial conditions from Glover and Mac Low (2007b; their Figure 1). Runs using a local shielding approximation with 64^3 (*dotted line*), 128^3 (*dash-dotted line*), 256^3 (*dashed line*), and 512^3 (*solid line*) zones are shown. All runs shown here use the local shielding approximation, which works quite well for turbulent initial conditions. Significant molecule formation occurs in 2 Myr, with some regions having already become fully molecular.

conversion from atomic to molecular form should take at least 10 Myr, longer than the entire lifetime of a transient cloud. However, estimates of this kind do not take account of dynamical processes such as supersonic turbulence or thermal instability.

Glover and Mac Low (2007a, b) used the ZEUS-MP MHD code (Norman 2000) modified to include a simplified chemical network to follow the non-equilibrium abundance of molecular hydrogen to study molecule formation in a turbulent, self-gravitating flow with and without magnetic fields. They found that initially uniform gas does indeed take tens of megayears to form molecules, but that initial supersonic turbulence produces density enhancements that allow regions to become fully molecular within 3 Myr (Figure 9.13), consistent with short timescales for MC formation.

One crucial issue is the possibility that MCs may actually contain sizable amounts of atomic gas interspersed within the molecular phase. Certainly, this is the impression that one gathers from density fields like that shown in Figure 9.7, where the dense gas appears to have a fractal structure, with substructure observed essentially down to the resolution limit. Observations appear to show atomic gas intermixed with the molecular gas (Williams *et al.* 1995; Li & Goldsmith 2003). Hennebelle and Inutsuka (2006) have suggested that some of this gas may be warm and proposed that it could be heated by dissipation of MHD waves. This issue remains open, but, if confirmed, it could have a deep impact on the dynamics and structure of MCs. Indeed, since the filling factor in MC is generally not greater than ~10%, the density of the interclump medium must be smaller than a few times $\simeq 10 \, \text{cm}^{-3}$. Otherwise, the mass in the interclump medium would be comparable to or greater than the mass within the clumps (Williams *et al.* (1995) estimate that the interclump particle density is lower than $10 \, \text{cm}^{-3}$). Thus, the assumption that the interclump medium has the same temperature than the gas inside the clumps leads to a thermal pressure much lower than the thermal pressure of the standard ISM.

9.4.4 Discussion and implications

The MC formation simulations discussed here are consistent with the scenario of Hartmann *et al.* (2001): the formation of a GMC involves accumulation of the gas from distances of up to hundreds of parsecs and takes up to 10–20 Myr to complete. However, most of this process occurs in the atomic phase, with molecular gas only forming late in the evolution, when a sufficiently large column density of dense gas has been collected (see also Franco & Cox 1986). The formation of molecular gas occurs roughly simultaneously with the onset of star formation for solar-neighbourhood conditions.

If this scenario proves correct, it has two important implications. First, the mechanism of cloud formation by accumulation and condensation implies, in the presence of magnetic fields, that the mass-to-flux ratio of the dense gas must *increase* with time. This is consistent with the observation that diffuse CNM clouds are generally magnetically subcritical (Heiles & Troland 2005), while MCs (which in this scenario begin their evolution as diffuse CNM clouds; see Vázquez-Semadeni *et al.* 2006) are generally magnetically critical or supercritical (Bourke *et al.* 2001; Crutcher 2004). Hartmann *et al.* (2001) suggest that the clouds in the solar neighbourhood should become supercritical roughly simultaneously with them becoming molecular. The precise timing of the transition from subcritical to supercritical, in relation to the time of becoming molecular and the onset of star formation, may be crucial in determining the SFE. Moreover, the global cloud contraction observed in the non-magnetic simulations may not take over until the clouds become supercritical in the magnetic case.

Second, the domination of the evolution by global gravitational contraction suggests a return to the scenario, originally proposed by Goldreich and Kwan (1974), of global gravitational contraction for MCs. This proposal was quickly dismissed by Zuckerman and Palmer (1974) through the argument that if all MCs converted all their mass into stars in roughly one free-fall time, the resulting star formation rate would be at least ten times larger than that presently observed in the galaxy. Zuckerman and Evans (1974) then proposed that the supersonic linewidths in the clouds are produced primarily by local motions (the hypothesis of microturbulence), which has been widely accepted until recently.

However, several considerations argue against the microturbulent picture. Turbulence is a regime of fluid flow characterized by having the largest velocities at the driving scale, as indicated by the negative slope of the turbulent energy spectrum in both the incompressible (Kolmogorov 1941) and compressible (Passot & Pouquet 1987; Kritsuk *et al.* 2007) cases. Comparisons of observations with numerical simulations in various contexts show that the dominant motions (and thus the driving scales) occur at scales the size of the whole cloud or larger (Ossenkopf & Mac Low 2002; Heyer & Brunt 2007). This contradicts the microturbulent picture, which requires small-scale driving.

Furthermore, evidence is beginning to accumulate that MCs or their clumps may, in fact, be gravitationally collapsing. This has been claimed for the Orion MC by Hartmann and Burkert (2007) and for NGC 2264 by Peretto *et al.* (2007). If gravitational contraction turns out to be a general feature of MCs, then the Goldreich and Kwan (1974; see also Field *et al.* 2008) suggestion may turn out to be correct after all, at least for a subset of the clouds or for the star-forming regions of MCs (a large fraction of the volume of an MC is devoid of star formation; e.g., Krumholz *et al.* 2006; Elmegreen 2007). In this case, the regulation of the SFE may be accomplished by the combined effects of magnetic support, of the turbulence produced in the clouds during their formation and of the dispersive action of stellar feedback.

The initial turbulence in the cloud caused by its formation produces non-linear density fluctuations that can collapse before the global cloud collapse is completed. The feedback from these first star formation events may be able to suppress, or at least reduce, subsequent events before all the gas of the cloud is turned into stars (Franco *et al.* 1994). Moreover, in the presence of magnetic field fluctuations, parts of the clouds may remain magnetically subcritical and thus supported against collapse, while other parts may become supercritical as they incorporate material from the surrounding WNM and go into collapse. In this case, the SFE in the locally supercritical regions may be large, while the global average over whole GMCs may be low, because most of their mass remains subcritical. The feedback from the active star-forming regions may then shred the clouds and leave subcritical fragments that may collapse later or even disperse away (Elmegreen 2007).

However, it is presently a matter of strong debate whether regulation of the SFE by stellar energy input occurs by dispersal of the star-forming clumps or by quasi-equilibrium support of the clouds. The recent study by Nakamura and Li (2007) suggests that near-equilibrium can be achieved between driving by stellar outflows and the self-gravity, but this study has a deep potential well and furthermore uses periodic boundary conditions, so that the cloud cannot be dispersed. Observationally, clusters older than several million years are generally observed to be devoid of gas, suggesting that they have been able to disperse (or consume) their parent cloud (Leisawitz *et al.* 1989; Hartmann *et al.* 2001).

These issues will hopefully be resolved in the near future both through observations aimed at distinguishing these two scenarios and by numerical simulations of the entire evolution of MCs, from formation to dispersal, including the feedback from stellar sources and magnetic fields.

References

Audit, E., & Hennebelle, P. 2005, *A&A* **433**, 1

Audit, E., & Hennebelle, P. 2008, in preparation

Ballesteros-Paredes, J., Hartmann, L., & Vázquez-Semadeni, E. 1999a, *ApJ.* **527**, 285

Ballesteros-Paredes, J., Vázquez-Semadeni, E., & Scalo, J. 1999b, *ApJ.* **515**, 286

Ballesteros-Paredes, J., & Mac Low, M.-M. 2002, *ApJ.* **570**, 734

Blitz, L., Fukui, Y., Kawamura, A., Leroy, A., Mizuno, N., & Rosolowsky, E. 2007, in *Protostars and Planets V*, eds. B. Reipurth, D. Jewitt, & K. Keil (Tucson: University of Arizona Press) 81

Blitz, L., & Shu, F. H. 1980, *ApJ.* **238**, 148

Blitz, L., & Williams, J. P. 1999, in *Physics of Star Formation and Early Stellar Evolution II*, eds. N. D. Kylafis, & C. J. Lada (Dordrecht: Kluwer)

Bourke, T. L., Myers, P. C., Robinson, G., & Hyland, A. R. 2001, *ApJ.* **554**, 916

Crutcher, R. M. 2004, *Ap&SS* **292**, 225

Dalgarno, A., & McCray, R. A. 1972, *ARA&A* **10**, 375

de Avillez, M., & Breitschwerdt, D. 2005, *A&A* **436**, 585

Dickey, J., & Lockman, F. 1990, *ARA&A* **28**, 215

Elmegreen, B. G. 1993, *ApJ.* **419**, L29

Elmegreen, B. G. 2007, *ApJ.* **668**, 1064

Elmegreen, B. G., & Palouš, J. (editors), 2007, *Triggered Star Formation in a Turbulent ISM* (Cambridge: Cambridge University Press)

Field, G. 1965, *ApJ.* **142**, 531

Field, G. B., Blackman, E. G., & Keto, E. R. 2008, *MNRAS* **385**, 181

Field, G., Goldsmith, D., & Habing, H. 1969, *ApJ. Lett.* **155**, 149

Folini, D., & Walder, R. 2006, *A&A* **459**, 1

Franco, J., & Cox, D. P. 1986, *PASP* **98**, 1076

Franco, J., Shore, S. N., & Tenorio-Tagle, G. 1994, *ApJ.* **436**, 795

Gammie, C. F. 1992, *The Formation of Giant Molecular Clouds.* PhD Thesis, Princeton University

Gazol, A., Vázquez-Semadeni, E., & Kim, J. 2005, *ApJ.* **630**, 911

Gazol, A., Vázquez-Semadeni, E., Sánchez-Salcedo, F., & Scalo, J. 2001, *ApJ.* **557**, L124

Gazol, A., Kim, J., Vázquez-Semadeni, E., & Luis, L. 2007, in *SINS – Small Ionized and Neutral Structures in the Diffuse Interstellar Medium*, eds. M. Haverkorn, & W. M. Goss (San Francisco: Astronomical Society of the Pacific) 154

Glover, S. C. O. G., & Mac Low, M.-M. 2007a, *ApJ. Supp.* **169**, 239

Glover, S. C. O. G., & Mac Low, M.-M. 2007b, *ApJ.* **659**, 1317

Goldreich, P., & Kwan, J. 1974, *ApJ.* **189**, 441

Goldreich, P., & Lynden-Bell, D. 1965, *MNRAS* **130**, 97

Hartmann, J., Ballesteros-Paredes, J., & Bergin, E. 2001, *ApJ*. **562**, 852

Hartmann, L., & Burkert, A. 2007, *ApJ*. **654**, 988

Heiles, C. 1997, *ApJ*. **481**, 193

Heiles, C. 2001, *ApJ*. **551**, L105

Heiles, C., & Troland, T. 2003, *ApJ*. **586**, 1067

Heiles, C., & Troland, T. 2005, *ApJ*. **624**, 773

Heithausen, A., Bensch, F., Stutzki, J., Falgarone, E., & Panis, J.-F. 1998, *A&A* **331**, L65

Heitsch, F., Burkert, A., Hartmann, L., Slyz, A., & Devriendt, J. 2005, *ApJ*. **633**, 113

Heitsch, F., Slyz, A., Devriendt, J., Hartmann, L., & Burkert, A. 2006, *ApJ*. **648**, 1052

Heitsch, F., Slyz, A., Devriendt, J., Hartmann, L., & Burkert, A. 2007, *ApJ*. **665**, 445

Heitsch, F., Zweibel, E., Slyz, A., & Devriendt, J. 2004, *ApJ*. **603**, 165

Hennebelle, P., & Audit, E. 2007, *A&A* **465**, 431

Hennebelle, P., Audit, E., & Miville-Deschênes, M.-A. 2007, *A&A* **465**, 445

Hennebelle, P., & Inutsuka, S.-I. 2006, *ApJ*. **647**, 404

Hennebelle, P., & Passot, T. 2006, *A&A* **448**, 1083

Hennebelle, P., & Pérault, M. 1999, *A&A* **351**, 309

Hennebelle, P., & Pérault, M. 2000, *A&A* **359**, 1124

Heyer, M., & Brunt, C. 2007, in *Triggered Star Formation in a Turbulent ISM*, eds. B. G. Elmegreen, & J. Palouš (Cambridge: Cambridge University Press) 9

Hollenbach, D. J., Werner, M. W., & Salpeter, E. E. 1971, *ApJ*. **163**, 165

Inoue, T., Inutsuka, S.-I., & Koyama, H. 2006, *ApJ*. **652**, 1331

Inoue, T., Inutsuka, S.-I., & Koyama, H. 2007, *ApJ*. **658**, L99

Jappsen, A.-K., Klessen, R. S., Larson, R. B., Li, Y., & Mac Low, M.-M. 2005, *A&A* **435**, 611

Joung, M. K. R., & Mac Low, M.-M. 2006, *ApJ*. **653**, 1266

Kennicutt, R. C., Jr. 1998, *ApJ*. **498**, 541

Kim, J., Franco, J., Hong, S. S., Santillán, A., & Martos, M. A. 2000, *ApJ*. **531**, 873

Kim, W.-T., & Ostriker, E. 2006, *ApJ*. **646**, 213

Kramer, C., Stutzki, J., Rohrig, R., & Corneliussen, U. 1998, *A&A* **329**, 249

Kravtsov, A. V. 2003, *ApJ*. **590**, L1

Kolmogorov, A. 1941, *Dokl. Akad. Nauk SSSR* **30**, 301

Koyama, H., & Inutsuka, S. 2000, *ApJ*. **532**, 980

Koyama, H., & Inutsuka, S. 2002, *ApJ*. **564**, L97

Koyama, H., & Inutsuka, S. 2006, *ApJ*., submitted (astro-ph/0605528)

Kritsuk, A. G., & Norman, M. L. 2002, *ApJ*. **569**, L127

Kritsuk, A. G., Norman, M. L., Padoan, P., & Wagner, R. 2007, *ApJ*. **665**, 416

Krumholz, M. R., Matzner, C. D., & McKee, C. F. 2006, *ApJ*. **653**, 361

Kulkarni, S. R., & Heiles, C. 1987, in *Interstellar Processes*, eds. D. J. Hollenbach, & H. Thronson (New York: Springer) 87

Lada, C. J., & Lada, E. A. 2003, *ARA&A* **41**, 57

Leisawitz, D., Bash, F. N., & Thaddeus, P. 1989, *ApJS* **70**, 731

Li, D., & Goldsmith, P. F. 2003, *ApJ*. **585**, 823

Li, Y., Mac Low, M.-M., & Klessen, R. S. 2005, *ApJ*. **626**, 823

Li, Y., Mac Low, M.-M., & Klessen, R. S. 2006, *ApJ*. **639**, 879

Mac Low, M.-M., & Klessen, R. S. 2004, *Rev. Mod. Phys.* **76**, 125

Matzner, C. D., & McKee, C. F. 2000, *ApJ*. **545**, 364

Mizuno, N., Kawamura, A., Onishi, T., Mizuno, A., & Fukui, Y., 2007, in *Triggered Star Formation in a Turbulent ISM*, eds. B. G. Elmegreen, & J. Palouš (Cambridge: Cambridge University Press) 128

Mouschovias, T. 1978, in *Protostars & Planets*, ed. T. Gehrels (Tucson: University of Arizona) 209

Myers, P. C., Dame, T. M., Thaddeus, P., Cohen, R. S., Silverberg, R. F., Dwek, E., & Hauser, M. G. 1986, *ApJ.* **301**, 398

Nakamura, F., & Li, L.-Y. 2007, *ApJ.* **656**, 721

Norman, C., & Silk, J. 1980, *ApJ.* **238**, 158

Norman, M. L. 2000, *Rev. Mex. Astron. Astrof. Ser. Conf.* **9**, 66

Ossenkopf, V., & Mac Low, M.-M. 2002, *A&A* **390**, 307

Ostriker, E. C., Stone, J. M., & Gammie, C. F. 2001, *ApJ.* **546**, 980

Padoan, P., Nordlund, Å. 1999, *ApJ.* **526**, 279

Passot, T., & Pouquet, A. 1987, *J. Fluid Mech.* **181**, 441

Passot, T., & Vázquez-Semadeni, E. 2003, *A&A* **398**, 845

Passot, T., Vázquez-Semadeni, E. & Pouquet, A., 1995, *ApJ.* **455**, 536

Peretto, N., Hennebelle, P., & André, P. 2007, *A&A* **464**, 983

Pittard, J. M., Dobson, M. S., Durisen, R. H., Dyson, J. E., Hartquist, T. W., & O'Brien, J. T. 2005, *A&A* **438**, 11

Rafikov, R. R. 2001, *MNRAS* **323**, 445

Roberts, W. W. 1969, *ApJ.* **158**, 123

Sánchez-Salcedo, F. J., Vázquez-Semadeni, E., & Gazol, A. 2002, *ApJ.* **577**, 768

Sántillan, A., Kim, J., Franco, J., Martos, M., Hong, S. S., & Ryu, D. 2000, *ApJ.* **545**, 353

Sasao, T. 1973, *PASJ* **25**, 1

Shu, F., Adams, F., & Lizano, S. 1987, *ARA&A* **25**, 23

Springel, V., Yoshida, N., White, S. D. M. 2001, *New Astron.* **6**, 79

Troland, T., & Heiles, C. 1986, *ApJ.* **301**, 339

Vázquez-Semadeni, E., Ballesteros-Paredes, J., & Rodriguez, L. 1997, *ApJ.* **474**, 292

Vázquez-Semadeni, E., Gazol, A., Passot, T., & Sánchez-Salcedo, F.J. 2003, in *Turbulence and Magnetic Fields in Astrophysics*, eds. E. Falgarone, & T. Passot (Dordrecht: Springer)

Vázquez-Semadeni, E., Gómez, G. C., Jappsen, A. K., Ballesteros-Paredes, J., González, R. F., & Klessen, R. S. 2007, *ApJ.* **657**, 870

Vázquez-Semadeni, E., Passot, T., & Pouquet, A. 1995, *ApJ.* **441**, 702

Vázquez-Semadeni, E., Ryu, D., Passot, T., González, R., & Gazol, A. 2006, *ApJ.* **643**, 245

Vishniac, E. T. 1994, *ApJ.* **428**, 186

Walder, R., & Folini, D. 1998, *A&A* **330**, 21L

Walder, R., & Folini, D. 2000, *ApSS* **274**, 343

Williams, J., Blitz, L., & Stark, A. 1995, *ApJ.* **451**, 252

Wolfire, M.G., Hollenbach, D., & McKee, C.F. 1995, *ApJ.* **443**, 152

Wolfire, M.G., Hollenbach, D., & McKee, C.F. 2003, *ApJ.* **587**, 278

Yang, C.-C., Gruendl, R. A., Chu, Y.-H., Mac Low, M.-M., & Fukui, Y. 2007, *ApJ.* **671**, 374Y

Zuckerman, B., & Evans, N. J. 1974, *ApJ.* **192**, L149

Zuckerman, B., & Palmer, P. 1974, *ARA&A* **12**, 279

10

The formation of distributed and clustered stars in molecular clouds

S. T. Megeath, Z.-Y. Li and Å. Nordlund

10.1 Introduction

During the last two decades, the focus of star formation research has shifted from understanding the collapse of a single dense core into a star to studying the formation of hundreds to thousands of stars in molecular clouds. In this chapter, we overview recent observational and theoretical progress towards understanding star formation on the scale of molecular clouds and complexes, i.e. the macrophysics of star formation (McKee & Ostriker 2007). We begin with an overview of recent surveys of young stellar objects (YSOs) in molecular clouds and embedded clusters, and we outline an emerging picture of cluster formation. We then discuss the role of turbulence to both support clouds and create dense, gravitationally unstable structures, with an emphasis on the role of magnetic fields (in the case of distributed stars), and feedback (in the case of clusters) to slow turbulent decay and mediate the rate and density of star formation. The discussion is followed by an overview of how gravity and turbulence may produce observed scaling laws for the properties of molecular clouds, stars and star clusters and how the observed, star formation rate (SFR) may result from self-regulated star formation. We end with some concluding remarks, including a number of questions to be addressed by future observations and simulations.

10.2 Observations of clustered and distributed populations in molecular clouds

Our knowledge of the distribution and kinematics of young stars, protostars and dense cores in molecular clouds is being rapidly improved by wide-field observations at X-ray, optical, infrared and (sub)millimeter wavelengths (Allen *et al.* 2007; Feigelson *et al.* 2007). In this section, we present multiwavelength portraits of some of the most active star-forming regions near our Sun and discuss some implications of these results.

Structure Formation in Astrophysics. ed. G. Chabrier. Published by Cambridge University Press.
© Cambridge University Press 2009.

10.2.1 Molecular cloud surveys

The Spitzer space telescope, with its unparalleled ability to detect IR excesses from young stellar objects (YSOs) with disk and envelopes over wide fields, is now producing the most complete censuses to date of YSOs in nearby molecular clouds. One example is the survey of the Orion A cloud, the most active star-forming cloud within 450 pc of the Sun (Megeath *et al.* 2008). The observed distribution of YSOs in this cloud exhibits structure on a range of spatial scales, tracing both the overall filamentary morphology of the clouds and smaller filamentary and clumpy sub-structures within the clouds, as shown in Figure 10.1. This figure also shows a range of observed stellar densities; YSOs can be found in large clusters containing OB stars, in smaller groups and in relative isolation.

Is there a preferred environment for star formation in giant molecular clouds (GMCs)? Do most low-mass stars form in rare, large clusters, in the more numerous small groups or in a distributed population of more isolated stars? To some

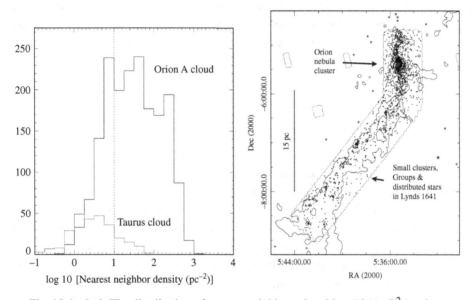

Fig. 10.1. *Left*: The distribution of nearest neighbour densities ($10/(\pi R_{10}^2)$, where R_{10} is the distance to the 10th nearest neighbour) for the Orion A GMC and the Taurus dark cloud complex (K. Luhman, personal communication). The *dotted line* gives the adopted threshold density for separating distributed stars from stars in groups and clusters. *Right*: The distribution of Spitzer-identified YSOs with infrared excesses in the Orion molecular clouds. The *grey line* gives the outline of the survey fields, including the large hockey stick-shaped field and three reference fields. The *black lines* are the $A_V = 3$ contour determined from an extinction map constructed from the 2MASS point source catalogue (R. Gutermuth, personal communication). The star symbols give the positions of OB stars from Brown *et al.* (1994).

extent, this depends on the criteria used to distinguish isolated and clustered stars; as shown in Orion A, the clusters are often not distinct objects but belong to an extended distribution of stars (Figure 10.1). One approach is to adopt a critical surface density for clustered stars, although there is no clear motivation for a particular threshold density as demonstrated by the continuous distribution of YSO surface densities shown in Figure 10.1. In the Orion cloud complex (Orions A and B), Megeath *et al.* (in preparation) decomposed the observed distribution of YSOs with IR excesses into groups of ten or more sources, where each member is in a contiguous region with a local surface density of 10 stars per pc^{-2} or higher. They find that 44% are in the 1000-member Orion Nebula Cluster (ONC), 18% are in 7 clusters with 30–100 YSOs and 9% are in 12 groups of 10–30 YSOs. They find that 28% of the YSOs with IR excesses (639 objects in total) form a distributed population outside the groups and clusters (also see Allen *et al.* 2007). The number of YSOs in the ONC is underestimated due to incompleteness, and consequently, the actual fraction of stars in the distributed population is somewhat lower (Megeath *et al.* in preparation). Despite this bias, the surveys are showing that even in GMCs forming massive stars, a substantial number of YSOs are found in relative isolation. The observed range of densities and environments must be explained by models of cloud fragmentation and evolution.

10.2.2 A gallery of embedded clusters

In Figures 10.2–10.5, we show the latest compilations of YSOs in four of the nearest embedded (or partially embedded) clusters to the Sun. The most striking result is the diversity of configurations apparent in the clusters. The dense gas in the Serpens cluster is concentrated in a 0.5-pc long filament divided into two main clumps; the protostars are primarily found in the two clumps. In contrast, the pre-main sequence stars are more evenly distributed over a 1-pc diameter region with only a modest concentration in the central region; this may be the result of dynamical evolution due to gravitational interactions between YSOs and to the non-spherically symmetric cloud potential. The dense gas in NGC 1333 forms a network of clumps and filaments spread over a region more than 1 pc in diameter. Both the protostars and pre-main sequence stars follow this structure, with the protostars concentrated in the dense gas and pre-main sequence stars found in the immediate vicinity of the dense gas; this implies little dynamical evolution in this region. Unlike the previous two clusters, the IC348 cluster shows a circularly symmetric and centrally condensed distribution. The molecular gas has been largely cleared from this region; the protostars are concentrated in a filamentary gas structure on the edge of the cluster.

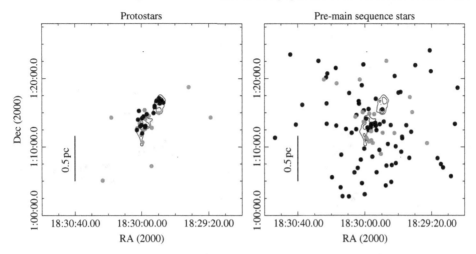

Fig. 10.2. The distribution of YSOs in the Serpens cloud core cluster at a distance of 260 pc (Winston *et al.* 2007). The contours show the dense gas detected in an 850 μm SCUBA map (Davis *et al.* 2000). The *left panel* shows the distribution of protostars separated into flat spectrum (grey markers) and Class I/0 (*black markers*) sources. The *right panel* shows the pre-main sequence stars with disks (*black markers*) and without disks (*grey markers*). The pre-main sequence stars without disks were identified by their elevated X-ray emission (Giardino *et al.* 2007); see Winston *et al.* (2007) for a discussion of the completeness of the displayed source selection. Note that the SCUBA map and Chandra images do not cover the entire displayed region.

The massive ONC shows a particularly complex morphology (Figures 10.1 and 10.5). In the centre of the ONC is the densest known region of young stars in the nearest 500 pc. This central condensation is elongated and aligned with the axis of the filamentary molecular cloud. To the north of the central condensation, a filamentary distribution of protostars follows the molecular cloud. It is surrounded by a more extended distribution of pre-main sequence stars. To the south, the molecular cloud appears to have been partially swept up into a shell by the massive stars in the centre of the ONC. A more extended distribution of YSOs in this region fills the shell, indicating that these stars probably originated in the expanding shell.

10.2.3 The distribution of protostars in embedded clusters

The distribution and spacing of protostars is an important constraint on the physics of fragmentation and the potential for subsequent interactions between protostars. As shown previously, the protostars trace the clumpy and filamentary distribution of the dense molecular gas. The protostars in embedded clusters are often closely packed, with median-projected nearest neighbour spacings ranging from 5000 to

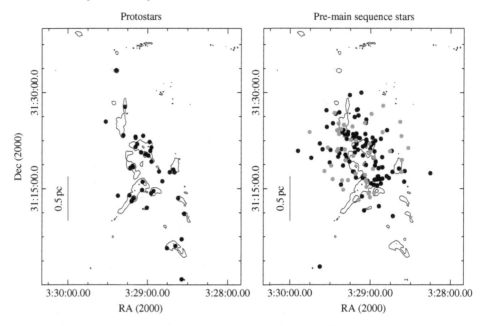

Fig. 10.3. The distribution of YSOs in the NGC 1333 cluster at a distance of 250 pc. The contours show the dense gas traced in the 850 μm SCUBA map (Sandell & Knee 2001). The *left panel* shows the distribution of protostars (Gutermuth *et al.* 2008); the *right panel* shows the distribution of pre-main sequence stars with disks (*black markers*; Gutermuth *et al.* 2008) and diskless pre-main sequence stars with elevated X-ray emission (*grey markers*; Getman *et al.* 2002). Note that the Chandra survey of Getman *et al.* does not cover the entire displayed field.

20 000 AU (Muench *et al.* 2007; Winston *et al.* 2007). In a few cases, peaks are found in the distribution of projected nearest neighbour distances for protostars (Gutermuth *et al.* 2008; Teixeira *et al.* 2006) and for dense cores (Stanke *et al.* 2006). These preferred separations for the protostars are within a factor of three of the local Jeans length, suggesting that the protostars may result from gravitational fragmentation (Teixeira *et al.* 2006). The detection of extremely dense groups of protostars with ∼7 objects in regions ∼10 000 AU diameter suggests that hierarchical fragmentation operates in some cases (Teixeira *et al.* 2007; Winston *et al.* 2007).

Can the protostars in the observed dense groups interact? In the Serpens region, the close spacing of protostars implies that the volumes from which these protostars accrete are often densely packed, if not overlapping (Winston *et al.* 2007). In a study of a dense group of protostars in Serpens, Winston *et al.* (2007) argued that the high gas density (Olmi & Testi 2002) and the subvirial RMS velocity of the protostars (as measured in the BIMA N_2H^+ detections of the protostellar cores;

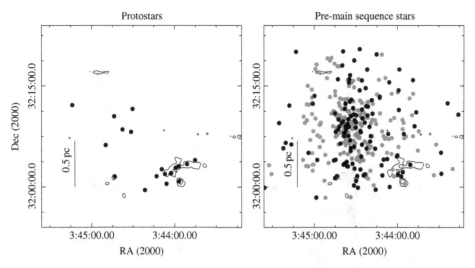

Fig. 10.4. The distribution of YSOs in the IC 348 Cluster. The contours show the dense gas as traced by the 850 μm SCUBA map (J. DiFrancesco, personal communication). The *left panel* shows the distribution of protostars (*black markers*); the *right panel* shows the distribution of pre-main sequence stars with (*black markers*) and without (grey markers) disks. The sample was taken from Muench *et al.* (2007).

Williams and Myers 2000) are consistent with the protostars accreting a significant portion of their mass through Bondi–Hoyle accretion. Bondi–Hoyle accretion in such a closely packed group may lead to the competitive accretion predicted in some numerical models of turbulent clouds (Bonnell & Bate 2006). Furthermore, if such groups stay bound for 400 000 years (the estimated duration of the protostellar phase Hatchell *et al.* 2007), each large (1000 AU) protostellar envelope would typically experience one collision. Further studies are needed to find direct evidence for interactions and to ascertain the fraction of stars that form in dense groups where the potential for interactions exists.

10.2.4 The evolution of embedded clusters: an observational perspective

A common trait of embedded clusters is a lack of circular symmetry: they are often elongated and in many cases show substructure (Gutermuth *et al.* 2005; Allen *et al.* 2007). This indicates that clusters retain the imprint of the filamentary, clumpy structure of the molecular clouds from which they formed; consequently, they do not appear to be relaxed, virialized clusters of stars. This is supported by observations of the kinematics of protostellar cores, which show that their velocity dispersions are subvirial (from observation of the NGC1333 and NGC2264 clusters; Peretto *et al.* 2007; Walsh *et al.* 2007). In the Orion Nebula, Furesz *et al.* (2008)

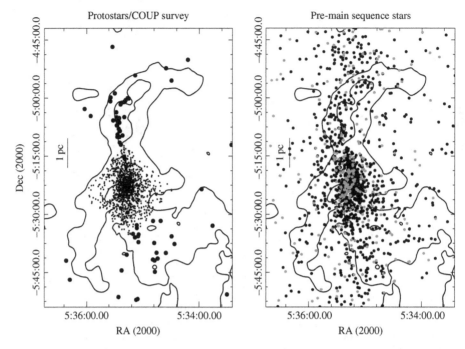

Fig. 10.5. The distribution of YSOs in the ONC. The contours show an extinction map constructed from the 2MASS PSC (R. Gutermuth, personal communication). *Left panel*: the large *black circles* are protostars identified in the Spitzer data (*Megeath et al.* in preparation); no protostars are identified in the centre of the Orion nebula where the 24-μm data is saturated. The small *black dots* are the detected X-ray sources from the Chandra Orion Ultradeep Project (Feigelson *et al.* 2005); this survey covered a diamond-shaped region defined by the extent of the X-ray sources. *Right panel*: the distribution of Spitzer-identified pre-main sequence stars with disks (*black markers*; Megeath *et al.* in preparation) and pre-main sequence stars without disks (*grey markers*) identified by variability (Carpenter *et al.* 2001).

found that the velocities of the pre-main sequence stars largely follow the overall velocity gradient in the gas. This shows that the more evolved pre-main sequence stars also retain the primordial velocity structure of their parental cloud.

Observations of 3–5 Myr clusters show that the parental gas is largely dispersed (see discussion of examples in Allen *et al.* 2007); at this time, the cluster may become unbound since the majority of the binding mass has been ejected from the cluster (Lada *et al.* 1984; Kroupa *et al.* 2001). The question is whether this timescale allows for the cluster to relax. Adopting radii of 0.5 pc for the clusters (encompassing the dense central regions of the clusters) and using the estimated number of stars (420, 156, 182 and 1000 stars for IC 348, NGC 1333, Serpens and ONC), we estimate relaxation times of 2.4, 1.8, 1.8 and 6 Myr, respectively. The actual relaxation times will be longer since the mass in the clusters is dominated by the gas (Adams & Myers 2001). This suggests that the relaxation time is approximately equal to or less than the gas dispersal time. Star formation continues in the residual cloud as the gas is being dispersed, as shown in IC 348 and the ONC (Figures 10.4 and 10.5). Muench *et al.* (2007) estimate that the current SFR in IC 348 is consistent with a constant SFR over the entire lifetime of the cluster. Thus, the timescales for star formation, gas dispersal and relaxation are similar (cf. also Allen *et al.* 2007).

A picture is emerging where clusters form in filamentary, clumpy clouds. During the first few million years of their evolution, clusters form stars over regions ~1 pc in diameter. During this time, the surrounding gas is being dispersed by winds and radiation, eventually leading to the cessation of star formation in 3–5 Myr. Although the stars may form with a low-velocity dispersion, at least relative to local structures in the gas, large scale collapse and velocity gradients in clouds may stir up the cluster (Furesz *et al.* 2008; Peretto *et al.* 2007). In extreme cases, this may lead to violent relaxation, as suggested for the ONC by Feigelson *et al.* (2005). Dynamical interactions between stars and between the stars and the cloud potential will scatter stars in random directions; however, given that the cloud mass (which typically embodies three-fourth of the total mass) is being ejected on a timescale similar to the relaxation time, it is unlikely that most embedded clusters will be able to achieve a relaxed equilibrium. One counter example may be IC 348, where the circularly symmetric, centrally condensed configuration of this cluster suggests that it is relaxed (Muench *et al.* 2007); however, circular symmetry could also result from dynamical expansion (Gutermuth *et al.* 2005). N-body simulations are needed to fully understand what the observed cluster morphologies imply about the dynamic state of the cluster.

10.3 Local theory of distributed and clustered star formation

Stars form in turbulent, magnetized clouds. The relative importance of magnetic fields and turbulence in controlling star formation is a matter of debate. Early quantitative models have concentrated on the formation and evolution of individual (quiescent, low-mass) cores and the role of magnetic fields (Shu *et al.* 1987; Mouschovias & Ciolek 1999). More recent studies have concentrated on the role of turbulence in cloud dynamics and core formation, as reviewed in, e.g., Mac Low and Klessen (2004), Elmegreen and Scalo (2004), Ballesteros-Paredes *et al.* (2007) and McKee and Ostriker (2007). Ultimately, the debate can only be settled by direct measurement of the flux-to-mass ratio for the bulk of the molecular gas, which is currently not available (Crutcher 1999). In the absence of such measurements, we have to rely on indirect evidence and theoretical arguments.

McKee (1999) argued that self-gravitating GMCs should be magnetically super-critical by a factor of $\sim$2 as a whole if turbulence and ordered magnetic fields provide comparable cloud support (Novak *et al.* 2007). Even in such a globally (moderately) supercritical GMC, the magnetic field may still play a key role in regulating star formation, because the formation of a typical star or of small stellar groups involves only a small sub-piece of the cloud that contains a few to tens of solar masses. Such a region can be locally subcritical, with a mass less than the magnetic critical mass, which can be hundreds of solar masses for typical parameters (McKee 1999), unless (i) a large fraction (perhaps $\sim$1/2 or more) of the material along a flux tube has collected in the region or (ii) the region is part of a substructure (i.e. a dense clump) that is already highly supercritical to begin with. In the former case, it would take a long time (perhaps more than 10 Myr) for the required mass accumulation to occur in turbulent GMCs of tens of parsecs in size and a few kilometers per second in rms speed, if it happens at all. If not, locally super-critical sub-pieces of stellar masses must be created by some other means, most likely through ambipolar diffusion. In the latter case, ambipolar diffusion is not expected to play a decisive role in star formation. The two cases may correspond to, respectively, distributed and clustered modes of star formation (Shu *et al.* 1987).

10.3.1 Ambipolar diffusion and distributed star formation

The best-studied region of distributed star formation is the Taurus molecular clouds. Although it is not clear whether these clouds are magnetically supercritical or subcritical as a whole, the dynamics of the more diffuse regions is probably magnetically dominated. The best evidence comes from thin strands of ^{12}CO emission that are aligned with the directions of the local magnetic field (Heyer *et al.* 2008, preprint), where the field is apparently strong enough to induce a measurable difference between the turbulent velocities along and perpendicular to the field direction. The filamentary morphology is strikingly similar to that observed in the nearby Riegel–Crutcher HI cloud, mapped recently by McClure-Griffiths *et al.* (2006) using 21-cm absorption against the strong continuum emission towards the galactic centre. Its filaments are also along the directions of the local magnetic field, which has an estimated strength of $\sim$30 μG. Kazes and Crutcher (1986) nearby location and found that $B_{\mathrm{los}} \sim 18\,\mu$G, which is consistent with the above estimate if a correction of $\sim$2 is applied for projection effect. The inferred strong magnetization of this cloud may not be too surprising, in view of the result that cold neutral HI structures are strongly magnetically subcritical in general (Heiles & Troland 2005). If diffuse molecular clouds such as the Taurus clouds are formed out of such HI gas, it is not difficult to imagine that at least some of them will be subcritical as well.

If the bulk of a molecular cloud is indeed magnetically subcritical, then the well-known low efficiency of star formation can be naturally explained: the formation of dense, star-forming, magnetically supercritical cores is regulated by ambipolar diffusion, which is generally a slow process, with a timescale an order of magnitude (or more) longer than the local free-fall time (Shu *et al.* 1987; Mouschovias & Ciolek 1999).

For relatively diffuse molecular gas with an A_V of a few, the problem is that the ambipolar diffusion timescale is too long to allow for significant star formation in a reasonable time due to ionization from UV background (McKee 1989; Myers & Khersonsky 1995). In order to form stars in a reasonable timescale, the rate of ambipolar diffusion must be enhanced (Fatuzzo & Adams 2002; Zweibel 2002). This is where turbulence can help greatly. Supersonic turbulence naturally creates dense regions where the background UV photons are shielded and where the gradient in magnetic field is large, thereby accelerating ambipolar diffusion. The strong magnetic field, on the contrary, prevents too large a fraction of the cloud mass being converted into stars in a turbulence crossing time. In this hybrid scenario of distributed star formation in relatively diffuse molecular clouds, both magnetic fields and turbulence play crucial roles: the star formation is accelerated by turbulent compression but regulated by magnetic fields through ambipolar diffusion (Li & Nakamura 2004).

The above scenario is illustrated in Nakamura and Li (2005), using 2D simulations of a sheet-like, magnetically subcritical cloud (with an initial dimensionless flux-to-mass ratio $\Gamma = 1.2$ everywhere). It is stirred with a (compressive) supersonic turbulence of rms Mach number $\mathcal{M} = 10$ at $t = 0$. The strong initial magnetic field prevents the bulk of the strongly shocked material from collapsing promptly. The shocked material is altered permanently; however, its flux-to-mass ratio Γ is reduced through enhanced ambipolar diffusion, which produces filamentary supercritical structures that are generally long-lived. Only the densest parts of the filaments – the dense cores – are directly involved in star formation. Even the dense cores are significantly magnetized, with flux-to-mass ratios about half the critical values or more. The strong magnetization ensures that ambipolar diffusion continues to play a role in the core evolution, reducing their internal velocity dispersions to typically subsonic levels, as observed (Myers 1995). Since only a fraction of the cloud material is magnetically supercritical (and thus capable of forming stars in the first place) and only a small fraction of the supercritical material is directly involved in star formation, the efficiency of star formation is naturally low.

An attractive feature of the 'turbulence-accelerated, magnetically regulated' star formation in diffuse, subcritical clouds (or sub-regions) is that the stars are expected to form at distributed locations. This is because the self-gravity, although important on the scale of individual supercritical cores, is cancelled out to a large extent, if not completely, by magnetic forces on large scales. The cancellation makes the clustering of dense cores difficult, unless the cores happen to be created close together by converging flows. By the same token, star clusters are more likely formed in dense, self-gravitating clumps that are magnetically supercritical as a whole, as we discuss in p. 240.

Even though the principle of the 'turbulence-accelerated, magnetically regulated' star formation is straightforward and well illustrated by 2D simulations, much work remains to be done to firm it up. Among the needed refinements is the extension of the 2D calculations to 3D. A first step in this direction has been taken by Kudoh *et al.* (2007) and Nakamura and Li (2008). Another improvement would be in the maintenance of turbulence, which will be more difficult, given our limited understanding of the origins of turbulence in molecular clouds in general, and in diffuse regions of distributed star formation in particular.

A region of inefficient, distributed star formation of considerable current interest is the Pipe nebula. Alves *et al.* (2007) found more than 150 dense, quiescent cores with a mass distribution that resembles the stellar initial mass function (IMF). The cores are distributed along a filament of more than 10 pc in length. None of the cores except the most massive one has yet collapsed and formed stars. This is difficult to understand, given their rather short free-fall times, unless the core creation is well synchronized. The synchronization is most naturally done by a large-scale shock. In such a case, the strong compression is expected to induce fast motions inside the filaments and cores unless the cores are magnetically cushioned. The Pipe nebula may turn out to be a good example of compression-induced, magnetically regulated star formation.

10.3.2 Turbulence, gravity and cluster formation

The most detailed simulations of cluster formation are performed using the smoothed particle hydrodynamic (SPH) technique, which generally does not include magnetic fields (see, however, Price & Bate 2007). These simulations are well suited for studying the interaction between turbulence compression, gravitational collapse and especially mass accretion onto collapsed objects (see Bonnell *et al.* 2007 for a recent review). They fall into two categories, with either a freely decaying (Bate *et al.* 2003; Bonnell *et al.* 2003) or a constantly driven turbulence (Klessen 2001). In the former, the initial turbulence typically generates several dense clumps, in which small stellar groups (Bate *et al.* 2003) or sub-clusters (Bonnell *et al.* 2003) form, depending on the number of Jeans masses initially contained in the cloud. The most massive member of each sub-cluster typically sits near the bottom of the gravitational potential well of the local collection of gas and stars, gaining mass preferentially through competitive accretion. The sub-clusters of stars and gas merge together at later times, making deeper gravitational potentials where the more massive stars near the bottom of the potential well can accrete preferentially, growing to higher masses.

The most attractive feature of the non-magnetic SPH calculations of cluster formation is that an IMF that broadly resembles the observed one is produced. In particular, the characteristic mass near the knee of the IMF is attributed to the initial cloud Jeans mass (Bonnell *et al.* 2006), and the steeper, Salpeter-like mass distribution above the knee is mostly shaped by competitive accretion. A potential difficulty is that the Jeans mass (and thus the characteristic mass of IMF) is expected to vary from region to region because of variations in both density and temperature. There is little observational evidence to support this expectation. The difficulty can be alleviated to some extent by the heating and cooling behaviours of molecular gas, which tend to set a characteristic density below (above) which the temperature decreases (increases) with density (Jappsen *et al.* 2005; Bonnell *et al.* 2006). Whether the thermal regulation of Jeans mass works for very dense clusters where the average Jeans mass is expected to be very small remains to be seen (McKee & Ostriker 2007).

A problem with cluster formation in decaying turbulence is that star formation may be too rapid. According to Klessen (2001), $\sim$20–30% of the cloud mass is accreted by sink particles in one free-fall time t_{ff}. The value of t_{ff} depends on the density (and thus the initial Jeans mass M_J for 10-K gas). For $M_J = 1 M_\odot$ (needed to reproduce the knee of IMF), $t_{ff} \approx 10^5$ year. This may be short compared to the lifetimes (typically 1 Myr or more) estimated for nearby embedded clusters, particularly those that include a substantial population of relatively evolved Class II and Class III objects. Increasing the free-fall time to 1 Myr would increase the Jeans mass to $10 M_\odot$, which would produce too high a mass for the knee of the IMF, unless the conversion of dense, self-gravitating gas into individual stars is inefficient (Shu *et al.* 1999; Matzner & McKee 2000). Alternatively, the rate of star formation can be reduced significantly if the turbulence is constantly driven on a small enough scale (Klessen 2001). The small-scale driving may, however, modify the hierarchical cluster formation observed in the simulations of the decaying turbulence case and affect competitive accretion and thus the high mass end of the IMF. Indeed, Klessen (2001) and Vázquez-Semadeni *et al.* (2003) suggested that turbulence driven on small scales may correspond to distributed star formation whereas that on large scales to clustered star formation. According to these works, large-scale driving does not slow down star formation significantly, so the problem of too rapid star formation in clusters may remain (but see Section 10.4). The problem can be alleviated somewhat by the inclusion of a strong magnetic field (Heitsch *et al.* 2001; Vázquez-Semadeni *et al.* 2005; Tilley & Pudritz 2007). In Section 10.3.3, we examine the role of protostellar outflows in slowing star formation by replenishing the turbulence in localized regions of active cluster formation.

10.3.3 Cluster formation in protostellar turbulence

The possibility of outflows replenishing the energy and momentum dissipated in a star-forming cloud was first examined in detail by Norman & Silk (1980). They envisioned the star-forming clouds to be constantly stirred up by the winds of optically revealed T Tauri stars. The idea was strengthened by the discovery of molecular outflows (McKee 1989), which point to even more powerful outflows from the stellar vicinity during the embedded, *protostellar* phase of star formation (Lada 1985; Bontemps *et al.* 1996). Shu *et al.* (1999) estimated the momentum output from protostellar outflows based on the galactic SFR and concluded that it is sufficient to sustain a level of turbulence of $\sim 1 - 2 \, \mathrm{km \, s^{-1}}$, similar to the line widths observed in typical GMCs. If the majority of stars are formed in local-ized parsec-scale dense clumps that occupy a small fraction of the GMC volume (Lada *et al.* 1991), their ability to influence the dynamics of the bulk of the GMC material will probably be reduced; other means of turbulence maintenance may be needed in regions of relatively little star formation, as concluded by Walawender *et al.* (2005) in the case of the Perseus molecular cloud. The concentration of star formation should, however, make the outflows more important in the spatially lim-ited, but arguably the most interesting regions of a GMC – the regions of cluster formation, where the majority of stars are thought to form.

The importance of outflows on cluster formation can be illustrated using a simple estimate. Let the masses of the cluster-forming dense clump and the stars formed in it be M_c and M_*, respectively. If the outflow momentum per unit stellar mass is P_*, then there is enough momentum to move all of the clump mass to a speed

$$v \sim \frac{M_* P_*}{M_c} \sim \epsilon P_* = 5 \, \mathrm{km \, s^{-1}} \left(\frac{\epsilon}{0.1} \right) \left(\frac{P_*}{50 \, \mathrm{km \, s^{-1}}} \right), \tag{10.1}$$

where $\epsilon = M_*/M_c$ is the star formation efficiency (SFE). For embedded clusters, the SFE can be 10% or more (Lada & Lada 2003). The value of P_* is somewhat uncertain. For low-mass stars, Nakamura & Li (2007) estimated a plausible range between 10 and $100 \, \mathrm{km \, s^{-1}}$. For the best-studied CO outflow-driving early B stars, the value is estimated at 50–$100 \, \mathrm{km \, s^{-1}}$ (D. Shepherd, personal communication). For the fiducial values of ϵ and P_* adopted in Eq. (10.1), the outflow-driven aver-age velocity is comfortably above the typical turbulence velocity of 1–$2 \, \mathrm{km \, s^{-1}}$ observed in nearby cluster-forming clumps. Indeed, if all of the cluster members were to form simultaneously, the clump would quickly become unbound. If the stars are formed more gradually (as evidenced by the presence of objects in a wide

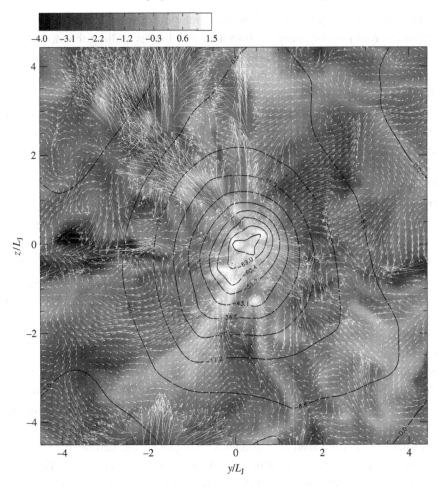

Fig. 10.6. Snapshots of a 3D magnetohydrodynamic (MHD) simulations of cluster formation including feedback from protostellar outflows. Plotted are the volume density map, velocity vectors, and contours of gravitational potential on a slice through the simulation box. For details, see Nakamura & Li (2007).

range of evolutionary states, from prestellar cores to Class III sources), then there is the possibility for the outflows to replenish the dissipated turbulence, keeping the cluster formation going for a time longer than the global free-fall time.

Quasi-equilibrium cluster formation in outflow-driven turbulence is illustrated in Figure 10.6. It is clear that dense cores tend to collect near the bottom of the gravitational potential well, where most of the stars form. Most of the outflow momentum is coupled, however, into the envelope, where most of the clump mass resides. The coupling of momentum into the envelope is facilitated by outflow collimation, which enables the outflows to propagate further away from the central region and drive motions on a larger scale that would decay more slowly. Gravity also plays an important role in generating the turbulent motions by pulling the slowed down material towards the centre and setting up a mass circulation between the core and the envelope. For parameters appropriate for nearby embedded clusters such as NGC 1333 ($\sim 10^3 M_\odot$ in mass and ~ 1 pc in size), the gravitationally induced infall and outflow-driven expansion are roughly balanced, creating a quasi-static environment in which the clump material is converted into stars at a relatively leisurely pace. For the particular example shown, the average SFE per global free-fall time is about 3%. This value is in the range inferred for NGC 1333 (Nakamura & Li 2007) and other objects (Krumholz & Tan 2007).

The protostellar turbulence in regions of active cluster formation is a special type of driven turbulence. It is driven anisotropically over a range of scales, as the collimated outflow propagates progressively further away from the driving source (Cunningham *et al.* 2006; Shang *et al.* 2006; Banerjee *et al.* 2007). There is a break in the energy spectrum near the characteristic length scale of the outflows, as predicted analytically by Matzner (2007). The break may provide a way to distinguish the protostellar turbulence from other types of turbulence.

It is interesting to speculate how massive stars may form in protostellar outflow-driven turbulence. One possibility, pointed out in Li and Nakamura (2006), is that, when the central region of the cluster-forming clump becomes dense enough, the outflows from the low-mass stars may become trapped. The trapping of outflows reduces the stellar feedback into the massive envelope, which leads to more mass falling to the central region, which in turn makes it more difficult for the outflow to get out. It may lead to a runaway collapse of the central region, with a large mass infall rate that is conducive to massive star formation (McKee & Tan 2002). The massive stars formed in this scenario will naturally be located near the bottom of the gravitational potential well, for which there is now growing observational evidence (Cesaroni 2005; Garay 2005). The trapping of outflows as a cluster-forming clump condenses may also provide a natural explanation for the accelerating star formation inferred for the ONC (Huff & Stahler 2006) and others (Palla & Stahler 2000). Detailed numerical simulations are needed to firm up this supposition.

10.4 Global theory of star formation in turbulent clouds

GMCs and their smaller-scale constituents, star-forming clumps and prestellar cores obey several remarkable scaling relations, which indicate that star formation to a significant extent exhibits a generic behaviour. The most important of these scaling relations are as follows:

1. Larson's velocity–size relation (Larson 1979), which expresses that the (3D) rms velocity dispersion σ_v, 3D of structures of size L scales as σ_v, 3D $\sim L^{p_v}$, where $p_v \approx 0.4$ (Larson 1981). Recently, Heyer and Brunt (2007) have shown that the relation is very tight when measured in a consistent manner over a limited range of scales and that *the normalization is essentially the same* in violently star-forming and nearly quiescent molecular clouds (Heyer *et al.* 2006).
2. Larson's velocity–mass relations (Larson 1981), which expresses that the densest structures on each scale are in near-virial equilibrium.
3. The cluster mass function, which is a power law with slope of dN/dM slightly less steep than -2 (Dowell *et al.* 2006 and references therein) – so stellar birth is distributed nearly evenly over mass, although a slightly larger number of stars are born in rich clusters.
4. The core mass function (CMF) and the IMF, which are approximate power laws for large enough masses. Individual examples may show deviations due to statistics or other influences, but the general power-law trends are well established.

These structures thus appear to be members of a hierarchy of scales, characterized by approximate scaling laws. Two additional, and crucially important, properties of this hierarchy are as follows:

1. On each scale, the actual lifetimes of the structures appear to be longer than their estimated turbulent decay times (as given by the turbulence crossing times). To maintain their (rather precise; cf. Heyer *et al.* 2006) velocity normalization in the hierarchy, the motions on each scale thus need significant driving.
2. A related but largely independent property is that the mass drain due to the star formation process is approximately constant across scales, at a level of only a few percent per unit free-fall time (cf. Krumholz & Tan 2007 and references therein).

The various properties enumerated above are linked in ways that are likely not to be coincidences: items (1) and (2) above imply that the draining of mass from each scale per unit crossing time is quite small. This property is in principle independent of the property that the structures appear to be long-lived; one could imagine the structures to have a small draining rate but still be quite short-lived. The opposite could also be the case; the structures could be drained more rapidly by star formation but could still be long-lived if the geometrical shapes of structures were somehow maintained as they were being drained.

10.4.1 Questions and difficulties

Most of these properties are difficult to explain: the small rate of star formation per free-fall time (Krumholz & Tan 2007), the driving source of the motions, the regularities of the cluster mass function the core mass function and the stellar IMF are all long-standing problems.

One can attempt to 'explain' Larson's density and velocity relations as a consequence of 'virialization', since they are roughly consistent with virial equilibrium. With velocity scaling as $L^{-1/2}$ (rather than the observed exponent $p_v \approx 0.4$) and density as L^{-1}, one could imagine that gravity maintains the velocity spectrum, as structures shrink in size and descend into ever deeper potential wells. That, however, still begs the question 'why would that happen?'; specifically, why would this happen in such a way that the lifetimes remain long and the draining rates at each level in the hierarchy remain nearly constant (and low)? One would rather expect that the characteristic timescale of such a cascade would be similar to the local gravitational free-fall time. The driving-by-gravity explanation is also very difficult to reconcile with the fact that potential wells actually get *shallower* with decreasing size, rather than deeper!

Likewise, it is very difficult to explain how local feedback could, without exceptional fine-tuning, both maintain a power-law scaling of the velocity with size – the energy input rate would have to match the turbulent dissipation rate and its variation with scale over a large range of scales – and maintain the proper amount of support against gravitational collapse to keep the draining rate constant and low over a range of structure sizes.

One is led to conjecture then that local feedback, rather than *maintaining* the observed scaling laws over a range of scales, instead may be responsible for *breaking* those scaling laws at local sites with rapid star formation, as observed in, e.g., CS clumps (Krumholz & Tan 2007), where the mass density (and possibly also draining rates) exceed values expected from the scaling relations. But that still leaves the questions of explaining the power-law scalings that hold at most scales and locations, of explaining the maintenance of these motions against dissipation and of explaining the constancy of draining rates over all other scales.

In fact, most of the scaling and lifetime properties may be explained, in a consistent and complete manner, by essentially 'turning the explanation upside down'; it is not the gravity that is in the lead but the kinetic energy, cascading from large to small scales, with gravity picking up those exceptional regions of space where the local gravitational potential energy dominates the local kinetic energy (i.e. the local virial parameter, as given in Eq. (10.2), happens to be small – cf. Figure 10.7).

10.4.2 A turbulent cascade origin of the scaling

A suitable entry point for discussion of the various scaling properties is Larson's velocity–size relation. As shown by Boldyrev (2002) and Boldyrev *et al.* (2002), such a scaling follows inevitably from the scaling relations of supersonic turbulence. A number of observationally well-defined diagnostics are in agreement

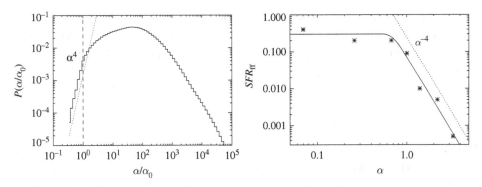

Fig. 10.7. The mass-weighted probability distribution of α (*left*), normalized to the virial ratio for the whole domain, in a numerical simulation of driven Mach = 18 supersonic turbulence with self-gravity (Padoan & Nordlund in preparation; resolution 1000^3) and the dependence of the SFR on the large-scale α (*right*) for a number of such simulations (stars – the broken power-law curve is only suggestive).

with the corresponding diagnostics extracted from simulations of supersonic (and super-Alfvénic) turbulence (Padoan *et al.* 1997, 1998, 2001, 2004). One can even measure, using the method of Lazarian and Pogosyan (2000), the velocity power spectrum exponent of the turbulence; the result is a wavenumber exponent ≈ -1.8, consistent with a velocity–size scaling with an exponent $p_v \approx 0.4$ (Padoan *et al.* 2006).

So, what we are observing as Larson's velocity–size scaling law is, most likely, just the inevitable and robust consequence of a cascade of kinetic energy from the largest (injection) scales, which occur at scales of the order of the galactic disk thickness, where kinetic energy may be contributed by several sources (de Avillez & Breitschwerdt 2005, 2007; Kim *et al.* 2006; McKee & Ostriker 2007; Ostriker 2007). No matter how that energy is fed into the medium, it must cascade to very small scales before it is dissipated. This occurs, in supersonic turbulence as well as in subsonic turbulence, via a turbulent cascade across an 'inertial range', where the denomination 'inertial' signifies that the motions are maintained by inertia.

This explains both the power-law distribution of velocity (and hence Larson's velocity–size relation) and the source of apparent 'driving' over a range of scales. As already shown by Larson (1979, cf. his Figure 1), the power-law scaling continues to sizes comparable to the thickness of the galactic disk, where it exhibits a break (at larger scales, the velocity dispersion is mainly associated with fluctuations in the galactic disk rotation rate). The title of Larson's (1981) paper was indeed, and very appropriately, '*Turbulence and Star Formation in Molecular Clouds*'!

The super-Alfvénic turbulence scenario may also be used to derive the form of the stellar IMF analytically (Padoan & Nordlund 2002) and to explain how brown dwarfs can form by the same mechanism as normal stars (Padoan & Nordlund 2004).

10.4.3 Interaction of turbulence and self-gravity

One might object to the comparisons of diagnostics from simulations of supersonic turbulence without self-gravity with observations of molecular clouds since these, according to the previous discussion, are often close to virial equilibrium, and it would thus seem that self-gravity cannot be ignored.

One could, on the contrary, also argue that the success of such exercises might indicate that self-gravity is only important in local regions and that these are sufficiently small not to disturb the comparisons significantly. Such a conjecture can actually be confirmed from an analysis of locally evaluated virial ratios ('virial parameters'), which we may define as (cf. Krumholz & McKee 2005)

$$\alpha = \frac{5v_{rms}^2}{6G\rho L^2},\tag{10.2}$$

where v_{rms} is a local (3D) rms velocity dispersion, measured in some neighbourhood of size L around comoving, Lagrangian tracer points where the local average density is ρ.

The left-hand side panel in Figure 10.7 shows an example of the mass-weighted probability distribution of the local virial ratio, using values of v_{rms}^2 and ρ evaluated locally. It is obvious that a large fraction of the mass has local values of α much larger than the virial ratio α_0 based on the rms velocity and average density for the whole domain and that only a small fraction of the mass ($\sim$1% in this case) has a smaller value.

A consideration of how local shearing motions are modified by compression explains this behaviour. Suppose we have identified a region of size L with a small density enhancement (and thus potentially a self-gravitating 'object'), which initially has a value of α close to the value for the whole box and a density close to the mean density. If we choose L to be, e.g., one-fourth of the size of the full box, then v must likewise be one-fourth of the average rms value to obtain the same α, so the initial selection is already limited to a subset of the volume (in the rest of the volume, a local α evaluated in this fashion would be larger, even initially).

In a local neighbourhood, and for a limited time, one can assume the local, comoving dynamics to be subsonic. As can be confirmed by data-browsing, the velocity field in supersonic turbulence consists of patches with smooth flow, separated (ideally) by discontinuities of the velocity and the mass density. With smooth velocity across a patch, one can subtract off the mean speed and consider the local velocity field, which can be split into a solenoidal and a compressive part. In a sufficiently small neighbourhood in space, the flow remains subsonic with respect to the mean, and in a sufficiently small neighbourhood in time – until the patch hits a shock – one may thus consider the local dynamics to be subsonic, smooth and continuous.

One consequence of this is that, within a neighbourhood thus limited in space and time, it makes sense to think in terms of approximate conservation of local angular momentum. The subsonic nature of the local dynamics means, e.g., that convergent motions couple to shear and rotations in the well-known sense that contraction tends to cause spin-up.

Now suppose the region in question is compressed to a tenth of its linear size, with compression taking place in two spatial directions. The cross section decreases with a factor of 100, but the density increases correspondingly, so the denominator in Eq. (10.2) is unchanged. Because of the tendency to locally conserve angular momentum, the v^2 in the numerator tends to increase by a large factor, and one finds that the α characterizing this local region has grown tremendously, which prevents the object from becoming self-gravitating.

In general, one finds that in regions of compressions, which are otherwise the locations that are most favourable for creating large mass densities, the tendency for growth of vorticity more than counters the growing influence of self-gravity, which makes it hard to achieve collapse under self-gravity, even when the virial parameter measured on the large scale is of the order unity.

There is thus a purely dynamical mechanism available to explain why structures on all scales in GMCs tend to survive longer than would be indicated by estimates based on the large-scale virial parameter (such an estimate is exactly what is being used when one says 'it is remarkable that these structures live longer than a few free-fall times'). The mechanism is even *unavoidable*; there is no way that this could *not* happen, in cases where the virial parameter based on the large scale is larger than unity!

When, on the contrary, the large-scale virial parameter is less than unity, opportunities rapidly open up for local collapse; the fraction of compressive, converging flows that already initially have low α increases, and a rapidly growing number of compressed regions find themselves to have local virial ratios low enough for collapse. To a first approximation, one can see the dependence of the draining rate on α (*right panel* of Figure 10.7) as resulting from a shift of the probability distribution function (PDF) of α (*left panel* of Figure 10.7, considered as a function of α rather than α/α_0).

Such an argument may be applied recursively. Structures that are much smaller than the initial size but that still have low enough local virial ratios for collapse can be taken as a basis for 'renormalization'. One concludes that, for any large-scale virial ratio, a hierarchy of smaller structures exists, with local virial ratios of the order unity and with internal structures consisting of yet smaller structures with local virial ratios of the order of unity. The hierarchy must be such as to conserve the draining rate at each scale; since the smaller-scale free-fall times are shorter than the larger-scale free-fall times, the smaller scales are 'slaved' to the larger scales.

10.4.4 Self-regulated star formation

One very important aspect of the difficulty to achieve locally small virial ratios when the large-scale virial ratio is larger than unity is that the rate of gravitational collapse of mass out of this hierarchy (which presumably is a good proxy of the SFR) turns out to be extremely sensitive to the large-scale virial ratio.

The *right panel* in Figure 10.7 illustrates the point, by showing a summary of the SFRs achieved in a number of simulations where the large-scale α was varied, both by changing the mean density and by changing the Mach number (Padoan & Nordlund in preparation). The functional behaviour is such that the SFR is approximately proportional to α^{-4} for α larger than unity, while for smaller alpha, the SFR is roughly constant and very large (~ 30 % per free fall time).

Krumholz & McKee (2005) arrived at an analytical estimate of the dependence of the SFR on α that is somewhat less steep but qualitatively similar, based on results from numerical simulations by Klessen (2001) and Vázquez-Semadeni *et al.* (2003, 2005).

In the regime where the SFR is nearly constant, it is significantly larger than values characterizing star formation in the Milky Way (Krumholz & Tan 2007). We thus conclude that galactic star formation is operating in a regime where α is on the steep, power law part of the dependence.

Does that then not require an unlikely fine-tuning? Why would the large-scale α happen to have a value such that the SFR agrees with the observed one? Well, suppose that it did not: if the value of α were too large, there would be practically no star formation, the driving of the interstellar medium (ISM) from massive stars and supernovae would cease, the level of turbulence would be reduced, the scale height of the ISM would drop and the large-scale α would decrease – both because of reduced velocities and because of increased densities.

Conversely, if α were too small, the SFR would be too high (relative to the level required to sustain the ISM turbulence) and the situation would rectify itself via the same feedback mechanism.

We conclude that the mechanism that determines the level of the star formation rate is large-scale stellar feedback. Note also that this only works if the driving of the ISM turbulence is, to a significant but not necessarily exclusive extent, due to large-scale feedback from star formation. That this is so is already abundantly established by elaborate numerical simulations, in particular the ones by de Avillez & Breitschwerdt (2005, 2007). Such a scenario is also entirely consistent with the threshold assumptions about star formation generally made in models of galaxy formation (Conroy *et al.* 2007; Nagai *et al.* 2007) and with the Kennicut–Schmidt law (Kennicutt 1998, 2007).

10.5 Concluding remarks

The last decade has brought significant observational and theoretical progress to the study of star formation in molecular clouds, particularly in observations of embedded clusters and simulations of turbulent clouds. With large-scale surveys of both molecular clouds and their populations of protostars and pre-main sequence stars now in hand and with increasingly sophisticated numerical models, there is excellent potential to rapidly advance our understanding of the macrophysics of star formation.

To do so will require better means for comparing observations and simulations. Observers and modellers need to devise a set of common 'observable' statistics that may be used to perform quantitative comparisons. Given the advances in measuring the distribution, evolutionary state and even kinematics of YSOs and dense cores, it is of great interest to relate the properties of the sink particles in simulations of star formation to the observed properties of dense cores and YSOs. In particular, it is important to determine whether the observed spatial distributions of protostars and more evolved pre-main sequence stars can be reproduced. The evolution of clusters should also be explored through N-body simulations with realistic and time-variable gas potentials to better understand the dynamical states implied by their observed morphologies.

There are areas where significant progress is needed on the observational front. Much progress is still needed to understand the ages and age spreads of embedded populations of YSOs. These are important constraints on the lifetimes of clouds and the duration of star formation. Significant advances have been made in the measurements of magnetic fields; however, since these measurements often target regions of active star formation, the field structure in the bulk of cloud material is poorly constrained. Large-scale mapping of the strength and direction of the magnetic field is needed to ascertain its dynamical role, particularly in relatively diffuse regions of clouds, where the field may play a dominant role.

Theoretically, there is a need to tie local simulations of star formation to global simulations of structure formation on GMC scales and beyond. On the GMC and smaller scales, there are additional effects not included in this chapter. These include the generation of small-scale structure from global cloud collapse

(Burkert & Hartmann 2004), generation of structure by thermal instabilities at the interfaces between the warm ISM and the molecular clouds (Heitsch *et al.* 2007), self-shielding and the transition from atomic to molecular phase in the interior of molecular clouds. In order to avoid the need for ad hoc initial and boundary conditions, such modelling may be performed as sub-sets of larger-scale and longer-duration global simulations, as often done in the study of galaxy formation in the context of cosmological structure formation simulations.

On larger scales, simulations need to account for the life cycle of molecular gas and determine whether 'lifetime' is even a meaningful concept in the context of molecular clouds. Relevant questions to investigate include: 'Given a snapshot of a molecular cloud, how long does it take before half of the molecular gas present in the snapshot is recycled?'; 'How much of that went into stars?'; 'How much was turned into warm and hot ISM, respectively', 'How much new molecular gas was added in the mean time?'; 'How much of the feedback is due to winds, UV-radiation and supernovae and on what scales is this feedback deposited?'. Such questions may be answered by realistic high-resolution global ISM simulations (such as the ones by de Avillez & Breitschwerdt 2007), which include the vertical structure of the galactic disk, realistic cooling functions and UV heating, self-gravity and magnetic fields. A next step in such simulations could be to let the SFR be determined self-consistently, by including approximations of how the SFR depends on the virial ratios of star-forming structures and their average magnetic fields.

Acknowledgements

ÅN acknowledges support from the Danish Natural Science Research Council, the Danish Center for Scientific Computing. The work of ZYL is supported in part by NASA (NNG05GJ49G) and NSF (AST-0307368) grants. This work is based in part on observations made with the Spitzer Space Telescope, which is operated by the Jet Propulsion Laboratory, California Institute of Technology, under a contract with NASA. Support for the work of STM was provided by NASA through an award issued by JPL/Caltech. All authors participated in the KITP programme 'Star Formation Through Cosmic Time', supported in part by the National Science Foundation under Grant No. PHY05-51164.

References

Adams, F. C. and Myers, P. C. 2001, *ApJ.* **553**, 744
Allen, L., Megeath, S. T., Gutermuth, R., Myers, P. C., Wolk, S., Adams, F. C., Muzerolle, J., Young, E., and Pipher, J. L. 2007, in B. Reipurth, D. Jewitt, and K. Keil (eds.), *Protostars and Planets V*, University of Arizona Press, Tucson, pp. 361–376
Alves, J., Lombardi, M., and Lada, C. J. 2007, *AA* **462**, L17

Ballesteros-Paredes, J., Klessen, R. S., Mac Low, M.-M., and Vazquez-Semadeni, E. 2007, in B. Reipurth, D. Jewitt, and K. Keil (eds.), *Protostars and Planets V*, University of Arizona Press, Tucson, pp. 63–80

Banerjee, R., Klessen, R. S., and Fendt, C. 2007, *ApJ.* **668**, 1028

Bate, M. R., Bonnell, I. A., and Bromm, V. 2003, *MNRAS* **339**, 577

Boldyrev, S. 2002, *ApJ.* **569**, 841

Boldyrev, S., Nordlund, Å., and Padoan, P. 2002, *ApJ.* **573**, 678

Bonnell, I. A. and Bate, M. R. 2006, *MNRAS* **370**, 488

Bonnell, I. A., Bate, M. R., and Vine, S. G. 2003, *MNRAS* **343**, 413

Bonnell, I. A., Clarke, C. J., and Bate, M. R. 2006, *MNRAS* **368**, 1296

Bonnell, I. A., Larson, R. B., and Zinnecker, H. 2007, in B. Reipurth, D. Jewitt, and K. Keil (eds.), *Protostars and Planets V*, University of Arizona Press, Tucson, pp. 149–164

Bontemps, S., Andre, P., Terebey, S., and Cabrit, S. 1996, *AA* **311**, 858

Brown, A. G. A., de Geus, E. J., and de Zeeuw, P. T. 1994, *AA* **289**, 101

Burkert, A. and Hartmann, L. 2004, *ApJ.* **616**, 288

Carpenter, J. M., Hillenbrand, L. A., and Skrutskie, M. F. 2001, *AJ* **121**, 3160

Cesaroni, R. 2005, *APSS* **295**, 5

Conroy, C., Wechsler, R. H., and Kravtsov, A. V. 2007, *ApJ.* **668**, 826

Crutcher, R. M. 1999, *ApJ.* **520**, 706

Cunningham, A. J., Frank, A., and Blackman, E. G. 2006, *ApJ.* **646**, 1059

Davis, C. J., Chrysostomou, A., Matthews, H. E., Jenness, T., and Ray, T. P. 2000, *ApJL* **530**, L115

de Avillez, M. A. and Breitschwerdt, D. 2005, *AA* **436**, 585

de Avillez, M. A. and Breitschwerdt, D. 2007, *ApJ.* **665**, L35

Dowell, J. D., Buckalew, B. A., and Tan, J. C. 2008, *Astron. J.* **135**, 823–835

Elmegreen, B. G. and Scalo, J. 2004, *ARAA* **42**, 211

Fatuzzo, M. and Adams, F. C. 2002, *ApJ.* **570**, 210

Feigelson, E., Townsley, L., Güdel, M., and Stassun, K. 2007, in B. Reipurth, D. Jewitt, and K. Keil (eds.), *Protostars and Planets V*, University of Arizona Press, Tucson, pp. 313–328

Feigelson, E. D., Getman, K., Townsley, L., Garmire, G., Preibisch, T., Grosso, N., Montmerle, T., Muench, A., and McCaughrean, M. 2005, *ApJS* **160**, 379

Furesz, G., Hartmann, L. W., Megeath, S. T., Szentgyorgyi, A. H., and Hamden, E. T. 2008, *ApJ.* **676**, 1109

Garay, G. 2005, in R. Cesaroni, M. Felli, E. Churchwell, and M. Walmsley (eds.), *Massive Star Birth: A Crossroads of Astrophysics*, Vol. 227 of *IAU Symposium*, Cambridge University Press, pp. 86–91

Getman, K. V., Feigelson, E. D., Townsley, L., Bally, J., Lada, C. J., and Reipurth, B. 2002, *ApJ.* **575**, 354

Giardino, G., Favata, F., Micela, G., Sciortino, S., and Winston, E. 2007, *AA* **463**, 275

Gutermuth, R. A., Megeath, S. T., Pipher, J. L., Williams, J. P., Allen, L. E., Myers, P. C., and Raines, S. N. 2005, *ApJ.* **632**, 397

Gutermuth, R. A., Myers, P. C., Megeath, S. T., Allen, L. E., Pipher, J. L., Muzerolle, J., Porras, A., Winston, E., and Fazio, G. 2008, *ApJ.* **674**, 336–356

Hatchell, J., Fuller, G. A., Richer, J. S., Harries, T. J., and Ladd, E. F. 2007, *AA* **468**, 1009

Heiles, C. and Troland, T. H. 2005, *ApJ.* **624**, 773

Heitsch, F., Mac Low, M.-M., and Klessen, R. S. 2001, *ApJ.* **547**, 280

Heitsch, F. *et al.* 2008, *ApJ* **683**, 786–795

Heyer, M. H. and Brunt, C. 2007, in B. G. Elmegreen and J. Palous (eds.), *IAU Symposium*, Vol. 237 of *IAU Symposium*, pp 9–16

Heyer, M. H., Williams, J. P., and Brunt, C. M. 2006, *ApJ.* **643**, 956

Heyer, M., Gong, H., Ostriker, E., and Brunt, C. 2008, *ApJ.* **680**, 420

Huff, E. M. and Stahler, S. W. 2006, *ApJ.* **644**, 355

Jappsen, A.-K., Klessen, R. S., Larson, R. B., Li, Y., and Mac Low, M.-M. 2005, *AA* **435**, 611

Kazes, I. and Crutcher, R. M. 1986, *AA* **164**, 328

Kennicutt, Jr., R. C. 1998, *ApJ.* **498**, 541

Kennicutt, R. C. 2007, in B. G. Elmegreen and J. Palous (eds.), *IAU Symposium*, Vol. 237 of *IAU Symposium*, Cambridge University Press, pp. 311–316

Kim, C.-G., Kim, W.-T., and Ostriker, E. C. 2006, *ApJL* **649**, L13

Klessen, R. S. 2001, *ApJL* **550**, L77

Kroupa, P., Aarseth, S., and Hurley, J. 2001, *MNRAS* **321**, 699

Krumholz, M. R. and McKee, C. F. 2005, *ApJ.* **630**, 250

Krumholz, M. R. and Tan, J. C. 2007, *ApJ.* **654**, 304

Kudoh, T., Basu, S., Ogata, Y., and Yabe, T. 2007, *MNRAS* **380**, 499

Lada, C. J. 1985, *ARAA* **23**, 267

Lada, C. J. and Lada, E. A. 2003, *ARAA* **41**, 57

Lada, C. J., Margulis, M., and Dearborn, D. 1984, *ApJ.* **285**, 141

Lada, E. A., Bally, J., and Stark, A. A. 1991, *ApJ.* **368**, 432

Larson, R. B. 1979, *MNRAS* **186**, 479

Larson, R. B. 1981, *MNRAS* **194**, 809

Lazarian, A. and Pogosyan, D. 2000, *ApJ.* **537**, 720

Li, Z.-Y. and Nakamura, F. 2004, *ApJL* **609**, L83

Li, Z.-Y. and Nakamura, F. 2006, *ApJL* **640**, L187

Mac Low, M.-M. and Klessen, R. S. 2004, *Rev. Mod. Phys.* **76**, 125

Matzner, C. D. 2007, *ApJ.* **659**, 1394

Matzner, C. D. and McKee, C. F. 2000, *ApJ.* **545**, 364

Megeath, S. T., Gutermuth, R. A., Muzerolle, E., Allgaier, J., Hora, J. R., Allen, L. E., Hartmann, L. W., Flaherty, K., Myers, P. C., Stauffer, J. R., Young, E. T., and Fazio, G. G. 2008, in preparation

McClure-Griffiths, N. M., Dickey, J. M., Gaensler, B. M., Green, A. J., and Haverkorn, M. 2006, *ApJ.* **652**, 1339

McKee, C. F. 1989, *ApJ.* **345**, 782

McKee, C. F. 1999, in C. J. Lada and N. D. Kylafis (eds.), *NATO ASIC Proc. 540: The Origin of Stars and Planetary Systems*, pp 29

McKee, C. F. and Ostriker, E. C. 2007, *ARAA* **45**, 565

McKee, C. F. and Tan, J. C. 2002, *Nature* **416**, 59

Mouschovias, T. C. and Ciolek, G. E. 1999, in C. J. Lada and N. D. Kylafis (eds.), *NATO ASIC Proc. 540: The Origin of Stars and Planetary Systems*, pp 305

Muench, A. A., Lada, C. J., Luhman, K. L., Muzerolle, J., and Young, E. 2007, *AJ* **134**, 411

Myers, P. C. 1995, in C. Yuan and J.-H. You (eds.), *Molecular Clouds and Star Formation*, pp 47

Myers, P. C. and Khersonsky, V. K. 1995, *ApJ.* **442**, 186

Nagai, D., Kravtsov, A. V., and Vikhlinin, A. 2007, *ApJ.* **668**, 1

Nakamura, F. and Li, Z.-Y. 2005, *ApJ.* **631**, 411

Nakamura, F. and Li, Z.-Y. 2007, *ApJ.* **662**, 395

Nakamura, F. and Li, Z.-Y. 2008, *ApJ.* October 20 issue

Norman, C. and Silk, J. 1980, *ApJ.* **238**, 158

Novak, G., Dotson, J. L., and Li, H. 2007, *arXiv:0707.2818*

Olmi, L. and Testi, L. 2002, *AA* **392**, 1053

Ostriker, E. C. 2007, in B. G. Elmegreen and J. Palous (eds.), *IAU Symposium*, Vol. 237 of *IAU Symposium*, Cambridge University Press, pp. 70–75

Padoan, P., Jimenez, R., Juvela, M., and Nordlund, Å. 2004, *ApJL* **604**, L49

Padoan, P., Jones, B. J. T., and Nordlund, A. P. 1997, *ApJ.* **474**, 730

Padoan, P., Juvela, M., Bally, J., and Nordlund, A. 1998, *ApJ.* **504**, 300

Padoan, P., Juvela, M., Goodman, A. A., and Nordlund, Å. 2001, *ApJ.* **553**, 227

Padoan, P., Juvela, M., Kritsuk, A., and Norman, M. L. 2006, *ApJ.* **653**, L125

Padoan, P. and Nordlund, Å. 2002, *ApJ.* **576**, 870

Padoan, P. and Nordlund, Å. 2004, *ApJ.* **617**, 559

Palla, F. and Stahler, S. W. 2000, *ApJ.* **540**, 255

Peretto, N., Hennebelle, P., and André, P. 2007, *AA* **464**, 983

Price, D. J. and Bate, M. R. 2007, *MNRAS* **377**, 77

Sandell, G. and Knee, L. B. G. 2001, *ApJL* **546**, L49

Shang, H., Allen, A., Li, Z.-Y., Liu, C.-F., Chou, M.-Y., and Anderson, J. 2006, *ApJ.* **649**, 845

Shu, F. H., Adams, F. C., and Lizano, S. 1987, *ARAA* **25**, 23

Shu, F. H., Allen, A., Shang, H., Ostriker, E. C., and Li, Z.-Y. 1999, in C. J. Lada and N. D. Kylafis (eds.), *NATO ASIC Proc. 540: The Origin of Stars and Planetary Systems*, pp 193

Stanke, T., Smith, M. D., Gredel, R., and Khanzadyan, T. 2006, *AA* **447**, 609

Teixeira, P. S., Zapata, L. A., and Lada, C. J. 2007, *ApJL* **667**, L179

Teixeira, P. S. *et al.* 2006, *ApJL* **636**, L45

Tilley, D. A. and Pudritz, R. E. 2007, *MNRAS*, **382**, 73

Vázquez-Semadeni, E., Ballesteros-Paredes, J., and Klessen, R. S. 2003, *ApJL* **585**, L131

Vázquez-Semadeni, E., Kim, J., and Ballesteros-Paredes, J. 2005, *ApJL* **630**, L49

Walawender, J., Bally, J., and Reipurth, B. 2005, *AJ* **129**, 2308

Walsh, A. J., Myers, P. C., Di Francesco, J., Mohanty, S., Bourke, T. L., Gutermuth, R., and Wilner, D. 2007, *ApJ.* **655**, 958

Williams, J. P. and Myers, P. C. 2000, *ApJ.* **537**, 891

Winston, E., Megeath, S. T., Wolk, S. J., Muzerolle, J., Gutermuth, R., Hora, J. L., Allen, L. E., Spitzbart, B., Myers, P., and Fazio, G. G. 2007, *ApJ.* **669**, 493

Zweibel, E. G. 2002, *ApJ.* **567**, 962

11

The formation and evolution of prestellar cores

P. André, S. Basu and S. Inutsuka

11.1 Introduction: dense cores and the origin of the IMF

Stars form from the gravitational collapse of dense cloud cores in the molecular interstellar medium of galaxies. Studying and characterizing the properties of dense cores is thus of great interest to gain insight into the initial conditions and initial stages of the star formation process.

Our observational understanding of low-mass dense cores has made significant progress in recent years, and three broad categories of cores can now be distinguished within nearby molecular clouds, which possibly represent an evolutionary sequence: starless cores, prestellar cores and 'Class 0' protostellar cores. Starless cores are possibly transient concentrations of molecular gas and dust without embedded young stellar objects (YSOs), typically observed in tracers such as $C^{18}O$ (Onishi *et al.* 1998), NH_3 (Jijina *et al.* 1999) or dust extinction (Alves *et al.* 2007), and which do not show evidence of infall. Prestellar cores are also starless ($M_\star = 0$) but represent a somewhat denser and more centrally concentrated population of cores which are self-gravitating, hence unlikely to be transient. They are typically detected in (sub)millimetre dust continuum emission and dense molecular gas tracers such as NH_3 or N_2H^+ (Benson & Myers 1989; Ward-Thompson *et al.* 1994; Caselli *et al.* 2002), are often seen in absorption at mid- to far-infrared wavelengths (Bacmann *et al.* 2000; Alves *et al.* 2001) and frequently exhibit evidence of infall motions (Gregersen & Evans 2000). Conceptually, all prestellar cores are starless but only a subset of the starless cores evolve into prestellar cores; the rest are presumably 'failed' cores that eventually disperse and never form stars. In practice, prestellar cores are characterized by large density contrasts over the local background medium. Specifically, the mean densities of observed prestellar cores exceed the mean densities of their parent clouds by a factor $\gtrsim 5$–10, while their mean column densities exceed the background column densities by a factor $\gtrsim 2$. For comparison, a critical self-gravitating Bonnor–Ebert isothermal spheroid

Structure Formation in Astrophysics. ed. G. Chabrier. Published by Cambridge University Press.

has a mean density contrast $\bar{\rho}_{BE}/\rho_{ext} \sim 2.4$ (Lombardi & Bertin 2001) and a mean column density contrast $\overline{\Sigma}_{BE}/\Sigma_{ext} \sim 1.5$ over the external medium.

Finally, Class 0 cores/objects are young accreting protostars observed early after point mass formation while most of the mass of the system is still in the form of a dense core/envelope as opposed to a YSO ($M_\star \ll M_{env}$) (André et al. 1993). They are believed to result from the gravitational collapse of prestellar cores. Class 0 protostars themselves evolve into Class I objects with $M_{env} < M_\star$ (Lada 1987; André & Montmerle 1994) as the protostellar envelope dissipates through accretion and ejection of circumstellar material. Class I objects subsequently evolve into Class II and Class III pre-main sequence stars surrounded by a circumstellar disk (optically thick and optically thin in the near-/mid-IR, respectively), but lacking a dense circumstellar envelope ($M_{env} \sim 0$).

Improving our understanding of the formation and evolution of dense cores in molecular clouds is crucially important since there is now good evidence that these early stages largely control the origin of the stellar initial mass function (IMF). Indeed, observations indicate that the prestellar core mass function (CMF) resembles the IMF (see Section 11.2), suggesting that the effective reservoirs of mass required for the formation of individual stars are already selected at the prestellar core stage. Furthermore, it is at the very end of the prestellar stage that multiple systems are believed to form and during the protostellar stage that a fraction of the prestellar core mass reservoir is accreted by the central protostellar components as a result of the accretion/ejection process.

11.2 Link between the prestellar CMF and the IMF

Wide-field (sub)millimetre dust continuum mapping is a powerful tool to take a census of prestellar dense cores and young protostars within star-forming clouds. The advent of large-format bolometer arrays on (sub)millimetre radiotelescopes such as the IRAM 30 m and the JCMT has led to the identification of numerous cold, compact condensations that do not obey the Larson (1981) self-similar scaling laws of molecular clouds and are intermediate in their properties between diffuse CO clumps and infrared YSOs (cf. André et al. 2000; Ward-Thompson et al. 2007 for reviews). As an example, Figure 11.1 shows the condensations found by Motte et al. (2001) at 850 μm in the NGC 2068 protocluster (Orion B). Such highly concentrated (sub)millimetre continuum condensations are denser by at least 3 to 6 orders of magnitude than typical CO clumps (Kramer et al. 1998) and feature large ($\gg$50%) mean column density contrasts over their parent background clouds, strongly suggesting that they are self-gravitating. The latter is directly confirmed by line observations in a number of cases. When available, the virial masses of the condensations indeed agree within a factor of $\sim$2 with the masses derived from the

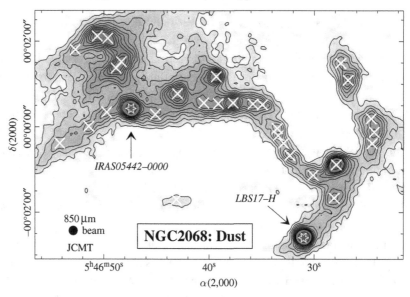

Fig. 11.1. SCUBA 850 μm dust continuum map of the NGC 2068 protocluster extracted from the mosaic of Orion B by Motte *et al.* (2001). A total of 30 compact prestellar condensations (marked by crosses), with masses between ∼0.4 and ∼4.5$M_{\odot}$, are detected in this ∼1 pc × 0.7 pc field.

(sub)millimetre dust continuum (André *et al.* 2007). A small fraction of these condensations lie at the base of powerful jet-like outflows and correspond to Class 0 objects. However, the majority of them are starless/jetless and appear to be the immediate prestellar progenitors of individual protostars or protostellar systems.

In particular, as first pointed out by Motte *et al.* (1998) in the case of the ρ Ophiuchi (L1688) cloud, the mass distribution of these starless dust continuum condensations is remarkably similar in shape to the stellar IMF (Figure 11.2). This was consistently found by a number of independent groups in the past few years (Testi & Sargent 1998; Johnstone *et al.* 2000, 2001; Motte *et al.* 2001; Enoch *et al.* 2006; Stanke *et al.* 2006; Nutter & Ward-Thompson 2007) in nearby star-forming regions such as ρ Ophiuchi, Serpens, Orion A and B and Perseus. In all of these clouds, the observed prestellar CMF is consistent with the Salpeter (1955) power-law IMF at the high-mass end ($dN/d\log M \propto M^{-1.35}$) and thus significantly steeper than the mass distribution of diffuse CO clumps ($dN/d\log M \propto M^{-0.6}$ over at least three decades in mass – e.g. Blitz 1993; Kramer *et al.* 1998). The difference presumably arises because CO clumps are primarily structured by supersonic turbulence (Elmegreen & Falgarone 1996) while prestellar condensations are largely free of supersonic turbulence and clearly shaped by self-gravity (Motte *et al.* 2001; André *et al.* 2007 and Section 11.3.2). The slope of the observed CMF becomes

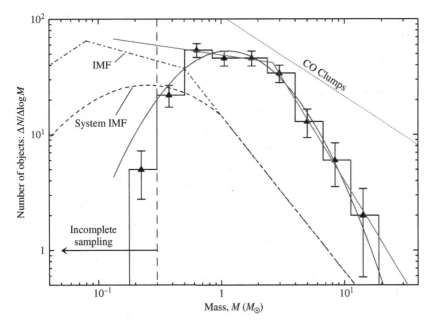

Fig. 11.2. Differential ($dN/d \log M$) mass distribution of the 229 starless dust continuum condensations detected at 850 μm with SCUBA in the Orion A/B cloud complex excluding the crowded OMC1 and NGC 2024 regions (histogram with error bars – adapted from Motte *et al.* 2001; Johnstone *et al.* 2001; and Nutter and Ward-Thompson 2007). This prestellar core sample is estimated to be complete down to ∼$0.3M_\odot$. A two-segment power-law fit and a log-normal fit are shown for comparison. The log-normal fit peaks at ∼$1.1M_\odot$ and has a standard deviation of ∼0.41 in $\log_{10} M_\odot$. For reference, the *dash-dotted curve* shows the shape of the single-star IMF (Kroupa 2001) and the *dashed curve* corresponds to the IMF of multiple systems (Chabrier 2005). (The log-normal part of the latter peaks at $0.25M_\odot$ and has a standard deviation of 0.55 in $\log_{10}M_\odot$.) The *dotted line* shows a $dN/d \log M \propto M^{-0.6}$ power-law distribution corresponding to the typical mass spectrum found for low-density CO clumps (see Blitz 1993; Kramer *et al.* 1998).

shallower than the Salpeter power law and more similar to the slope of the typical CO clump mass distribution at the low-mass end. Based on the results of present core surveys, the entire prestellar CMF can generally be fit equally well with either a two-segment broken power-law or a log-normal distribution down to the completeness limit of the observations. This is illustrated in Figure 11.2 which shows the differential CMF in log–log format for a sample of 229 starless submillimetre continuum cores in the Orion A/B complex identified in the SCUBA surveys of Motte *et al.* (2001) and Johnstone *et al.* (2001, 2006).

Note that there is some discussion in the literature (Reid & Wilson 2006) as to whether the differential or the cumulative form of the CMF should be used. The differential form is more intuitive to interpret as it can be fit using the standard least-squares technique after assigning simple Poisson error bars. It is, however, inadequate when the total number of objects is small ($\lesssim 100$) as it is significantly affected by the arbitrary choice of mass bins. The cumulative form, which is independent of binning, is preferable when dealing with small samples, but some care must be taken when comparing it with model distributions based on least-squares fits (cf. Reid & Wilson 2006). In practice, using the non-parametric Kolmogorov–Smirnov (K–S) test to compare the cumulative form of the CMF with model distributions is more reliable than least-squares fitting. The K–S test confirms that, within statistical uncertainties, the prestellar CMF observed in the above-mentioned nearby clouds is indistinguishable in shape from the stellar IMF, although the significance remains limited by the relatively small size of present prestellar core samples. The K–S test also shows that the observed CMF differs from the shallow power-law mass distribution of CO clumps at a very high significance level (for instance, in the Orion case illustrated in Figure 11.2, the probability that the two mass distributions are statistically similar in shape is only $P \approx 2 \times 10^{-6}$).

In addition, the median prestellar core mass observed in regions such as ρ Ophiuchi and Orion (~ 0.2–$1.5 M_\odot$) is only slightly larger than the characteristic $\sim 0.5 M_\odot$ set by the peak of the IMF in $dN/d\log M$ format. Such a close resemblance of the CMF to the IMF in both shape and mass scale is consistent with the view that the prestellar condensations identified in (sub)millimetre dust continuum surveys are about to form stars on a one-to-one basis, with a fixed and relatively high local efficiency, i.e. $\epsilon_{\mathrm{core}} \equiv M_\star / M_{\mathrm{core}} \gtrsim 30$–$50\%$.

Interestingly, in a recent near-IR extinction imaging study of the Pipe dark cloud, Alves *et al.* (2007) found a population of 159 starless cores whose mass distribution similarly follows the shape of the IMF. This finding is reminiscent of the CMF results obtained from the (sub)millimetre dust continuum, although it is important to stress that most of the starless cores in the Pipe Nebula are gravitationally unbound objects confined by external pressure (Lada *et al.* 2008). Hence, they do not qualify as prestellar cores and a large fraction of them may never evolve into stars. Assuming nevertheless that most of them will evolve into self-gravitating prestellar cores and subsequently collapse into stars, the Alves *et al.* (2007) result suggests that the IMF may be determined even earlier than the prestellar stage.

As appealing a direct connection between the prestellar CMF and the IMF might be, several caveats should be kept in mind. First, although core mass estimates based on optically thin (sub)millimetre dust continuum emission are straightforward, they rely on uncertain assumptions about the *dust (temperature and emissivity) properties* (Stamatellos *et al.* 2007a). Second, current determinations of the CMF are limited by small-number statistics in any given cloud and may be affected by incompleteness at the low-mass end (Johnstone *et al.* 2000).

With *Herschel*, the future submillimetre space telescope to be launched in 2009, it will be possible to dramatically improve on the statistics and to largely eliminate the mass uncertainties through direct measurements of the dust temperatures (cf. André & Saraceno 2005). Third, in some regions such as the ρ Ophiuchi cloud, the shape of the CMF is in better agreement with the IMF of individual field stars than with the IMF of multiple systems (André *et al.* 2007). This is surprising since the spatial resolution of current surveys for prestellar cores ($\sim$2000 AU at best) is not sufficient to probe core multiplicity. Furthermore, multiple systems are believed to form *after* the prestellar stage by subsequent dynamical fragmentation during the collapse phase, close to the time of protostar formation (Goodwin *et al.* 2007). Thus, one would expect the masses of prestellar cores to be more directly related to the masses of multiple systems than to the masses of individual stars. It is possible that a fraction at least of the cores observed at masses lower than the peak of the CMF in dN/d log M format (e.g., $\sim$1.1$M_\odot$ in Figure 11.2) are not gravitationally bound, hence not prestellar in nature (cf. André *et al.* 2007).

Last but not least, there is a potential timescale problem. As pointed out by Clark *et al.* (2007), if the lifetime of prestellar cores depends on their mass, then the observed mass distribution is not necessarily representative of the intrinsic CMF (see also Elmegreen 2000). This is due to the fact that an observer is more likely to detect long-lived cores than short-lived cores. In practice, however, the mean densities of prestellar cores are essentially uncorrelated with their masses, so that there is no systematic dependence of the dynamical timescale on the mass. The importance of the potential timescale bias can be assessed by considering a *weighted* core mass function in which each core is assigned a weight equal to $\langle t_{\rm ff}\rangle/t_{\rm ff} = \bar{\rho}^{1/2}/\langle\bar{\rho}^{1/2}\rangle$ (instead of 1), where $\langle t_{\rm ff}\rangle$ is the average free-fall time of the sampled cores. Such a weighting makes it possible to recover the intrinsic shape of the CMF if the lifetime of each core is proportional to its free-fall time. André *et al.* (2007) applied this technique to the sample of 57 starless dust continuum condensations identified by Motte *et al.* (1998) at 1.2 mm in the ρ Ophiuchi cloud. In this case, the above-mentioned weighting does not change the high-mass end of the CMF and only affects the low-mass end: it renders the entire ρ Oph CMF derived above the $\sim$0.1$M_\odot$ completeness level remarkably consistent with a single, Salpeter power-law mass distribution (see figure 8 of André *et al.* 2007). We conclude that the steep, Salpeter-like slope of the CMF at the high-mass end is robust but that the departure from a single power-law distribution at the low-mass end, in the form of a break near the median core mass (cf. Figure 11.2), is less robust.

Despite these limitations, the observational findings summarized in this section are very encouraging as they support scenarios according to which the bulk of *the IMF is partly determined by pre-collapse cloud fragmentation* (Larson 1985, 2005; Elmegreen 1997; Padoan & Nordlund 2002). The finding that the high-mass end of the prestellar CMF is substantially steeper than the $dN/d \log M \propto M^{-0.6}$ mass distribution of low-density CO clumps is very significant: while most of the mass of the low-density CO medium is contained in the largest, most massive clumps, most of the prestellar mass destined to evolve into stars is in small, low-mass cores. Clearly, one of the keys to the problem of the origin of the IMF lies in a good understanding of the processes responsible for the formation of prestellar cores/condensations out of low-density structures within molecular clouds. However, it is likely that additional processes, such as subfragmentation into binary/multiple systems and, more generally, mechanisms controlling the value of the star formation efficiency at the core level (ϵ_{core}), also play an important role and, in particular, are required to generate the low-mass ($M_\star < 0.3 M_\odot$) end of the IMF (cf. Bate *et al.* 2003; Ballesteros-Paredes *et al.* 2006). In the following sections, we discuss core formation and core subfragmentation models in turn.

11.3 Core formation models versus observational constraints

The mechanisms by which prestellar cores form and evolve in molecular clouds are the subject of a major theoretical debate at the moment. There is little doubt that self-gravity ultimately plays a dominant role and it has even been proposed that dense cores may form by purely gravitational fragmentation (Larson 1985; Hartmann 2002). However, the respective roles of magnetic fields and interstellar turbulence in regulating the core/star formation process are highly controversial. In particular, the classical picture of slow, quasi-static core formation by ambipolar diffusion in magnetically supported clouds (Mouschovias 1987; Shu *et al.* 1987, 2004) has been seriously challenged by a new, much more dynamic paradigm, which emphasizes the role of supersonic turbulence in supporting clouds on large scales and generating density fluctuations on small scales (Padoan & Nordlund 2002; Mac Low & Klessen 2004). Conceptually, core formation models may be conveniently divided into four categories, depending on whether linear or nonlinear (turbulent) perturbations initiate core formation and on whether magnetic fields are dynamically dominant or not (Table 11.1). In this classification, the 'standard' picture and the new turbulent paradigm correspond to two extreme models, according to which cores form by magnetically regulated gravitational fragmentation and super-Alfvénic turbulent fragmentation, respectively. In all turbulent models, cores are initially formed by cloud material compressed by shocks arising from supersonic turbulence. There are, however, several versions of the dynamic picture of core/star formation which mainly differ in the way they explain the origin of the IMF. In the Padoan and Nordlund (2002) scenario, the IMF is almost entirely set by the properties of interstellar turbulence at the prestellar stage, while in the alternative model proposed by Bate and Bonnell (2005), turbulence is largely

Table 11.1. *Summary of main features of core formation models*

	Core formation scenarios	
Models	Gravitational fragmentation (linear perturbations)	Turbulent fragmentation (non-linear perturbations)
Weak B	Short (few Myr) timescale infall mildly supersonic ordered curved field lines initial CMF very narrow	Very short (<Myr) timescale infall highly supersonic field lines distorted initial CMF is broad, IMF-like
Strong B	Long (∼10 Myr) timescale subsonic infall small field line curvature initial CMF very narrow	Short (few Myr) timescale subsonic relative infall and supersonic systematic speeds ordered field lines initial CMF is broad, IMF-like

irrelevant for the IMF which originates from competitive accretion and dynamical interactions at the protostellar stage. The Klessen and Burkert (2000) scenario is intermediate between these two extremes in that both turbulence and dynamical interactions play a role in shaping the IMF. The various turbulent models also differ depending on whether the turbulence is freely decaying (Tilley & Pudritz 2007) or continuously driven (Vázquez-Semadeni *et al.* 2005).

11.3.1 Theoretical description of cloud fragmentation models

The classical problems of fragmentation of a sheet-like layer or a cylinder are relevant to star formation because they show that there is a *preferred* scale of gravitational fragmentation as soon as one considers a structured (i.e. flattened or filamentary) region with some scale length H (see Larson 1985). This is not the case for a uniform medium, in which case the Jeans analysis shows that the largest possible scale has the fastest growth rate. Sheets and filaments can be easily generated by the formation process of the molecular cloud itself or by subsequent internal turbulent or gravitational motions. For isothermal sheets, $H = c_s^2/(\pi G \Sigma)$, where c_s is the isothermal sound speed and Σ is the column density of the sheet. For highly flattened sheets, the preferred fragmentation scale is $\lambda_m = 2\pi H$, while for a layer with the extended isothermal atmosphere calculated by Spitzer (1942), $\lambda_m = 4.4\pi H$ (Simon 1965). These two length scales likely bracket the possibilities for more realistic isothermal non-magnetic sheet-like configurations. The growth time of the fragmentation instability is essentially the dynamical timescale $t_d \simeq H/c_s$ for the cases described above and is more generally identified with the free-fall timescale $t_{ff} \simeq 1/\sqrt{G\rho}$ that applies to both unpressured sheets and those confined by a strong external pressure.

However, purely gravitational fragmentation instability of an entire molecular cloud on the dynamical timescale given by the mean column density and temperature is ruled out (Zuckerman & Palmer 1974), due to the observed low efficiency of galactic star formation. Nevertheless, it remains a relevant concept to understanding fragmentation in the cluster-forming subregions of molecular clouds. A proposed explanation for the low efficiency of star formation is the so-called standard model of star formation (Mouschovias 1987; Shu *et al.* 1987), in which cores are formed on a diffusive timescale (much longer than a dynamical timescale), due to the ambipolar drift of neutrals past dynamically dominant magnetic fields. This mode of star formation is often erroneously labelled as 'isolated' star formation. In reality, ambipolar diffusion-driven core formation is also a fragmentation process as surely as its non-magnetic counterpart – the difference is primarily in the timescales of evolution. One may, however, make the following distinction based on the mass-to-magnetic flux ratio, M/Φ, compared to the critical value, $(M/\Phi)_{crit}$, necessary for support against gravitational collapse. Molecular cloud envelopes may be magnetically dominated, with subcritical to transcritical mass-to-flux ratios, i.e. $\mu_{env} \equiv (M/\Phi)/(M/\Phi)_{crit} \leq 1$. Cloud envelopes also have low mean density and relatively high levels of ionization and thereby evolve on such a long timescale for fragmentation that no runaway has essentially occurred by the time we observe the clouds. Conversely, the cluster-forming cores (and also regions of weak clustering found in the Taurus molecular cloud) may be magnetically supercritical with $\mu_{clus} > 1$ and also have greater mean density, resulting in fragment formation on relatively shorter timescales and lengthscales.

For the above reasons and because Zeeman measurements (see the compilation of Crutcher 1999 and Section 11.3.2.3) establish that mass-to-flux ratios are clustered about the critical value, it is important to study cloud fragmentation including the effect of dynamically significant magnetic fields and also ambipolar diffusion, if possible. Magnetized sheets can be studied most easily in two limits. If the formation process of the sheet is by dynamical compression, say from stellar winds or supernovae, the ambient interstellar magnetic field may be swept up into the expanding shell and be oriented primarily in the plane of the sheet. Conversely, if self-gravity plays an important role in the formation of the sheet, then it will be flattened along the mean direction of the magnetic field.

When the magnetic field is in the plane of the sheet, the character of fragmentation and the fate of the cloud can be further categorized into two subcases (Nagai *et al.* 1998). If the geometrical thickness of the sheet is comparable to (or larger than) the natural scale height $\left(\sim c_s/\sqrt{G\rho_c}\right)$, or equivalently, if the ambient pressure is much smaller than the midplane pressure, compressional motions along the magnetic field lines result in the formation of filamentary clouds elongated perpendicular to the field lines. The line-mass (mass per unit length) of the filament is larger than that of the (isothermal) equilibrium filament so that collapse towards the axis of the filament continues until temperature increases (Inutsuka & Miyama 1992). In this case, the characteristic (i.e. minimum) mass scale for fragmentation is determined by the final fragmentation of the filamentary cloud and significantly smaller than the initial Jeans mass of the sheet (Inutsuka & Miyama 1997). The resulting mass scale can be called 'the minimum Jeans mass' and its dependence on the initial temperature and metallicity of the cloud was discussed by Masunaga and Inutsuka (1999). The expected CMF was analytically derived by Inutsuka (2001) using the Press–Schechter formalism. The overall timescale of the whole process is of the order of the initial free-fall time $\left(t_{\rm ff} \sim 1/\sqrt{G\rho_{\rm c}}\right)$.

In contrast, if the sheet-like cloud is confined by strong external pressure due to an ambient warm (or hot) medium, the thickness of the sheet is smaller than $\sim c_{\rm s}/\sqrt{G\rho_{\rm c}}$. The timescale for fragmentation is still of the order of the gravitational free-fall time $t_{\rm ff}$, but sound waves can now propagate many times across the (thin) sheet within this timescale. Thus, the fragmentation is in an incompressible mode and the axes of the resulting filaments are parallel to the direction of the magnetic field. In this case, the filaments may fragment into cores whose masses can be smaller than the Jeans mass. In effect, this second case provides a mechanism for generating gravitationally stable cores (Inutsuka & Miyama 1997).

When the magnetic field is oriented perpendicular to the sheet, recent simulations of cloud fragmentation which include ambipolar diffusion have allowed an extensive parameter study of the interplay of magnetic fields, ambipolar diffusion and turbulence in the formation of cores (Basu & Ciolek 2004; Basu *et al.* 2008a). Given the standard cosmic-ray-induced ionization fraction $x_i \simeq 10^{-7}(n/10^4\,{\rm cm}^{-3})^{-1/2}$ (Elmegreen 1979; Nakano 1979), there is an observationally distinguishable difference between subcritical and supercritical fragmentation. Subcritical fragmentation leads to subsonic infall motions onto cores and within them, while supercritical cloud fragmentation leads to extended supersonic infall on the core scale ~ 0.1 pc and even beyond.

Figure 11.3 shows a comparison of column density structure and velocity vectors for four different models (from Basu *et al.* 2009a,b). Each image is shown at the time of runaway collapse of the first supercritical dense core that forms in the simulation, with a column density enhancement of a factor of 10. The top two images are for subcritical (*left*) and supercritical (*right*) clouds evolving from linear initial perturbations. The bottom two images are for subcritical (*left*) and supercritical (*right*) clouds evolving from highly non-linear (turbulent) initial conditions. Each image represents a physically distinct path to core/star formation. Velocity vectors are overlaid, although the normalizations may differ (see caption). The subcritical

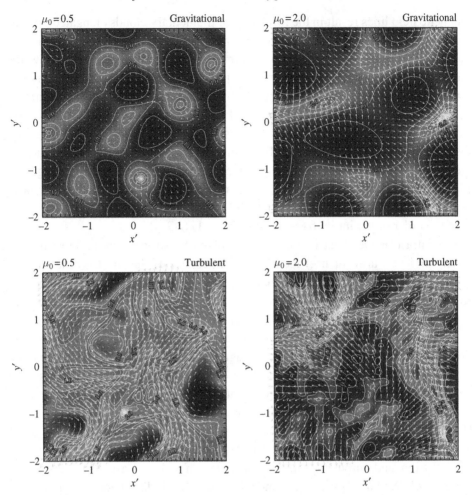

Fig. 11.3. Image and contours of column density and velocity vectors of neutrals for four different models at the time of runaway collapse of the first supercritical dense core (Basu *et al.* 2009a,b). *Top left*: gravitational fragmentation for a subcritical model $\mu_0 = 0.5$. *Top right*: the same for a supercritical model with $\mu_0 = 2.0$. *Bottom left*: turbulent fragmentation for $\mu_0 = 0.5$. *Bottom right*: turbulent fragmentation for $\mu_0 = 2.0$. The image is of the logarithm of column density and the contour lines represent values of column density enhancement in multiplicative increments of $2^{1/2}$, having the values $[0.7, 1.0, 1.4, 2, 2.8, 4.0, \ldots]$. The horizontal or vertical distance between foot points of velocity vectors corresponds to a speed $0.5\,c_s$ in the top panels, to $1.0\,c_s$ in the bottom left panel and to $3.0\,c_s$ in the bottom right panel. Spatial coordinates are normalized to $2\pi H$, the wavelength of maximum growth rate in the limit of no magnetic field and external pressure.

models have $\mu_0 = 0.5$ (i.e. the mass-to-flux ratio is half the critical value) and the supercritical models have $\mu_0 = 2$. Amongst the gravitational fragmentation models (linear initial perturbations), there is extended and mildly supersonic motions only in the supercritical model, whereas the subcritical model has a maximum speed of only $\sim 0.4 c_s$. For the subcritical gravitational fragmentation model, the runaway collapse to the first core occurs at a time 7.4 Myr, whereas for the supercritical model it occurs at 0.83 Myr, assuming a background column density 10^{22} cm^{-2} and temperature 10 K. Fragmentation of a transcritical ($\mu_0 \simeq 1$) cloud is qualitatively unique in that the fragmentation spacing is much larger than the typical value $\simeq 2\pi H$ that is valid for both highly supercritical and highly subcritical clouds (Ciolek & Basu 2006). The timescale of transcritical fragmentation is intermediate to the subcritical and supercritical cases, but closer to the former. Here, we focus primarily on the distinction between decidedly subcritical and supercritical clouds. The *bottom panels* of Figure 11.3 show the corresponding $\mu_0 = 0.5$ (*left*) and $\mu_0 = 2$ (*right*) models, now with turbulent initial perturbations with rms amplitude $v_a = 2 c_s$ in each of v_x and v_y. The power spectrum is such that $v_k^2 \propto k^{-4}$, so most of the energy is in the largest scale modes: the result is immediate large-scale compressive motions in the cloud. The $\mu_0 = 0.5$ model undergoes rapid ambipolar diffusion during the first compression, but still rebounds due to stored magnetic energy. It subsequently oscillates several times before continuing ambipolar diffusion causes runaway collapse in the highest density region. The total time for this process is 1.2 Myr, similar to that of supercritical gravitational fragmentation. In contrast, the $\mu_0 = 2$ model goes into prompt collapse during the first compression, on a timescale of merely $4.0 \times 10^4 = 0.04$ Myr. Systematic supersonic motions exist in both turbulent models, but the relative infall is generally subsonic in the subcritical model. A comparison of the two *bottom panel* images shows that the supercritical model is extremely filamentary, while the subcritical model is much less so since it has had a chance to rebound from the initial compression. These turbulent simulations are done in the thin-disk approximation and are consistent with the earlier results of Li and Nakamura (2004) and Nakamura and Li (2005), also done using the same approximation. Recent fully 3D simulations (Kudoh *et al.* 2007; Kudoh & Basu 2008) also confirm the results. A summary of main outcomes of the four different modes of core/star formation is given in Table 11.1.

The results described above are used to generate statistics of core masses, sizes, shapes, etc. Each model with a unique set of parameters is run ~ 100 times in order to generate a meaningful measure of the various outcomes arising from different realizations of the initial random perturbations. Thresholding techniques are used to identify cores – for details of the technique, see Basu *et al.* (2009a). Figure 11.4 shows a comparison of the initial CMF for supercritical models with $\mu_0 = 2$ but

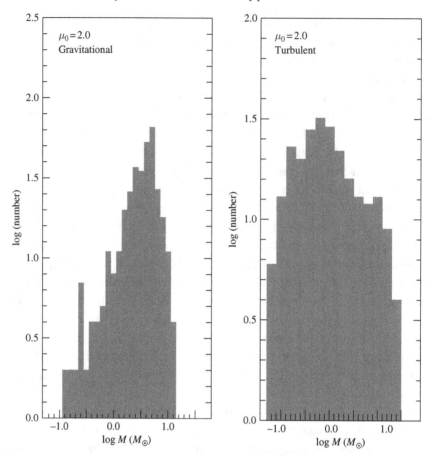

Fig. 11.4. Histograms of masses contained within regions with column density enhancement above a factor of 2, measured at the end point of simulations with $\mu_0 = 2.0$ (Basu *et al.* 2009a,b). The *left panel* corresponds to gravitational fragmentation and the *right panel* corresponds to turbulent fragmentation. See text for details. Each figure is obtained from the compilation of results of a large number of simulations. The bin width is 0.1.

without and with turbulent initial perturbations, respectively. We emphasize that these are *initial* CMFs because they reflect the status of cores at the time that the first supercritical dense core undergoes runaway collapse. The clear distinction to be made between the two models is that the linear initial perturbation case (*left panel*) has an extremely sharp peak (consistent with a preferred mass scale for fragmentation in the linear theory), although there is a broader tail at the low-mass end due to some cores that are very young and just emerging above the threshold. In contrast, the turbulent initial condition case develops a broad tail of high-mass cores, due to much of the turbulent power being on large scales. However, the high-mass cores often have quite disturbed structures (see Figure 11.3, *lower panels*), and it is not clear that they would subsequently collapse monolithically.

The narrow initial CMF seems to be a problem for purely gravitational fragmentation models. However, a broader distribution may yet be possible if the newly formed cores continue to accrete from their environment. If the accretion from the environment is non-uniform from core to core, an initially narrow and/or log-normal CMF may become skewed so that a high-mass tail develops. For example, Basu and Jones (2004) have generalized an earlier result of Myers (2000) and shown that an initially log-normal CMF and an exponential distribution of subsequent accretion times from the cloud results in a CMF that is like a log-normal at low masses but has a power-law tail at high masses. The model formally requires that the accretion rate onto a core is linearly proportional to the instantaneous core mass. Additionally, Basu and Jones (2004) show that a different accretion law can also lead to a broadened near-power-law tail.

11.3.2 Observational diagnostics

In principle, it should be possible to discriminate between these various core formation scenarios based on detailed observational studies of the characteristics of prestellar cores and young protostars.

11.3.2.1 Core formation efficiency and spatial distribution from surveys

A number of very large submillimetre continuum surveys of nearby cloud complexes have been completed recently which provide the spatial distribution of cores within these complexes and set constraints on the efficiency of the core formation process (Johnstone *et al.* 2004; Hatchell *et al.* 2005; Enoch *et al.* 2006; Motte *et al.* 2007). As an illustration, Figure 11.5 shows part of the 3 deg^2 SCUBA 850-μm survey of the Perseus cloud complex by Hatchell *et al.* (2005). Such extensive surveys show that prestellar cores and Class 0 protostars are found in localized subregions within molecular clouds which occupy only a very small fraction of their volume. These localized active subregions often correspond to cluster-forming clumps associated with embedded near-IR clusters. On this basis, it has been suggested that there may be a threshold in background column density (or, equivalently, visual extinction at $A_V \sim 5$–10) for core formation (Onishi *et al.* 1998; Johnstone *et al.* 2004). Observationally, however, establishing the presence (or absence) of such a threshold is difficult since there are also detection thresholds. According to Hatchell *et al.* (2005), there is no real threshold but the probability of forming a prestellar core is a steeply rising function of background column density. In any case, the results of these wide-field surveys for cores clearly demonstrate the global inefficiency of the core formation process: the fraction of cloud mass observed in

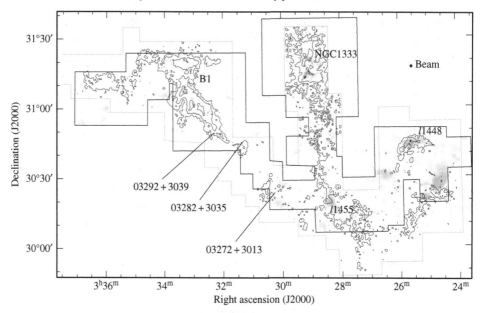

Fig. 11.5. SCUBA 850-μm dust continuum map (greyscale) of the western part of the Perseus cloud complex. Contours of $C^{18}O(1-0)$ integrated intensity are overlaid. The solid black boundary indicates the total area mapped with SCUBA, while the grey boundary marks the area mapped in $C^{18}O$. Note that dust continuum cores are detected only in localized subregions, such as NGC∼1333, within the complex. (Adapted from Hatchell *et al.* 2005.)

the form of prestellar cores is very low (≲1 − 20% – Hatchell *et al.* 2005; Nutter *et al.* 2006). Furthermore, there is some evidence that core/star formation in the observed cluster-forming clumps has been induced by external triggers (Kirk *et al.* 2006; Nutter *et al.* 2006). These results are broadly consistent with the view that the low-density envelopes of molecular clouds are supported against collapse by magnetic fields, as in the classical ambipolar diffusion picture (Mouschovias 1987; Shu *et al.* 1987; McKee 1989).

11.3.2.2 Core lifetimes

In the turbulent paradigm, cloud cores are always dynamically evolving (Dib *et al.* 2007) and survive for at most a few free-fall times. Observationally, a rough estimate of the lifetime of starless cores can be obtained from the number ratio of cores with and without embedded YSOs in a given core sample. Using this technique, Lee and Myers (1999) found that the typical lifetime of starless cores with average volume density $\sim 10^4\,cm^{-3}$ was $\sim 1-1.5 \times 10^6$ yr. Furthermore, by considering several samples of isolated cores spanning a range of core densities, Jessop and Ward-Thompson (2000) established that the typical core lifetime decreased as the

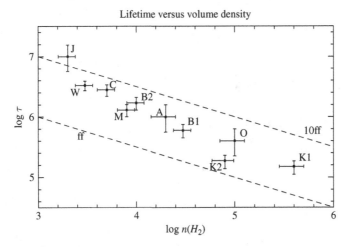

Fig. 11.6. Plot of inferred core lifetime against mean volume density for various samples of starless and prestellar cores. The two *dashed lines* correspond to one and ten free-fall times, respectively. (Adapted from Jessop & Ward-Thompson 2000 and Kirk *et al.* 2005.)

mean volume density in the core sample increased (Figure 11.6 – see also Kirk *et al.* 2005). As can be seen in Figure 11.6, all of the observed lifetimes lie between one free-fall time, which is the timescale expected in free-fall collapse, and ten free-fall times, the timescale expected for highly subcritical cores undergoing ambipolar diffusion. The observed timescales are typically longer than the free-fall time by a factor of $\sim$2–5 in the density range of 10^4–10^5 cm^{-3}. This suggests that starless cores cannot all be rapidly evolving, at variance with purely hydrodynamic scenarios of core formation (Ballesteros-Paredes *et al.* 2003) but in agreement with numerical simulations of moderately supercritical, turbulent molecular clouds (Galván-Madrid *et al.* 2007). The observed timescales also appear to be too short for all of the cores to form from a highly subcritical state. These statistical estimates of core lifetimes are, however, quite uncertain since they assume that all of the observed cores follow the same evolutionary path and that the core/star formation rate is constant.

11.3.2.3 Magnetic field measurements

In principle, observations of magnetic fields can provide a strong discriminator between the two extreme paradigms of the core/star formation process. Two main observational techniques have been used to estimate the magnetic field strength in cloud cores. First, the Zeeman effect in, e.g., the 18-cm lines of OH, gives a direct measurement of the line-of-sight component of the magnetic field, but relatively few positive detections have been obtained. For existing detections, a

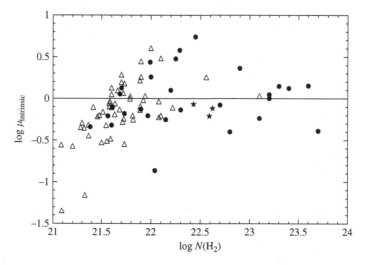

Fig. 11.7. Plot of observed mass-to-magnetic flux ratio, in units of the critical value and divided by 3 to correct for projection bias, against cloud/core column density. Dots are for Zeeman detections of $B_\parallel$ (above 3σ); stars are for Chandraskehar–Fermi estimates of $B_\perp$; open triangles are lower limits (corresponding to observed upper limits to the field strength). (Adapted from Heiles and Crutcher 2005 – see also Crutcher 2004 for details.)

very good correlation is found between the magnetic field strength B_{los} and $n^{0.5}\,\sigma_v$, where n is the gas density and σ_v is the line-of-sight velocity dispersion observed in optically thin molecular line tracers (Crutcher 1999; Basu 2000). This shows that magnetic fields play a dynamically important role during the contraction of at least some cloud cores and is consistent with cloud turbulence being MHD in character. Second, maps of linearly polarized dust emission at submillimetre wavelengths, combined with the Chandrasekhar and Fermi method, provide an indirect estimate of the plane-of-the-sky component of the magnetic field (Ward-Thompson *et al.* 2000; Crutcher *et al.* 2004). The available B-field measurements based on either the Zeeman technique or the Chandrasekhar and Fermi method are summarized in Figure 11.7, which plots the inferred mass-to-magnetic flux ratio corrected for projection effects against the column density of each core (see Crutcher 2004; Heiles & Crutcher 2005). It can be seen that all cloud cores are scattered around the critical mass-to-flux ratio in this plot, suggesting that, on average, cores are close to magnetically critical. Furthermore, there is some hint in Figure 11.7 that the mass-to-flux ratio may be systematically subcritical at column densities lower than $\sim 3 \times 10^{21}$ cm^{-2}. This tentative trend is weak, however, as it relies only on the locations of Zeeman non-detections in the diagram. Thus, existing magnetic field measurements do not allow a definite conclusion to be drawn, even if they seem to favour pictures of the core formation process in which the magnetic field does play an important role.

11.3.2.4 Radial density structure

The density profiles of isolated prestellar cores are now fairly well known. Two methods have been used: (i) mapping the optically thin (sub)millimetre continuum *emission* from the cold dust contained in the cores and (ii) mapping the same cold core dust in *absorption* against the background infrared emission (originating from warm cloud dust or remote stars).

Ward-Thompson *et al.* (1994, 1999) and André *et al.* (1996) employed the first approach to probe the structure of prestellar cores (see also Shirley *et al.* 2000). Under the simplifying assumption of spatially uniform dust temperature and emissivity properties, they concluded that the radial density profiles of isolated prestellar cores were flatter than $\rho(r) \propto r^{-1}$ in their inner regions (for $r \leq R_{\text{flat}}$) and approached $\rho(r) \propto r^{-2}$ only beyond a typical radius $R_{\text{flat}} \sim 2500$–$5000$ AU.

More recently, the use of the *absorption* approach, both in the mid-IR from space (Bacmann *et al.* 2000) and in the near-IR from the ground (Alves *et al.* 2001), made it possible to confirm and extend the (sub)millimetre emission results, essentially independently of any assumption about the dust temperature distribution. In some cases, such as L1689B, the absorption studies indicate that isolated prestellar cores feature sharp edges defining outer radii $R_{\text{out}} \sim 0.1$ pc (cf. Bacmann *et al.* 2000).

The circularly averaged column density profiles can often be fit remarkably well with models of pressure-bounded Bonnor–Ebert spheres, as first demonstrated by Alves *et al.* (2001) for B68. This is also the case of cores detected in the submillimetre continuum such as L1689B (Kirk *et al.* 2005). The quality of the fits shows that equilibrium Bonnor–Ebert spheroids provide a good, first-order model for the structure of isolated prestellar cores. In detail, however, there are problems with this model. First, the inferred density contrasts (from centre to edge) are generally larger (i.e. $\gtrsim 20$–80 – cf. Bacmann *et al.* 2000; Kirk *et al.* 2005) than the maximum contrast of ~ 14 for stable Bonnor–Ebert spheres. Second, the effective gas temperature needed in the fits is often significantly larger than measured core temperatures (Ward-Thompson *et al.* 2002; Lai *et al.* 2003). These arguments suggest that prestellar cores are either already contracting (see Lee *et al.* 2001 and Section 11.3.2.5) or experiencing extra support from static or turbulent magnetic fields (Curry & McKee 2000). As shown by Bacmann *et al.* (2000), one way to account for large-density contrasts and high effective temperatures is to consider models of cores initially supported by a static magnetic field and evolving through ambipolar diffusion (Basu & Mouschovias 1994; Ciolek & Mouschovias 1994).

However, good Bonnor–Ebert fits can often be found for dynamically evolving 'cores' produced by turbulent compression (Ballesteros-Paredes *et al.* 2003;

Gómez *et al.* 2007). Thus, the observed density profiles do not provide a very strong diagnostic of proposed core formation models.

11.3.2.5 Velocity structure

Observing the velocity field within and around dense cores is probably the most effective way to discriminate between quasi-static and dynamic core formation scenarios. If cores are produced by shocks in large-scale supersonic flows, large-velocity gradients and local maxima of the line-of-sight velocity dispersion are expected in the immediate vicinity of cores (Ballesteros-Paredes *et al.* 2003; Klessen *et al.* 2005). A much more quiescent ambient velocity field is expected in the magnetically controlled picture (Nakamura & Li 2005 – see also Section 11.3.1 and Figure 11.3).

Briefly, isolated starless cores are characterized by subsonic levels of internal turbulence (Myers 1983; Goodman *et al.* 1998; Caselli *et al.* 2002), small rotational velocity gradients (Goodman *et al.* 1993; Caselli *et al.* 2002) and extended, subsonic infall motions (Tafalla *et al.* 1998; Lee *et al.* 2001). Moreover, the environment of isolated prestellar cores is also extremely quiescent. This has been shown recently through deep mapping of several cores in low-density molecular gas tracers such as $^{13}CO(1-0)$ and $C^{18}O(1-0)$. Figure 11.8a shows a grey-scale

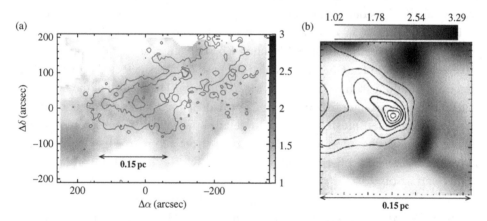

Fig. 11.8. (a) Grey-scale image of the line-of-sight velocity dispersion (in units of the isothermal sound speed) derived from deep $^{13}CO(1-0)$ line observations towards and around the prestellar core L1689B. (Adapted from André, *et al.* 2009) Column density contours from Bacmann *et al.* (2000) are superimposed and delineate the core boundaries. The maximum observed value of σ_{los} is only $\sim 0.34\,km s^{-1}$ or $\sim 1.7 c_s$ (b) Grey-scale image of the line-of-sight velocity dispersion superimposed on column density contours for a model core obtained in SPH numerical simulations of gravoturbulent fragmentation. (Adapted from Klessen *et al.* 2005.) Note that these simulations produce localized maxima where $\sigma_{los} \gtrsim 3 c_s$ at the shock positions.

map of the line-of-sight velocity dispersion σ_{los} observed in ^{13}CO(1–0) towards the prestellar core L1689B in the Ophiuchus complex (André *et al.* in preparation). It can be seen that σ_{los} remains at most transonic ($\sigma_{los} < 2c_s$, where c_s is the isothermal sound speed) everywhere in a region of more than 0.25 pc in diameter around the column density peak. Such a quiescent velocity field is at variance with purely hydrodynamic models of gravoturbulent fragmentation (Klessen *et al.* 2005). While these models successfully produce a fair amount ($\sim 25\%$) of cores with subsonic internal velocity dispersions, corresponding to dense, post-shock stagnation points at the intersection of converging flows, they also produce highly supersonic maxima of σ_{los} in the low column density gas around the cores (cf. Figure 11.8b), which are not observed. Current observations therefore provide strong indirect evidence that the evolution of isolated prestellar cores is magnetically controlled.

The environment of individual cores in cluster-forming regions is known to be more turbulent (Caselli & Myers 1995), so that hydrodynamic core formation models may be more appropriate in this case. Indeed, observations of Class 0 objects indicate that protostellar collapse is more dynamic, with supersonic infall velocities and large mass accretion rates ($> 10\, c_s^3/G$), in cluster-forming clumps (Di Francesco *et al.* 2001; Belloche *et al.* 2006). Evidence of coherent, supersonic contraction motions over more than 0.5 pc has even been found in some proto-clusters (Motte *et al.* 2005; Peretto *et al.* 2006). On small scales, however, the compact (~ 0.03 pc) prestellar condensations of the Ophiuchus, Serpens, Perseus and Orion protoclusters are themselves characterized by subsonic levels of internal turbulence (Myers 2001, André *et al.* 2007) reminiscent of the thermal cores of Taurus. Furthermore, the condensation-to-condensation velocity dispersion measured in these cluster-forming regions is small and only subvirial (André *et al.* 2007). This is consistent with the view that protoclusters often start their evolution from 'cold', out-of-equilibrium initial conditions (cf. Adams *et al.* 2006), perhaps as a result of external perturbations (cf. Nutter *et al.* 2006). The observed small relative velocity dispersion also implies that collisions between condensations in a low-mass protocluster such as L1688 are relatively rare and that, in general, prestellar condensations do not have time to interact with one another before evolving into pre-main sequence objects (André *et al.* 2007). Furthermore, in such protoclusters, the mass accretion rate expected from competitive, Bondi-like accretion of background gas onto the condensations (Bonnell *et al.* 2001) is estimated to be at least a factor ~ 3 lower than the mass infall rate resulting from gravitational collapse at the Class 0 and Class I stages (André *et al.* 2007 – see also Krumholz *et al.* 2005). Therefore, competitive accretion cannot play a dominant role once individual protostellar collapse sets in. However, Bondi-like accretion of unbound gas is more effective before protostellar collapse and may possibly govern the growth of starless, self-gravitating condensations initially produced by gravitational fragmentation (cf. Section 11.3.1 and Figure 11.4) towards a Salpeter-like IMF mass spectrum (cf. Myers 2000; Basu & Jones 2004; Clark & Bonnell 2005).

11.4 Collapse and subfragmentation of prestellar cores

11.4.1 Core collapse models: thermodynamics

The evolution of gravitationally collapsing cores and the formation of protostars are radiation-magnetohydrodynamical processes. Although we need to model these processes by solving equations of radiative transfer and magnetohydrodynamics simultaneously in multi-dimensions, a direct calculation of all the equations remains challenging. The most sophisticated theoretical models so far are either non-magnetic radiation hydrodynamical calculations based on the (flux-limited) diffusion approximation or magnetohydrodynamical calculations with some prescribed equations of states.

Here, we first explain the thermodynamics of gravitational collapse revealed by dynamical modelling with detailed radiative transfer in *spherical* symmetry (Larson 1969; Narita *et al.* 1970; Stahler *et al.* 1980; Winkler & Newman 1980; Masunaga *et al.* 1998; Masunaga & Inutsuka 2000a). The classical results based on grey-approximation were confirmed by recent work that solved the frequency-dependent RHD equations by the 'Variable Eddington Factor Method' (Masunaga & Inutsuka 2000a). The latter provides the time evolution of the apparent spectrum of the radiation field (Spectral Energy Distribution – SED) in addition to the detailed dynamical evolution of the protostar.

Figure 11.9 shows the time evolution of the central temperature (as a function of the central density) in a collapsing cloud. Once compressional heating dominates over radiative cooling, the central temperature increases gradually above the low value (~ 10 K) found in molecular clouds. At first, the slope of the temperature curve corresponds to a ratio of specific heats $\gamma_{\mathrm{eff}} = 5/3$: $T(\rho) \propto \rho^{2/3}$ for 10 K $<T<10^2$ K. This apparently monoatomic gas behaviour is due to the fact that the rotational degree of freedom of molecular hydrogen is not excited in this low-temperature regime ($E(J = 2-0)/k_{\mathrm{B}} = 512$ K). When the temperature rises above $\sim 10^2$ K, the slope becomes that of diatomic molecules ($\gamma_{\mathrm{eff}} = 7/5$). In both cases, the effective ratio of specific heats is larger than the critical value for gas pressure support against self-gravity: $\gamma_{\mathrm{eff}} > \gamma_{\mathrm{crit}} \equiv 4/3$. Thus, the collapsing velocity is decelerated and forms a shock at the surface of a quasi-adiabatic hydrostatic object, the 'first core'. Its radius is about 1 AU in spherically symmetric calculations (but is larger by an order of magnitude in 2D/3D calculations with rotation). It mainly consists of H_2. The increase in density and temperature inside the first core is slow but

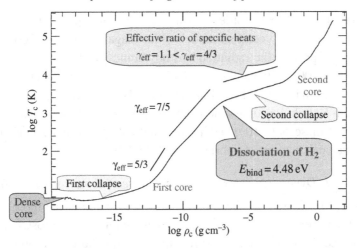

Fig. 11.9. Temperature evolution at the centre of a gravitationally collapsing cloud obtained by Masunaga and Inutsuka (2000a) in their radiation hydrodynam-ical calculation of protostellar collapse in spherical symmetry. The *first collapse* phase corresponds to the formation of the *first protostellar core* that consists mainly of hydrogen molecules. The dissociation of hydrogen molecule triggers the *second collapse* that eventually produces the *second core*, i.e. a protostellar object. Each of these phases in the temperature evolution is characterized by a distinct value of the effective ratio of specific heats, γ_{eff}.

monotonic. When the temperature becomes $>10^3$ K, the dissociation of H_2 starts. The binding energy of H_2 is about 4.5 eV which is much larger than the thermal energy per hydrogen molecule in this temperature regime. Therefore, the dissocia-tion of H_2 acts as an efficient coolant of the gas, which reduces the effective ratio of specific heats below the critical value ($\gamma_{\mathrm{eff}} < 4/3$) and triggers the second dynami-cal collapse.[1] In this 'second collapse' phase, the collapsing velocity becomes very large and engulfs the first core. As a result, the first core lasts only for $\sim 10^3$ yr. In the course of the second collapse, the central density reaches a stellar value ($\rho_\star \sim 1\,\mathrm{g\,cm}^{-3}$) and a truly hydrostatic protostellar object forms in the centre.

Radiation hydrodynamic (RHD) calculations automatically produce the time evolution of the accretion luminosity and SED of the collapsing object (cf. Masunaga & Inutsuka 2000a). The resultant luminosity evolution has a sharp growth at the formation of the first core and a peak around the formation of the second protostellar core in the case of dynamical initial conditions, while it only shows a gradual growth in the case of hydrostatic equilibrium initial conditions. This difference in the time evolution of the accretion luminosity may provide a useful observational diagnostic.

---

[1] Note that the role of H_2 dissociation in this second collapse is analogous to that of the photo-disintegration of Fe as the cause of the pre-supernova gravitational collapse.

Molecular emission line profiles of various important species were also calculated in self-consistent dynamical models (Masunaga & Inutsuka 2000b). Recent 3D modelling of protostellar radiation hydrodynamics can be found, e.g. in Whitehouse and Bate (2006), Stamatellos *et al.* (2007b), and Krumholz *et al.* (2007).

Further evolution includes the T-Tauri phase on Hayashi tracks, for which the relevant (Kelvin–Helmholtz) timescale is about two orders of magnitude larger than the dynamical timescale ($\sim 10^5$ yr) of protostellar collapse, which is only accessible by steady-state calculations (Chabrier & Baraffe 2000).

11.4.1.1 Dynamical roles of the first and second protostellar cores

The successive formations of the short-lived first core and of the more stable second core play important roles in the dynamical evolution of protosteller collapse. This is because the associated pressure support helps the formation of a rapidly rotating disk-like structure around the central object in the presence of non-zero initial angular momentum (Saigo *et al.* 2000).

11.4.1.1.1 Rotationally driven fragmentation into multiple systems. The mechanisms responsible for the formation of multiple systems have been the subject of extensive theoretical work (see, e.g., Goodwin *et al.* 2007 for a review). The essence of the results may be summarized as follows. First, gravitational subfragmentation within a collapsing cloud core is not expected during the early isothermal phase of evolution, unless the mass of the core is much larger than the initial Jeans mass. This result has been found in numerical simulations and is supported by semi-analytic calculations in the case of an initially uniform cloud with solid rotation (Tsuribe & Inutsuka 1999a, b). Initial central concentration in the density profile makes core fragmentation even more difficult. However, if the evolution is followed to the higher density regime where the gas becomes adiabatic, a disk-like structure forms which allows another mode of binary formation to develop, i.e. disk fragmentation around the central protostar. For example, calculations based on a piecewise polytropic equation of state show that the central portion of a collapsing core becomes adiabatic and forms a disk-like structure around the central object, which subsequently fragments into 'satellite' objects (Matsumoto & Hanawa 2003). The study of realistic radiative transfer effects on these modes of binary fragmentation remains an important task for future work. The above phenomena correspond to the fragmentation of/around the first core, which might account for the formation of binaries with separations $\gtrsim 1$ AU. Obviously, formation of binaries with even shorter separations is also expected in the disk around the second protostellar core. These multiple epochs of core fragmentation may result in multiple peaks in the separation distribution of binary stars as proposed by Machida *et al.* (2008b).

The effects of initial turbulence on binary fragmentation have also been studied extensively (see, e.g., Goodwin *et al.* 2007). Purely hydrodynamic SPH simulations of rotating cloud core collapse show that a very low level of initial core turbulence (e.g., $E_{turb}/E_{grav} \sim 5\%$) leads to the formation of a multiple system (Goodwin *et al.* 2004; Hennebelle *et al.* 2004). In such hydrodynamic simulations, fragmentation is driven by a combination of rotation/turbulence and occurs in large ($\gtrsim 100$ AU) disk-like structures or 'circumstellar accretion regions' (CARs – cf. Goodwin *et al.* 2007). These CARs are highly susceptible to spiral instabilities which *always* fragment them into small-*N multiple systems with N* > 2 and typically *N* ~ 3–4 within a radius ~150 AU (Goodwin *et al.* 2004). However, recent MHD simulations of *magnetized* core collapse (Price & Bate 2007; Hennebelle & Teyssier 2008) show that the presence of an even moderate magnetic field strongly modifies angular momentum transport during collapse and at least partly suppresses core fragmentation. Therefore, it is unclear at the present time whether the collapse of an individual prestellar core typically produces one, two or more stars.

11.4.1.1.2 Implications for the CMF–IMF connection. If each prestellar core is the progenitor of *N* ~ 2–4 stars, then the IMF of individual stars cannot be the direct product of the CMF but results instead from the convolution of the CMF with the typical distribution of object masses produced by binary fragmentation for one mass of core (cf. Delgado-Donate *et al.* 2003; Goodwin *et al.* 2008). For realistic binary fragmentation scenarios, the IMF will still follow the CMF at the high-mass end (because the majority of each core's mass can still end up in one stellar component) but may differ substantially from the CMF at the low-mass end. Such a picture is consistent with present determinations of the CMF (see Section 11.2). However, it is presently unclear whether the collapse of an individual prestellar core typically produces one, two or more stars. Accordingly, the origin of the low-mass end ($\lesssim 0.1 M_{\odot}$) of the IMF is highly uncertain. Observationally, this is an area where the future large-millimetre interferometer ALMA (Atacama large millimeter array) will yield key progress.

11.4.1.1.3 Driving MHD outflows. Another effect of the formation of the first core occurs in the MHD evolution of the self-gravitating collapsing core. A magnetically supercritical core whose rotation axis is parallel to the mean direction of the magnetic field lines leads to self-similar collapse as long as the equation of state is isothermal (Basu & Mouschovias 1994). Once the first core is formed, a rapidly rotating disk-like structure develops owing to the change in the effective equation of state and its rapid rotation winds up the field lines creating a significant amount of toroidal magnetic field. This enhanced toroidal magnetic field produces a bipolar outflow driven by magnetic pressure (Tomisaka 2002). Thus, the formation of the first hydrostatic core plays a critical role in launching the protostellar outflow. A similar process happens again around the second core, with higher ejection velocities reminiscent of the observed optical jets and high-velocity neutral winds (Machida *et al.* 2006, 2007, 2008a).

The observational detection of the first core would not only confirm the predictions of RHD modelling but would also set strong constraints on MHD models of protostellar outflows as described in Section 11.4.2.

11.4.2 Resistive MHD effects and onset of outflows

11.4.2.1 MHD modelling with resistivity

The full modelling of magnetohydrodynamical processes in star formation must include non-ideal MHD effects (Nakano *et al.* 2002; Tassis & Mouschovias 2007). Ambipolar diffusion is important at early times during the low-density core formation phase, but Ohmic dissipation is more important at later times in the high-density collapse phase. The Hall current term can also be important in an intermediate regime (Wardle 2004). To account for the dissipation of magnetic fields during the formation of protostars, Machida *et al.* (2006, 2007, 2008a) used the resistive MHD equation with prescribed resistivity in their 3D nested grid code simulations. Their basic equations are as follows:

$$\frac{\partial \rho}{\partial t} + \nabla \cdot (\rho \boldsymbol{v}) = 0, \tag{11.1}$$

$$\rho \frac{\partial \boldsymbol{v}}{\partial t} + \rho (\boldsymbol{v} \cdot \nabla) \boldsymbol{v} = -\nabla P - \frac{1}{4\pi} \boldsymbol{B} \times (\nabla \times \boldsymbol{B}) - \rho \nabla \phi, \tag{11.2}$$

$$\frac{\partial \boldsymbol{B}}{\partial t} = \nabla \times (\boldsymbol{v} \times \boldsymbol{B}) + \eta \nabla^2 \boldsymbol{B}, \tag{11.3}$$

$$\nabla^2 \phi = 4\pi G \rho, \tag{11.4}$$

where ρ, $\boldsymbol{v}$, P, $\boldsymbol{B}$, η and ϕ denote the density, velocity, pressure, magnetic flux density, resistivity and gravitational potential, respectively. Machida *et al.* estimated the resistivity η in Eq. (11.3) according to Nakano *et al.* (2002) and assumed that η was a function of density and temperature. They further assumed a barotropic equation of state to mimic the temperature evolution shown in Figure 11.9. Hence, η could be expressed as a function of density only: $\eta = c_\eta \, \eta_0(\rho)$, where $\eta_0(\rho)$ is a function of the central density. The initial conditions adopted by Machida *et al.* correspond to a spherical cloud with a critical Bonnor–Ebert density profile having a central (number) density $\rho_{c,0} = 3.841 \times 10^{-20}$ g cm^{-3} ($n_{c,0} = 10^4$ cm^{-3}). In this case, the critical Bonnor–Ebert sphere radius, $R_c = 6.45 \, c_s [4\pi G \rho_{\mathrm{BE}}(0)]^{-1/2}$,

Table 11.2. *Model parameters and results*

Model	ω	B_{init} (μG)	Ω_0 (s^{-1})	α_0	β_0	B_{fin} (kG)	P_{fin} (days)
SR	0.003	17	7.0×10^{-16}	0.5	3×10^{-5}	2.18	3.0
MR	0.03	17	7.0×10^{-15}	0.5	3×10^{-3}	0.40	2.1
RR	0.3	17	7.0×10^{-14}	0.5	3×10^{-1}	–	–

Note: Representing the thermal, rotational and gravitational energies as U, K and W, the relative factors against the gravitational energy are defined as $\alpha_0 = U/|W|$ and $\beta_0 = K/|W|$.

corresponds to $R_c = 4.58 \times 10^4$ AU. The total mass inside the critical radius was $M_0 = 7.6 M_\odot$. Initially, the cloud was in solid body rotation around the z-axis (at a rate Ω_0) and had a uniform magnetic field ($B_{init} = 17\,\mu$G) parallel to the z-axis (or rotation axis). To promote contraction, the density was increased by 70% starting from the critical Bonnor–Ebert sphere.

The various models investigated by Machida *et al.* can be characterized by a single non-dimensional parameter ω, related to the cloud's initial rotation rate and defined using the central density ρ_0 as $\omega = \Omega_0/(4\pi\, G\, \rho_0)^{1/2}$. The parameter ω, the initial magnetic field strength, the initial angular velocity Ω_0, the ratio of the thermal (α_0) and rotational (β_0) energies to the gravitational energy, the final magnetic field strength at the centre and the rotation period of the protostar at the final snapshot of the calculation are summarized in Table 11.2, where SR, MR and RR stand for (initially) slow, medium and rapid rotator, respectively.

11.4.2.2 Protostellar outflows and jets

As a result of their calculations, Machida *et al.* found that two distinct flows (low- and high-velocity flows) are driven by the first and second cores. They proposed that the low-velocity flow from the first core corresponds to observed molecular outflows, while the high-velocity flow from the protostar corresponds to observed optical jets. As an illustration of their simulations, a snapshot of model RR is shown in Figure 11.10.

The results of Machida *et al.* show that the flow driven by the first core has a slow speed and a wide opening angle, while the flow driven by the protostar has a high speed and a well-collimated structure. The flow speed roughly corresponds to the escape speed of the driving object. The difference in the depth of the gravitational potential between the first and the second core therefore causes the difference in flow speed. Typically, observed molecular outflows and optical jets have speeds $v_{out,obs} \simeq 30$ km s^{-1} and $v_{jet,obs} \simeq 100$ km s^{-1}, respectively. At the end of the calculations, the low- and high-velocity flows of Machida *et al.* only have speeds $v_{LVF} \simeq 3$ km s^{-1} and $v_{HVF} \simeq 30$ km s^{-1}, respectively. However, the first

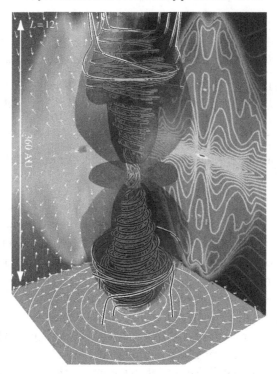

Fig. 11.10. Bird's-eye view of model RR ($l = 12$). The structure of high-density region ($n > 10^{12}$ cm^{-3}; iso-density surface) and magnetic field lines (black-and-white streamlines) are plotted. The structure of the outflow is shown by the iso-velocity surface inside which the gas is outflowing from the centre. The density contours and velocity vectors (thin arrows) on the mid-plane of $x = 0$, $y = 0$ and $z = 0$ are, respectively, projected in each wall surface.

and second cores only have masses $M_{\text{first core}} \simeq 0.01 M_{\odot}$ and $M_{\text{second core}} \simeq 10^{-3} M_{\odot}$, respectively, at the end of the calculations. Each core grows in mass by at least 1–2 orders of magnitude in the subsequent gas accretion phase. Since the escape speed increases as the square root of the mass of the central object at a fixed radius, the speeds of the low- and high-velocity flows may increase by a factor of ~10, and reach $v_{\text{LVF}} \simeq 30$ km s^{-1} and $v_{\text{HVF}} = 300$ km s^{-1}, respectively, which correspond to typical observed values.

The difference in collimation between the two flows is caused both by different configurations of the magnetic field lines around the driving object and by different driving mechanisms (see also Banerjee & Pudritz 2006; Hennebelle & Fromang 2008). The magnetic field lines around the first core have an hourglass configuration because they converge to the cloud centre as the cloud collapses, and Ohmic dissipation is ineffective before first core formation. Furthermore, the

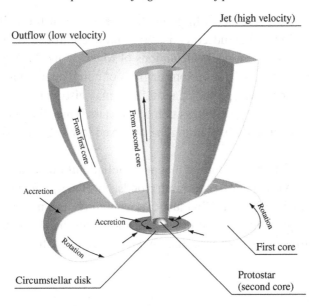

Fig. 11.11. Schematic picture proposed by Machida *et al.* (2008a) for the jet and outflow driven from the protostar and the fist core, respectively.

flow emerging near the first core is mainly a disk wind driven by the magneto-centrifugal wind mechanism. The centrifugal force is dominant in the low-velocity flow, whereas the magnetic pressure gradient force is more effective in driving the high-velocity flow. The magnetic field lines are straightened by the magnetic tension force near the protostar because the magnetic field is decoupled from the neutral gas. However, the magnetic field lines are strongly twisted in the region in close proximity to the protostar, where the magnetic field is coupled with the neutral gas again. Thus, the strong toroidal field generated around the protostar can drive a high-velocity flow, which is guided by the straight configuration of the magnetic field.

Figure 11.11 summarizes the main features of the outflows and jets modelled by Machida *et al.* (2008a). Note that the calculations of Machida *et al.* cover only the very early phase of protostar evolution. Further longer-term calculations are needed to better understand the correspondence between the models and the observed protostellar outflows.

11.4.2.3 Effects of outflows/jets on star-forming cores/clouds

In principle, outflows and jets from new-born stars may have strong dynamical effects on their environments. It has been proposed that the violent impact of fast outflows may result in the destruction of the cores from which the driving protostars are born (Nakano *et al.* 1995; Matzner & McKee 2000), or even in the dispersal of the parental molecular clouds. In particular, the local star formation efficiency at the level of individual cloud cores is most likely controlled by the effects of protostellar outflows: The model calculations of Matzner and McKee (2000) give $\epsilon_{core} \equiv M_\star/M_{core} \sim 25$–$75\%$ for various degrees of core flattening and magnetization. On a more global level, protostellar outflows may also significantly reduce the fraction of total molecular cloud mass going into stars. Thus, this process provides an interesting possibility for explaining the observed low star formation efficiency in the galaxy.

Li and Nakamura (2006) proposed that protostellar outflows could also sustain a high level of turbulence in cluster-forming clouds, but Banerjee *et al.* (2007) argued that this was actually difficult. Obviously, a quantitative examination of all these processes will require appropriate dynamical modelling of protostellar outflows, which remains to be done both theoretically and observationally.

11.5 Conclusions: proposed view for the star formation process

The prestellar CMF appears to be consistent with the stellar IMF between ~ 0.1 and $\sim 5 M_\odot$, although large uncertainties remain especially at the low- and high-mass ends (cf. Section 11.2). Small internal and relative motions are measured for these prestellar cores, implying that they are much less turbulent than their parent cloud and generally do not have time to interact before collapsing to (proto)stars (cf. Section 11.3.2.5). These results strongly support scenarios according to which the IMF is largely determined at the prestellar stage.

None of the extreme scenarios proposed for the formation of prestellar cores can explain all observations. Pure ambipolar diffusion is too slow to be the main core formation mechanism for typical levels of cloud ionization. Purely hydrodynamic pictures have trouble accounting for the inefficiency of core formation and the detailed velocity structure of individual cores. A mixed scenario (cf. Nakamura & Li 2005; Basu *et al.* 2009b) may be the solution: supersonic MHD turbulence in a molecular cloud close to magnetic criticality generates seed cores, which grow in mass until they become gravitationally unstable and collapse in a magnetically controlled fashion while decoupling from their turbulent environment. Such a mixed picture has several advantages. On the one hand, relatively strong magnetic fields prevent global collapse and lead to inefficient core/star formation on large (GMC) scales. On the other hand, the rate of ambipolar diffusion is enhanced by turbulence in shock-compressed regions (Fatuzzo & Adams 2002; Zweibel 2002; Nakamura & Li 2005). The difference between the clustered and the isolated or distributed modes of star formation may result from local variations around the critical mass-to-flux ratio and/or from the presence or absence of external triggers.

Cluster-forming clumps such as L1688 in Ophiuchus or NGC2264-C in Monoceros are likely (slightly) magnetically supercritical. There is some evidence that these clumps are in a state of global collapse induced by large-scale external triggers (Peretto *et al.* 2006; Nutter *et al.* 2006). In this case, the local star formation efficiency within each condensation is high ($\gtrsim 50\%$) and a promising core formation mechanism is purely gravitational, Jeans-like fragmentation of compressed cloud layers (cf. Palous 2007; Peretto *et al.* 2007), possibly followed by subsequent core growth (Basu & Jones 2004).

Regions of more distributed and less efficient star formation, such as Taurus or the Pipe nebula, are likely slightly subcritical or nearly critical. The size of individual cores is then larger (Motte *et al.* 1998), the local star formation efficiency is lower (~ 15–30% – Onishi *et al.* 2002; Alves *et al.* 2007) and the feedback from protostellar outflows may be more important in limiting accretion and defining stellar masses (Shu *et al.* 2004).

To fully understand how stars form and how the IMF comes about, a better knowledge of the presently poorly known initial conditions for molecular cloud formation is required, a subject of growing interest (see, e.g., Chapter 9 by Hennebelle *et al.*). It is also crucial to further investigate the processes by which prestellar cores form, evolve, and eventually collapse and fragment into multiple protostellar systems. With present submillimetre instrumentation, observational studies are limited by small-number statistics and restricted to the nearest regions. The advent of major new facilities in the coming years should yield several breakthroughs in this field. With an angular resolution at 75–300 µm comparable to, or better than, the largest ground-based millimetre-wave radiotelescopes, *Herschel*, the far infraRed and submillimetre telescope to be launched by ESA in 2009 (cf. Pilbratt 2005), will make possible complete surveys for prestellar cores down to the proto-brown dwarf regime in the cloud complexes of the Gould Belt (cf. André & Saraceno 2005). High-resolution ($0.01''$–$0.1''$) studies with the ALMA (becoming partly available in 2011, fully operational in 2013 – cf. Bachiller and Cernicharo 2008) at ~ 450 µm–3 mm will allow us to probe the kinematics of individual condensations in distant, massive protoclusters. Complementing each other nicely, *Herschel* and ALMA will tremendously improve our global understanding of the initial stages of star formation in the galaxy.

References

Adams, F. C., Proszkow, E. M., Fatuzzo, M., & Myers, P. C. 2006, *ApJ.*, **641**, 504
Alves, J. F., Lada, C. J., & Lada, E. A. 2001, *Nature*, **409**, 159
Alves, J. F., Lombardi, M., & Lada, C. J. 2007, *A&A*, **462**, L17
André, P., Belloche, A., Motte, F., & Peretto, N. 2007, *A&A*, **472**, 519
André, P., & Montmerle, T. 1994, *ApJ.*, **420**, 837

André, P. , & Saraceno, P. 2005, in *The Dusty and Molecular Universe: A Prelude to Herschel and ALMA*, ESA SP-577, p. 179

André, P., Ward-Thompson, D., & Barsony, M. 1993, *ApJ.*, **406**, 122

André, P., Ward-Thompson, D., & Barsony, M. 2000, in *Protostars and Planets IV*, ed. V. Mannings, A. Boss, & S. Russell (Tucson: University of Arizona Press), 59

André, P., Pety, J., & Bacmann, A. 2009, in preparation

André, P., Ward-Thompson, D., & Motte, F. 1996, *A&A*, **314**, 625

Bachiller, R., & Cernicharo, J. (ed.) 2008, *Science with ALMA: A new era for Astrophysics* (Berlin: Springer), *Ap&SS*, **313**, 1–3

Bacmann, A., André, P., Puget, J.-L., Abergel, A., Bontemps, S., & Ward-Thompson, D. 2000, *A&A*, **361**, 555

Ballesteros-Paredes, J., Gazol, A., Kim, J., Klessen, R. S., *et al.* 2006, *ApJ.*, **637**, 384

Ballesteros-Paredes, J., Klessen, R. S., & Vázquez-Semadeni, E. 2003, *ApJ.*, **592**, 188

Banerjee, R., Klessen, R. S., & Fendt, C. 2007, *ApJ.*, **668**, 1028

Banerjee, R., & Pudritz, R. E. 2006, *ApJ.*, **641**, 949

Basu, S. 2000, *ApJ.*, **540**, L103

Basu, S., & Ciolek, G. E. 2004, *ApJ.*, **607**, L39

Basu, S., Ciolek, G. E., & Wurster, J. 2009a, NewA, **14**, 221

Basu, S., Ciolek, G. E., Dapp, W., & Wurster, J. 2009b, submitted to NewA (arXiv:0810.0783)

Basu, S., & Jones, C. E. 2004, *MNRAS*, **347**, L47

Basu, S., & Mouschovias T. Ch. 1994, *ApJ.*, **432**, 720

Bate, M. R., & Bonnell, I. A. 2005, *MNRAS*, **356**, 1201

Bate, M. R., Bonnell, I. A., & Bromm, V. 2003, *MNRAS*, **339**, 577

Belloche, A., Hennebelle, P., & André, P. 2006, *A&A*, **453**, 145

Benson, P. J., & Myers, P. C. 1989, *ApJS*, **71**, 89

Blitz, L. 1993, in *Protostars & Planets III*, eds. E. H. Levy & J. I. Lunine (Tucson: University of Arizona Press), p. 125

Bonnell, I. A., Bate, M. R., Clarke, C. J., & Pringle, J. E. 2001, *MNRAS*, **323**, 785

Caselli, P., Benson, P. J., Myers, P. C., & Tafalla, M. 2002, *ApJ.*, **572**, 238

Caselli, P., & Myers, P. C. 1995, *ApJ.*, **446**, 665

Chabrier, G. 2005, in *The Initial Mass Function 50 years later*, eds. E. Corbelli *et al.* (Dordrecht: Springer), Astrophysics and Space Science Library, **327**, p. 41

Chabrier, G., & Baraffe, I. 2000, *ARA&A*, **38**, 337

Ciolek, G. E., & Basu, S. 2006, *ApJ.*, **652**, 442

Ciolek, G. E., & Mouschovias, T. Ch. 1994, *ApJ.*, **425**, 142

Clark, P. C., & Bonnell, I. A. 2005, *MNRAS*, **361**, 2

Clark, P. C., Klessen, R. S., & Bonnell, I. A. 2007, *MNRAS*, **379**, 57

Crutcher, R. M. 1999, *ApJ.*, **520**, 706

Crutcher, R. M. 2004, *Ap&SS*, **292**, 225

Crutcher, R. M., Nutter, D. J., Ward-Thompson, D., & Kirk, J. 2004, *ApJ.*, **600**, 279

Curry, C. L., & McKee, C. F. 2000, *ApJ.*, **528**, 734

Delgado-Donate, E. J., Clarke, C. J., & Bate, M. R. 2003, *MNRAS*, **342**, 926

Dib, S., Kim, J., Vázquez-Semadeni, E., Burkert, A., & Shadmehri, M. 2007, *ApJ.*, **661**, 262

Di Francesco, J., Myers, P. C., Wilner, D. J., Ohashi, N., & Mardones, D. 2001, *ApJ.*, **562**, 770

Elmegreen, B. G. 1979, *ApJ.*, **232**, 729

Elmegreen, B. G. 1997, *ApJ.*, **486**, 944

Elmegreen, B. G. 2000, in *Star Formation from the Small to the Large Scale*, eds. F. Favata, A. A. Kaas & A. Wilson (Noordnuijk, The Netherlands: European Space Agency Publications Division, ESTEC), ESA SP-445, p. 265

Elmegreen, B. G., & Falgarone, E. 1996, *ApJ.*, **471**, 816

Enoch, M. L., Young, K. E., Glenn, J., Evans, N. J., *et al.* 2006, *ApJ.*, **638**, 293

Fatuzzo, M., & Adams, F. C. 2002, *ApJ.*, **570**, 210

Galván-Madrid, R., Vázquez-Semadeni, E., Kim, J., & Ballesteros-Paredes, J. 2007, *ApJ.*, **670**, 480

Gómez, G. C., Vázquez-Semadeni, E., Shadmehri, M., & Ballesteros-Paredes, J. 2007, *ApJ.*, **669**, 1042

Goodman, A. A., Barranco, J. A., Wilner, D. J., & Heyer, M. H. 1998, *ApJ.*, **504**, 223

Goodman, A. A., Benson, P. J., Fuller, G. A., & Myers, P. C. 1993, *ApJ.*, **406**, 528

Goodwin, S., Whitworth, A., & Ward-Thompson, D. 2004, *A&A*, **414**, 633

Goodwin, S. P., Kroupa, P., Goodman, A., & Burkert, A. 2007, in *Protostars and Planets V*, ed. B. Reipurth, *et al.* (Tucson: University of Arizona Press), p. 133

Goodwin, S. P., Nutter, D., Kroupa, P., Ward-Thompson, D., & Whitworth, A. 2008, *A&A*, **477**, 823

Gregersen, E. M., & Evans, N. J. II. 2000, *ApJ.*, **538**, 260

Hartmann, L. 2002, *ApJ.*, **578**, 914

Hatchell, J., Richer, J. S., Fuller, G. A., Qualtrough, C. J., Ladd, E. F., & Chandler, C. J. 2005, *A&A*, **440**, 151

Heiles, C., & Crutcher, R. M. 2005, in *Cosmic Magnetic Fields*, eds. R. Wielebinski & R. Beck (Berlin: Springer), Lecture Note in Physics, **664**, p. 137

Hennebelle, P., & Fromang, S. 2008, *A&A*, **477**, 9

Hennebelle, P., & Teyssier, R. 2008, *A&A*, **477**, 25

Hennebelle, P., Whitworth, A., Cha, S.-H, & Goodwin, S. 2004, *MNRAS*, **348**, 687

Inutsuka, S. 2001, *ApJ.*, **559**, L149

Inutsuka, S., & Miyama, S. M. 1992, *ApJ.*, **388**, 392

Inutsuka, S., & Miyama, S. M. 1997, *ApJ.*, **480**, 681

Jessop, N. E., & Ward-Thompson, D. 2000, *MNRAS*, **311**, 63

Jijina, J., Myers, P. C., & Adams, F. C. 1999, *ApJS*, **125**, 161

Johnstone, D., Di Francesco, J., & Kirk, H. 2004, *ApJ.*, **611**, L45

Johnstone, D., Fich, M., Mitchell, G. F., Moriarty-Schieven, G. 2001, *ApJ.*, **559**, 307

Johnstone, D., Matthews, H., & Mitchell, G. F. 2006, *ApJ.*, **639**, 259

Johnstone, D., Wilson, C. D., Moriarty-Schieven, G., Joncas, G., Smith, G., Gregersen, E., & Fich, M. 2000, *ApJ.*, **545**, 327

Kirk, H., Johnstone, D., & Di Francesco, J. 2006, *ApJ.*, **646**, 1009

Kirk, J. M., Ward-Thompson, D., & André, P. 2005, *MNRAS*, **360**, 1506

Klessen, R. S., Ballesteros-Paredes, J., Vázquez-Semadeni, E., & Durán-Rojas, C. 2005, *ApJ.*, **620**, 786

Klessen, R. S., & Burkert, A. 2000, *ApJS*, **128**, 287

Kramer, C., Stutzki, J., Rohrig, R., & Corneliussen, U. 1998, *A&A*, **329**, 249

Kroupa, P. 2001, *MNRAS*, **322**, 231

Krumholz, M. R., Klein, R. I., & McKee, C. F. 2007, *ApJ.*, **656**, 959

Krumholz, M. R., McKee, C. F., & Klein, R. I. 2005, *Nature*, **438**, 332

Kudoh, T., & Basu, S. 2008, *ApJ.*, **679**, L97

Kudoh, T., Basu, S., Ogata, Y., & Yabe, T. 2007, *MNRAS*, **380**, 499

Lada, C. J. 1987, in *Star Forming Regions*, eds. M. Peimbert & J. Jugaku (Dordrecht: Reidel), IAU Symp. 115, 1

Lada, C. J., Muench, A. A., Rathborne, J. M., Alves, J., & Lombardi, M. 2008, *ApJ.*, **672**, 410

Lai, S.-P., Velusamy, T., Langer, W. D., & Kuiper, T. B. H., *et al.* 2003, *ApJ.*, **126**, 311

Larson, R. B. 1969, *MNRAS*, **145**, 271

Larson, R. B. 1981, *MNRAS*, **194**, 809

Larson, R. B. 1985, *MNRAS*, **214**, 379

Larson, R. B. 2005, *MNRAS*, **359**, 211

Lee, C. W., & Myers, P. C. 1999, *ApJS*, **123**, 233

Lee, C. W., Myers, P. C., & Tafalla, M. 2001, *ApJS*, **136**, 703

Li, Z.-Y., & Nakamura, F. 2004, *ApJ.*, **609**, L83

Li, Z.-Y., & Nakamura, F. 2006, *ApJ.*, **640**, L187

Lombardi, M., & Bertin, G. 2001, *A&A*, **375**, 1091

Machida, M. N., Inutsuka, S., & Matsumoto, T. 2006, *ApJ.*, **647**, L151

Machida, M. N., Inutsuka, S., & Matsumoto, T. 2007, *ApJ.*, **670**, 1198

Machida, M. N., Inutsuka, S., & Matsumoto, T. 2008a, *ApJ.*, **676**, 1088

Machida, M. N., Tomisaka, K, Matsumoto, T., & Inutsuka, S. 2008b, *ApJ.*, **677**, 327

Mac Low, M.-M., & Klessen, R. S. 2004, *RvMP*, **76**, 125

McKee, C. F. 1989, *ApJ.*, **345**, 782

Masunaga, H., & Inutsuka, S. 1999, *ApJ.*, **510**, 822

Masunaga, H., & Inutsuka, S. 2000a, *ApJ.*, **531**, 350

Masunaga, H., & Inutsuka, S. 2000b, *ApJ.*, **536**, 406

Masunaga, H., Miyama, S. M. & Inutsuka, S. 1998, *ApJ.*, **495**, 346

Matsumoto, T., & Hanawa, T. 2003, *ApJ.*, **595**, 913

Matzner, C., & McKee, C. 2000, *ApJ.*, **545**, 364

Motte, F., André, P., & Neri, R. 1998, *A&A*, **365**, 440

Motte, F., André, P., Ward-Thompson, D., & Bontemps, S. 2001, *A&A*, **372**, L41

Motte, F., Bontemps, S., Schilke, P., Lis, D., Schneider, N., & Menten, K. 2005, in *Massive Star Birth: A Crossroads of Astrophysics*, IAU Symp. 227, eds. R. Cesaroni et al. (Cambridge: Cambridge University Press), 151

Motte, F., Bontemps, S., Schilke, P., Schneider, N., Menten, K. M., & Broguière, D. 2007, *A&A*, **476**, 1243

Mouschovias, T. Ch. 1987, in *Physical Processes in Interstellar Clouds*, eds. G. E. Morfill & M. Scholer (Dordrecht: Reidel), 453

Myers, P. C. 1983, *ApJ.*, **270**, 105

Myers, P. C. 2000, *ApJ.*, **530**, L119

Myers, P. C. 2001, in *From Darkness to Light*, eds. T. Montmerle & P. André, *ASP Conf. Ser.*, **243**, 131

Nagai, T., Inutsuka, S.(-I.), & Miyama, S. M. 1998, *ApJ.*, **506**, 306

Nakamura, F., & Li, Z.-Y. 2005, *ApJ.*, **631**, 411

Nakano, T. 1979, *PASJ*, **31**, 697

Nakano, T., Hasegawa, T., & Norman, C. 1995, *ApJ.*, **450**, 183

Nakano, T., Nishi, R., & Umebayashi, T. 2002, *ApJ.*, **573**, 199

Narita, S., Nakano, T., & Hayashi, C. 1970, *Prog. Theor. Phys.*, **43**, 942

Nutter, D., & Ward-Thompson, D. 2007, *MNRAS*, **374**, 1413

Nutter, D., Ward-Thompson, D., & André, P. 2006, *MNRAS*, **368**, 1833

Onishi, T., Mizuno, A., Kawamura, A., Ogawa, H., Fukui, Y. 1998, *ApJ.*, **502**, 296

Onishi, T., Mizuno, A., Kawamura, A., Tachihara, K., & Fukui, Y. 2002, *ApJ.*, **575**, 950

Padoan, P., & Nordlund, A. 2002, *ApJ.*, **576**, 870

Palous, J. 2007, in *Triggered Star Formation in a Turbulent ISM*, IAU Symp. 237, eds. B. G. Elmegreen & J. Palous (Cambridge: Cambridge University Press), 114

Peretto, N., André, P., & Belloche, A. 2006, *A&A*, **445**, 979

Peretto, N., Hennebelle, P., & André, P. 2007, *A&A*, **464**, 983

Pilbratt, G. 2005, in *The Dusty and Molecular Universe: A Prelude to Herschel and ALMA*, ESA SP-577, p. 3

Price, D. J., & Bate, M. R. 2007, *MNRAS*, **377**, 77

Reid, M. A., & Wilson, C. D. 2006, *ApJ.*, **650**, 970

Saigo, K., Matsumoto, T., & Hanawa, T. 2000, *ApJ.*, **531**, 971

Salpeter, E. E. 1955, *ApJ.*, **121**, 161

Shirley, Y., Evans II, N. J., Rawlings, J. M. C., & Gregersen, E. M. 2000, *ApJS*, **131**, 249

Shu, F. H., Adams, F. C., & Lizano, S. 1987, *ARA&A*, **25**, 23

Shu, F. H., Li, Z.-Y., & Allen, A. 2004, *ApJ.*, **601**, 930

Simon, R. 1965, *Annales d'Astrophysique*, **28**, 40

Spitzer, L., Jr. 1942, *ApJ.*, **95**, 329

Stahler, S. W., Shu, F. H., & Taam, R. E. 1980, *ApJ.*, **241**, 637

Stamatellos, D., Whitworth, A. P., & Ward-Thompson, D. 2007a, *MNRAS*, **379**, 1390

Stamatellos, D., Whitworth, A. P., Bisbas, T., & Goodwin, S. 2007b, *A&A*, **475**, 37

Stanke, T., Smith, M. D., Gredel, R., & Khanzadyan, T. 2006, *A&A*, **447**, 609

Tafalla, M., Mardones, D., Myers, P. C., Caselli, P., Bachiller, R., & Benson, P. J. 1998, *ApJ.*, **504**, 900

Tassis, K., & Mouschovias, T. Ch. 2007, *ApJ.*, **660**, 370

Testi, L., & Sargent, A. I. 1998, *ApJ.*, **508**, L91

Tilley, D. A., & Pudritz, R. E. 2007, *MNRAS*, **382**, 73

Tomisaka, K. 2002, *ApJ.*, **575**, 306

Tsuribe, T., & Inutsuka, S. 1999a, *ApJ.*, **523**, L155

Tsuribe, T., & Inutsuka, S. 1999b, *ApJ.*, **526**, 307

Vázquez-Semadeni, E., Kim, J., Shadmehri, M., & Ballesteros-Paredes, J. 2005, *ApJ.*, **618**, 344

Wardle, M. 2004, *Ap&SS*, **292**, 317

Ward-Thompson, D., André, P., Crutcher, R., Johnstone, D., Onishi, T., & Wilson, C. 2007, *Protostars and Planets V*, eds. B. Reipurth, D. Jewitt, & K. Keil (Tucson: University of Arizona Press), 33

Ward-Thompson, D., André, P., & Kirk, J. M. 2002, *MNRAS*, **329**, 257

Ward-Thompson, D., Kirk, J. M., Crutcher, R. M., Greaves, J. S., Holland, W. S., & André, P. 2000, *ApJ.*, **537**, L135

Ward-Thompson, D., Motte, F., & André, P. 1999, *MNRAS*, **305**, 143

Ward-Thompson, D., Scott, P. F., Hills, R. E., & André, P. 1994, *MNRAS*, **268**, 276

Whitehouse, S. C., & Bate, M. R. 2006, *MNRAS*, **367**, 32

Winkler, K.-H. A., & Newman, M. J. 1980, *ApJ.*, **236**, 201; 1980, *ApJ.*, **238**, 311

Zuckerman, B., & Palmer, P. 1974, *ARA&A*, **12**, 279

Zweibel, E. G. 2002, *ApJ.*, **567**, 962

12

Models for the formation of massive stars

M. R. Krumholz and I. A. Bonnell

Abstract

The formation of massive stars is currently an unsolved problem in astrophysics. Understanding the formation of massive stars is essential because they dominate the luminous, kinematic and chemical output of stars. Furthermore, their feedback is likely to play a dominant role in the evolution of molecular clouds and any subsequent star formation therein. Although significant progress has been made observationally and theoretically, we still do not have a consensus as to how massive stars form. There are two contending models to explain the formation of massive stars: core accretion and competitive accretion. They differ primarily in how and when the mass that ultimately makes up the massive star is gathered. In the core accretion model, the mass is gathered in a pre-stellar stage due to the overlying pressure of a stellar cluster or a massive pre-cluster cloud clump. In contrast, competitive accretion envisions that the mass is gathered during the star formation process itself, being funnelled to the centre of a stellar cluster by the gravitational potential of the stellar cluster. Although these differences may not appear overly significant, they involve significant differences in terms of the physical processes involved. Furthermore, the differences also have important implications in terms of the evolutionary phases of massive star formation and ultimately that of stellar clusters and star formation on larger scales. Here, we review the dominant models and discuss prospects for developing a better understanding of massive star formation in the future.

12.1 Introduction

The difficulties in understanding massive star formation lie in that we are not able to fully ascertain the properties of a cloud in which massive stars form or to

Structure Formation in Astrophysics. ed. G. Chabrier. Published by Cambridge University Press.
© Cambridge University Press 2009.

determine whether the properties we do observe are that of the pristine initial conditions or those due to subsequent evolution. For example, the central condensation often observed in massive cores could be an indication of the central condensation required in the core accretion model or equally represent the evolved state from a dynamical collapsing model. It is also equally difficult to determine the census of young stars in these cores and to what degree their dynamics may thus be affected. We are therefore left in a state of some ambiguity in terms of constraining the initial conditions for the models.

There is also considerable ambiguity regarding the end state of the massive star formation process, particularly as regards the multiplicity and clustering of massive stars. Observations unambiguously show that massive stars have a much higher companion fraction than low-mass stars (Duchêne *et al.* 2001; Preibisch *et al.* 2001; Shatsky & Tokovinin 2002; Lada 2006), even at early, still-embedded stages of star formation (Apai *et al.* 2007). A significant fraction of massive stars appear to be twins, systems with mass ratios near unity (Pinsonneault & Stanek 2006). However, the full mass and period distribution of massive stars is much less well-determined than it is for low-mass stars, and the statistical significance of the Pinsonneault & Stanek results has been disputed due to the small sample size and the potential for selection bias produced by using a sample of eclipsing binaries (Lucy 2006).

Similarly, the clustering properties of massive stars are debated. While the great majority of massive stars are either part of star clusters or runaways ejected from clusters, since on statistical grounds one would expect most massive stars to be in clusters, it is unclear if this observation implies that O stars form only, or preferentially, in clusters. Statistical analyses of cluster data reach contradictory conclusions, with some concluding that the stellar initial mass function (IMF) is truncated at a value that is an increasing function of the mass of the cluster in which that star is found (Weidner & Kroupa 2004, 2006) and others concluding that the data are consistent with random and independent sampling from stellar and cluster mass functions, with a cluster mass function that is continuous down to a single star (Oey *et al.* 2004; Elmegreen 2006; Parker & Goodwin 2007). Observationally, $4 \pm 2\%$ of O stars do not appear to be surrounded by clusters and are not runaways (de Wit *et al.* 2004, 2005), which would seem to imply that O stars can form in isolation. However, it remains possible that these O stars formed via a different mechanism than most massive stars.

On the theoretical side, a traditional problem in massive star formation is how the mass is actually accreted by the protomassive star and how this accretion circumvents the radiation pressure produced by the high stellar luminosity and the opacity of dust grains in the accretion flow. A number of potential solutions have recently been advanced based on the physics of disk accretion and the ways of circumventing the effects of the photon pressure such that we can conclude massive star formation is unlikely to be halted solely due to radiation pressure. However, these conclusions are still preliminary, since to date no simulation of massive star formation including radiative effects has successfully demonstrated the formation of stars up to $\sim 80 M_\odot$, the largest stellar mass that has been securely measured (the eclipsing spectroscopic binary WR 20a; Bonanos *et al.* 2004; Rauw *et al.* 2005).

In this chapter, we begin by describing the two primary models of massive star formation in Sections 12.2–12.3, and we then focus on the challenges that both models face in Section 12.4. We then suggest directions in which we can make progress on those challenges and in distinguishing between the two models in Section 12.5.

12.2 The core accretion model

12.2.1 The model

The basic premise of the core accretion model is that all stars, low and high mass, form by a top-down fragmentation process in which a molecular cloud breaks up into smaller and smaller pieces under the combined influence of turbulence, magnetic fields and self-gravity. This process continues down to some smallest scale, called a core, which does not undergo significant internal subfragmentation before collapse to stellar densities. Thus, a core represents the object out of which an individual star or a bound stellar system (for example a binary) forms, and the mass of the core determines the mass reservoir available to form the star. A massive star must therefore form from a massive core (McKee & Tan 2002, 2003). Cores also represent the scale in the star formation process at which gas becomes unstable to a global gravitational collapse. On all larger scales, objects have gravitationally unstable parts within them, but they are not in a state of overall collapse, and they turn a relatively small fraction of their mass into stars per dynamical time.

This model leads immediately to several predictions that may be tested against observations. First, in the core accretion scenario, a core is the mass reservoir available for accretion onto the star that forms within it, so there must be a direct relationship between the core and the stellar mass functions. Indeed, the core mass distribution will effectively set the stellar mass distribution. Second, for systems too young for the stars to have moved significantly from their birth sites, the spatial and velocity distributions of cores and stars should be similar – properties such as clustering strength, degree of mass segregation and kinematics should be similar for cores and for the youngest stars. Third, since the individual cores are gravitationally collapsing but the bulk of the gas is not, the overall rate of star formation in clouds much larger than an individual core should be small, in the sense that $\lesssim 10\%$ of the mass should be converted into stars per cloud dynamical time.

Finally, it is worth noting that the origin of the core mass function and core kinematics need not be specified as part of the core accretion model. Although a number of theories have been advanced to explain these quantities (Padoan & Nordlund 2002), these models and the core accretion model are independent of one another and must be tested separately.

12.2.2 Observational tests

12.2.2.1 Core and stellar mass functions

Thus far, the predictions of the core accretion model appear to hold up well against observations, although there are several significant caveats and areas of uncertainty. Observations of many different star-forming regions in dust continuum emission (Motte *et al.* 1998; Testi & Sargent 1998; Johnstone *et al.* 2001; Onishi *et al.* 2002; Reid & Wilson 2005, 2006) and near-infrared extinction mapping (Alves *et al.* 2007) essentially all find that the core mass distribution has a functional form strikingly similar to the stellar IMF. The core mass function has a power law tail of Salpeter slope at high masses and a flattening at low masses. The most recent and sensitive study, that of Alves *et al.* (2007), even hints at a turn-down in the mass function below the peak, exactly as is seen in the stellar IMF. The sole significant difference between the core and the star mass functions appears to be that the core mass function is shifted to higher masses by a factor of ~3, exactly what one would expect if a significant fraction of pre-stellar cores were ejected by outflows rather than accreting onto a star (Matzner & McKee 2000).

There is a significant caveat that the core mass function has only been reliably measured up to masses $\lesssim 20 M_\odot$ on the Salpeter side and down to $\gtrsim 0.5 M_\odot$ on the brown dwarf side. The latter limitation comes from the difficulty of detecting such low-mass objects, while the former comes from the difficulty of resolving objects at the large distances to which observations must reach to find such massive cores. While there are observations of cores up to a few hundred $M_\odot$, massive enough to form the largest stars (Beuther & Schilke 2004; Beuther *et al.* 2005, 2006; Sridharan *et al.* 2005; Reid & Wilson 2006), these objects are both expected and observed to have internal structure, and so one might worry that they are simply blended collections of low-mass cores. High-resolution interferometric observations show that at least some of these massive cores are very strongly centrally concentrated, however, which argues against this possibility (Beuther *et al.* 2007). That said, it cannot be ruled out definitively without higher resolution observations or a larger sample of massive cores.

12.2.2.2 Core spatial and kinematic distributions

The prediction of the core accretion model that young stars and star-forming cores should have similar spatial distributions also appears to be consistent with observations. In young clusters, stars with masses $\gtrsim 5 M_\odot$ are preferentially located in cluster centres, but at lower stellar masses, the mass function appears to be independent of position within a cluster (Hillenbrand & Hartmann 1998). There is some dispute about whether this segregation is a function of birthplace (Bonnell & Davies 1998; Huff & Stahler 2006) or is produced dynamically in young clusters (Tan *et al.* 2006; McMillan *et al.* 2007), but to the extent that it is primordial it should be reflected in the spatial distribution of cores. In at least one star-forming cloud, ρ-Ophiuchus, exactly such a pattern of core segregation is observed: massive cores are found only in the centre, but cores $\lesssim 5M_\odot$ show a position-independent mass function (Elmegreen & Krakowski 2001; Stanke *et al.* 2006). The major caveat regarding this observation is that it has not been replicated yet in more distant and more massive star-forming clouds, so it is unclear to what extent it is a generic feature of massive star-forming regions.

On the question of kinematics, observations also appear consistent with the core accretion model. Observations of populations of pre-stellar cores and cores around Class 0 protostars (Goodman *et al.* 1998; Walsh *et al.* 2004; André *et al.* 2007; Kirk *et al.* 2007; Rathborne *et al.* 2008) consistently find four kinematic signatures for low-mass cores across a wide range of cluster-forming clouds. First, the linewidths along sightlines that pass through core centres are essentially always sonic, with no significant turbulent component. Second, the linewidth increases only slowly as one examines sightlines progressively further from core centres, becoming only transonic (i.e. Mach numbers ~ 2) at projected distances of ~ 0.1 pc from core centres. Third, the mean velocity difference between cores and their envelopes are also small. Fourth, if one computes a centroid velocity for each core, the dispersion of core centroid velocities is generally smaller than the virial velocity in the observed region.

Simulations of cores forming in the context of driven, virialized turbulent flows in which the large-scale region is not globally collapsing are able to reproduce most but not all of these observations, while simulations in which the region is in global collapse have much greater difficulty (Offner *et al.* 2008). In non-collapsing regions, cores have subsonic or transonic velocity dispersions along the line of sight because the collapse is limited and localized, although the linewidths are still somewhat larger than observed ones, probably due to the absence of magnetic fields in the simulations. Because collapse is localized, cores are not bounded by strong infall shocks and the velocity dispersion at their edges is fairly small. Cores move subvirially with respect to one another because they are born from the densest, shocked gas that is preferentially at low-velocity dispersion (Padoan *et al.* 2001; Elmegreen 2007).

In contrast, simulations in which the turbulence is not driven and the entire region enters a process of global collapse have much greater difficulty (Klessen *et al.* 2005; Ayliffe *et al.* 2007). These simulations also find relatively low differences in mean velocity between cores and envelopes, but, due to the large-scale infall that occurs in them, they generally also find supersonic velocity dispersions, either at core centres or in their outskirts. They also produce core-to-core velocity dispersions that are about the same size as those seen in the gas (Offner *et al.* 2008), rather than smaller as the observations demand. This feature occurs because the cores are effectively dissipationless and retain the velocity dispersions they have at birth. The gas, however, continually loses energy through shocks, and thus after a while its velocity dispersion decreases to the point where it matches that of the cores.

12.2.2.3 The star formation rate

The final test of the core accretion model is the overall rate of star formation. Core accretion requires that, at any instant, collapse in cloud be confined to discrete regions containing a small fraction of the mass. Thus, the star formation rate (SFR) per cloud dynamical time must be small, $\lesssim 10\%$. In contrast, competitive accretion models require that clouds convert their mass into stars quickly, with an SFR per dynamical time $\gtrsim 10\%$, at least by the end of the star formation process or when averaged over many clouds.

This prediction can be tested in two ways: statistical and object-by-object. Statistically, one can estimate SFRs in a large sample of clouds by observing a galaxy in a tracer of star formation such as infrared emission and then compare it to the total mass of dense, cluster-forming gas by observing the same galaxy in a dense gas tracer such as the HCN($1 \rightarrow 0$) line, which is emitted by gas at densities $\sim 10^5\,\mathrm{cm}^{-3}$ (Gao & Solomon 2004a,b; Wu *et al.* 2005). If the dense gas mass divided by the free-fall time at the mean density of that gas is much larger than the measured SFR in the galaxy, then the clouds must, on average, convert their mass into gas relatively slowly. In contrast, if the cloud mass divided by the free-fall time is comparable to the galactic SFR, then the clouds must convert into stars rapidly.

Krumholz and Tan (2007) perform this exercise using a variety of tracers of dense gas and find that, with the exception of one tracer that only yields an upper limit, all the data require that clouds convert to stars at a rate well under 10% of the mass per dynamical time. This appears to be consistent with the requirements of core accretion. However, it is worth mentioning the caveat that this technique averages over a large number of star forming clouds and assumes that all of them are active star formers. If there are a significant number of clouds that do not form stars, then the SFR in the active star formers could be significantly higher (Elmegreen 2007). This picture would be inconsistent with core accretion although it would also demand an explanation for what mechanism prevents gas at densities of $10^5\,\mathrm{cm}^{-3}$ from forming stars.

For individual objects, one can estimate the SFR by computing the age spread within a star cluster and then estimating the efficiency of conversion of gas into stars based on cluster dynamics. The age divided by the dynamical time of the parent cloud multiplied by the efficiency yields the fraction of the mass that must have been converted into stars per dynamical time. This method is considerably less certain, since for clusters where it is possible to estimate an age spread using pre-main sequence tracks, the parent cloud has generally already been dispersed. One must therefore estimate the mass of that cloud (in order to determine the overall efficiency) and its dynamical time without observing it directly. Moreover, even the age spreads themselves are uncertain, since the pre-main sequence tracks used to age-date stars are of limited accuracy.

Given these uncertainties, it is not surprising that interpretations of the data vary widely. Elmegreen (2000b, 2007) argues that star formation in a cluster generally ends in only 1–2 crossing times, which, together with a typical cluster formation efficiency of 20–50% (Lada & Lada 2003), implies rapid star formation. Tan *et al.* (2006) and Huff and Stahler (2006), analyzing much the same data set, conclude instead that the typical duration of star formation is 4–5 crossing times, implying a much lower rate of star formation. Resolving this controversy will require both better observations of still partially embedded young clusters and improvements in the pre-main sequence models.

It is worth noting, however, that several groups recently obtained parallax measurements of the distance Orion Nebula Cluster (ONC), the object about which much of the debate regarding the duration of cluster formation has taken place. These observations correct the distance from roughly 500 to about 400 pc (Sandstrom *et al.* 2007; Caballero 2008; Hirota *et al.* 2007; Menten *et al.* 2007), which will reduce the inferred luminosity for all the Orion stars by $\sim$30%. This will in turn increase the age spread inferred from the pre-main sequence tracks. This revision obviously favours the extended formation hypothesis, although determining by exactly how much the revised distance raises the age spread will require a full re-analysis of the locations of ONC stars on pre-main sequence tracks using the revised luminosities.

12.2.3 *Massive stars in the core accretion model*

12.2.3.1 *Initial fragmentation*

In the context of the core accretion model, a massive star forms from a core that is similar to those that form low-mass stars. The one challenge that is particular to the core accretion model is how a massive star, an object that is potentially hundreds of Jeans masses in size, can form by direct collapse. Why does the fragmentation process not always continue down to objects that are $\sim$1 Jeans mass in size? Indeed, purely hydrodynamic simulations of core collapse and fragmentation find exactly this behaviour (Dobbs *et al.* 2005).

The simulations and analytic work of Krumholz (2006) and Krumholz *et al.* (2007b,c) provide a likely answer: at very early times in the star formation process, radiation feedback powered by the gravitational potential energy of collapsing gas will modify the effective equation of state of the gas on scales of ~0.01–0.1 pc, roughly the size of observed pre-stellar cores. The effective equation of state is critical to determining whether or not gas fragments (Larson 2005; Jappsen *et al.* 2005): equations of state that are isothermal or softer favour fragmentation, while stiffer equations of state prevent it. In the simulations of Krumholz *et al.*, the accretion luminosity coming from the first ~0.1–1$M_\odot$ star in the simulation gives the gas an equation of state with an effective ratio of specific heats $\gamma = 1.1$–1.2, stiffer than isothermal and thus unfavourable to fragmentation. This is truly not an equation of state, since the gas temperature depends on position relative to the illuminating source as well as density, but it appears to have the same effect on fragmentation as a true equation of state. In simulations with radiation, Krumholz *et al.* find that a typical massive core forms only a handful of fragments and most of these form out of gravitationally unstable protostellar disks that are shielded from radiation by their high column densities. Figure 12.1 illustrates this result, showing a comparison between a non-radiative and a radiative simulation starting from identical initial conditions. The strong suppression of fragmentation found in these simulations suggests that in the dense regions where massive stars form gas *cannot*

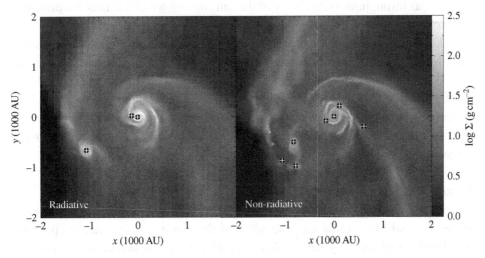

Fig. 12.1. The plots show the column density through a 4000 AU box around the most massive protostars in two simulations of the collapse and fragmentation of a massive core by Krumholz *et al.* (2007b). The simulation shown on the *left* includes radiative transfer and feedback while the one on the *right* omits these effects, but they are otherwise identical in initial conditions, evolutionary time and resolution. The plus signs indicate the locations of protostars.

fragment on size scales below a few hundredths of a parsec, so monolithic collapse is the only possibility.

The major remaining caveat on this work is that the simulations begin from somewhat unrealistic initial conditions in which cores possess turbulent velocity structures but not turbulent density structures. In reality, even at the time when a first protostar forms, there should be density structure present in the gas, and some overdense regions may be sufficiently shielded against radiation to collapse despite significant external illumination. Future simulations using radiative transfer on larger scales will have to examine this effect.

12.2.3.2 Massive protostellar disks and companion formation

Since the collapsing core has non-zero angular momentum, a protostellar disk naturally results. The sizes of these disks are typically ~ 1000 AU, a value imposed by the core size and the typical angular momentum carried by core's turbulent velocity fields. These disks are quite different from those around lower-mass protostars in low-density regions. Due to the high density in massive cores, the accretion rate onto these disks is extremely high. Early in their lifetimes, rapid accretion drives massive protostellar disks to a regime of gravitational instability characterized by Toomre Q values near unity and disk-to-star mass ratios ~ 0.5 (Kratter & Matzner 2006; Krumholz *et al.* 2007b; Kratter *et al.* 2008). While in this stage, the disks have strong spiral arms and are subject to gravitational fragmentation. This condition persists throughout the majority of the time during which the massive protostar is accreting.

Disks in this evolutionary phase are also subject to fragmentation. Fragments that form in the disk intend to migrate inwards as the disk accretes. The fate of these fragments is unknown, and most of them likely merge with the central protostar. However, it seems likely that at least some of them survive to become close companions, explaining the high companion fraction of massive stars. Those which are closer than ~ 1 AU are likely to gain considerable mass via mass transfer from the primary, which passes through a phase of rapid expansion to large radii on its way to the main sequence. Such systems are likely to become massive twins (Krumholz & Thompson 2007).

Massive, gravitationally unstable disks are potentially observable, and their properties could be an important signpost of the massive star formation process. Simulated observations by Krumholz *et al.* (2007a) indicate that these disks should be visible out to distances of ~ 0.5 kpc using EVLA observations of molecular lines in the 20–30 GHz range, and out to several kpc using ALMA observations of lines in the 200–300 GHz range. Figure 12.2 illustrates some simulated observations of massive disks. Dust continuum observations, which provide morphological but not kinematic information, should be possible at considerably larger distances. These

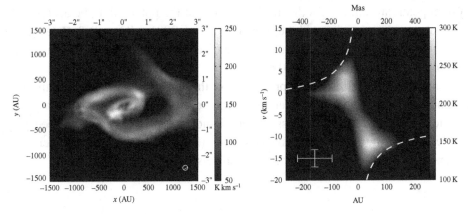

Fig. 12.2. The plots show simulated observations of a massive protostellar disk in the CH$_3$CN 220.7472 GHz line using ALMA (*left*) and of the inner part of the same disk using the NH$_3$ 26.5190 GHz line (the (8, 8) inversion transition) using the EVLA (*right*). The left image shows velocity-integrated brightness temperature as a function of position, while the right image shows brightness temperature as a function of position and velocity for a line along the disk plane. The beam sizes are indicated in both plots, and the colour scales are chosen so that their bottom values correspond to the 3σ sensitivity of the assumed observation. The *dashed line* shows a Keplerian rotation curve for an 8.3$M_\odot$ star, the mass of the star at the centre of the disk. Note the presence of very strong spiral structure in the position–position plot and that the disk is offset from zero velocity in the position–velocity plot. Both of these are observational signposts of strong gravitational instability in the disk. (Adapted from Krumholz *et al.* (2007a)).

observations should reveal several indicators of gravitational instability, including strong spiral arms, hot spots produced by accreting protostars formed in the disk, rotation profiles that are super-Keplerian in their outer parts relative to their inner parts (due to the non-trivial disk-to-star mass ratio) and velocity offsets between the disk and the star (due to motion of the star under the influence of the disk's gravity).

12.3 The competitive accretion model

The primary difference between the core accretion model and the competitive accretion model is the initial conditions and the physical process invoked to gather the mass of the most massive star. Models of competitive accretion are derived from the requirement to form a full cluster of stars in addition to forming the few massive stars in the system (Klessen *et al.* 1998; Bate *et al.* 2003; Bonnell *et al.* 2003). As such, the initial conditions are dynamically unrelaxed allowing for large-scale fragmentation into the many objects needed to populate a stellar cluster. In this scenario, fragmentation produces lower-mass stars while the few higher-mass stars form subsequently due to continued accretion in stellar clusters where the overall system potential funnels gas down to the centre of the potential, to be accreted by the protomassive stars located there (Bonnell *et al.* 2001*a*; Bonnell & Bate 2006). This model not only produces massive stars, but a full IMF and mass-segregated stellar clusters (Bonnell *et al.* 2007).

Competitive accretion relies primarily on the physics of gravity and the formation of massive stars in clustered environments. That gravity is the dominant physical process in forming massive stars is not surprising, as gravity need play a role in both the larger-scale formation of stellar cluster and the smaller-scale fragmentation and collapse of individual objects.

12.3.1 Initial fragmentation

The initial fragmentation stage of a stellar cluster in a turbulent cloud involves the rapid generation of structure due to the turbulence, followed by the collapse of individual fragments at Jeans length separations (≈ 0.05 pc) in the cloud (Larson 1984). Such fragmentation is unaffected by any heating from other newly formed stars due to this relatively large separation. Numerical simulations have repeatedly shown that the gravitational fragmentation of a turbulent medium results in a range of stellar masses based on the thermal Jeans mass of the cloud (Klessen *et al.* 1998; Klessen & Burkert 2000; Klessen *et al.* 2000; Bate *et al.* 2003; Bonnell *et al.* 2004; Bate & Bonnell 2005). In the context of cloud core mass functions that are similar to the stellar IMF, this can easily be understood as being due to the subfragmentation of the higher-mass cores, while the lower-mass ones are never gravitationally bound (Clark & Bonnell 2004, 2006). This is just what has recently been observed in the Pipe nebula (Lada *et al.* 2008).

Thus, turbulence drives structures into the molecular cloud, while the thermal physics determines the fragmentation scale and thus the characteristic mass for star formation (Larson 2005; Jappsen *et al.* 2005; Bonnell *et al.* 2006). Such fragmentation does not produce stars with masses ten or more times larger than the Jeans mass, implying that massive stars need to form through an alternative mechanism. Turbulent fragmentation is sometimes invoked to explain the formation of high-mass fragments (Padoan *et al.* 2001), but such fragments need to be very widely separated and cannot account for the clustered environments of massive star formation.

Once individual fragments have formed, they fall together to form small N-systems which grow through accretion of gas and other fragments. These systems merge with other small N-clusters, generating a hierarchical merger scenario for stellar cluster formation much as is invoked in cosmology. A hierarchial fragmentation process is most straightforward in order to explain the origin of stellar clusters.

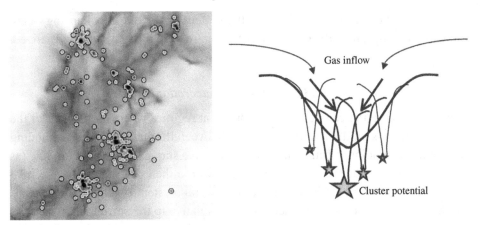

Fig. 12.3. The location of the most massive stars in a simulation of cluster formation (*left*) shows their preferential location in the centres of clusters due to competitive accretion. The *right panel* shows a schematic view of the competitive accretion process, whereby the cluster potential funnels gas down to the cluster core. The stars located there are therefore able to accrete more gas and become higher-mass stars. The gas reservoir can be replenished by infall into the large-scale cluster potential. (Adapted from Bonnell *et al.* 2007).

The infalling gas is crucial in providing the material to form the more massive stars in the clusters.

12.3.2 Accretion in a cluster potential

Competitive accretion provides just such a mechanism as it relies on the continued accretion onto a lower-mass star due to the large-scale gravitational potential of a stellar cluster (Bonnell *et al.* 2001*a*). Such a potential naturally funnels gas down towards the centre of the system, to be accreted by the soon-to-be massive stars located there.

Competitive accretion relies on the inefficiency of fragmentation such that there is a large common reservoir of gas from which the protostars can accrete (Bonnell *et al.* 2001*a*, 2004). Observations of both pre-stellar structures and young stellar clusters support this view with the large majority of the total mass being in a distributed gaseous form (Motte *et al.* 1998; Johnstone *et al.* 2000, 2004; Lada & Lada 2003; Bastian & Goodwin 2006). The second requirement is that the gas be free to move under the same gravitational acceleration as the stars. If the gas is fixed in place due to magnetic fields, then accretion will be limited. When these two requirements are filled, the dynamical timescales for accretion and evolution are similar such that a significant amount of gas can be accreted. The necessary condition is that the system (cluster) is bound such that the overall potential can have a significant effect. It is not necessary for the system to be cold or collapsing as the potential will still funnel the gas down to the centre even if the system is virialized.

12.3.3 Dynamics of accretion

Competitive accretion can occur in two different regimes where the physics of the accretion depends on the relative velocities between the stars and the surrounding gas. First, at the point of fragmentation, the relative star–gas velocities are low (as are the star–star velocities). The accretion rates are then determined by the tidal limits of each star's potential relative to the overall system potential. This means that the stars themselves do not need to move relative to the surrounding gas. It is their ability to gravitationally attract the gas, which is governed by their tidal radii in the cluster potential, that determines the accretion rates and therefore the final mass distribution.

Second, once stars dominate the potential, they will virialize and have random velocities relative to the gas. Thus, in general, the relative velocities between gas and stars are large and the accretion is determined by a Bondi–Hoyle-type accretion. This produces a steeper mass spectrum due to the strong mass dependence of the accretion rates (Zinnecker 1984; Bonnell *et al.* 2001*b*). Simulations and analytical arguments have shown that this type of process results in a Salpeter-like IMF with mass spectra of $\mathrm{d}N \propto m^{-\gamma}\mathrm{d}m$ with γ in the range of 2–2.5 (Bonnell *et al.* 2001*b*, 2003). It should be noted that even in a virialized cluster, there are exceptions to a Bondi–Hoyle accretion formalism. For example, the most massive star sits in or near the centre of the potential. The relative velocity of the gas to this star is then small, resulting in near-spherical infall limited more by the tidal radii of the cluster core. Even stars that are further out and have higher velocities will periodically have velocities more aligned to that of the infalling gas and thus a much lower relative velocity than would be naively predicted.

12.3.4 Limitations of competitive accretion

Recently, Krumholz *et al.* (2005c) have cast some doubt in this process by claiming that accretion in such environments cannot significantly increase a star's mass and therefore does not play a role in establishing the stellar IMF. In part, this is correct as with competitive accretion, most stars do not continue to accrete. It is the few which do that are important in terms of forming higher-mass stars and the IMF. What the Krumholz analysis overlooks is that Bondi–Hoyle accretion,

$$\dot{M}_* \approx 4\pi\rho\frac{(GM_*)^2}{v^3},$$

(12.1)

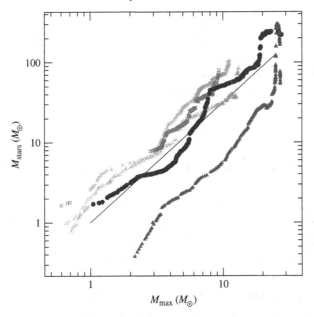

Fig. 12.4. The total mass in stars is plotted against the mass of the most massive star in the system. Competitive accretion predicts that as the total mass grows, the mass of the most massive star also increases as $M_{\mathrm{max}} \propto M_{\mathrm{stars}}^{2/3}$ (Bonnell *et al.* 2004).

is very dependent on the exact values of the gas density ρ, stellar mass M_* and velocity dispersion v, such that with slight variations in these properties, the accretion can vary from 10^{-9} to $10^{-4} M_\odot$ year^{-1}. The higher values take more typical gas densities found in the cores of stellar clusters and velocity dispersion of $0.5\,\mathrm{km\,s^{-1}}$, in keeping with the expected level of turbulence and stellar velocity dispersion expected on these scales (Bonnell & Bate 2006).

The large range in accretion rates due to small differences in the gas density, the relative velocity and the initial stellar (fragment) mass highlights the strength of competitive accretion in explaining the full stellar IMF. It is only the rare stars which gain sufficient mass due to competitive accretion that attain high-mass status. As importantly, competitive accretion limits the growth of the bulk of the stars keeping a low median stellar mass.

12.3.5 Predictions

There are several observational predictions that can be gathered from the numerical simulations of competitive accretion. The most striking of these is the relationship between the cluster properties and the mass of the most massive star. Stellar clusters form when newly formed stars fall together to produce small N-groups. These groups grow to form larger clusters by accreting gas and stars from the surrounding environment. The gas accompanies the stars into the cluster potential with a significant fraction being accreted by the more massive stars. This naturally produces a correlation between the number of stars in the cluster (or the total mass in stars) and the mass of the most massive star therein (Bonnell *et al.* 2004). This prediction, where

$$M_{\text{max}} \propto M_{\text{cluster}}^{2/3},$$ (12.2)

agrees remarkably well with the analysis of observed young stellar clusters by Weidner and Kroupa (2006).

Competitive accretion also provides an explanation for the frequency and properties of binary systems among massive stars (Bonnell & Bate 2005). Binary systems form readily through fragmentation and three-body capture in the small N-systems in the centre of the forming clusters. The continued accretion hardens binary systems while increasing the masses of the component. Thus, wide (100 AU) low-mass binary systems that undergo continued accretion evolve into high-mass close binary systems. The frequency of such high-mass binary systems is 100% in these systems with the majority of these being close with massive companions. They generally are found to have fairly high eccentricities which, along with their small separations, imply that some of these should merge to form higher-mass stars. Stellar densities required to perturb such a binary system and force it to merge are estimated to be of order 10^6 stars pc^{-3}.

12.3.6 Observational comparisons

Competitive accretion requires a bound stellar system in order for the large-scale potential to funnel the gas down to the forming massive stars. This does not mean that the system is cold and globally collapsing, and in fact the system will continually appear to be nearly virialized as any contraction will engender more transverse motions in the turbulent gas. On small scales, the initial fragmentation produces cores which are internally subsonic as is expected in any turbulent cloud (Padoan *et al.* 2001) as they form due to the shock removal of the turbulent motions (Padoan *et al.* 2001). They also have a low velocity relative to each other and to the nearby interfragment gas (Ayliffe *et al.* 2007). Once they collapse into stars and move due to the large-scale potential, they gain significant velocity dispersions.

The fact that the accreting stellar clusters are always quasi-virial means that the star formation process occurs over several free-fall times of the original system (Bonnell *et al.* 2004). This is a lower limit as it neglects the formation time of the system which must be larger than its localized free-fall time. A further complication to this arises from using final dynamical times to compute the SFR. In

Bonnell *et al.* (2003), the final cluster dynamical time is of order three times smaller than the initial crossing time of the cloud. Using the final cluster timescale thus underestimates the SFR per free-fall time by a factor of 3 or more such that the ONC estimates are very compatible to those from the simulation (Krumholz & Tan 2007). On large scales, the SFR is lower still as much of the gas on such scales need not be bound (Clark *et al.* 2005, 2007).

12.4 Challenges in both models: feedback

While the core accretion and competitive accretion models differ on how mass is gathered, they both face similar problems in explaining how that mass, once it reaches the vicinity of the star, gets to the stellar surface. The primary barrier is the radiation of the massive protostar, which will be extremely powerful because massive stars have very short Kelvin–Helmholtz times that allow them to reach the main sequence while still accreting. For a 50-$M_\odot$ ZAMS star, $t_{KH} = GM^2/(RL) \approx 2 \times 10^4$ year (using the ZAMS models of Tout *et al.* 1996), while the formation time of such a star is a few $\times 10^5$ year (McKee & Tan 2002). Radiation can interfere with accretion both directly via the force it exerts and indirectly by ionizing the gas and raising its temperature above the escape speed from the star-forming region.

12.4.1 Radiation pressure

The problem of radiation pressure in massive star formation can be viewed in terms of the Eddington limit. In spherical symmetry, a star of luminosity L and mass M shining on a gas of opacity κ at a distance r exerts a force per unit mass $\kappa L/(4\pi r^2 c)$, while the gravitational force per unit mass is GM/r^2. The gravitational force will be stronger only if $L/M < 4\pi Gc/\kappa = 1.3 \times 10^4 \kappa_1 (L_\odot/M_\odot)$, where κ_1 is κ measured in units of cm^2 g^{-1}. For ZAMS stars, the light-to-mass ratio reaches $10^4 L_\odot/M_\odot$ by a mass of 35–40$M_\odot$, and the opacity of dust gas at solar metallicity varies from hundreds of cm^2 g^{-1} at visible and UV wavelengths to a few cm^2 g^{-1} in the far-infrared. This would seem to imply that accretion of dusty gas onto a protostar should be impossible due to radiation pressure for stars larger than a few tens of solar masses, a point made by a series of 1D calculations (Larson & Starrfield 1971; Kahn 1974; Wolfire & Cassinelli 1987).

Since more massive stars are observed to exist, something in this argument must fail. One possibility is that massive stars form not by accretion but by mergers of lower-mass stars, which will not be affected by radiation pressure (Bonnell *et al.* 1998; Bonnell & Bate 2005). However, while this process may take place, the high stellar densities required to produce an appreciable merger rate suggest that it is unlikely to be the dominant formation mechanism under galactic conditions.

A more likely route around the problem of radiation pressure is to recognize that, in an extremely optically thick protostellar envelope, the gas can reshape the radiation field by beaming it preferentially in certain directions. There is a certain subtlety to this point, since non-spherical *accretion* is not by itself sufficient to overcome radiation pressure. If accretion were non-spherical but the radiation field remained spherical, then the net force on matter would still be away from the star at all points. However, if the opaque gas can force the radiation to flow around certain structures and travel away from the star non-spherically, then the radiation flux within those structures will be reduced and the net force can be inwards, allowing accretion.

The most obvious candidate for a structure that can exclude the radiation field is a protostellar disk (Nakano 1989; Nakano *et al.* 1995, 2000; Jijina & Adams 1996). In such a disk, direct stellar radiation will be absorbed in a thin layer within which the dust is near its sublimation temperature, roughly 1000–1500 K depending on the exact composition of the grains (Ossenkopf & Henning 1994; Pollack *et al.* 1994). Inside this layer, the opacity of the gas is small and radiation streams outwards freely and spherically without exerting much force on the infalling matter. At the dust destruction front, radiation is rapidly down-converted to infrared wavelengths and then diffuses outwards through the dusty envelope. Since the disk is much more opaque than its surroundings, this diffusion is anisotropic and much less flux passes through the disk than through the region around it. Matter that reaches the disk is shielded and is able to accrete. It will briefly feel a net outward acceleration as it passes through the dust destruction front and encounters the spherical radiation field within it, but if it has enough momentum inwards, its flow will not be reversed before the dust sublimes and it ceases feeling much radiative acceleration (Wolfire & Cassinelli 1987). 2D multi-group radiation-hydrodynamic simulations of this process starting from non-turbulent initial conditions demonstrate that it allows the formation of stars up to $43M_\odot$ (Yorke & Sonnhalter 2002). The upper mass limit is set in this simulation because eventually the radiation field is able to reverse the infall of matter which has not yet reached the disk, choking off further inflow to the disk.

Expanding to a 3D view and including additional physical processes both open up additional possibilities for anisotropizing the radiation field. As it happens in two dimensions, 3D simulations show that radiation blows bubbles above and below the accretion disk, preventing gas from falling onto the star except through the disk. However, in three dimensions these bubbles prove to be subject to an instability analogous to Rayleigh–Taylor instability, which causes the bubbles to break up into low-density chimneys filled with outward-moving radiation and high-density walls that channel gas inwards to the disk (Krumholz *et al.* 2005*a*). If there are substantial magnetic fields present in the infalling gas, this separation of the gas into dense inflowing channels and diffuse radiation chimneys can be significantly accelerated by photon bubble instability (Turner *et al.* 2007). Finally, both observations (Beuther *et al.* 2002, 2003, 2004; Beuther & Shepherd 2005) and simulations (Banerjee & Pudritz 2007) show that massive protostars produce hydromagnetic outflows just like their low-mass counterparts. These outflows are launched near the star, where their dust is sublimed away and the dust does not have time to re-form until the outflowing gas is well away from the star. As a result, the cavities that protostellar outflows punch into massive cores are optically thin, providing an escape channel for radiation that provides significant anisotropy and reduces the radiation pressure force near the equatorial plan by up to an order of magnitude (Krumholz *et al.* 2005b).

Together, these calculations suggest that the radiation pressure problem is considerably less daunting than was suggested by early spherically symmetric estimates. However, this problem is not completely closed and a number of outstanding issues remain. First, all of the multi-dimensional simulations and analytic calculations to date have assumed that dust and gas are perfectly well-coupled and that, except for subliming at high temperatures, grains do not change their properties as gas accretes. However, 1D radiation-hydrodynamic simulations show that neither of these assumptions are strictly true (Suttner *et al.* 1999), and it is not clear how this will affect the radiation pressure problem in more dimensions.

Second, how magnetic fields will affect the radiation pressure problem is still largely an open question. Only one radiation-magnetohydrodynamic (MHD) simulation of massive star formation has been reported in the literature (Turner *et al.* 2007), and this focused solely on magnetic effects in a small 2D patch of a protostellar envelope. The role of magnetic fields on larger scales and in three dimensions remains unexplored.

Third, essentially all work on radiation pressure to date has taken place in the context of the core accretion model rather than the competitive accretion model. The initial conditions generally start from a single gravitationally bound massive core. The sole exception to this is Edgar and Clarke (2004), who simulate Bondi–Hoyle accretion onto a massive star and find that radiation pressure completely halts accretion onto stars larger than about $10 M_\odot$. Since the calculations begin with a smooth flow that is not undergoing gravitational collapse, it is unclear whether the results will continue to hold in the context of a more sophisticated competitive accretion model. However, this result does point out that whether radiation pressure can be overcome in the absence of an initial bound core remains an open question.

12.4.2 Ionization

The second form of feedback that might prevent accretion is ionizing radiation, which can heat gas to temperatures of 10^4 K, with sound speeds of $10 \, \mathrm{km \, s^{-1}}$. If the ionized region expands to include gas for which the escape speed from the massive star and its surrounding gas is smaller than $10\,\mathrm{km\,s^{-1}}$, the ionized gas will begin to escape, potentially choking off the supply of mass to the protostar. Whether or not this happens depends on the interaction between the ionizing radiation field and the gas inflow. Walmsley (1995) first pointed out that in a spherically symmetric accretion flow, ionizing radiation can be trapped against the stellar surface simply by the flux of inflowing matter. The mass flux required to accomplish this is relatively modest, $\lesssim 10^{-4} M_\odot\ \mathrm{year}^{-1}$ even for mid-O stars. The accretion rates for such massive objects expected from analytic models and simulations (McKee & Tan 2003; Bonnell *et al.* 2004; Krumholz *et al.* 2007b), or simply inferred on dimensional grounds from the observed properties of massive star-forming regions, are of comparable size.

Keto (2002, 2003, 2007) has generalized this model to cases where the accretion flow does not completely quench the HII region and where the accretion flow is cylindrically symmetric and has non-zero angular momentum. He finds that even after an HII region is able to lift off a stellar surface, it may be confined for an extended period because the continual supply of new gas keeps the ionized region confined close enough to the star so that even gas with a sound speed of $10\,\mathrm{km\,s^{-1}}$ remains bound and accretes onto the star. In this case, ionization continues despite the presence of an HII region. Depending on the relative sizes of the ionization radius and the gas circularization radius, some of the accretion disk may fall within the ionized region, leading to formation of an ionized, accreting disk as observed in G10.6-0.4 (Keto & Wood 2006). The general conclusion of this work is that the ionizing radiation cannot significantly inhibit the formation of a massive star. As long as there is sufficient material left to accrete, the ionized region can be trapped so that gas can continue to accrete through it.

On the numerical side, Dale *et al.* (2005) has simulated the expansion of an HII region into a turbulent, non-magnetized parsec-sized cluster of gas and stars produced by a previous simulation of cluster formation via competitive accretion (Bonnell & Bate 2002). In agreement with Keto, Dale *et al.* find that ionizing radiation feedback cannot significantly impede the formation of massive stars. Although an ionizing source deposits in a protocluster gas cloud an amount of energy significantly larger than its binding energy, this energy does not succeed in unbinding the cluster or preventing the mass in dense structures from accreting. Instead, it simply blows off a relatively small fraction of the low-density gas. Overall, the behaviour of the gas with respect to the ionizing radiation has much in common with its behaviour with respect to the non-ionizing radiation: an inhomogeneous gas flow shapes the radiation field, forcing radiation to flow out through low-density channels while removing relatively little mass and at the same time allowing dense, shielded gas to move inwards under the influence of gravity.

Based on this combination of numerical and analytic work, it seems unlikely that the ionizing radiation poses a serious challenge to either the core accretion model or the competitive accretion model of massive star formation.

12.5 Future directions

12.5.1 Observations

There is considerable room for future observations to improve our models of how massive stars work. Such measurements are difficult to use directly in discriminating between the competitive and the core accretion models, because the models differ primarily in how the mass is gathered, a process that occurs on timescales far too long to observe. The challenge, therefore, is to design observations that can address the secondary conditions implied or required by the models. Here, we present some potentially powerful observational tests for this purpose. Most of these are submillimetre measurements that will require ALMA, although some may be possible earlier using the current generation of telescopes, such as PdBI, SMA and CARMA. Others involve observations in the radio, infrared and X-rays. These observations can be roughly divided into three scales: the 'micro' scale of individual star systems and their immediate gaseous progenitors, the 'meso' scale of individual star clusters and the 'macro' scale of star formation on galactic scales.

12.5.1.1 Micro scales

On the micro scale, one significant advance could come through observations of a large sample of disks around massive protostars. Both competitive accretion and core accretion models predict that massive stars should have disks $\sim$1000 AU in size. However, competitive accretion models also predict that the typical massive star has a large number of close encounters with other stars which will destroy or greatly truncate the disk at regular intervals (Scally & Clarke 2001; Bonnell *et al.* 2003; Pfalzner *et al.* 2006; Moeckel & Bally 2007). Continuing accretion causes the disk to grow back, but this model implies that in a sample of massive, embedded stars without significant HII regions, there should be some fraction at any given time that lack disks. In the core accretion model, encounters are much rarer, and a disk should always be present until accretion stops and the remnant disk material is removed by ionizing radiation from the massive star. The fraction of diskless, still-embedded massive stars predicted by competitive accretion models is not yet certain, but an observation of whether such stars exist is a potentially powerful discriminator between models.

A second very useful micro-scale observation would be a better determination of the properties of massive cores – the basic building blocks of the core accretion model. This model requires the similarity between the shapes of the core and the initial stellar mass, and spatial distributions observed for lower-mass cores continue up to at least a few hundred $M_\odot$. Agreement between these properties has only been established in nearby low- and intermediate-mass regions, and evidence that this mapping does not hold for massive cores in more distant regions would rule out the simple core accretion model.

We also like to find how centrally concentrated massive cores are and to what extent do they break up into collections of lower-mass clumps at high resolution, since both of these will have an impact on how and whether they fragment. Observations today hint at a high degree of central concentration (Beuther *et al.* 2007), but this is only for a single core, and even in this case, the observations lack the resolution to address the degree of clumping.

The question of clumping also leads to the problem of determining to what extent massive cores already contain low-mass stars. Motte *et al.* (2007) find that even infrared-quiet massive cores harbour $SiO(2 \rightarrow 1)$ emission indicative of outflows from low-mass stars. They infer that any truly starless phase for massive cores must last $\lesssim 1000$ year. Given these cores' masses and mean densities, this is not surprising, since any reasonable density distribution would suggest that they contain at least $\sim M_\odot$ of gas at densities high enough to collapse in 1000 years. However, the degree of low-mass stellar content is an important discriminator between theories. Core accretion models predict that massive cores fragment weakly, so one should essentially never find a massive core that has converted a significant fraction of its mass to stars but has formed only low-mass stars. Any low-mass stars that exist in such a core should grow to become massive stars as more gas falls onto them. Competitive accretion models, on the contrary, predict that an outcome in which massive cores fragment and produce only low-mass stars should not be uncommon. For this reason, determining the low-mass stellar content of IR-quiet massive cores would be extremely useful. This might be possible using X-ray observations, which are very sensitive to the presence of low-mass stars (Feigelson *et al.* 2005).

A final micro-scale observational question concerns the apparently isolated O stars identified by de Wit *et al.* (2004, 2005). The existence of these stars already seems to suggest that the presence of a cluster is not absolutely required to form massive stars. It would be very helpful to know if these stars are unusual in any way other than being isolated. For example, are their rotation rates of magnetic properties different than those of O stars found in clustered environments? This would help address the question of whether these stars formed by a different process than typical O stars.

12.5.1.2 Meso scales

On the meso scale, the most useful observations would be better constraints on the dynamical state of cluster-forming gas clumps and on the formation history of the star clusters that result. As discussed earlier, the question of how massive stars form is intimately tied to the question whether star clusters form in a process of global collapse in which the parent cloud never reaches any sort of equilibrium configuration or whether cluster-forming clouds are quasi-equilibrium objects within which collapse occurs only in localized regions. The data that exist now can be interpreted in conflicting ways (Elmegreen 2000b; Tan *et al.* 2006; Elmegreen 2007; Huff & Stahler 2007; Krumholz & Tan 2007) and cannot definitively settle which model more closely approximates reality.

For gaseous star-forming clumps, one can approach this problem by searching for signatures of supersonic collapse via line profiles. There is some evidence for such collapse in certain regions of star cluster formation (Peretto *et al.* 2006), but in others infall is either completely absent or subsonic (Garay 2005). The question of whether the typical region of massive star formation is in a state of global collapse, however, cannot be answered without more systematic searches covering large number of star-forming clouds.

For revealed star clusters, observations can help by pinning down the star formation history in more detail, hopefully settling the question of the true age spreads in clusters. This project will require both improvements in observational data and in the pre-main sequence models used to convert observed stellar luminosities and temperatures into ages. Observationally, the current surveys suffer from potential problems with incompleteness, interlopers and dust obscuration. Incompleteness preferentially removes the oldest and lowest mass members from a sample, making clusters appear younger than their true age, while interlopers tend to be older stars whose inclusion in the sample can artificially age clusters. Obscuration by dust in the vicinity of the cluster can push the age in either direction, since it worsens incompleteness but, by dimming stars, it also makes them appear older than their true age. Future surveys should be able to do considerably better than the current generation in minimizing these effects.

For the pre-main sequence tracks, improvements are likely to come from empirical calibrations of the models against binary systems, where masses can be obtained dynamically and the stars are likely to be coeval (Palla & Stahler 2001; Stassun *et al.* 2004; Boden *et al.* 2005). As more pre-main sequence binaries are discovered and their masses and radii determined, it should be possible to progressively improve the models and reduce the uncertainties in the ages they produce.

A final point to remember is that distance uncertainty is a constant concern, since, as the case of the ONC shows, even a $\lesssim 20\%$ change in the estimate distance can significantly skew estimates of the star formation history. Radio parallax measurements of the distances to many more star-forming regions should be forthcoming in the next few years and may force significant revisions in our models of star formation histories.

12.5.1.3 Macro scales

Galactic- and larger-scale observations are powerful tools for massive star formation studies because they offer the opportunity to obtain statistical constraints, something generally impossible on smaller scales due to the rarity of massive stars. Such observations already indicate the existence of an upper mass cutoff of $\sim 150 M_\odot$, beyond which stars appear to be very rare or non-existent (Elmegreen 2000a; Weidner & Kroupa 2004; Figer 2005). It is possible that this limit is a result of instabilities in stars after they form rather than something set at formation, but if it is set at formation, then it is a result that massive star formation models should be able to explain. An observational determination of whether there is any variation of this maximum mass from galaxy to galaxy, or within our galaxy, might provide a valuable clue to the origin of the cutoff and thus to the correct picture of how massive stars form. More generally, any convincing evidence for variation in the IMF from region to region, beyond that expected simply from statistical fluctuations, would be helpful.

Galactic-scale observations can also be useful in constraining the star formation rate as a function of density. Krumholz and McKee (2005) define the SFR per free-fall time, $\mathrm{SFR_{ff}}$, as the fraction of a gaseous object's mass that is incorporated into stars per free-fall time of that object. Krumholz and Tan (2007) show that, when averaged over galactic scales, $\mathrm{SFR_{ff}}$ is typically a few percent for objects ranging in density from entire giant molecular clouds, $\sim 10^2 \, \mathrm{cm}^{-3}$ in density, to sub-pc size clumps traced by HCN($1 \rightarrow 0$) emission, which is excited only at densities $\sim 10^5 \, \mathrm{cm}^{-3}$ or higher. They argue that this implies that cluster-forming gas clumps cannot be rapidly collapsing, a result required by the core accretion model and incompatible with competitive accretion. However, systematic uncertainties limit the determination of $\mathrm{SFR_{ff}}$ to factors of ~ 3–4, meaning that a rise of $\mathrm{SFR_{ff}}$ with density cannot be definitively ruled out. The largest such uncertainty is in the conversion of observed molecular line luminosities to masses, and this can be significantly reduced by better observational calibrations of the 'X' factor for the various molecules used in surveys.

Observations to date also cannot rule out the possibility that the galactic-averaged value of $\mathrm{SFR_{ff}}$ is low because it averages over a significant amount of the emitting gas that is not associated with star-forming regions. In this picture, star-forming clumps would collapse rapidly, as required by competitive accretion, but their high values of $\mathrm{SFR_{ff}}$ would be diluted by the inclusion in galactic averages of a significant amount of gas that emits in a given line but that is not bound and is not forming stars at all. Resolved observations of a large sample of line-emitting regions in nearby galaxies, combined with resolved infrared observations to estimate SFRs in those regions, could determine whether such sterile clouds exist. A preliminary survey of the Milky Way finds that large HCN-emitting gas clouds generally have the same values of SFR$_{ff}$ inferred from galactic averages (Wu *et al.* 2005), but there is considerable scatter in the result, and since the sample was selected based on indicators of massive star formation, the survey does not address the possible existence of sterile HCN-emitting clouds.

12.5.2 Simulations

There is also considerable room for improvement in simulations of massive star formation. The simulations that have been performed to date can be roughly placed into a continuum, running from those with large dynamic ranges in three dimensions but rather limited input physics to those that include extremely detailed physics, but with lower dimensionality, lower dynamic range, or both. Including all or even most of the physical processes relevant to massive star formation in a simulation that has the same dynamic range as that achieved by calculations including only hydrodynamics plus gravity is not feasible on current computers. However, it is possible to push from both ends of the spectrum towards the middle. We end this chapter by highlighting what we believe are two of the most promising directions for progress of this sort.

12.5.2.1 Cluster formation with feedback

Much of the debate between competing models concerns whether the collapse process that forms a massive star is localized to a single massive core or occurs globally within a much larger cloud that forms an entire cluster. However, the dynamics of clusters are certainly modified by feedback from embedded protostars, and the role of feedback has been subjected only to a very limited exploration in simulations.

One obvious place for improvement would be to extend the work of Krumholz *et al.* (2007b) on radiative suppression of fragmentation to larger scales. The Krumholz *et al.* simulations follow the collapse of $\sim$0.1 pc-size cores down to $\sim$10 AU scales; the cores are turbulent initially, but the turbulence is allowed to decay freely. It should be possible to repeat a simulation of this sort on large scales, following an entire cluster-forming gas clump $\sim$0.5 pc in size and $\sim$5000$M_\odot$ in mass. Such a simulation would address the question of whether the dynamics of competitive accretion, which inevitably happen in a simulation containing many thermal Jeans masses where the turbulence is not driven, are significantly modified by radiative feedback. The challenge here is whether the calculations could be done at high enough resolution to allow meaningful comparison to earlier work

and still run in a reasonable amount of time. Simulations involving radiative transfer are generally approximately five times as expensive as ones involving only
hydrodynamics and gravity, and their resolution is correspondingly worse. New
radiation-hydrodynamic algorithms that are potentially much faster than those used
in the Krumholz *et al.* simulations may help with this problem (Whitehouse & Bate
2004; Whitehouse *et al.* 2005; Shestakov & Offner 2008).

Another important priority is to improve on the pioneering work of Li and Nakamura (2006) and Nakamura and Li (2007) on the effect of protostellar outflows on
cluster dynamics. They simulate magnetized clouds in which forming protostars
generate outflows and find that these outflows drive turbulence strongly enough to
keep the SFR to $SFR_{ff} \sim 5\%$ and to prevent the cloud from going into overall collapse. If these results are valid, then outflows qualitatively change the dynamics of
star cluster formation in a way that appears on its face inconsistent with the competitive accretion model in which massive stars form by accreting gas over large
distances from a region in global rather than local collapse. However, there are
significant caveats. The simulations use a fairly low-resolution fixed grid with no
adaptivity. Low resolution could artificially isotropize the energy from outflows,
making them more effective at driving turbulence than more collimated outflows
would be. The simulations also use a very simple implementation of sink particles to represent forming protostars; these sink particles all have the same mass,
and they instantaneously appear on the computational grid rather than accreting gas over time. As a result, they cannot directly study the impact of outflows
on the dynamics of accretion. Clearly, it is important to repeat these simulations
using higher-resolution adaptive mesh or Lagrangian simulations and using a more
sophisticated implementation of sink particles in which it is possible to follow
accretion over time (Bate *et al.* 1995; Krumholz *et al.* 2004).

12.5.2.2 Fragmentation and feedback with magnetic fields

Another almost completely unexplored area is the effect of magnetic fields on
fragmentation, accretion and feedback in regions of massive star formation. The
first MHD simulations of low-mass star formation with enough dynamic range to
study fragmentation, using three very different numerical techniques, all find that
magnetic fields significantly reduce fragmentation compared to the purely hydrodynamical case (Hosking & Whitworth 2004; Price & Bate 2007; Hennebelle &
Teyssier 2008; Machida *et al.* 2008). To date no simulations of magnetic fields in
the context of massive star or star cluster formation have been performed. It would
be very valuable to repeat earlier hydrodynamic or radiation-hydrodynamic simulations (Bonnell *et al.* 2004 or Krumholz *et al.* 2007b) with MHD to see how
magnetic fields change fragmentation and initial accretion.

Magnetic fields may change how feedback from embedded protostars interact with the environment, an effect already hinted at by a number of simulations. Turner *et al.* (2007) show that magnetic fields make it easier to overcome the radiation pressure barrier in massive star formation because they allow photon bubble instabilities to develop, while Nakamura and Li (2007) show that magnetic fields, by providing a means to redistributed energy from protostellar outflows throughout a cluster, enhance outflows' ability to drive turbulent motions and affect cloud dynamics. Krumholz *et al.* (2007d) find that magnetic fields can help confine HII regions and also transmit their effects via fast magnetosonic and Alfvén waves to distant parts of a flow. This could potentially alter the findings of Dale *et al.* (2005) that because HII regions expand asymmetrically into low-density parts of a flow, they cannot unbind clusters or greatly reduce the SFR within them. In summary, all the simulations of massive star formation that have been performed to date could be significantly altered by the inclusion of magnetic fields. With the advent of adaptive mesh refinement (AMR) and smoothed particle hydrodynamics (SPH) MHD codes (Fromang *et al.* 2006; Price & Bate 2007) capable of achieving significant dynamic ranges in MHD problems, it is important to repeat these calculations with MHD.

Acknowledgements

We thank S. S. R. Offner for helpful discussions. This work was supported by NASA through Hubble Fellowship grant HSF-HF-01186 awarded by the Space Telescope Science Institute, which is operated by the Association of Universities for Research in Astronomy, Inc., for NASA, under contract NAS 5-26555; the National Science Foundation under Grant No. PHY05-51164; the Arctic Region Supercomputing Center; the National Energy Research Scientific Computing Center, which is supported by the Office of Science of the US Department of Energy under contract DE-AC03-76SF00098, through ERCAP grant 80325; the NSF San Diego Supercomputer Center through NPACI program grant UCB267 and the US Department of Energy at the Lawrence Livermore National Laboratory under contract W-7405-Eng-48.

References

Alves, J., Lombardi, M. & Lada, C. J. 2007 The mass function of dense molecular cores and the origin of the IMF. *A&A* **462**, L17–L21

André, P., Belloche, A., Motte, F. & Peretto, N. 2007 The initial conditions of star formation in the Ophiuchus main cloud: Kinematics of the protocluster condensations. *A&A* **472**, 519–535

Apai, D., Bik, A., Kaper, L., Henning, T. & Zinnecker, H. 2007 Massive binaries in high-mass star-forming regions: A multi-epoch radial velocity survey of embedded O-stars. *ApJ.* **655**, 484–491

Ayliffe, B. A., Langdon, J. C., Cohl, H. S. & Bate, M. R. 2007 On the relative motions of dense cores and envelopes in star-forming molecular clouds. *MNRAS* **374**, 1198–1206

Banerjee, R. & Pudritz, R. E. 2007 Massive star formation via high accretion rates and early disk-driven outflows. *ApJ.* **660**, 479–488

Bastian, N. & Goodwin, S. P. 2006 Evidence for the strong effect of gas removal on the internal dynamics of young stellar clusters. *MNRAS* **369**, L9–L13

Bate, M. R. & Bonnell, I. A. 2005 The origin of the initial mass function and its dependence on the mean Jeans mass in molecular clouds. *MNRAS* **356**, 1201–1221

Bate, M. R., Bonnell, I. A. & Bromm, V. 2003 The formation of a star cluster: Predicting the properties of stars and brown dwarfs. *MNRAS* **339**, 577–599

Bate, M. R., Bonnell, I. A. & Price, N. M. 1995 Modelling accretion in protobinary systems. *MNRAS* **277**, 362–376

Beuther, H., Leurini, S., Schilke, P., Wyrowski, F., Menten, K. M. & Zhang, Q. 2007 Interferometric multi-wavelength (sub)millimeter continuum study of the young high-mass protocluster IRAS 05358+3543. *A&A* **466**, 1065–1076

Beuther, H. & Schilke, P. 2004 Fragmentation in massive star formation. *Science* **303**, 1167–1169

Beuther, H., Schilke, P. & Gueth, F. 2004 Massive molecular outflows at high spatial resolution. *ApJ.* **608**, 330–340

Beuther, H., Schilke, P., Gueth, F., McCaughrean, M., Andersen, M., Sridharan, T. K. & Menten, K. M. 2002 IRAS 05358+3543: Multiple outflows at the earliest stages of massive star formation. *A&A* **387**, 931–943

Beuther, H., Schilke, P. & Stanke, T. 2003 Multiple outflows in IRAS 19410+2336. *A&A* **408**, 601–610

Beuther, H. & Shepherd, D. 2005 Precursors of UCHII regions and the evolution of massive outflows. In *Cores to Clusters: Star Formation with Next Generation Telescopes* (eds M. S. Kumar, M. Tafalla & P. Caselli), Astrophysics and Space Science Library, Vol. 324, Springer, New York, pp. 105–119

Beuther, H., Sridharan, T. K. & Saito, M. 2005 Caught in the act: The onset of massive star formation. *ApJ.* **634**, L185–L188

Beuther, H., Zhang, Q., Sridharan, T. K., Lee, C.-F. & Zapata, L. A. 2006 The high-mass star-forming region IRAS 18182-1433. *A&A* **454**, 221–231

Boden, A. F., Sargent, A. I., Akeson, R. L., Carpenter, J. M., Torres, G., Latham, D. W., Soderblom, D. R., Nelan, E., Franz, O. G. & Wasserman, L. H. 2005 Dynamical masses for low-mass pre-main-sequence stars: A preliminary physical orbit for HD 98800 B. *ApJ.* **635**, 442–451

Bonanos, A. Z., Stanek, K. Z., Udalski, A., Wyrzykowski, L., Żebruń, K., Kubiak, M., Szymański, M. K., Szewczyk, O., Pietrzyński, G. & Soszyński, I. 2004 WR 20a is an eclipsing binary: Accurate determination of parameters for an extremely massive Wolf-Rayet system. *ApJ.* **611**, L33–L36

Bonnell, I. A. & Bate, M. R. 2002 Accretion in stellar clusters and the collisional formation of massive stars. *MNRAS* **336**, 659–669

Bonnell, I. A. & Bate, M. R. 2005 Binary systems and stellar mergers in massive star formation. *MNRAS* **362**, 915–920

Bonnell, I. A. & Bate, M. R. 2006 Star formation through gravitational collapse and competitive accretion. *MNRAS* **370**, 488–494

Bonnell, I. A., Bate, M. R., Clarke, C. J. & Pringle, J. E. 2001a Competitive accretion in embedded stellar clusters. *MNRAS* **323**, 785–794

Bonnell, I. A., Bate, M. R. & Vine, S. G. 2003 The hierarchical formation of a stellar cluster. *MNRAS* **343**, 413–418

Bonnell, I. A., Bate, M. R. & Zinnecker, H. 1998 On the formation of massive stars. *MNRAS* **298**, 93–102

Bonnell, I. A., Clarke, C. J. & Bate, M. R. 2006 The Jeans mass and the origin of the knee in the IMF. *MNRAS* **368**, 1296–1300

Bonnell, I. A., Clarke, C. J., Bate, M. R. & Pringle, J. E. 2001b Accretion in stellar clusters and the initial mass function. *MNRAS* **324**, 573–579.

Bonnell, I. A. & Davies, M. B. 1998 Mass segregation in young stellar clusters. *MNRAS* **295**, 691

Bonnell, I. A., Larson, R. B. & Zinnecker, H. 2007 The origin of the initial mass function. In *Protostars and Planets V* (eds. B. Reipurth, D. Jewitt, & K. Keil), University of Arizona Press, Tucson, pp. 149–164

Bonnell, I. A., Vine, S. G. & Bate, M. R. 2004 Massive star formation: Nurture, not nature. *MNRAS* **349**, 735–741

Caballero, J. A. 2008 Dynamical parallax of sigma Ori AB: Mass, distance and age. *MNRAS* **383**, 750–754

Clark, P. C. & Bonnell, I. A. 2004 Star formation in transient molecular clouds. *MNRAS* **347**, L36–L40

Clark, P. C. & Bonnell, I. A. 2006 Clumpy shocks and the clump mass function. *MNRAS* **368**, 1787–1795

Clark, P. C., Bonnell, I. A., Zinnecker, H. & Bate, M. R. 2005 Star formation in unbound giant molecular clouds: The origin of OB associations? *MNRAS* **359**, 809–818

Clark, P. C., Klessen, R. S. & Bonnell, I. A. 2007 Clump lifetimes and the initial mass function. *MNRAS* **379**, 57–62

Dale, J. E., Bonnell, I. A., Clarke, C. J. & Bate, M. R. 2005 Photoionizing feedback in star cluster formation. *MNRAS* **358**, 291–304

de Wit, W. J., Testi, L., Palla, F., Vanzi, L. & Zinnecker, H. 2004 The origin of massive O-type field stars. I. A search for clusters. *A&A* **425**, 937–948

de Wit, W. J., Testi, L., Palla, F. & Zinnecker, H. 2005 The origin of massive O-type field stars. II. Field O stars as runaways. *A&A* **437**, 247–255

Dobbs, C. L., Bonnell, I. A. & Clark, P. C. 2005 Centrally condensed turbulent cores: Massive stars or fragmentation? *MNRAS* **360**, 2–8

Duchêne, G., Simon, T., Eislöffel, J. & Bouvier, J. 2001 Visual binaries among high-mass stars. An adaptive optics survey of OB stars in the NGC 6611 cluster. *A&A* **379**, 147–161

Edgar, R. & Clarke, C. 2004 The effect of radiative feedback on Bondi–Hoyle flow around a massive star. *MNRAS* **349**, 678–686

Elmegreen, B. G. 2000a Modeling a high-mass turn-down in the stellar initial mass function. *ApJ.* **539**, 342–351

Elmegreen, B. G. 2000b Star formation in a crossing time. *ApJ.* **530**, 277–281.

Elmegreen, B. G. 2006 On the similarity between cluster and galactic stellar initial mass functions. *ApJ.* **648**, 572–579

Elmegreen, B. G. 2007 On the rapid collapse and evolution of molecular clouds. *ApJ.* **668**, 1064–1082

Elmegreen, B. G. & Krakowski, A. 2001 A search for environmental influence on stellar mass. *ApJ.* **562**, 433–439

Feigelson, E. D., Getman, K., Townsley, L., Garmire, G., Preibisch, T., Grosso, N., Montmerle, T., Muench, A. & McCaughrean, M. 2005 Global X-ray properties of the Orion Nebula region. *ApJS* **160**, 379–389.

Figer, D. F. 2005 An upper limit to the masses of stars. *Nature* **434**, 192–194

Fromang, S., Hennebelle, P. & Teyssier, R. 2006 A high order Godunov scheme with constrained transport and adaptive mesh refinement for astrophysical magnetohydrodynamics. *A&A* **457**, 371–384

Gao, Y. & Solomon, P. M. 2004a HCN Survey of normal spiral, infrared-luminous, and ultraluminous galaxies. *ApJS* **152**, 63–80

Gao, Y. & Solomon, P. M. 2004b The star formation rate and dense molecular gas in galaxies. *ApJ.* **606**, 271–290

Garay, G. 2005 Massive and dense cores: The maternities of massive stars. In *IAU Symposium 227: Massive Star Birth: A Crossroads of Astrophysics* (eds R. Cesaroni, M. Felli, E. Churchwell & M. Walmsley), Cambridge University Press, Cambridge, pp. 86–91

Goodman, A. A., Barranco, J. A., Wilner, D. J. & Heyer, M. H. 1998 Coherence in dense cores. II. The transition to coherence. *ApJ.* **504**, 223

Hennebelle, P. & Teyssier, R. 2008 Magnetic processes in a collapsing dense core. II Fragmentation. Is there a fragmentation crisis? *A&A* **477**, 25–34

Hillenbrand, L. A. & Hartmann, L. W. 1998 A preliminary study of the Orion Nebula Cluster structure and dynamics. *ApJ.* **492**, 540

Hirota, T., Bushimata, T., Choi, Y. K., Honma, M., Imai, H., Iwadate, K., Jike, T., Kameno, S., Kameya, O., Kamohara, R., Kan-ya, Y., Kawaguchi, N., Kijima, M., Kim, M. K., Kobayashi, H., Kuji, S., Kurayama, T., Manabe, S., Maruyama, K., Matsui, M., Matsumoto, N., Miyaji, T., Nagayama, T., Nakagawa, A., Nakamura, K., Oh, C. S., Omodaka, T., Oyama, T., Sakai, S., Sasao, T., Sato, K., Sato, M., Shibata, K. M., Shintani, M., Tamura, Y., Tsushima, M. & Yamashita, K. 2007 Distance to Orion KL measured with VERA. *PASJ* **59**, 897–903

Hosking, J. G. & Whitworth, A. P. 2004 Fragmentation of magnetized cloud cores. *MNRAS* **347**, 1001–1010

Huff, E. M. & Stahler, S. W. 2006 Star formation in space and time: The Orion Nebula Cluster. *ApJ.* **644**, 355–363

Huff, E. M. & Stahler, S. W. 2007 Cluster formation in contracting molecular clouds. *ApJ.* **666**, 281–289

Jappsen, A.-K., Klessen, R. S., Larson, R. B., Li, Y. & Mac Low, M.-M. 2005 The stellar mass spectrum from non-isothermal gravoturbulent fragmentation. *A&A* **435**, 611–623

Jijina, J. & Adams, F. C. 1996 Infall collapse solutions in the inner limit: Radiation pressure and its effects on star formation. *ApJ.* **462**, 874

Johnstone, D., Di Francesco, J. & Kirk, H. 2004 An extinction threshold for protostellar cores in Ophiuchus. *ApJ.* **611**, L45–L48

Johnstone, D., Fich, M., Mitchell, G. F. & Moriarty-Schieven, G. 2001 Large area mapping at 850 microns. III. Analysis of the clump distribution in the Orion B molecular cloud. *ApJ.* **559**, 307–317

Johnstone, D., Wilson, C. D., Moriarty-Schieven, G., Joncas, G., Smith, G., Gregersen, E. & Fich, M. 2000 Large-area mapping at 850 microns. II. Analysis of the clump distribution in the ρ Ophiuchi molecular cloud. *ApJ.* **545**, 327–339

Kahn, F. D. 1974 Cocoons around early-type stars. *A&A* **37**, 149–162

Keto, E. 2002 On the evolution of ultracompact HII regions. *ApJ.* **580**, 980–986.

Keto, E. 2003 The formation of massive stars by accretion through trapped hypercompact HII regions. *ApJ.* **599**, 1196–1206

Keto, E. 2007 The formation of massive stars: Accretion, disks, and the development of hypercompact HII regions. *ApJ.* **666**, 976–981

Keto, E. & Wood, K. 2006 Observations on the formation of massive stars by accretion. *ApJ.* **637**, 850–859

Kirk, H., Johnstone, D. & Tafalla, M. 2007 Dynamics of dense cores in the Perseus molecular cloud. *ApJ.* **668**, 1042–1063

Klessen, R. S., Ballesteros-Paredes, J., Vázquez-Semadeni, E. & Durán-Rojas, C. 2005 Quiescent and coherent cores from gravoturbulent fragmentation. *ApJ.* **620**, 786–794

Klessen, R. S. & Burkert, A. 2000 The formation of stellar clusters: Gaussian cloud conditions. I. *ApJS* **128**, 287–319

Klessen, R. S., Burkert, A. & Bate, M. R. 1998 Fragmentation of molecular clouds: The initial phase of a stellar cluster. *ApJ.* **501**, L205

Klessen, R. S., Heitsch, F. & Mac Low, M. 2000 Gravitational collapse in turbulent molecular clouds. I. Gasdynamical turbulence. *ApJ.* **535**, 887–906

Kratter, K. M. & Matzner, C. D. 2006 Fragmentation of massive protostellar discs. *MNRAS* **373**, 1563–1576

Kratter, K. M., Matzner, C. D. & Krumholz, M. R. 2008 Global models for the evolution of embedded, accreting protostellar disks. *ApJ.* **681**, 375

Krumholz, M. R. 2006 Radiation feedback and fragmentation in massive protostellar cores. *ApJ.* **641**, L45–L48

Krumholz, M. R., Klein, R. I. & McKee, C. F. 2005a Radiation pressure in massive star formation. In *IAU Symposium 227: Massive Star Birth: A Crossroads of Astrophysics* (eds R. Cesaroni, M. Felli, E. Churchwell & M. Walmsley), Cambridge University Press, Cambridge, pp. 231–236

Krumholz, M. R., Klein, R. I. & McKee, C. F. 2007a Molecular line emission from massive protostellar disks: Predictions for ALMA and EVLA. *ApJ.* **665**, 478–491

Krumholz, M. R., Klein, R. I. & McKee, C. F. 2007b Radiation-hydrodynamic simulations of collapse and fragmentation in massive protostellar cores. *ApJ.* **656**, 959–979

Krumholz, M. R., Klein, R. I., McKee, C. F. & Bolstad, J. 2007c Equations and algorithms for mixed-frame flux-limited diffusion radiation hydrodynamics. *ApJ.* **667**, 626–643

Krumholz, M. R. & McKee, C. F. 2005 A general theory of turbulence-regulated star formation, from spirals to ultraluminous infrared galaxies. *ApJ.* **630**, 250–268

Krumholz, M. R., McKee, C. F. & Klein, R. I. 2004 Embedding Lagrangian sink particles in eulerian grids. *ApJ.* **611**, 399–412

Krumholz, M. R., McKee, C. F. & Klein, R. I. 2005b How protostellar outflows help massive stars form. *ApJ.* **618**, L33–L36

Krumholz, M. R., McKee, C. F. & Klein, R. I. 2005c The formation of stars by gravitational collapse rather than competitive accretion. *Nature* **438**, 332–334

Krumholz, M. R., Stone, J. M. & Gardiner, T. A. 2007d Magnetohydrodynamic evolution of HII regions in molecular clouds: Simulation methodology, tests, and uniform media. *ApJ.* **671**, 518–535

Krumholz, M. R. & Tan, J. C. 2007 Slow star formation in dense gas: Evidence and implications. *ApJ.* **654**, 304–315

Krumholz, M. R. & Thompson, T. A. 2007 Mass transfer in close, rapidly accreting protobinaries: An origin for massive twins? *ApJ.* **661**, 1034–1041.

Lada, C. J. 2006 Stellar multiplicity and the initial mass function: Most stars are single. *ApJ.* **640**, L63–L66

Lada, C. J. & Lada, E. A. 2003 Embedded clusters in molecular clouds. *ARA&A* **41**, 57–115

Lada, C. J., Muench, A. A., Rathborne, J. M., Alves, J. F. & Lombardi, M. 2008 The nature of the dense core population in the Pipe Nebula: Thermal cores under pressure. *ApJ.* **672**, 410–422

Larson, R. B. 1984 Gravitational torques and star formation. *MNRAS* **206**, 197–207

Larson, R. B. 2005 Thermal physics, cloud geometry and the stellar initial mass function. *MNRAS* **359**, 211–222

Larson, R. B. & Starrfield, S. 1971 On the formation of massive stars and the upper limit of stellar masses. *A&A* **13**, 190

Li, Z.-Y. & Nakamura, F. 2006 Cluster formation in protostellar outflow-driven turbulence. *ApJ.* **640**, L187–L190

Lucy, L. B. 2006 Spectroscopic binaries with components of similar mass. *A&A* **457**, 629–635

Machida, M. N., Tomisaka, K., Matsumoto, T. & Inutsuka, S.-I. 2008 Formation scenario for wide and close binary systems. *ApJ,* **677**, 327–347

Matzner, C. D. & McKee, C. F. 2000 Efficiencies of low-mass star and star cluster formation. *ApJ.* **545**, 364–378

McKee, C. F. & Tan, J. C. 2002 Massive star formation in 100,000 years from turbulent and pressurized molecular clouds. *Nature* **416**, 59–61

McKee, C. F. & Tan, J. C. 2003 The formation of massive stars from turbulent cores. *ApJ.* **585**, 850–871

McMillan, S. L. W., Vesperini, E. & Portegies Zwart, S. F. 2007 A dynamical origin for early mass segregation in young star clusters. *ApJ.* **655**, L45–L49

Menten, K. M., Reid, M. J., Forbrich, J. & Brunthaler, A. 2007 The distance to the Orion Nebula. *A&A* **474**, 515–520

Moeckel, N. & Bally, J. 2007 Binary capture rates for massive protostars. *ApJ.* **661**, L183–L186

Motte, F., Andre, P. & Neri, R. 1998 The initial conditions of star formation in the rho Ophiuchi main cloud: Wide-field millimeter continuum mapping. *A&A* **336**, 150–172

Motte, F., Bontemps, S., Schilke, P., Schneider, N., Menten, K. M. & Broguière, D. 2007 The earliest phases of high-mass star formation: A 3 square degree millimeter continuum mapping of Cygnus X. *A&A,* 476, 1243–1260

Nakamura, F. & Li, Z.-Y. 2007 Protostellar turbulence driven by collimated outflows. *ApJ.* **662**, 395–412

Nakano, T. 1989 Conditions for the formation of massive stars through nonspherical accretion. *ApJ.* **345**, 464–471

Nakano, T., Hasegawa, T., Morino, J. & Yamashita, T. 2000 Evolution of protostars accreting mass at very high rates: Is Orion IRc2 a huge protostar? *ApJ.* **534**, 976–983

Nakano, T., Hasegawa, T. & Norman, C. 1995 The mass of a star formed in a cloud core: Theory and its application to the Orion A cloud. *ApJ.* **450**, 183

Oey, M. S., King, N. L. & Parker, J. W. 2004 Massive field stars and the stellar clustering law. *AJ* **127**, 1632–1643

Offner, S. S. R., Krumholz, M. R., Klein, R. I. & McKee, C. F. 2008 The dynamics of molecular cloud cores in the presence of driven and decaying turbulence. *ApJ.* **136**, 404–420

Onishi, T., Mizuno, A., Kawamura, A., Tachihara, K. & Fukui, Y. 2002 A complete search for dense cloud cores in Taurus. *ApJ.* **575**, 950–973

Ossenkopf, V. & Henning, T. 1994 Dust opacities for protostellar cores. *A&A* **291**, 943–959

Padoan, P., Juvela, M., Goodman, A. A. & Nordlund, Å. 2001 The turbulent shock origin of proto-stellar cores. *ApJ.* **553**, 227–234

Padoan, P. & Nordlund, Å. 2002 The stellar initial mass function from turbulent fragmentation. *ApJ.* **576**, 870–879

Palla, F. & Stahler, S. W. 2001 Binary masses as a test for pre-main-sequence tracks. *ApJ.* **553**, 299–306

Parker, R. J. & Goodwin, S. P. 2007 Do O-stars form in isolation? *MNRAS* **380**, 1271–1275

Peretto, N., André, P. & Belloche, A. 2006 Probing the formation of intermediate- to high-mass stars in protoclusters. A detailed millimeter study of the NGC 2264 clumps. *A&A* **445**, 979–998

Pfalzner, S., Olczak, C. & Eckart, A. 2006 The fate of discs around massive stars in young clusters. *A&A* **454**, 811–814

Pinsonneault, M. H. & Stanek, K. Z. 2006 Binaries like to be twins: Implications for doubly degenerate binaries, the type Ia supernova rate, and other interacting binaries. *ApJ.* **639**, L67–L70

Pollack, J. B., Hollenbach, D., Beckwith, S., Simonelli, D. P., Roush, T. & Fong, W. 1994 Composition and radiative properties of grains in molecular clouds and accretion disks. *ApJ.* **421**, 615–639

Preibisch, T., Weigelt, G. & Zinnecker, H. 2001 Multiplicity of massive stars. In *IAU Symposium* 200 (eds H. Zinnecker & R. Mathieu), Astronomical Society of the Pacific, San Francisco, p. 69

Price, D. J. & Bate, M. R. 2007 The impact of magnetic fields on single and binary star formation. *MNRAS* **377**, 77–90

Rathborne, J. M., Lada, C. J., Muench, A. A., Alves, J. F. & Lombardi, M. 2008 The nature of the dense core population in the Pipe Nebula: A survey of NH3, CCS, and HC5N molecular line emission. *ApJS* **174**, 396–425

Rauw, G., Crowther, P. A., De Becker, M., Gosset, E., Nazé, Y., Sana, H., van der Hucht, K. A., Vreux, J.-M. & Williams, P. M. 2005 The spectrum of the very massive binary system WR 20a (WN6ha + WN6ha): Fundamental parameters and wind interactions. *A&A* **432**, 985–998

Reid, M. A. & Wilson, C. D. 2005 High-mass star formation. I. The mass distribution of submillimeter clumps in NGC 7538. *ApJ.* **625**, 891–905

Reid, M. A. & Wilson, C. D. 2006 High-mass star formation. II. The mass function of submillimeter clumps in M17. *ApJ.* **644**, 990–1005

Sandstrom, K. M., Peek, J. E. G., Bower, G. C., Bolatto, A. D. & Plambeck, R. L. 2007 A parallactic distance of 389^{+24}_{-21} parsecs to the Orion Nebula Cluster from very long baseline array observations. *ApJ.* **667**, 1161–1169

Scally, A. & Clarke, C. 2001 Destruction of protoplanetary discs in the Orion Nebula Cluster. *MNRAS* **325**, 449–456

Shatsky, N. & Tokovinin, A. 2002 The mass ratio distribution of B-type visual binaries in the Sco OB2 association. *A&A* **382**, 92–103

Shestakov, A. I. & Offner, S. S. R. 2008 A multigroup diffusion solver using pseudo transient continuation for a radiation-hydrodynamic with patch-based AMR. *JCP* **227**, 2154–2186

Sridharan, T. K., Beuther, H., Saito, M., Wyrowski, F. & Schilke, P. 2005 High-mass starless cores. *ApJ.* **634**, L57–L60

Stanke, T., Smith, M. D., Gredel, R. & Khanzadyan, T. 2006 An unbiased search for the signatures of protostars in the ρ Ophiuchi molecular cloud . II. Millimetre continuum observations. *A&A* **447**, 609–622

Stassun, K. G., Mathieu, R. D., Vaz, L. P. R., Stroud, N. & Vrba, F. J. 2004 Dynamical mass constraints on low-mass pre-main-sequence stellar evolutionary tracks: An eclipsing binary in orion with a $1.0 M_\odot$ primary and a $0.7 M_\odot$ Secondary. *ApJS* **151**, 357–385

Suttner, G., Yorke, H. W. & Lin, D. N. C. 1999 Dust coagulation in infalling protostellar envelopes. I. Compact grains. *ApJ.* **524**, 857–866

Tan, J. C., Krumholz, M. R. & McKee, C. F. 2006 Equilibrium star cluster formation. *ApJ.* **641**, L121–L124

Testi, L. & Sargent, A. I. 1998 Star formation in clusters: A survey of compact millimeter-wave sources in the serpens core. *ApJ.* **508**, L91–L94

Tout, C. A., Pols, O. R., Eggleton, P. P. & Han, Z. 1996 Zero-age main-seqence radii and luminosities as analytic functions of mass and metallicity. *MNRAS* **281**, 257–262

Turner, N. J., Quataert, E. & Yorke, H. W. 2007 Photon bubbles in the circumstellar envelopes of young massive stars. *ApJ.* **662**, 1052–1058

Walmsley, M. 1995 Dense cores in molecular clouds. In *Revista Mexicana de Astronomia y Astrofisica Series* de conferencias, Vol. **1**, *Circumstellar Disks, Outflows and Star Formation* (eds S. Lizano & J. M. Torrelles), p. 137

Walsh, A. J., Myers, P. C. & Burton, M. G. 2004 Star formation on the move? *ApJ.* **614**, 194–202

Weidner, C. & Kroupa, P. 2004 Evidence for a fundamental stellar upper mass limit from clustered star formation. *MNRAS* **348**, 187–191

Weidner, C. & Kroupa, P. 2006 The maximum stellar mass, star-cluster formation and composite stellar populations. *MNRAS* **365**, 1333–1347

Whitehouse, S. C. & Bate, M. R. 2004 Smoothed particle hydrodynamics with radiative transfer in the flux-limited diffusion approximation. *MNRAS* **353**, 1078–1094

Whitehouse, S. C., Bate, M. R. & Monaghan, J. J. 2005 A faster algorithm for smoothed particle hydrodynamics with radiative transfer in the flux-limited diffusion approximation. *MNRAS* **364**, 1367–1377

Wolfire, M. G. & Cassinelli, J. P. 1987 Conditions for the formation of massive stars. *ApJ.* **319**, 850–867

Wu, J., Evans, N. J., Gao, Y., Solomon, P. M., Shirley, Y. L. & Vanden Bout, P. A. 2005 Connecting dense gas tracers of star formation in our galaxy to high-z star formation. *ApJ.* **635**, L173–L176

Yorke, H. W. & Sonnhalter, C. 2002 On the formation of massive stars. *ApJ.* **569**, 846–862

Zinnecker, H. 1984 Star formation from hierarchical cloud fragmentation – A statistical theory of the log-normal initial mass function. *MNRAS* **210**, 43–56

Part IV

Protoplanetary Disks and Planet Formation

13

Observational properties of disks
and young stellar objects

G. Duchene, F. Menard, J. Muzerolle and S. Mohanty

13.1 Introduction

The collapse of a molecular cloud core leads to the creation of a newborn star and
attendant circumstellar accretion disk. The action of disk accretion during and after
collapse likely controls the initial mass and angular momentum of these young
stellar objects (YSOs). It also has a hand in shaping the conditions under which
planetary systems are born. Determining the observational properties of YSOs,
particularly their variations with environment, mass and time, can therefore place
stringent constraints on the physical processes at play during both stellar and planetary formation. In this chapter, we give an overview of the current knowledge of
YSO properties, with an emphasis on their implications for both processes.

Circumstellar disks are a ubiquitous outcome of the star formation process, making them a powerful probe of YSO evolution. In addition, disks are the birthsite of
planetary systems and can therefore be used to constrain the overall planet formation process (via statistical analyses) and some of the key physical mechanisms,
such as grain growth, vertical settling or radial migration (via detailed studies of
individual objects). We focus our discussion on the analysis of young, optically
thick, gas-rich protoplanetary disks. In particular, we focus on their dust component which, although it amounts to only a tiny fraction (on the order of 1%) of the
total mass, represents the building blocks of planetesimals and planets. We also
address some observations of more evolved debris disks.

We then explore observational diagnostics of the accretion process in YSOs.
The current paradigm of magnetically mediated accretion involves the disruption
and channelling of circumstellar disk gas by the stellar magnetic field. This process
allows accretion onto the star while simultaneously producing significant mass loss
from magnetocentrifugal or stellar winds. The early stellar angular momentum
evolution may be controlled by the star–disk magnetic interaction. UV radiation
from stellar accretion shocks and X-rays from coronal field lines irradiate the disk
and affect its chemistry and evolution. Constraining the physical properties of this

Structure Formation in Astrophysics. ed. G. Chabrier. Published by Cambridge University Press.
© Cambridge University Press 2009.

complex region is thus important for understanding star formation, early stellar evolution and the conditions for planet formation in disks.

Finally, we discuss our observations of disk accretion in young brown dwarfs (BDs). The formation mechanism of BDs presents a challenge to theorists. The central dilemma is easily framed: since the average thermal Jeans mass in molecular clouds is usually of order a solar mass, how does one form free-floating objects ten to a hundred times less massive? The two hypotheses most often appealed to, 'ejection' and turbulent fragmentation, make specific predictions that can be tested by observations of disks and accretion activity in young BDs. The search for disks, and a classical T Tauri-like disk accretion phase, around young BDs has consequently been a major area of research in the last few years. Surveys have now firmly established that disks are very common in young BDs, just as in stars. Here, we focus on the evidence for accretion activity.

13.2 Circumstellar disks

13.2.1 Observational approaches

Historically, circumstellar disks have first been studied through the thermal emission of their dust component via analyses of the spectral energy distribution (SED) of YSOs. While the disk structure and composition is in principle imprinted in its SED, the high optical depth of the disks is such that its interpretation remains ambiguous and model-dependent (Chiang *et al.* 2001). Nonetheless, the optically thin millimetre regime allows for direct estimate of the total dust mass in disks, while mid-infrared silicate emission feature spectroscopy is a probe of the hot surface layers, providing insight into grain size and composition.

In the last decade, and at a rapidly accelerating pace, spatially resolved observations of disks have opened a new era of disk analysis, where previously ambiguous or unconstrained quantities can be directly determined. The outer radii and surface layer dust content of many disks have been determined by scattered light images in the visible and near-infrared; the surface density profile of disks can be constrained via continuum millimetre interferometric mapping, while the disk inner radius and the star–disk interface region are now accessible, thanks to the advent of long-baseline optical interferometry. Multi-wavelength scattered light imaging, which takes advantage of wavelength-dependent dust opacity, represents a powerful approach to study non-uniform dust properties. In effect, comparing visible to near- and mid-infrared images allows to 'peel off' the surface layers and start probing disk layering.

It must be emphasized, however, that each of these techniques is sensitive to grains of different sizes (μm- versus mm-sized, typically) and to different regions of the disk (midplane/surface and inner/outer regions). Therefore, no single method provides a deep enough understanding of disks, and only a multi-technique approach allows to constrain the basic physical processes relevant to planet formation. As a consequence, most of our current knowledge of circumstellar disks stems from a combination of these various approaches.

13.2.2 Disks as tracers of star formation

Let us first consider protoplanetary disks as a 'collateral' outcome of star formation. Disk frequencies as high as 90% have been observed in the youngest populations, with little evidence of a dependence on environment or stellar mass up to intermediate-mass stars and down to BDs (Hillenbrand *et al.* 1998; Haisch *et al.* 2001; Luhman *et al.* 2007a). This suggests that most, if not all, low- and intermediate-mass objects form through a common mechanism, even across the stellar–substellar boundary. On the contrary, the formation history of higher-mass O- and B-stars remains much debated, in part because of their small number, larger typical distance to the Sun and the very short timescales associated to their evolution. While several independent lines of evidence support the frequent presence of circumstellar disks, a similar formation scenario as for lower-mass stars, around 10–$20 M_\odot$ stars, disks around higher-mass objects have remained elusive and the jury is still out concerning their formation mechanism (see Zinnecker & Yorke 2007 for a recent review).

At the next level of detail, the orientation of circumstellar disks can be an indication on the leading physical processes associated to star formation. For instance, several studies have found that disks within multiple systems are preferentially aligned with each other, albeit with many cases of clear non-alignment (Monin *et al.* 2007). Turbulent fragmentation or misaligned rotation and elongation axes of initial cores can account for these results. On a larger scale, while prestellar cores in Taurus tend to be preferentially prolate and oriented perpendicularly to the cloud magnetic field (Hartmann 2002), the orientation of young star–disk systems is independent of that of the surrounding molecular cloud, indicating that the magnetic field, while it may play a major role during the pre-collapse phase, has little influence on the final stages of star formation (Ménard & Duchêne 2004).

13.2.3 Disks and planet formation: initial conditions and overall evolution

Although disk masses estimated from millimetre fluxes are subject to caution induced by uncertain dust opacities and gas-to-dust ratio, it is now widely accepted that the mass of protoplanetary disks around YSOs are in the range of 0.1–10% of the central object's mass (see Natta *et al.* 2000 for a review). In many cases, the disk is at least as massive as the minimum mass solar nebula. Through both thermal emission mapping and scattered light images, the outer radius of circumstellar disk is usually found to exceed 100 AU (Dutrey *et al.* 1996; Kitamura *et al.* 2002), leaving ample room to form full-fledged planetary systems as our own Solar System. In summary, there is little doubt that a large fraction of circumstellar disks are both massive enough and large enough to provide adequate initial conditions to form planetary systems, in qualitative agreement with the relatively high frequency of extrasolar planetary systems (see contribution by M. Mayor & S. Udry, this volume).

Not only are circumstellar disks equally frequent around YSOs over a wide range of stellar masses, they also generally share a common geometry. SED analysis of BDs, TTSs and (the so-called group I) HAeBes disks are consistent with hydrostatic flared structures out to radii of several hundred astronomical units (AUs), with the gaseous and dust component well mixed. This is further reinforced by spatially resolved images of objects across the mass range (Burrows *et al.* 1996; Perrin *et al.* 2006; Luhman *et al.* 2007b). The main exception to this basic geometry is the group of the so-called group II HAeBes, whose SED is interpreted as that of a flat disk whose surface is not directly irradiated by the central star. This could be the result of self-shadowing by a thermally supported puffed-up inner rim at the dust sublimation radius (Dullemond *et al.* 2001; Natta *et al.* 2001) or of global grain growth that results in a much lower dust opacity (Dullemond & Dominik 2004). It remains unclear whether a simple evolutionary scheme can link the normal flared disks and the shadowed ones, however.

The average survival time for the inner region of protoplanetary disks (a few AU from the central star), as traced by near- and mid-infrared excesses, is on the order of a few Myr (Strom *et al.* 1989; Haisch *et al.* 2001). There is, however, a large dispersion in individual disk lifetimes, from less than 1 Myr to over 10 Myr (Hillenbrand 2005). This latter estimate represents the longest possible timescale for planet formation within a few astronomical units of the central star. Interestingly, the inner disk clearing timescale, between a full-fledged optically thick structure to almost undetected levels, is much shorter, on the order of 0.1 Myr only (McCabe *et al.* 2006). The physical processes that may be at play during this clearing are discussed in Section 13.2.5. Since the proportion of objects with such holes, dubbed 'transition disks', and large mass reservoirs at large radii is relatively small, it seems that the outer regions of the disk are then cleared on an equally short timescale, on the order of 0.1 Myr (Duvert *et al.* 2000), a potential challenge for disk evolution models. Definitive conclusions regarding that later phase still await sensitive millimetre surveys to complement the many transition objects uncovered with the Spitzer telescope.

13.2.4 Disks and planet formation: first stages

One key process in the early evolution of protoplanetary disks is the aggregation of small dust grains into increasingly larger bodies via sticking collisions. This process, which has been experimentally verified, is theoretically and numerically expected to be very efficient in circumstellar disks (Dominik *et al.* 2007 and references therein) and is a continuous focus of empirical studies. Shallow opacity laws in the radio regime suggest the presence of millimetre-sized particles, for instance (Beckwith & Sargent 1991; Acke *et al.* 2004). The observed continuum of silicate feature strength and shape reveals a smooth evolution from disks whose surface layers are dominated by 'small', interstellar-like grains (0.1 μm) to those where 'large' grains (1 μm and larger) have taken over (van Boekel *et al.* 2003; Kessler-Silacci *et al.* 2006). Finally, direct analyses of scattered light images have conclusively shown that grains of a few microns in size are present in disks but that the dust grain size distribution in the surface layer does not extend monotonously to millimetre-sized particles (see Watson *et al.* 2007 for a recent review). Recent Spitzer data suggest that such grains, while absent from the interstellar medium, are also present in molecular clouds (Indebetouw *et al.* 2005), and it is difficult to disentangle dust processing occurring within disks from pre-collapse evolution. The existence of millimetre-sized bodies in disks, however, is most likely the results of local growth.

While these independent techniques have built a convincing case for the ongoing evolution of dust grains in disks, one must remember that they sample spatially different dust populations. The apparent contradiction between millimetre spectral indices and analysis of scattered light images could therefore be interpreted as evidence for a stratified disk structure, with large grains numerous in the midplane but absent in the surface layers (Duchêne *et al.* 2003). Similarly, disks whose SED is characterized by a shallow spectral index and strong 10- and 20-μm silicate features can only be explained if dust grains in the surface and midplane layers have very different grain size distributions (Figure 13.1). This differential settling phenomenon is expected to take place in disks (Barrière-Fouchet *et al.* 2005; Dullemond & Dominik 2005; Ciesla 2007) since submicron grains are small enough to be well-mixed with the gaseous component, whereas larger grains rapidly sink towards the midplane. In one particular system, Duchêne *et al.* (2004) found that scattering at 3.7 μm occurs in a layer that is twice as close to the disk midplane than at visible wavelengths. Since larger grains are required to account for the ring-scattered light morphology at the longest wavelength, this supports the picture of vertically stratified, possibly settled, disk, as predicted in hydrodynamical simulations (Pinte *et al.* 2007).

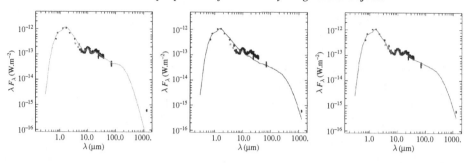

Fig. 13.1. Observed and model SED for the IM Lup circumstellar disk. (Adapted from Pinte *et al.*, 2008) (*Left*) the model includes only small grains and underpredicts millimetre fluxes. (*Centre*) the model, that includes fully mixed large grains, fails to reproduce the strong mid-infrared silicate features. (*Right*) the model, which also includes large grains along with differential settling, simultaneously reproduces these features and is a very good fit to the observations.

Another physical phenomenon predicted by numerical simulations is the radial migration of grains, which is dependent on the grain size with a maximum efficiency around grain radii of 1 mm–1 cm (Barrière-Fouchet *et al.* 2005; Garaud 2007). Current millimetre interferometers (required to probe such grain sizes) do not provide sufficient spatial resolution to test this evolution, but this will become one of the key results for ALMA surveys of known disks. In the meantime, long baseline interferometry has offered the first direct evidence of the radial dependence of dust grain properties. By comparing the spectra arising from the inner AU and the entire mid-infrared-emitting region (on the order of 10 AU in radius), van Boekel *et al.* (2004) showed that the crystallinity rate is highest in the innermost regions but that the observed rate in the 'outer' regions requires some form of outward radial migration of the smallest grains. A similar behaviour has been recently observed for a TTS disk (Schegerer *et al.* 2008), suggesting that the same phenomena are at play across a wide range of stellar masses.

13.2.5 Disks and planet formation: final stages

If planets indeed form in just a few million years, there might be some YSOs whose circumstellar disk already contains planetary-mass bodies. Unfortunately, the largest bodies we can detect through their thermal emission have radii on the order of 1 cm (Testi *et al.* 2003) and only indirect evidence can therefore be proposed for the presence of large bodies within circumstellar disks. Embedded planets are expected to clear gaps in disk (Lin & Papaloizou 1993), but it is virtually impossible to detect the signature of a planet-induced gap in the object's SED (Wolf & D'Angelo 2005; Varnière *et al.* 2006a). The high spatial resolution required to unambiguously detect such a gap is not achievable with current instrumentation although this will be possible in a few years with ALMA and/or long baseline interferometry imaging instruments.

Scattered light images and thermal emission maps of some protoplanetary disks have revealed large-scale structures (Grady *et al.* 2001; Fukagawa *et al.* 2004; Piétu *et al.* 2005), such as spiral arms, which could be interpreted as induced by a planetary-mass object within the disk (Bate *et al.* 2003b; Wolf & D'Angelo 2005). The large spatial scale associated with these structures, tens of astronomical units, rather favours an overall instability or a companion-induced perturbation, however (Clampin *et al.* 2003). The so-called 'transition disks', with their cleared inner regions (see Najita et al. 2007 and references therein), are also frequently interpreted as having formed a planet at the outer edge of the inner hole (Rice *et al.* 2006; Varnière *et al.* 2006b; Crida & Morbidelli 2007). However, such inner holes can also be opened via photoevaporation (Alexander et al. 2006; Maddison *et al.* 2007), so that it is not yet possible to confirm the presence of protoplanets around YSOs.

While it remains currently impossible to test the possibility that gas-rich disks around YSOs have already formed planets, disk sculpting has been observed in several gas-depleted debris disks surrounding stars in the 10–1000 Myr age range. The fact that some debris disks show narrow ring geometries could be naturally explained by the shepherding effect of planets, although alternative explanations have been proposed (see Besla & Wu 2007 and references therein). Arguably, the most convincing evidence of planet sculpting is the off-centre ring surrounding Fomalhaut (Kalas *et al.* 2005), which is most easily explained by the presence of an eccentric planet, a scenario that could also account for asymmetries observed in other debris disks (Kuchner & Holman 2003).

The scattering properties of optically thin debris disks, which are dominated by the small grains resulting from collisions between undetectable parent bodies, can also inform on the properties of early counterparts to our current Solar System. While measured albedos span a wide range of values, revealing no obvious trend, the observed scattering phase function seems to generally favour small grains that scatter isotropically. Strongly forward-throwing dust populations have been found in some cases, maybe suggesting different evolutionary paths (Graham *et al.* 2007; Kalas *et al.* 2007). Complex grain structure is also suggested in cases where rich multi-techniques data sets are available. For instance, the SED, scattered light colours and phase function of the HD 181327 debris disk cannot be accounted for with a simple power-law size distribution (Schneider et al. 2006). While radius-dependent dust properties and/or wavy size distributions (Thébault & Augereau 2007) may account for the problems encountered in that study, an intriguing, though untested, alternative hypothesis is the possibility that dust grains in debris disks are not compact and spherical, but rather elongated aggregates made of small individual elements. Such aggregates could have scattering and extinction properties that simultaneously mimic small and large grains, depending on the observational approach being used. Interestingly, such a complex grain structure would be consistent with the finding that dust grains in the AU Mic debris disks must be highly porous to account for scattered light intensity and linear polarization levels (Graham *et al.* 2007).

13.3 Accretion-related phenomena: probing the central engine

13.3.1 Accretion and winds

13.3.1.1 T-Tauri stars

The magnetospheric accretion paradigm in classical T Tauri (CTT) stars has been established through many different observational diagnostics. Permitted emission lines such as Hα are ubiquitous and likely trace gas in accretion flows as well as winds. Hα emission is seen in both CTTs and their non-accreting (and mostly disk-less) counterparts, weak T-Tauri (WTT) stars. However, it is typically stronger when accretion is present; an Hα equivalent width (EW) threshold EW[Hα] >10 Å traditionally has been adopted as the baseline signature of ongoing accretion in CTTs, while smaller values signify only chromospheric emission characteristic of WTTs. The difference is more starkly revealed in velocity-resolved line profiles, which are very broad (FWHM $\gtrsim 200\,\mathrm{km\,s^{-1}}$) and asymmetric in CTTs but narrow (FWHM $\lesssim 100\,\mathrm{km\,s^{-1}}$) and symmetric in WTTs. The occasional redshifted absorption seen in some CTTs, along with the large line widths, provided the first hints of ballistic infall of gas in these systems. However, the presence of blueshifted absorption also suggested a wind component. Subsequent radiative transfer modelling showed that magnetospheric infall was able to explain much of the observed line emission (Hartmann *et al.* 1994; Muzerolle *et al.* 1998a), although winds also contribute in very active objects (Alencar *et al.* 2005; Kurosawa *et al.* 2006).

The other ubiquitous tracer of accretion activity in CTTs is continuum excess in the UV and optical regimes. The former is typically observed as Balmer continuum emission with a conspicuous Balmer jump, while the latter shows up as 'veiling' of photospheric absorption lines. The emission spectrum at shorter wavelengths is generally consistent with blackbody emission from ~ 8000 K hot spots on the stellar surface (although the excess does not continue to drop off at $\lambda \gtrsim 6000$ Å, possibly indicating additional components at lower temperatures; White and Hillenbrand 2004). This excess emission also varies with time, sometimes with periodic behaviour commensurate with the stellar rotation rate, although often stochastically varying as well. These characteristics indicate emission from an accretion shock where gas accreting from the disk falls onto the stellar surface at free-fall velocities. Detailed shock models have been successful in reproducing the observed spectral shape and depth of the Balmer jump (Calvet & Gullbring 1998), leading to estimates of small surface filling factors (typically of order 1%) and mass accretion rates.

These accretion diagnostics can all be used to make quantitative estimates of the accretion rate of material transferred from the disk to the star ($\dot{M}_*$). It is generally assumed that $\dot{M}_*$ is the same as the actual accretion rate in the disk, though this need not be the case. Neglecting possible radial dependencies, $\dot{M}_*$ can be used to indirectly trace the bulk gas content in disks. For example, in the simple picture of a steady-state viscously evolving disk, the gas surface density is proportional to the accretion rate. Comparisons of $\dot{M}_*$ with stellar mass, age and environment then give us clues on the global evolution of protoplanetary accretion disks. The most commonly used diagnostic is the UV excess, as it is the most straightforward to employ; the excess luminosity is directly proportional to the accretion luminosity, which can then be related to the accretion rate via $L_{\mathrm{acc}} \sim GM_*\dot{M}/R_*$. The main uncertainties in such a calculation include the bolometric correction for unobserved flux at shorter wavelengths, the mass and radius of the star (determined indirectly from theoretical evolutionary tracks) and the extinction to the source. This method is not applicable in all cases, particularly for faint or highly extincted sources. Other, more indirect indicators have thus been adopted, including line profile modelling and infrared emission line luminosities.

Using these techniques, mass accretion rates have now been measured for relatively large samples of CTTs. The values span a large range, from 10^{-10} to $10^{-6} M_\odot$ year^{-1} for 1-Myr-old solar-mass objects. Figure 13.2 shows the distribution of $\dot{M}$ for all Taurus CTTs with masses 0.2–1M$_\odot$ reported in the literature to date. We note that this is a fairly complete sample, including roughly 90% of all known members with masses in this range that exhibit infrared excess. Given that the peak of this distribution is nearly two orders of magnitude larger than the minimum measurable $\dot{M}$ for this mass range, there are likely few (if any) missing weak accretors. The spread of values is mostly intrinsic, since typical accretion variability is only $\sim$30% for stars in this mass range. Various investigations of this type have been carried out, including measurements for objects spanning a wide range of ages and masses, in order to constrain disk accretion and evolution. We will examine the mass relationship later in Section 13.3.4. Concerning age, several studies have shown that on average $\dot{M}_*$ decreases with time roughly as expected for simple viscous evolution (Hartmann *et al.* 1998; Calvet *et al.* 2005). However, the large intrinsic spread at any given age is still not fully understood. It may indicate a range of initial disk masses or sizes or possibly different accretion

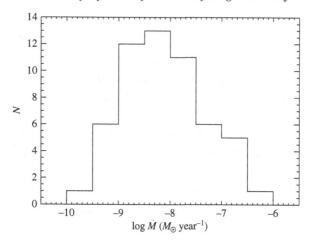

Fig. 13.2. The distribution of mass accretion rates for Taurus CTTs with masses $0.2–1M_\odot$. The data have been culled from Hartigan *et al.* (1995), Gullbring *et al.* (1998), Hartmann *et al.* (1998), White and Ghez (2001), and White and Hillenbrand (2005). We have attempted to account for differences in assumed geometrical factors, bolometric corrections and adopted mass tracks by scaling everything to Gullbring *et al.* (1998).

physics (e.g., gravitational instabilities versus the magneto-rotational instabilities and layered accretion).

A natural by-product of the accretion process is mass loss in winds and/or jets. Constraining the launching mechanism(s) and source region(s) is crucial to understanding the angular momentum evolution of young stars during and after the disk accretion phase. Three general models have come into vogue in the past 10–15 years: X-winds, disk winds and stellar winds. All of these models operate on similar principles, namely magnetically channelled mass loss powered by accretion energy. Their main difference lies in the launch point: X-winds from a specific location at the magnetic truncation point of the disk; disk winds from a larger range of radii in the inner disk and stellar winds from open field lines emanating from the stellar polar regions. The challenge has been to obtain observational diagnostics that can clearly distinguish between them. Forbidden line emissions such as [OI] $\lambda6300$ have been the mainstays for delineating the large-scale structure of the wind and jet, as well as measuring mass loss rates (Hartigan *et al.* 1995). High spatial resolution observations enabled by large telescopes with adaptive optics and HST now enable unprecedented measurements at $\sim10–50\,\text{AU}$ scales. Many jets have been laterally resolved, leading to estimates of collimation scales and possible evidence for rotation (Coffey *et al.* 2004). However, to really probe down to the launching region, observations on scales of 1 AU or less are needed. The technique of spectroastrometry has been used in some instances to follow line emission close to that scale. Takami *et al.* (2003) discovered extended Hα emission down to 5–10 AU suggestive of microjets in several T-Tauri stars. They also resolved counterjets in two objects, indirectly revealing the presence of AU-scale inner disk 'holes' (one of these objects, CS Cha, was recently confirmed by Spitzer, lacking dust emission at $\lambda \lesssim 10\,\mu$m; Espaillat *et al.* 2007).

Recent analyses of complicated He I line profiles show that all three processes may be involved to some degree, with stellar and disk winds possibly domi-nant in most actively accreting objects (Dupree *et al.* 2005; Edwards *et al.* 2006; Kwan *et al.* 2007). Preliminary modelling by Kwan *et al.* (2007) indicates that many of the He I 10830 profiles are consistent with a stellar polar wind origin. Yet, time-series observations of He I from AA Tau (Bouvier *et al.* 2007) show anti-correlations with accretion signatures that favour a disk wind origin. More observations and modelling of this very useful diagnostic are needed to better constrain which scenario may dominate.

13.3.1.2 Herbig Ae/Be stars and protostars

There have been many observations of similar phenomena in Herbig Ae/Be stars that have been linked to magnetically mediated accretion analogous to that of T-Tauri stars. However, the overall evidence is rather mixed, not the least of which is the still scant evidence of magnetic fields of the required strength (see Section 13.3.2). By definition, these objects exhibit strong permitted line emission, particularly Hα; however, line profiles typically show P Cygni shapes indicative of winds rather than infalling accretion flows (Catala *et al.* 1999). A small subset of objects (of which UX Orionis is the prototype) do exhibit redshifted absorption components indicative of infall (Muzerolle *et al.* 2004). Vink *et al.* (2005) showed that most Herbig Ae stars exhibit Hα line polarization patterns similar to those of T-Tauri stars and suggestive of scattering of emission from a compact region of an exterior rotating circumstellar disk. (Herbig Be stars do not show the same effect, instead exhibiting line depolarization consistent with an undisrupted disk; Mottram *et al.* 2007.) Moreover, UV excess most likely produced by accretion shocks on the stellar surface has been detected in some Herbig stars (Grady *et al.* 2004). Lopez *et al.* (2006) used measurements of Paβ and Brγ emission lines to estimate mass accretion rates for ∼24 Herbig Ae/Be stars, deriving values in the range of 10^{-8}–$10^{-6} M_\odot$ year^{-1}, slightly higher on average than typical CTT rates.

Interestingly, recent VLT interferometry has spatially resolved Brackett γ line emission in the Herbig Ae star HD 104237 (one of the jet sources), indicating an origin in a disk wind launched near the inner dust rim of the disk (Tatulli *et al.* 2007). Jets and outflows detected in Lyα, Hα and [SII] emission (Devine *et al.* 2000; Grady *et al.* 2004) have been observed emanating from roughly a half dozen intermediate-mass objects (Devine *et al.* 2000; Grady *et al.* 2004; McGroarty *et al.* 2004). The jet opening angles, shock properties and lengths are comparable to the better-studied counterparts in low-mass YSOs. Based on STIS corona-graphic observations, Wassell *et al.* (2006) recently estimated a mass loss rate of $\sim 10^{-8} M_\odot$ year^{-1} for the Herbig Ae star HD 163296, on the high end of the range observed for T-Tauri stars. The jet properties all point to a common launching and collimation mechanism over a large range in stellar mass, namely magnetic inter-action near the star–disk interface. Further work is required to assess the Herbig population as a whole, which in general consists of rather heterogeneous circum-stellar properties. More detailed radiative transfer models including the effects of both winds and accretion flows need to be calculated to better understand the origin of the line emission.

What about the youngest objects? Large telescopes now permit similar obser-vations of optical diagnostics in many heavily reddened Class I protostars. White and Hillenbrand (2004) detected photospheric features in 11 Taurus Class I objects and hence were able to determine their stellar properties and measure continuum veiling. With this information, stellar accretion rate estimates were derived for protostars for the first time using a method commonly used for Class II T-Tauri stars. They were also able to estimate mass loss rates from forbidden emission. The results show that accretion and mass loss processes are very similar between the two presumed evolutionary stages, so much so that it is not clear that the two classes are truly separate. However, measurements of the most deeply embedded objects have not yet been determined, so characterization of disk accretion activity at the earliest stages of star formation is still lacking. Infrared spectrographs allow further investigation of protostellar properties and disk accretion activity (although this is somewhat limited by the often increased dust continuum in these objects). For example, Doppmann *et al.* (2005) did a study of stellar and accretion properties of Class I objects using high-resolution infrared spectroscopy. They found that a majority of objects observed exhibit Brγ emission, many with profile shapes con-sistent with magnetospheric accretion model predictions. More observations and modelling are needed to test potential accretion diagnostics and assess the relative contribution of winds, which are expected to be more substantial than in Class II T-Tauri stars since the mass loss rates are generally larger.

13.3.2 X-rays and magnetic fields

Measurements of magnetic fields in young stars provide a critical test of the magnetospheric accretion paradigm. Theoretical models, such as that of Königl (1991) and Shu *et al.* (1994), require sufficient stellar field strengths to disrupt the inner disk and channel accretion and wind flows, providing the requisite angular momentum sink for the system. Only in recent years, robust detections of magnetic activity in young stars, particularly T-Tauri stars, have been made. Most of these detections involve the measurement of Zeeman splitting in field-sensitive absorption lines in the near infrared (see Johns-Krull 2007 and references therein). Mean field strengths of 1–3 kG have now been measured in over a dozen T-Tauri stars. Such field strengths are in general consistent with magnetospheric accretion theory. However, the field geometry at the stellar surface appears to be more complex than a pure dipole. Indeed, there is a distinct lack of correlation, for particular objects given their known mass accretion rates, masses and radii, between the observed field strengths and those predicted by dipole magnetospheric accretion models (Johns-Krull 2007). Nevertheless, spectropolarimetric observations of the He I emission line, which likely traces the stellar accretion shock, have revealed a strong magnetic field in that region, suggesting that the magnetic field *is* ordered within the accretion flow itself, as expected in magnetospheric theories (Johns-Krull *et al.* 1999; Yang *et al.* 2007). To reconcile these observations, Mohanty and Shu (2008) have recently generalized the popular model of X-wind accretion to include *multi-pole* stellar magnetic fields. The predictions of this generalized model appear to be in much better agreement with the data (Johns-Krull & Gafford 2002; Mohanty & Shu 2008). Better constrained accretion rates and hot-spot field strengths will allow more stringent tests of the theory and show whether the star–disk magnetic interaction can indeed generally enforce co-rotation of the star with the inner edge of the disk, as the theory predicts. Another open question is the origin of the magnetic activity – the conventional dynamo cannot operate in most T-Tauri stars since they are fully convective. More recent theories that treat other types of dynamos are still inconsistent with observations (Johns-Krull 2007). Thus, it may be that the fields are primordial in origin (though more work on both the field diffusion timescales and the dynamo generation mechanisms needs to be done before firm conclusions can be drawn).

All of the above results concern young stars with a relatively narrow range of masses, $\sim 0.5 - 1 M_\odot$. As discussed above, young objects in a much larger mass range exhibit evidence of similar accretion activity. Measurements of magnetic fields and activity around high- and low-mass stars and BDs are needed to better constrain where magnetospheric accretion may operate and where it may break down. Recent investigations of line polarization in some Herbig Ae/Be stars have provided the first indications of magnetic fields in more massive PMS stars, with magnetic field strengths on the order of $\sim 0.1 - 1$ kG (Hubrig *et al.* 2006; Wade *et al.* 2007). Only $\sim 10\%$ of the Herbig stars observed exhibit measurable fields, similar to the fraction of older Ap stars (the field strengths are also similar, suggesting a fossil field origin for those objects). However, the upper limits for the roughly 90% lacking detections are too small to meet the requirements of most magnetospheric accretion theories (Wade *et al.* 2007), casting some doubt on the general applicability of the model to this mass range. More sensitive measurements, particularly using diagnostics of the hot-spot field strength such as the accretion shock emission lines seen in T-Tauri stars, will hopefully clarify the situation. Direct magnetic field measurements at the other end of the mass spectrum are very difficult owing to the faintness of such objects; however, such measurements are now starting to be made (Reiners & Basri, 2008, submitted) and should provide further constraints on accretion models.

Another manifestation of magnetic activity in young stars, and one that is ostensibly easier to observe and characterize, is X-ray emission. X-ray activity is universal in T-Tauri stars. The typical quiescent luminosities and temperatures are similar to main sequence stars, with a similar correlation between X-ray and stellar luminosity ($L_X/L_{bol} \lesssim 10^{-3}$). Recent long-term monitoring of T-Tauri stars such as the Chandra COUP survey of Orion have revealed frequent flaring activity with duty cycles of days to weeks and durations of hours to ~ 1 day and energies hundreds of times greater than solar events (Wolk *et al.* 2005). These bulk characteristics are consistent with X-ray emission from magnetic coronae.

There do appear to be some differences between CTTs and WTTs. CTTs appear to have systematically lower L_X (Preibisch *et al.* 2005; Telleschi *et al.* 2007). This may indicate that accretion somehow modifies coronal structure, or else CTTs X-rays are generated in a different region such as the accretion flow itself. X-ray spectra of some CTTs have revealed high gas densities consistent with an origin in accreting gas (Kastner *et al.* 2002). However, other evidence is not so clear. Analysis of time variability and flare events from the COUP data shows a lack of correlation between X-ray and optical variations (Wolk *et al.* 2005). Since the optical variability in CTTs traces emission from the accretion shock, the bulk of the X-ray emission must come from coronal gas separate from the accreting field lines. Simulations of magnetic field topology including multiple components, with the coronal fields (in a complex multi-polar geometry) inside the field connecting to the disk, show that the accreting gas can absorb X-rays from the coronal field (Gregory *et al.* 2007). A recent monitoring campaign of the CTTs AA Tau found no evidence of decreased X-ray emission during optical eclipse events, indicating that the X-ray emitting region is *outside* the accretion field, possibly at high latitudes; one X-ray flare occurred near the middle of optical eclipse (Grosso *et al.* 2007).

X-ray activity in objects with accretion disks are of particular interest since X-rays have been implicated as a main contributor to disk evolution. For example, X-rays can ionize disk gas, helping to drive accretion via the magneto-rotational instability and mass loss via photoevaporative winds. Glassgold *et al.* (1997) first proposed that X-ray ionization of the disk surface layers may be sufficient to allow coupling to the disk magnetic field. More recently, Chiang and Murray-Clay (2007) have suggested a similar process for 'transitional' disks, whose inner regions have been rendered optically thin by grain growth or giant planet formation. These theories predict some correlation between L_X and the disk mass accretion rate ($\dot{M}$). However, none has been seen to date. Telleschi *et al.* (2007) showed considerable scatter between L_X/L_{bol} and $\dot{M}$ for the well-studied CTTs in Taurus. Unfortunately, the lack of simultaneity between the two sets of measurements may mask any intrinsic correlation.

X-ray emission in higher-mass PMS stars and young embedded Class I objects provides a key tracer of magnetic activity in these systems, where direct magnetic field measurements are difficult or impossible. Detections of protostellar X-rays have now been made in a number of nearby star-forming regions. The number counts, L_X distribution and plasma temperatures for Class I objects are roughly similar to that of the CTTs, suggesting a similar origin from coronal or accretion fields (Güdel *et al.* 2007; Winston *et al.* 2007). Interestingly, a correlation between L_X and optical and near-infrared flux was observed during the outburst of the Class I source V1647 Ori (Kastner *et al.* 2006), contrary to CTTs and suggesting that perhaps different physics operate during such outbursting events than in quiescent accretion activity. Unfortunately, the X-ray properties of the most highly embedded (and presumably youngest) Class 0 objects remain elusive, as few, if any, convincing detections have been made to date. Meanwhile, X-ray emission observed from Herbig Ae stars has generally been attributed to unresolved low-mass companions. However, a few detections cannot be so easily dismissed (e.g. AB Aur; Stelzer *et al.* 2006), perhaps indicating magnetic activity is present in at least some Herbig stars (consistent with the few magnetic field measurements).

13.4 Accretion in brown dwarfs: clues to substellar formation

13.4.1 Introduction

BDs are by definition substellar objects: with a mass $\lesssim 80 M_{Jup}$ ($0.08 M_\odot$), they are incapable of sustaining stable hydrogen fusion, and thus, after an initial period of Deuterium fusion, they simply continue to cool down and grow fainter with time. Over the last decade, astronomers have discovered hundreds of these diminutive bodies, both in the field as well as in star-forming regions and young clusters, with masses ranging down to nearly (or perhaps even below) the planetary mass boundary (~ 12 M_{Jup}). While some BDs are companions to stars, the vast majority are isolated bodies.

Two main competing hypotheses have emerged to explain how such low-mass objects form. The 'ejection' scenario posits that BDs are in fact 'stellar embryos', born within molecular cloud cores but subsequently flung out from the cores by dynamical interactions with their neighbours before accumulating enough gas to become full-fledged stars (Reipurth & Clarke 2001; Bate *et al.* 2003a). A key prediction here is that disks around young BDs should be severely truncated by these interactions and hence far smaller than those girdling newborn stars. As a corollary, the disk accretion phase that is ubiquitous in CTTs may be short-lived and rare in BDs.

In the alternative 'turbulent fragmentation' picture (Padoan & Nordlund 2004), turbulent shocks in molecular clouds create gravitationally bound cores with a range of masses: from large ones that form stars to ultra-low mass ones – far below the cloud's mean thermal Jeans mass but *locally* unstable – that collapse directly into BDs. In this view, there is no fundamental distinction between the formation mechanism of stars and BDs: both form quietly within their own cores. Extended disks, and an associated CTT phase, should then be as common in BDs as in stars.

These two formation mechanisms have key implications for low-mass star formation in general as well. In the first case, gravitational interactions also set the final stellar masses, by limiting the amount of material that can accrete onto nascent protostars: through competitive accretion, repeated truncation of the primordial disks and outright ejection from the core. In the second case, the mass spectrum of star-forming cores is ultimately regulated by cloud turbulence. Thus, discriminating between these substellar formation scenarios is also vital for understanding the physics governing the *stellar* initial mass function (IMF).

It is worth pointing out at this juncture that simply the *presence* of disks and accretion, while providing constraints on BD formation, is *not* sufficient to unambiguously distinguish between the two formation scenarios discussed above. We return to this subject in Sections 13.5 and 13.6, where we review the constraints implied by current data and if and how future observations combined with current ones may finally settle the issue of BD formation.

13.4.2 Accretion and outflow signatures in brown dwarfs

The more straightforward measure of accretion in CTTs, namely the UV/optical continuum excess or veiling, is difficult to measure in BDs since they are faint and red and, as we shall see in Sections 13.3 and 13.4.1, have very small accretion rates. The most commonly used accretion indicator in BDs has been the $H\alpha$ emission line profile. In particular, White and Basri (2003) have noted that empirically, in high-resolution spectra, the full width of the $H\alpha$ line at 10% of the emission peak ($\equiv$ FW10) is always $\gtrsim 270\,\mathrm{km\,s^{-1}}$ in accretors regardless of spectral type, from solar-type stars down to the substellar limit (SpT $\sim$ M6 – M7). In subsequent work,

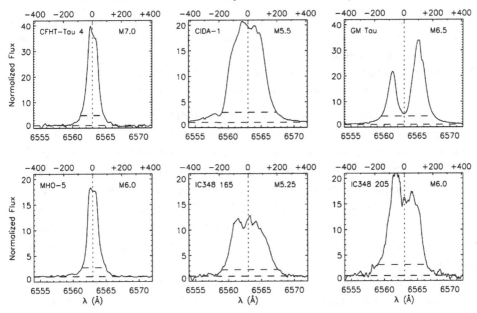

Fig. 13.3. *Hα* profiles in a sample of accreting BDs, exhibiting the broad line wings (except in a few pole-on cases), and line asymmetries due to geometrical and wind effects, associated with CTT-like magnetospheric accretion. (From Mohanty *et al.* 2005).

Jayawardhana *et al.* (2003) lowered this cutoff to FW10 $\gtrsim$200 km s^{-1} for the BD regime (SpT > M6), accounting for the lower infall velocities of these objects.

Various teams have now obtained high-resolution optical spectra of more than a hundred VLMS and BDs (SpT $\sim$ M5–M9.5: mass $\sim$0.1–0.015$M_\odot$) in nearby star-forming regions ($d \lesssim$ 400 pc, age: $\lesssim$1–10 Myr), covering a major fraction of the known BDs in these regions (Muzerolle *et al.* 2003, 2005; Mohanty *et al.* 2003, 2005; Barrado y Navascués *et al.* 2004). Using the FW10[Hα] diagnostic above, scores of these sources have been clearly identified as accretors. Some of the salient qualitative features observed are (Figures 13.3 and 13.4) as follows:

1. accretion is observed over the full range of BD masses, down to the lowest mass objects studied ($\sim$15M$_{Jup}$), approaching the planetary mass boundary;
2. the broad line-wings of the Hα accretion line-profiles indicate free-fall infall velocities, implying the same paradigm applicable to CTTs: magnetospherically channelled accretion flows from an inner disk edge at a few stellar radii;
3. the Hα line profiles in VLMS and BD accretors often evince asymmetries and time variability analogous to those seen in higher-mass CTTs (see Section 13.4.3);
4. while some emission lines, such as HeI (5876 Å), two of the CaII infrared triplet lines (8498, 8542 Å) and of course Hα itself, appear in both accreting and simply chromospherically active VLMS and BDs, others – HeI (6672 Å), [OI] (8446 Å) and one of the

CaII infrared triplet lines (8662 Å) – appear to be present only in accreting VLMS and BDs (though not in *all* the latter) and are thus good secondary indicators of accretion when they are present, in addition to the Hα FW10 diagnostic described above;

5. forbidden emission lines such as [OI] (6300 Å), indicative of mass outflow in jets and winds, are observed in a few accreting BDs (see further below).

Accretion signatures are observed in VLMS and BDs in near-infrared spectra as well. In particular, accreting VLMS and BDs often exhibit Paβ and Brγ in emission, while merely chromospherically active VLMS and BDs do not (Figure 13.4; Natta *et al.* 2004, 2006; Gatti *et al.* 2006). It is true that these lines are not as sensitive to accretion as Hα: their smaller optical depth results in their absence in some low-mass accretors that still exhibit broad Hα. Nevertheless, these near-infrared lines are especially important accretion diagnostics in regions of high extinction (e.g. ρ Oph), where high-resolution optical spectroscopy is prohibitively time-consuming.

Typical CTT jet/wind indicators, such as emission in forbidden lines of [OI], [NII] and [SII] and blueshifted absorption in Hα, have been identified in only a handful of VLMS and BDs (Figures 13.3 and 13.4; Barrado y Navascués *et al.* 2004; Mohanty *et al.* 2005; Muzerolle *et al.* 2005; Whelan *et al.* 2005, 2007). This is not too surprising: if the outflow rates are comparable to the accretion rates (as theory suggests), then the very small accretion rates in the VLMS/BD regime would generally result in outflow signatures near/below current sensitivity limits. Indeed, in the one BD in which both accretion and outflow rates have been estimated from high-resolution spectra (LS-RCrA-1: Barrado y Navascués *et al.* 2004; Mohanty *et al.* 2005), the outflow estimates range from an order of magnitude lower than to comparable to the accretion rate, roughly consistent with theory. The jets in two BDs (ρ Oph 102 and 2MASS 1207-3932) have now also been spectro-astrometrically resolved (Whelan *et al.* 2005, 2007); the data suggest scaled-down versions of CTT jets, in agreement with the idea that the outflows scale with accretion. However, no BD jet has yet been spatially resolved in imaging.

13.4.3 Measuring the accretion rates

Because the UV/optical excess is typically too weak to measure, most estimates of mass accretion rates in BDs have been made using the more indirect method of modelling the Hα emission line profiles (Muzerolle *et al.* 2000, 2003, 2005). In addition, two secondary methods have come into usage. The first relies on the empirically observed correlation between $\dot{M}$ and the emission flux in CaII infrared triplet lines. The correlation was first noted by Muzerolle *et al.* (1998b) for CTTs, using $\dot{M}$ measured from the UV/optical excess. Mohanty *et al.* (2005) then showed that it extended into the BD regime as well (with $\dot{M}$ now measured from modelling the Hα line), for the 8662 Å line of the triplet. The other secondary technique relies on the analogous correlation observed between $\dot{M}$ and the flux in the Paβ and Brγ lines (Natta *et al.* 2004, 2006; Gatti *et al.* 2006). Using these methods, $\dot{M}$ have now been derived for scores of VLMS and BDs. The accretion rates are in the range 10^{-10}–10^{-12} $M_{\odot}$year^{-1}: 2–4 orders of magnitude lower than the average in CTTs.

13.4.4 Dependence of $\dot{M}_{\mathrm{acc}}$ on central mass

The accretion rates derived so far point to a clear correlation between $\dot{M}$ and the mass of the central object (M_*), all the way from solar-mass stars down to the lowest mass BDs (Figure 13.4, for Taurus members only). In particular, it appears that roughly, $\dot{M} \propto M_*^2$ (albeit with considerable scatter), over ~4 orders of magnitude in $\dot{M}$ and 2 orders in M_* (Muzerolle *et al.* 2003, 2005; Natta *et al.* 2004, 2006; Mohanty *et al.* 2005).

The reasons for this relationship are unclear, though a few explanations have been offered. Padoan *et al.* (2005) argue that it results from Bondi–Hoyle accretion from the surrounding cloud onto the disk. In the latter picture, $\dot{M}_{\mathrm{BH}} \propto M[\rho_{\infty}/(c_{\infty}^2 + v_{\infty}^2)]$, where $\dot{M}_{\mathrm{BH}}$ is the accretion rate onto the star+disk system from the surrounding medium, M is the total mass of star+disk, ρ_{∞} and c_{∞} are, respectively, the gas density and sound speed in the surrounding medium, and v_{∞} is the velocity of the medium relative to the star+disk system. This directly gives the correlation observed between infall rate and central mass and explains the observed scatter in terms of the variability of stellar velocity and conditions in the cloud. The first drawback, however, is that Bondi–Hoyle accretion refers to infall from the cloud onto the star+disk, while the observed accretion rates are through the disk onto the star. It is not clear why the two should be related, except perhaps in a time-averaged sense. Moreover, there does not appear to be any systematic difference in $\dot{M}$, for stars of similar mass, between regions with very large differences in cloud properties and stellar velocities, whereas the Bondi–Hoyle scenario would predict a systematic variation of a few orders of magnitude (e.g. Trumpler 37 versus Taurus: Sicilia-Aguilar *et al.* 2005; Hartmann *et al.* 2006).

Dullemond *et al.* (2006) instead suggest that the relationship arises due to the rotational properties of the original star-forming cores (specifically, they assume that all cores regardless of size are rotating at the same fraction of breakup velocity), combined with properties of viscous accretion disks. Their adopted conditions essentially result in two relationships – $M_{\mathrm{disk}} \propto M_*^2$ and $\dot{M} \propto M_{\mathrm{disk}}$ – which in combination produce the observed $\dot{M} \propto M_*^2$ correlation. However, the data appear substantially inconsistent with the $\dot{M} \propto M_{\mathrm{disk}}$ relation they expect. On the contrary, current disk masses are very insecure due to large uncertainties in grain opacities;

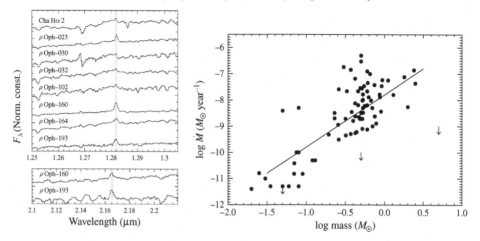

Fig. 13.4. *Left panels*: Paβ (*top*) and Brγ (*bottom*) emission detected in a sample of BDs. The dotted vertical lines show the emission position. (Adapted from Natta *et al.* (2004)). *Right panel*: mass accretion rate as a function of (sub)stellar mass, for all measured members of Taurus. The *solid line* is an arbitrarily placed $\dot{M} \propto M_\odot^2$ trend. Arrows indicate the minimum measurable accretion rate for three characteristic masses.

it is hence possible that this explanation for the $\dot{M}-M_*^2$ relation will appear more viable with improved disk masses.

Finally, Hartmann *et al.* (2006) suggest that the relationship arises due to a combination of factors that change with the mass regime: perhaps the lowest mass stars have the least massive disks that can be thoroughly magnetically active with accretion driven throughout by the magneto-rotational instability, while higher-mass stars have more massive disks, which have magnetically inactive regions (dead zones), where gravitational instabilities may drive accretion. The viability of this hypothesis, as that of Dullemond *et al.* (2006) above, depends on obtaining more robust disk masses. It also predicts that disk evolution should proceed much more rapidly in the lowest mass sources, with $\dot{M}$ dropping off significantly faster with age in BDs than in higher-mass CTTs. The sample of VLMS and BD accretors as a function of age is too limited at this stage to test this claim (see Section 13.4.2).

Finally, a proposed bias that is often brought up in this context deserves mention. Is it possible that the apparent $\dot{M}-M_*^2$ correlation results from an observational bias, wherein VLMS and BDs with very high accretion rates are not identified as such? Specifically, low-mass sources with very large $\dot{M}$ would be heavily veiled and masqueraded as higher-mass stars, thereby escaping inclusion in the VLMS/BD sample. Including such sources would reduce the steepness of the accretion rate–mass relationship and may remove the correlation altogether. However, this suggestion is probably not viable. A large fraction of young VLMS and

BDs known today were originally identified in infrared surveys, and the effects of accretion-related veiling should be much smaller at such wavelengths than in the optical wavelength. Thus, sources with high veiling should be apparent by a significant mismatch between the optical and the infrared spectral types. No such objects have been identified, implying that the $\dot{M}$ in VLMS and BDs is truly much lower than in CTTs and that the observed $\dot{M}-M_*^2$ correlation does not result from a bias against low-mass sources with very large accretion rates. It is also possible that sources with the lowest $\dot{M}$ in some mass regimes (particularly for the BDs) are not identified as accretors due to sensitivity limits. However, this is likely not the case at least for the Taurus solar-mass CTTs, which as we have indicated comprise a fairly complete sample. Thus, the correlation may be steeper than current observations suggest, but probably not shallower.

13.4.5 Implications for brown dwarf formation

To summarize, all the evidence above points to a very similar infancy for BDs and higher-mass stars. Disks and a disk accretion phase are equally ubiquitous in the stellar and substellar regimes, and the phenomenology and physics of the star–disk interaction – magnetospherically channelled inflow from an inner disk edge at a few (sub)stellar radii, the presence of outflowing winds/jets along with inflow, a relationship between the accretion rate and the central mass – are also the same in both regimes. Does this point to a similar formation process for BDs as for stars, in agreement with the 'turbulent fragmentation' scenario and in contrast to the 'ejection' hypothesis? While it is tempting to conclude so, the evidence for this is not yet ironclad.

In particular, the optical and infrared accretion and disk diagnostics arise within disk radii of <1 AU. While the presence of disks and disk accretion does rule out the most severe 'ejection' picture, wherein BD disks are completely sheared away, it does not rule out dynamical interactions which still allow some surrounding material to remain. In particular, current simulations imply that 'ejection' can still permit BD disks up to 10–20 AU in radic (Bate *et al.* 2003a). Given the factor of 10^2–10^4 smaller $\dot{M}$ in BDs compared to CTTs, this would also allow disk accretion in BDs to continue as long as in stars. What is required therefore is a measure of the true size of BD disks. Recent sub-mm/mm measurements do imply BD disk radii of at least ~10 AU (they *may* be much larger, but the data cannot determine whether this is so), which is nominally near the limit of the 'ejection' picture (Scholz *et al.* 2006). However, even if the disks were much smaller (~1 AU) initially, they would viscously expand to ~10 AU over Myr timescales. Only if BD disks are on average significantly larger than 10–20 AU can 'ejection' be ruled out confidently. Evidence for this does not yet exist. However, various investigations, some ongoing and others planned for the future, may help settle the question.

13.4.6 Further formation constraints: recent and future observations

One of the first tests proposed to test formation scenarios was to measure space velocities and spatial distribution: the initial 'ejection' model (Reipurth & Clarke 2001) suggested that young BDs should have larger mean velocities than higher-mass stars and should therefore also be more spatially dispersed than the latter in star-forming regions. Various surveys (Luhman 2006) have shown that this is not the case: BDs have similar space velocities and distributions as stars. However, subsequent simulations showed that BD velocities would remain comparable to the average stellar velocity dispersion even for 'ejection' (Bate *et al.* 2003a), so this is no longer a stringent test.

The existence of wide binaries is another possible test: the same dynamical interactions that would eject BDs and truncate their disks would also disrupt wide binaries. A few wide separation (∼40–250 AU) BD binaries have now been discovered (Chauvin *et al.* 2004; Luhman 2004); at least for these systems, the 'ejection' scenario appears highly unlikely. The discovery of a large sample of such systems would argue against 'ejection' being the primary mode of BD formation. On the contrary, there is increasing evidence that the binary component separation decreases continuously as one moves to lower masses and that BDs simply continue this trend (Kraus *et al.* 2005; Basri & Reiners 2006). A lack of frequent wide binaries among BDs might therefore simply reflect binary formation mechanisms in general, without shedding light on the viability of the 'ejection' scenario for BDs in particular.

Measuring the true size of BD disks provides another means of discriminating between formation scenarios: as mentioned earlier, BD disks much larger than 10–20 AU would strongly argue against 'ejection'. Current modelling, based on fitting SEDs to observed sub-mm/mm dust emission, already suggests that the disks are at least as large as ∼10 AU. However, the number of BDs with such measurements is few (Scholz *et al.* 2006), and the modelling is also dependent on very uncertain dust opacities. Moreover, as discussed, viscous spreading alone can produce ∼10 AU disks even if they started out much smaller. A much better test would be to directly resolve the disks. The unprecedented sensitivity and spatial resolution of ALMA will be a great advantage in this area. However, even ALMA might be insufficient: while disks larger than ∼10 AU in the nearest star-forming regions (∼150 pc) will nominally be resolved, materials beyond ∼30 AU in BD disks may be too cold to yield detectable sub-mm/mm emission, even with ALMA (Natta & Testi 2008).

Thus, ALMA may not be able to decide whether BD disks are significantly larger than implied by ejection combined with viscous spreading.

Finally, the discovery of Class 0 BDs, i.e. proto-BDs embedded in their own isolated cores just like protostars, would provide strong support for the 'turbulent fragmentation' picture and argue against 'ejection'. Indeed, Spitzer has recently identified a number of very low luminosity objects (VELLOs), embedded within cores previously thought to be starless. Initial modelling suggests that these Class 0 objects have masses squarely in BD regime (Young *et al.* 2004; Bourke *et al.* 2006; Huard *et al.* 2006). These masses are rather insecure; however, since they depend on deconvolving the stellar, disk and envelope luminosity contributions, the accretion rates onto the central objects are poorly constrained. Studies are now underway to get more precise masses by obtaining infrared spectral types for the VELLOs and also to constrain the accretion rates by measuring the Paβ and Brγ emission fluxes. Naively, one expects that if the central masses are currently in the BD regime and the measured accretion rates are also too low to allow them to eventually reach stellar masses over the normal lifetime of a core, then these must be true proto-BDs. However, reality may not be so simple. While the VELLO masses are themselves very low, the cores they are embedded in are of order a solar mass, i.e. not particularly small ones. It is difficult to imagine why a solar mass core would produce a BD-mass object and not a low-mass star. Instead, it is possible that the accretion onto the central object is episodic, with long periods of low accretion punctuated by short bursts of intense accretion (as suggested by FU Orionis outbursts); most of the mass accumulated would be during the short bursts. In this case, a BD mass allied with a very low accretion rate would perhaps be the usual state of a low-mass star during much of the Class 0 phase, with high accretion rates being statistically rare. Hence, currently known VELLOs may not necessarily be proto-BDs or be able to adjudicate between BD formation scenarios.

What the 'turbulent fragmentation' scenario really says is that BDs form out of gravitationally bound substellar-mass cores. Recent observations already indicate that cores with BD masses are indeed present in star-forming regions (Greaves 2005, Walsh *et al.* 2007). However, the bound nature of these cores is still uncertain; while the line-width data do suggest that some are gravitationally bound, the conclusion is not robust due to sensitivity issues. ALMA will provide a huge advance in this regard: it should be able to identify substellar-mass cores with ease and even spatially resolve the nearest ones (Natta & Testi 2008). The firm detection of isolated gravitationally bound BD-mass cores would be a clear indication that BDs can form just like stars, in agreement with the 'turbulent fragmentation' scenario. The further discovery of a proto-BD within one of these cores would of course be the final proof.

Acknowledgements

SM acknowledges the support of the Spitzer Fellowship for this work.

References

Acke, B., van den Ancker, M. E., Dullemond, C. P., van Boekel, R., & Waters, L. B. F. M. 2004, *A&A*, **422**, 621
Alencar, S. H. P., Basri, G., Hartmann, L., & Calvet, N. 2005, *A&A*, **440**, 595
Alexander, R. D., Clarke, C. J., & Pringle, J. E. 2006, *MNRAS*, **369**, 229
Barrado y Navascués, D., Mohanty, S., & Jayawardhana, R. 2004, *ApJ.*, **604**, 284
Barrière-Fouchet, L., Gonzalez, J.-F., Murray, J. R., Humble, R. J., & Maddison, S. T. 2005, *A&A*, **443**, 185
Basri, G. & Reiners, A. 2006, *AJ*, **132**, 663
Bate, M. R., Bonnell, I. A., & Bromm, V. 2003a, *MNRAS*, **339**, 577
Bate, M. R., Lubow, S. H., Ogilvie, G. I. & Miller, K. A. 2003b, *MNRAS*, **341**, 213
Beckwith, S. V. W. & Sargent, A. I. 1991, *ApJ.*, **381**, 250
Besla, G. & Wu, Y. 2007, *ApJ.*, **655**, 528
Bourke, T., *et al.* 2006, *ApJ. Lett*, **649**, L37
Bouvier, J., *et al.* 2007, *A&A*, **463**, 1017
Burrows, C. J., *et al.* 1996, *ApJ.*, **473**, 437
Calvet, N., Briceño, C., Hernández, J., Hoyer, S., Hartmann, L., Sicilia-Aguilar, A., Megeath, S. T., & D'Alessio, P. 2005, *AJ*, **129**, 935
Calvet, N. & Gullbring, E. 1998, *ApJ.*, **509**, 80
Catala, C., Donati, J. F., Böhm, T., *et al.* 1999, *A&A*, **345**, 884
Chauvin, G., *et al.* 2004, *A&AL*, **425**, L29
Chiang, E. I., Joung, M. K., Creech-Eakman, M. J., Qi, C., Kessler, J. E., Blake, G. A., & van Dishoeck, E. F. 2001, *ApJ.*, **547**, 1077
Chiang, E. I. & Murray-Clay, R. A. 2007, *Nat. Phys.*, **3**, 604
Ciesla, F. J. 2007, *ApJ. Lett*, **654**, L159
Clampin, M., *et al.* 2003, *AJ*, **126**, 385
Coffey, D., Bacciotti, F., Woitas, J., Ray, T. P., & Eislöffel, J. 2004, *ApJ.*, **604**, 758
Crida, A. & Morbidelli, A. 2007, *MNRAS*, **377**, 1324
Devine, D., Grady, C. A., Kimble, R. A., Woodgate, B., Bruhweiler, F. C., Boggess, A., Linsky, J. L., & Clampin, M. 2000, *ApJ.*, **542**, L115
Dominik, C., Blum, J., Cuzzi, J. N., & Wurm, G. 2007, in *Protostars and Planets V* (B. Reipurth, D. Jewitt, & K. Keil eds.), University of Arizona Press, Tucson 783
Doppmann, G. W., Greene, T. P., & Covey, K. R. 2005, *AJ*, **130**, 1145
Duchêne, G., McCabe, C., Ghez, A. M., & Macintosh, B. A. 2004, *ApJ.*, **606**, 969
Duchêne, G., Ménard, F., Stapelfeldt, K., & Duvert, G. 2003, *A&A*, **400**, 559
Dullemond, C., Natta, A., & Testi, L. 2006, *ApJ. Lett*, **645**, L69
Dullemond, C. P. & Dominik, C. 2004, *A&A*, **417**, 159
Dullemond, C. P. & Dominik, C. 2005, *A&A*, **434**, 971
Dullemond, C. P., Dominik, C., & Natta, A. 2001, *ApJ.*, **560**, 957
Dupree, A. K., Brickhouse, N. S., Smith, G. H., & Strader, J. 2005, *ApJ.*, **625**, L131
Dutrey, A., Guilloteau, S., Duvert, G., Prato, L., Simon, M., Schuster, K., & Menard, F. 1996, *A&A*, **309**, 493
Duvert, G., Guilloteau, S., Ménard, F., Simon, M., & Dutrey, A. 2000, *A&A*, **355**, 165
Edwards, S., Fischer, W., Hillenbrand, L., & Kwan, J. 2006, *ApJ.*, **646**, 319

Espaillat, C., Calvet, N., D'Alessio, P., *et al.* 2007, *ApJ.*, **664**, L111

Fukagawa, M., *et al.* 2004, *ApJ. Lett*, **605**, L53

Garaud, P. 2007, *ApJ.*, **671**, 2091

Garcia Lopez, R., Natta, A., Testi, L., & Habart, E. 2006, *A&A*, **459**, 837

Gatti, T., Testi, L., Natta, A., Randich, S., & Muzerolle, J. 2006, *A&A*, **460**, 547

Glassgold, A. E., Najita, J., & Igea, J. 1997, *ApJ.*, **480**, 344

Grady, C. A., *et al.* 2001, AJ, **122**, 3396

Grady, C. A., Woodgate, B., Torres, C. A. O., Henning. Th., Apai, D., Rodmann, J., Wang, H., Stecklum, B., Linz, H., Williger, G. M., Brown, A., Wilkinson, E., Harper, G. M., Herczeg, G. J., Danks, A., Vieria, G. L., Malumuth, E., Collins, N. R., & Hill, R. S. 2004, *ApJ.*, **608**, 809

Graham, J. R., Kalas, P. G., & Matthews, B. C. 2007, *ApJ.*, **654**, 595

Greaves, J. S. 2005, *Astr. Nach.*, **326**, 1044

Gregory, S. G., Wood, K., & Jardine, M. 2007, *MNRAS*, **379**, L35

Grosso, N., Bouvier, J., Montmerle, T., Fernández, M., Grankin, K., & Zapatero Osorio, M. R. 2007, *A&A*, **475**, 607

Güdel, M., Telleschi, A., Audard, M., Skinner, S. L., Briggs, K. R., Palla, F., & Dougados, C. 2007, *A&A*, **468**, 515

Gullbring, E., Hartmann, L., Briceno, C., & Calvet, N. 1998, *ApJ.*, **492**, 323

Haisch, K. E., Jr., Lada, E. A., & Lada, C. J. 2001, *ApJ. Lett.*, **553**, L153

Hartigan, P., Edwards, S., Ghandour, L. 1995, *ApJ.*, **452**, 736

Hartmann, L. 2002, *ApJ.*, **578**, 914

Hartmann, L., Calvet, N., Gullbring, E., & D'Alessio, P. 1998, *ApJ.*, **495**, 385

Hartmann, L., D'Alessio, P., Calvet, N., & Muzerolle, J. 2006, *ApJ.*, **648**, 484

Hartmann, L., Hewett, R., & Calvet, N. 1994, *ApJ.*, **426**, 669

Hillenbrand, L. A. 2005, Proceedings of *A Decade of Discovery: Planets Around Other Stars* SISCI Symposium (astro-ph/0511083)

Hillenbrand, L. A., Strom, S. E., Calvet, N., Merrill, K. M., Gatley, I., Makidon, R. B., Meyer, M. R., & Skrutskie, M. F. 1998, *AJ*, **116**, 1816

Huard, T., *et al.* 2006, *ApJ.*, **640**, 391

Hubrig, S., Yudin, R. V., Schöller, M., & Pogodin, M. A. 2006, *A&A*, **446**, 1089

Indebetouw, R., *et al.* 2005, *ApJ.*, **619**, 931

Jayawardhana, R., Mohanty, S., & Basri, G. 2003, *ApJ.*, **592**, 282

Johns-Krull, C. & Gafford, A. 2002, *ApJ.*, **573**, 685

Johns-Krull, C. M. 2007, *ApJ.*, **664**, 975

Johns-Krull, C. M., Valenti, J. A., Hatzes, A. P., & Kanaan, A. 1999, *ApJ.*, **510**, L41

Kalas, P., Duchene, G., Fitzgerald, M. P., & Graham, J. R. 2007, *ApJ. Lett.*, **671**, L161

Kalas, P., Graham, J. R., & Clampin, M. 2005, *Nature*, **435**, 1067

Kastner, J. H., Huenemoerder, D. P., Schulz, N. S., Canizares, C. R., & Weintraub, D. A. 2002, *ApJ.*, **567**, 434

Kastner, J. H., Richmond, M., Grosso, N., *et al.* 2006, *ApJ.*, **648**, L43

Kessler-Silacci, J., *et al.* 2006, *ApJ.*, **639**, 275

Kitamura, Y., Momose, M., Yokogawa, S., Kawabe, R., Tamura, M., & Ida, S. 2002, *ApJ.*, **581**, 357

Königl, A. 1991, *ApJL*, **370**, L39

Kraus, A., White, R., & Hillenbrand, L. 2006, *ApJ.*, **649**, 306

Kuchner, M. J. & Holman, M. J. 2003, *ApJ.*, **588**, 1110

Kurosawa, R., Harries, T. J., & Symington, N. H. 2006, *MNRAS*, **370**, 580

Kwan, J., Edwards, S., & Fischer, W. 2007, *ApJ.*, **657**, 897

Lin, D. N. C. & Papaloizou, J. C. B. 1993, in *Protostars and Planets III* (E. Levy & J. Lunine eds.), University of Arizona Press, Tucson 749

Luhman, K. 2004, *ApJ.*, **614**, 398

Luhman, K. 2006, *ApJ.*, **645**, 676

Luhman, K. L., Joergens, V., Lada, C., Muzerolle, J., Pascucci, I., & White, R. 2007a, in *Protostars and Planets V* (B. Reipurth, D. Jewitt, & K. Keil eds.), University of Arizona Press, Tucson 443

Luhman, K. L., *et al.* 2007b, *ApJ.*, **666**, 1219

Maddison, S. T., Fouchet, L., & Gonzalez, J.-F. 2007, *Ap&SS*, **311**, 3

McCabe, C., Ghez, A. M., Prato, L., Duchêne, G., Fisher, R. S., & Telesco, C. 2006, *ApJ.*, **636**, 932

McGroarty, F., Ray, T. P., & Bally, J. 2004, *A&A*, **415**, 189

Ménard, F. & Duchêne, G. 2004, *A&A*, **425**, 973

Mohanty, S., Jayawardhana, R., & Barrado y Navascués, D. 2003, *ApJ. Lett*, **593**, L109

Mohanty, S., Jayawardhana, R., & Basri, G. 2005, *ApJ.*, **626**, 498

Mohanty, S. & Shu, F. 2008, *ApJ.*, in press (astro-ph/0806.4869)

Monin, J.-L., Clarke, C. J., Prato, L., & McCabe, C. 2007, in *Protostars and Planets V* (B. Reipurth, D. Jewitt, & K. Keil eds.), University of Arizona Press, Tucson 395

Mottram, J. C., Vink, J. S., Oudmaijer, R. D., & Patel, M. 2007, *MNRAS*, **377**, 1363

Muzerolle, J., Calvet, N., & Hartmann, L. 1998a, *ApJ.*, **492**, 743

Muzerolle, J., D'Alessio, P., Calvet, N., & Hartmann, L. 2004, *ApJ.*, **617**, 406

Muzerolle, J., Hartmann, L., & Calvet, N. 1998b, *AJ*, **116**, 455

Muzerolle, J., Hillenbrand, L., Calvet, N., Briceño, C., & Hartmann, L. 2003, *ApJ.*, **592**, 266

Muzerolle, J., Luhman, K., Briceño, C., Hartmann, L., & Calvet, N. 2005, *ApJ.*, **625**, 906

Muzerolle, J., *et al.* 2000, *ApJ. Lett*, **545**, L141

Najita, J. R., Strom, S. E., & Muzerolle, J. 2007, *MNRAS*, **378**, 369

Natta, A., Grinin, V., & Mannings, V. 2000, in *Protostars and Planets IV* (V. Mannings, A. P. Boss, & S. S. Russell eds.), University of Arizona Press, Tucson 559

Natta, A., Prusti, T., Neri, R., Wooden, D., Grinin, V. P., & Mannings, V. 2001, *A&A*, **371**, 186

Natta, A. & Testi, L. 2008, *ApSS*, **313**, 113

Natta, A., Testi, L., Muzerolle, J., Randich, S., Comerón, F., & Persi, P. 2004, *A&A*, **424**, 603 [N04]

Natta, A., Testi, L., & Randich, S. 2006, *A&A*, **452**, 245

Padoan, P., Kritsuk, A., Norman, M., & Nordlund, Å. 2005, *ApJ. Lett.*, 622, L61

Padoan, P. & Nordlund, Å. 2004, *ApJ.*, **617**, 559

Perrin, M. D., Duchêne, G., Kalas, P., & Graham, J. R. 2006, *ApJ.*, **645**, 1272

Piétu, V., Guilloteau, S., & Dutrey, A. 2005, *A&A*, **443**, 945

Pinte, C., Fouchet, L., Ménard, F., Gonzalez, J.-F., & Duchêne, G. 2007, *A&A*, **469**, 963

Pinte, C., Padgett, D. L., Ménard, F., Stapelfeldt, K. R., Schneider, G., Olofsson, J., Panić, O., Augereau, J. C., Duchêne, G., Krist, J., Pontoppidan, K., Perrin, M. D., Grady, C. A., Kessler-Silacci, J., van Dishoeck, E. F., Lommen, D., Silverstone, M., Hines, D. C., Wolf, S., Blake, G. A., Henning, T., Stecklum, B. 2008, *A&A*, **489**, 633

Preibisch, T., Kim, Y.-C., Favata, F., *et al.* 2005, *ApJS*, **160**, 401

Reiners, A. & Basri, G. 2008, *ApJ.*, **684**, 1390

Reipurth,B. & Clarke,C., 2001, *AJ*, **122**, 432.

Rice, W. K. M., Armitage, P. J., Wood, K., & Lodato, G. 2006, *MNRAS*, **373**, 1619

Schegerer, A. A., Wolf, S., Ratzka, Th., & Leinert Ch. 2008, *A&A*, **478**, 779

Schneider, G., *et al.* 2006, *ApJ.*, **650**, 414

Scholz, A., Jayawardhana, R., & Wood, K. 2006, *ApJ.*, **645**, 1498

Shu, F. H., Najita, J., Ostriker, E., Wilkin, F., Ruden, S., Lizano, S. 1994, *ApJ.*, **429**, 781

Sicilia-Aguilar, A., Hartmann, L., Hernández, J., Briceño, C., & Calvet, N. 2005, *AJ*, **130**, 188

Stelzer, B., Micela, G., Hamaguchi, K., & Schmitt, J. H. M. M. 2006, *A&A*, **457**, 223

Strom, K. M., Strom, S. E., Edwards, S., Cabrit, S., & Skrutskie, M. F. 1989, *AJ*, **97**, 1451

Takami, M., Bailey, J., & Chrysostomou, A. 2003, *A&A*, **397**, 657

Tatulli, E., Isella, A., Natta, A., *et al.* 2007, *A&A*, **464**, 55

Telleschi, A., Güdel, M., Briggs, K. R., Audard, M., & Palla, F. 2007, *A&A*, **468**, 425

Testi, L., Natta, A., Shepherd, D. S., & Wilner, D. J. 2003, *A&A*, **403**, 323

Thébault, P. & Augereau, J.-C. 2007, *A&A*, **472**, 169

van Boekel, R., Waters, L. B. F. M., Dominik, C., Bouwman, J., de Koter, A., Dullemond, C. P., & Paresce, F. 2003, *A&A*, **400**, L21

van Boekel, R., *et al.* 2004, *Nature*, **432**, 479

Varnière, P., Bjorkman, J. E., Frank, A., Quillen, A. C., Carciofi, A. C., Whitney, B. A., & Wood, K. 2006a, *ApJ. Lett*, **637**, L125

Varnière, P., Blackman, E. G., Frank, A., & Quillen, A. C. 2006b, *ApJ.*, **640**, 1110

Vink, J. S., Drew, J. E., Harries, T. J., Oudmaijer, R. D., & Unruh, Y. 2005, *MNRAS*, **359**, 1049

Wade, G. A., Bagnulo, S., Drouin, D., Landstreet, J. D., & Monin, D. 2007, *MNRAS*, **376**, 1145

Walsh, A., *et al.* 2007, *ApJ.*, **655**, 958

Wassell, E. J., Grady, C. A., Woodgate, B., Kimble, R. A., & Bruhweiler, F. C. 2006, *ApJ.*, **650**, 985

Watson, A. M., Stapelfeldt, K. R., Wood, K., & Ménard, F. 2007, in *Protostars and Planets V* (B. Reipurth, D. Jewitt, & K. Keil eds.), University of Arizona Press, Tucson 523

Whelan, E. *et al.* 2005, *Nature*, **435**, 652

Whelan, E. *et al.* 2007, *ApJ. Lett*, **659**, L45

White, R.J. & Basri, G. 2003, *ApJ.*, **582**, 1109

White, R. J. & Ghez, A. M. 2001, *ApJ.*, **556**, 265

White, R. J. & Hillenbrand, L. A. 2004, *ApJ.*, **616**, 998

White, R. J. & Hillenbrand, L. A. 2005, *ApJ.*, **621**, L65

Winston, E., Megeath, S. T., Wolk, S. J., *et al.* 2007, *ApJ.*, **669**, 493

Wolf, S. & D'Angelo, G. 2005, *ApJ.*, **619**, 1114

Wolk, S. J., Hardner, F. R., Jr., Flaccomio, E., *et al.* 2005, *ApJS*, **160**, 423

Yang, H., Johns-Krull, C. M., & Valenti, J. A. 2007, *AJ*, **133**, 73

Young, C., *et al.* 2004, *ApJS*, **154**, 396

Zinnecker, H. & Yorke, H. W. 2007, *ARA&A*, **45**, 481

14

Structure and dynamics of protoplanetary disks

C. P. Dullemond, R. H. Durisen and J. C. B. Papaloizou

14.1 Introduction

The dust and gas disks surrounding many pre-main-sequence stars are thought to be the birthplaces of planets. They are therefore the locations of structure formation in the universe, albeit on small scales. In recent years, the topic of protoplanetary disks has gained an increasing amount of attention in astrophysics, which is in large part due to the enormous increase of high-quality observational data that have been published recently, from the Spitzer Space Telescope, from the Very Large Telescope, from millimetre interferometers and so forth. Moreover, studies of the properties of extrasolar planetary systems have put new constraints on the formation and migration of newly born planets in such disks. The topic of the structure and evolution of protoplanetary disks is wide open, and significant new developments, both from theory and from observations, can be expected in the near future.

This chapter gives an overview of some aspects of theoretical modelling of protoplanetary disks. The review consists of four parts. First, we discuss the overall formation, evolution and dispersal of protoplanetary disks. We show that much of what we know of the long timescale disk evolution is still based on relatively simple models. We discuss their vertical structure, as derived from radiative transfer models and their comparison to observations. We then show that magneto-rotational and gravitational instabilities (GIs) can introduce complex dynamics and non-linear structure to these disks, and we discuss an on-going debate about the role of GIs in planet formation. Finally, once a planet is formed, it will have interaction with the remaining disk and can migrate. This is the final topic of this chapter.

14.2 The birth and death of a protoplanetary disk

The fundamental reason why disks exist around young stars is that the typical specific angular momentum j of the molecular cloud cores that collapse to form stars

Structure Formation in Astrophysics. ed. G. Chabrier. Published by Cambridge University Press.

is about 10^{21} cm^2 s^{-1} orders of magnitude larger than the maximum for a star in radial force balance (see table 1 of Bodenheimer (1995)). On the contrary, $j \approx 10^{20}$ and $3 \cdot 10^{20}$ cm^2 s^{-1} for Jupiter's and Neptune's orbits, respectively. So a cloud collapse that forms a star is likely to form a circumstellar disk of Solar System size around that star. This much seems clear, but beyond that, disk formation is a topic involving great uncertainties. There is little conclusive evidence from observations, because of the high extinction in the collapsing circumstellar envelope that feeds the disk. Some models of real embedded systems (Osorio *et al.* 2003) suggest disk masses comparable to stellar masses in these early phases.

Much of what we understand about the disk formation phase comes from numerical modelling. Early models of rotating cloud collapse include the analytic work by Ulrich (1976), Cassen and Moosman (1981) and Terebey *et al.* (1984). But radiation-hydrodynamic work by Tscharnuter (1987) and Yorke *et al.* (1993) have shown that even with the limiting assumption of axisymmetry, the flow pattern of the collapsing gas is more complex than what the analytic models predict. When going to full 3D, the situation becomes even more interesting, with the possibility of non-axisymmetric modes and even fragmentation (Truelove *et al.* 1998; Matsumoto & Hanawa 2003; Banerjee *et al.* 2004; Vorobyov & Basu 2005). While these models describe in detail the formation and very early evolution of such disks, they are limited by computational power and various arbitrary assumptions about initial conditions and about the distribution and redistribution of angular momentum during and after collapse. A protoplanetary disk spans a very large range in radii (from within 0.01 AU out to a few 100 AU), meaning that a single outer orbit corresponds to a million inner orbits, all of which correspond to less than 10^{-3} of the life time of a disk, which is a few million years. A complete picture of the formation and evolution of a protoplanetary disk over its entire dynamic range is therefore prohibitively expensive when using these 3D computational techniques.

A much simpler approach to the formation and evolution of disks is chosen by Nakamoto and Nakagawa (1994) and Hueso and Guillot (2005). In their approach, the disk is modelled using the vertically integrated equations for axisymmetric viscous disk evolution coupled to a simple recipe for the infalling envelope that feeds the disk with fresh matter in the early (Lada-André Class O/I) phase. The disk is characterized by a radial surface density profile as a function of time $\Sigma(r, t)$ which obeys

$$\frac{\partial \Sigma}{\partial t} = \frac{3}{r} \frac{\partial}{\partial r} \left[\sqrt{r} \frac{\partial}{\partial r} \left(\Sigma \nu \sqrt{r} \right) \right], \tag{14.1}$$

(Shakura & Sunyaev 1973; Lynden-Bell & Pringle 1974) where $\nu \equiv \alpha c_{\rm s}^2 / \Omega_K$ with $\Omega_K = \sqrt{GM_*/r^3}$ is the anomalous viscosity of the disk, parameterized by a dimensionless α which is typically taken to be $\simeq 0.01$. The origin of this anomalous viscosity is still not fully clear, but is thought to be due to magneto-rotational turbulence, which will be discussed in Section 14.4.1. The symbol $c_s = \sqrt{kT_{mid}(r)/\mu}$ is the isothermal sound speed, where $T_{mid}(r)$ is the gas temperature at the midplane of that radius and $\mu = 2.3\ m_p$ is the mean molecular weight of the gas. If the midplane temperature $T_{mid}(r)$ is known, the solution of Eq. (14.1) is readily obtained numerically. For special cases, analytical solutions also exist (Lynden-Bell & Pringle 1974; see also Hartmann *et al.* 1998). The great advantage of these kinds of models over fully 3D hydrodynamic ones is that the viscous disk equations are 1D and they lend themselves ideally for implicit numerical integration. This allows for the evolution of the disk over millions of years using only very minor computational resources. It is for this reason that such models, in spite of their simplicity and limitations, are still extraordinarily useful today, both for theoretical studies of disks and for interpreting observational data.

Models of this kind show that the evolution of a protoplanetary disk involves accretion of the inner part of the disk and viscous spreading of the outer part of the disk (see Lynden-Bell & Pringle 1974). They also show that the later evolution appears to be strongly influenced by the conditions of the collapse process that formed the disk (Hueso & Guillot 2005). It was suggested by Dullemond *et al.* (2006) that such a link between molecular cloud core and protoplanetary disk may also explain the correlations that have been observed between the stellar mass and the measured accretion rates in their circumstellar disks (see, e.g., Muzerolle *et al.* 2003). Other disk evolution scenarios aimed at explaining this relation were presented by Hartmann *et al.* (2006) and Alexander and Armitage (2006).

The models described so far all assume a *given* global strength of the turbulence that drives the accretion process. On average, a value of $\alpha = 0.01$ appears to be consistent with observed characteristics of protoplanetary disks (Hartmann *et al.* 1998) as well as with values derived from detailed 3D simulations of magneto-rotationally driven turbulence in a shearing box (Stone *et al.* 1996) and of GIs (Boley *et al.* 2006). However, the magnetohydrodynamic (MHD) turbulence can only remain active if the disk is sufficiently ionized to couple the magnetic field to the gas. There may exist regions in the disk that are neither warm enough to thermally ionize heavy elements nor reachable by cosmic rays and/or X-rays because they have too high a column depth (Gammie 1996). Such regions are called *dead zones* as they are expected to be non-turbulent and therefore do not accrete. The extent of the dead zone, and whether it exists in the first place, is still heavily debated, in particular in the context of the abundance of free electrons (Sano *et al.* 2000; Desch 2004; Semenov *et al.* 2004; Ilgner & Nelson 2006) and hydrodynamic stirring of the dead zone by the active regions (Fleming & Stone 2003).

If dead zones exist, they may have interesting consequences for the evolution of a disk. In the model of Gammie, the dead zone exists only in a limited range in radii (outward of ~ 0.1 AU and inward of the radius where the total surface density of the disk drops below about $100\,\mathrm{g\,cm^{-2}}$) and only near the midplane. The surface is, in this model, sufficiently ionized through cosmic ray ionization. Since the maximum accretion rate is a given function of radius (set by the penetration depth of the cosmic rays), whereas the supply of matter from the outer disk may be higher, the matter piles up in the dead zone. As we shall discuss below, this may lead to episodic massive accretion events which may be related to the FU Orionis outburst phenomenon.

Typically, dead zones and FU Ori outbursts are expected for very young systems ($\lesssim 1\,\mathrm{Myr}$). As the disks mature, they become less violent and slowly fade away. During this 'protoplanetary' phase, it is believed that planets are formed. However, observational data from Haisch *et al.* (2001), Carpenter *et al.* (2005) and Sicilia-Aguilar *et al.* (2006) have shown that stars older than about 10 Myr only very rarely have dusty circumstellar disks, while stars of about 1 Myr very often have dusty disks. This means that gas giant planet formation must have taken place within 10 Myr, which puts interesting constraints on giant planet formation theories. It is also interesting that the simple viscous accretion and spreading of protoplanetary disks are too gradual for this very strong observational trend, so some other process must operate to remove the disk (or at least its dust) on a timescale of about 10 Myr. There is no conclusive evidence yet as to what this process is, but there are various natural possibilities. The most promising appears to be the photoevaporation of these disks caused by the extreme ultraviolet (EUV) irradiation of the disk by its own host star (Hollenbach 1994). Alexander *et al.* (2006a; see also Clarke *et al.* 2001) showed that when EUV photoevaporation is coupled to viscous disk evolution, then this process tends to drill a hole of about 1 AU radius in the inner disk within typically a few million years, after which the outer disk is rapidly destroyed. This model may explain both the observed disappearance of disks and the existence of a peculiar class of T Tauri stars for which infrared observations have shown that they have a huge inner hole in their disk (Calvet *et al.* 2002; Bouwman *et al.* 2003; D'Alessio *et al.* 2005). These 'transition disks' are rare, but this could be a natural consequence of the rapid dispersal of the disk once a hole is drilled.

However, transition disks also exist which clearly still have a signature of accretion of material onto the stellar surface. This is inconsistent with photoevaporation models. An explanation for this type of source could be the growth of dust grains in these inner holes (Tanaka *et al.* 2005). Since dust grains are the main opacity carriers, a rapid growth of grains through aggregation will lower their opacity and may render the disk optically thin. This may even be related to the more general problem of disk dispersal (Dullemond & Dominik 2004, 2005, Tanaka *et al.* 2005; D'Alessio *et al.* 2006; Nomura & Nakagawa 2006). At the present, however, many if not most of these questions are still open.

14.3 Vertical structure

14.3.1 Basic vertical structure

The drawback of describing the disk by a surface density $\Sigma(r, t)$ instead of a full 3D simulation is that the information about the vertical structure is lost. However, for a given $\Sigma(r, t)$, the vertical temperature and density structure can be approximately reconstructed. Combined with the time-dependent surface density $\Sigma(r, t)$ following from viscous disk theory, one obtains a fairly complete picture of the overall structure and evolution of a protoplanetary disk. The drawback of this approach is that the disk is a priori assumed to be axisymmetric, smooth and laminar, which is likely an over-simplification. However, as argued above, it is up to now the only feasible way to model the disk structure and evolution over its entire radial domain and over its entire life span. For that reason, these kinds of disk models are still one of the main pillars of the theory of protoplanetary disks.

The main challenge of such models is to reconstruct the vertical temperature structure. Once that is known, the assumption of vertical hydrostatic equilibrium combined with the given value of the surface density immediately yields the vertical density distribution as well. In its full complexity, computing the vertical temperature structure involves multi-dimensional dust continuum and molecular/atomic line transfer, gas chemistry, dust–gas thermal heat exchange, photoelectric heating of the gas, heating through viscous dissipation of gravitational energy, etc. At present, no model exists that covers this full complexity. Fortunately, the adoption of a few not unreasonable assumptions simplifies the problem considerably. The most far-reaching simplification, adopted in most current disk models, is that the gas temperature equals the dust temperature. This reduces the entire problem to a 2D/3D dust continuum radiative transfer problem, which is complicated but manageable. Even further simplified models such as the 1+1D irradiated accretion disk models of D'Alessio *et al.* (1998) and the 1+0D two-layer irradiated passive disk model of Chiang and Goldreich (1997) still capture much of the physics of circumstellar disks and reproduce the solid-state emission features and SED shape of T Tauri star disks reasonably well.

These models have shown that protoplanetary disks quite naturally tend to have a *flaring* geometry, i.e. a geometry in which the surface height H_s goes as $d(H_s/R)/dR > 0$ (Kenyon & Hartman 1987). A flaring geometry allows the stellar photons to irradiate the surface of the disk and heat the upper layers. The thermal emission from these layers subsequently keeps the equatorial disk regions warm. The resulting vertical density and temperature structure for such an irradiated disk are shown in Figure 14.1. The warm surface layer on top of a less warm (but not completely cold) disk interior gives the characteristic solid-state features in emission (Calvet *et al.* 1991), as observed for nearly all protoplanetary disks. Usually,

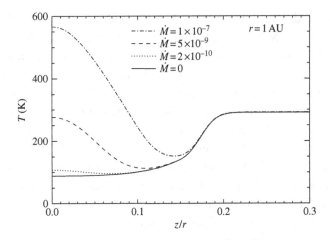

Fig. 14.1. Model computation of the vertical temperature profile at 1 AU of a pro-toplanetary disk around a T Tauri star. The solid (*bottom*) curve is the result when the disk is laminar (i.e. not accreting). The other curves show how the midplane region heats up when accretion is active. The models were made for turbulence parameter $\alpha = 0, 10^{-4}, 10^{-3}, 10^{-2}$ from bottom to top.

irradiational heating dominates the heating/cooling balance of the disk. However, at radii close enough to the central star and at high enough accretion rate, the release of heat through viscous dissipation of gravitational potential energy can start dominating the heating/cooling balance near the midplane, as seen in Figure 14.1 (see, e.g., D'Alessio *et al.* 1998; Davis 2007). When the accretion rate is high enough, this may even start dominating the effective temperature at the disk's surface and hence dominate the spectrum of the disk (Lachaume 2004). In principle, this could yield features in absorption even for face-on disks, but this is rarely seen.

In recent years, more and more efforts are being undertaken to investigate how the disk structure changes when the gas and dust temperatures decouple, as they are expected to do in the very tenuous upper layers of the disk. This area of research is still very much under development, but first results indicate that depending on the abundance of polycyclic aromatic hydrocarbons and small dust grains in the gas, the gas temperature could drastically exceed the dust temperature in these very upper layers (Jonkheid *et al.* 2004; Kamp & Dullemond 2004; Nomura & Millar 2005; Nomura *et al.* 2007). Dust evolution such as sedimentation and/or aggregation will therefore have a strong effect on the gas temperature.

14.3.2 *2D and 3D radiative transfer models*

Due to their relative simplicity and elegance, simple two-layer (1+0D) and the more detailed 1+1D disk models are quite popular. However, in recent years, modellers are gradually moving towards more realistic 2D/3D continuum radiative transfer models. Most current 2D/3D radiative transfer models are based on *parameterized* density structures, after which the radiative transfer algorithm computes the dust temperature, the SEDs and the images (Robitaille *et al.* 2006; Scholz *et al.* 2007; Visser *et al.* 2007, to name only a few). By now, this technique has become mainstream. Recently, there is also a gradual movement towards the inclusion of 1D vertical hydrostatic equilibrium into these models (Dullemond 2002; Nomura 2002; Dullemond & Dominik 2004; Walker *et al.* 2004).

The drawback of 2D/3D models is that they tend to have a large number of parameters. The most important of these are disk mass, surface density profile, disk height and flaring geometry, inner and outer radius, dust composition and grain size distribution. Any slight increase of complexity will add a few more. Clearly, to constrain all these parameters, we also need a very large variety of data: not only the SED/spectrum but also spatial information at different scales, such as infrared interferometric visibility curves, millimetre-maps, resolved infrared images, etc. Fitting such a large set of data for each source is a cumbersome activity, but an increasing number of sources are being studied in this way (Wolf *et al.* 2003; Pinte *et al.* 2007; Pontoppidan *et al.* 2007a,b). With the huge database of pre-computed radiative transfer models of Robitaille *et al.* (2006), a first quick fitting of data is drastically simplified and accelerated, though detailed fitting will always require further fine-tuning using radiative transfer codes. Watson *et al.* (2007) show how such radiative transfer models can be used to fit disk models to images, and what one can learn from this.

Summarizing the conclusions from all these radiative transfer studies would fill several more pages, but on the whole such models have shown that protoplanetary disks still have quite a large abundance of dust until shortly before they are destroyed, their geometries can have different degrees of flaring, they always have warm surface layers on top of cooler midplane regions (proving that they are energetically irradiation-dominated in most parts of the disk), some disks have large inner holes and there is evidence for at least some grain growth and sedimentation occurring in these disks.

14.4 Gravitational instabilities in protoplanetary disks

14.4.1 Instabilities and structure formation

The smooth axisymmetric models for the global structure and evolution of a protoplanetary disk discussed so far are useful tools for understanding various observational data of protoplanetary disks, but they are evidently highly simplified. The hydrodynamic structure of protoplanetary disks is likely to be much richer, with complex turbulent motion on the small scales and gravitationally induced structures on the large scale. Even though the appearance and vertical structure are strongly affected by local dissipation and stellar irradiation, the rotational and gravitational components dominate the energy. It is natural to ask whether large-scale features can be produced by tapping these mechanical energy reservoirs and what this might have to do with planet formation.

The disk evolution picture presented in Section 14.2 already requires non-uniform structure at least on small scales. Equation (14.1) assumes that turbulence in the flow produces hydrodynamic stresses accurately parametrized by α, with $\alpha \sim 10^{-2}$ required by observations. Various mechanisms have been proposed as the source for this turbulence (Papaloizou & Lin 1995; Balbus 2003; Gammie and Johnson 2005), and all involve the onset of instability due to linear or non-linear perturbations from laminar flow. Pure hydrodynamic instability of nearly Keplerian shear with high Reynolds number (Hawley *et al.* 1999; Lesur and Longaretti 2005; Ji *et al.* 2006) and convective instability due to vertical heat flow (Cabot 1996; Stone and Balbus 1996; Balbus 2000) do not appear to result in meaningful values of α. At present, three mechanisms are receiving the most attention: magneto-rotational instabilites (MRIs), gravitiational instabilities (GIs), and baroclinic instabilities (BIs). For MRIs and GIs, the ability to produce significant effective αs has been demonstrated (Balbus 2003; Durisen *et al.* 2007). Instabilities generated by pre-existing non-uniform structures, such as edges (Papaloizou and Pringle 1987), vortensity extrema (Papaloizou & Lin 1989) and other local extrema (Lovelace *et al.* 1999; Li *et al.* 2001), are also promising candidates under special conditions but are not self-consistent ways to create structure in an initially smooth disk.

Shallow turbulent flows with high shear can produce large-scale vortices which sustain themselves by swallowing smaller eddies with the same sign of vorticity, a phenomenon known as *negative eddy viscosity*. The classic astronomical example is Jupiter's Red Spot (Marcus 1993). BIs in particular, which involve simulation of inertial waves through buoyancy effects, are likely to lead to significant vorticity fluctuations. Numerical experiments, such as by Johnson and Gammie (2005), demonstrate that a thin Keplerian disk given small random vorticity fluctuations will develop large-scale anticyclonic vortices, similar to the high-pressure systems in the Earth's atmosphere. Anticyclones are favoured due to the retrograde epicycles executed by perturbed fluid elements, i.e. anticyclones roll with the Keplerian shear, while cyclonic eddies swirl counter to the shear and get torn apart (Adams and Watkins 1995). Whether true BIs arise due to entropy gradients in disks has been somewhat controversial (Klahr & Bodenheimer 2003; Johnson & Gammie 2006), but very recent simulations suggest that strong enough radial temperature gradients combined with fast enough radiative cooling can sustain BIs and large-scale vortex production in thin disks (Petersen *et al.* 2007a,b). If these vortices become asymmetric, e.g. tilted with respect to the azimuthal direction, large effective αs could result. However, in 3D disks, vortices penetrating to the midplane may be fragile (Barranco & Marcus 2005). Because large-scale stable vortices may accelerate planetesimal and gas giant planet formation (Adams & Watkins 1995; Klahr & Bodenheimer 2006), the possibility that these structures may arise in disks deserves further study.

14.4.2 GI basics and transport

Let us turn our attention now to a mechanism that has been demonstrated to produce dramatic structure. As reviewed in Durisen *et al.* (2007), simulations show that disk self-gravity causes spiral waves to erupt when the Toomre (1981) Q-parameter

$$Q = \frac{c_{\mathrm{gas}}\kappa}{\pi G \Sigma} \sim \frac{T^{1/2}\Omega}{\Sigma} \tag{14.2}$$

is less than about 1.5–1.7. Here, c_{gas} is the gas sound speed-averaged appropriately over the vertical structure and κ is the epicyclic frequency, which is $\approx \Omega_{\mathrm{K}}$ in a nearly Keplerian disk. The ratio embodied in Q tells us that a disk will become unstable to its own self-gravity once it becomes sufficiently cold (low T) and/or massive (high Σ). The instability manifests itself by spontaneous growth of noise into non-linear spiral wave disturbances on a dynamic timescale, i.e. on the order of the rotation period P_{rot}. Semi-analytic theory suggests that amplification can be by swing (Toomre 1981) or SLING (Adams *et al.* (1989); Shu *et al.* 1990), but, in real disks, swing seems to dominate for low disk mass (Pickett *et al.* 1998; Mayer *et al.* 2004). At large amplitude, the spiral waves steepen into shocks (Pickett *et al.* 1998, 2000; Nelson *et al.* 1998, 2000a), and there is non-linear mode coupling (Laughlin *et al.* 1998). Even if only a few well-defined modes grow at first, a broad spectrum of interacting spiral waves develops, with most power confined between inner and outer Lindblad resonances (Pickett *et al.* 2003; Mejía *et al.* 2005; Boley *et al.* 2006). Gammie (2001) refers to this as *gravitoturbulence*.

The ultimate fate of GIs in a disk is controlled by its thermodynamics, particularly radiative cooling and the equation of state (EOS). The predominately trailing spiral waves cause net outward transport of angular momentum (Larson 1984). So, a GI-active disk is heated not only by shocks but also through work done by gravity when torques induce net mass inflow inside the corotation radii of the dominant spiral waves (Mejía *et al.* 2005; Boley *et al.* 2006; Cai *et al.* 2008). Just enough cooling can maintain marginal instability with $Q \sim 1.4$–1.7. Simulations (Tomley *et al.* 1991; Gammie 2001; Rice *et al.* 2003; Mejía *et al.* 2005) confirm that a quasi-steady *asymptotic state* of non-linear GI activity can be achieved through a balance of heating and cooling (Goldreich & Lynden-Bell 1965; Pringle 1981). On the contrary, if the cooling is too fast, the spiral waves fragment into gravitationally bound clumps instead.

A lot can be learned by considering simple EOSs, such as ideal gases with constant ratios of specific heats γ, combined with idealized cooling, such as an imposed volumetric cooling rate ϵ/t_{cool}, where ϵ is the internal energy density and t_{cool} is an assumed cooling time. In local shearing-box simulations of infinitesimally thin disks, Gammie (2001) finds that

$$\alpha \approx \frac{4}{9} \left[\Gamma(\Gamma - 1)t_{cool}\Omega\right]^{-1} \tag{14.3}$$

in the asymptotic state, where Γ is a 2D version of γ obtained by vertical integration of the EOS and that disks fragment if $t_{cool}\Omega$ < about 3 (or t_{cool} < about $P_{rot}/2$) for $\Gamma = 2$, usually referred to as the *Gammie fragmentation criterion*. Let M_d be the disk mass and M_s the stellar mass. Global $t_{cool}\Omega$ = constant 3D simulations with $M_d \leq 0.25M_s$ and $\gamma = 5/3$ confirm both the fragmentation criterion (Rice *et al.* 2003) and Eq. (14.3) (Lodato & Rice 2004). Global 3D disk simulations with t_{cool} = constant everywhere also support the Gammie criterion but tend to produce somewhat larger αs than Eq. (14.3) (Mejía *et al.* 2005). Differences between GI transport and strict α-disk behaviour (Laughlin & Rozyczka 1996) may be due to the global character of GIs (Balbus & Papaloizou 1999). Lodato and Rice (2005) find that, when $M_d > 0.25M_s$, disks do not act locally even with $t_{cool}\Omega$ = constant. GIs can instead exhibit episodic eruptive behaviour. Global 3D simulations (Rice *et al.* 2005) show that the constant in Gammie's fragmentation criterion increases to about 10–12 for $\gamma = 7/5$ (or t_{cool} < about $2P_{rot}$). For an isothermal EOS ($\gamma \to 1$), disks fragment whenever Q < about 1.4 (Johnson & Gammie 2003). An example of disk fragmentation is shown in Figure 14.2a.

GIs in stratified disks are a 3D phenomenon (Pickett *et al.* 1998, 2000, 2003). Strong corrugations of the disk surface are caused by *shock bores*, spiral shock waves that take on characteristics of hydraulic jumps (Martos & Cox 1998; Boley *et al.* 2005; Boley & Durisen 2006). If the shocks are roughly adiabatic, the vertical pressure gradient forces in the post-shock region are much higher than in the pre-shock region. The gas goes out of vertical hydrostatic equilibrium and jumps upwards at roughly the post-shock c_{gas}. Jumps can be factors of two or more in H. Huge breaking waves result as the jumping gas spreads horizontally and curls over the shock front. GI-induced spiral waves in real stratified disks are thus not well described as just 'density waves'. The dynamics of shock bores may have important consequences for disk appearance, as well as for mixing and stirring of gas and solids.

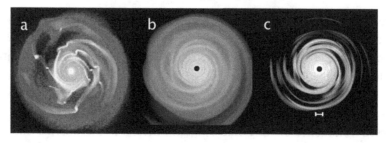

Fig. 14.2. Logarithmic grey scales. (a) Surface density of a fragmenting disk
$(40 \times 40\,\text{AU})$ with $M_s = 1\,M_\odot$, $M_d = 0.1\,M_\odot$ and initial $Q = 1.4$ evolved isother-
mally after 4.5 outer rotations (Mayer *et al.* 2002). (b) Midplane density of a
non-fragmenting disk $(80 \times 80\,\text{AU})$ with $M_s = 0.5\,M_\odot$, $M_d = 0.07\,M_\odot$ and initial
$Q_{\min} = 1.8$ evolved using D'Alessio opacities after 13.2 outer rotations (Boley
et al. 2006). (c) Intensity map at 1 mm of the disk in (b). The line shows 0.05
arcsec if the disk is placed at 150 pc.

14.4.3 Fragmentation of 'real' disks

If GIs are sufficiently violent and rapid, they might lead to disk fragmentation and
hence to the instantaneous formation of companion objects, such as brown dwarfs
or gas giant planets (Boss 1997, 2000). Whether this mechanism is viable is, however, a long-standing debate. It depends on whether a real disk can simultaneously
be cool enough to be unstable and have a short enough cooling time to fragment
(Rafikov 2005). The vertical thermal structure and radiative transport discussed
in Section 14.3 are thus critical. So far, all hydrosimulations assume equal dust
and gas temperature, and the high-T, high altitude regions in Figure 14.1 are usually not explicitly calculated, although some authors do include crude treatments of
irradiation. Nelson *et al.* (2000a) apply a radiative algorithm with realistic opacities
(Pollack *et al.* 1994) to global 2D disk simulations. Isentropic hydrostatic columns
in the disk are cooled by computing vertical radiative equilibrium. In simulations
of 50 AU radii disks with $M_s = 0.5 M_\odot$ and $M_d = 0.2 M_\odot$, they find $t_{\text{cool}} \sim 3\text{–}10$
P_{rot} (Durisen *et al.* 2007) and no fragmentation. Although the Gammie criterion
becomes complex with realistic opacities (Johnson & Gammie 2003), fragmentation still does not occur unless t_{cool} is short and Q is low. When Boss (2001,
2002a,b) used full 3D radiative diffusion in global simulations of a 20 AU radius
disk with $M_s = 1 M_\odot$ and $M_d = 0.09 M_\odot$, he finds fast cooling leading to fragmentation was independent of the magnitude of the dust opacity. Boss (2004a) attributes
this fast cooling to convection, because he detects upwellings and superadiabatic
gradients associated with the spiral arms.

Inconsistent results on disk fragmentation in the 'planet-forming region' (a
few to tens of astronomical units) persist to the present. The Boss treatment
of the disk photosphere assumes a constant T condition set at a radial optical
depth of 10. The Indiana University hydrogroup (IUHG) attempts a more detailed

treatment with two different algorithms. The Cai/Mejía scheme uses flux-limited diffusion (FLD) for vertical optical depths $\tau > 2/3$, explicitly treats cooling in $\tau < 2/3$ regions and fits the thick disk interior to the thin disk atmosphere with an Eddington solution (Mejía *et al.* 2005; Boley *et al.* 2006; Cai *et al.* 2006). The newer Boley scheme uses FLD only in the *r*- and ϕ-directions when τ is large and solves the wavelength-averaged radiative transfer equation for all τ in the *z*-direction using one ray (Boley *et al.* 2007a). Simulations of a 40 AU radius disk with $M_s = 0.5 M_\odot$ and $M_d = 0.07 M_\odot$ show neither fast cooling by convection nor fragmentation with either scheme (Figure 14.2b). The IUHG simulations yield an effective asymptotic $\alpha \sim$ few$\cdot 10^{-2}$ similar to Eq. (14.3) and with behaviour intermediate between $t_{cool}\Omega = $ constant and $t_{cool} = $ constant. Recent SPH simulations with radiative cooling by Mayer *et al.* (2007) and Stamatellos and Whitworth (2008) add to the controversy by finding fragmentation only in special cases. Figure 14.2c illustrates how a non-fragmenting GI-active disk might appear to ALMA at the distance of the Taurus star-forming region.

Before additional radiative hydrocodes are brought to bear on this problem, Boley *et al.* (2007a) suggest a suite of radiation and convection tests that codes should pass. They also give a numerical example showing, as argued by Rafikov (2007), that, even when vigorous convection occurs, it carries only tens of percent of the flux. It seems unlikely that t_{cool} can be reduced by factors of ten or more, as required by the Boss results. Boley *et al.* (2006, 2007a) speculate that vertical motions reported by Boss (2004a) and by Mayer *et al.* (2007) are shock bores, not convection, and they suspect that t_{cool} disagreements are more likely due to treatment of radiative boundary conditions. Other contributing factors may be resolution, accuracy and EOS (Boss 2007; Boley *et al.* 2007b). Detailed code comparisons are needed to elucidate these issues.

If convection does not cause fast cooling, then, contrary to Boss (2002a), we expect sensitivity to disk metallicity, grain growth and settling of solids, because these change the dust opacity. With the Cai/Mejía scheme, the strength of GIs in the planet-forming region decreases as metallicity Z and maximum grain radius a_{max} are increased (Cai *et al.* 2006). Over the range $1/4 < Z/Z_\odot < 2$ and 1 micron $< a_{max} < 1$ mm, notable differences are seen but no fragmentation. With the addition of infrared radiation shining down on the disk, presumably from a warm envelope, Cai *et al.* (2008) find that GIs can be weakened or suppressed. Mild irradiation selectively suppresses high-order structure in favour of global two-armed spirals. Thus, envelope and probably also stellar irradiation make fragmentation even more unlikely (Matzner & Levin 2005). The simultaneous action of MRIs and GIs also tends to weaken GI amplitudes (Fromang 2005). The only hope then left for planet formation by disk fragmentation may be extreme reduction of disk opacity by the growth of solids (see Section 14.4.4).

Despite controversy about $r < 10$s of AU, massive disks that are large enough may have fragmentation in their outer regions (Kratter & Matzner 2006). An example of fragmentation in the outer regions of a massive disk evolved with realistic opacities that can be found in Krumholz *et al.* (2007), but the objects formed by fragmentation in this case are stars and brown dwarfs, respectively, not planets.

An additional concern for the disk instability theory is whether clumps formed by fragmentation survive to become protoplanets. An array of numerical effects are worrisome (Truelove *et al.* 1997; Pickett *et al.* 2003; Mayer *et al.* 2004; Nelson 2006; Boss 2007; Pickett and Durisen 2007). Considerable light may be shed on this soon through a study led by L. Mayer, A. J. Gawryszczak and A. C. Boley using a wide variety of codes to evolve the same disk.

14.4.4 Unified and hybrid theories

Interesting areas for future investigation are connections between GIs and other disk phenomena, including FU Orionis outbursts, thermal processing of solids, settling and stirring of solids and planetesimal formation. It has been clear since Laughlin and Bodenheimer (1994) that strong inflow due to GIs can occur during the early phase of rapid disk accretion. This has also been seen, under very different assumptions about protostellar collapse, by Vorobyov and Basu (2005, 2006), where bursts of GI activity produce episodic mass transport with $\dot{M} \sim 10^{-4} M_\odot/\text{yr}^{-1}$, comparable to FU Ori levels. Individual strong bursts with similar $\dot{M}$s occur in disks that are cooled to instability from initially stable states (Mejía *et al.* 2005; Boley *et al.* 2006).

Repeating GI bursts (Armitage *et al.* 2001) are possible at later stages of disk evolution if a *dead zone* forms in the ~ 1–10 AU region. Without MRI, mass may accumulate in the dead zone until GIs erupt (Armitage *et al.* 2001; Hartmann *et al.* 2006). Following ideas by Boss and Durisen (2005a,b) and Boley (2007, personal communication) suggest that bursting dead zones may simultaneously trigger FU Ori outbursts in the inner disk, generate strong spiral waves capable of producing chondrules and annealing solids, produce vigorous mixing of gas and solids (Boss 2004b; Boley *et al.* 2005) and accelerate the formation of gas giant planets. High-resolution simulations of bursting dead zones at 4 AU show that FU Ori $\dot{M}$s are possible, but spiral shocks capable of melting chondrule precursors seem to require low enough dust opacity that the disk is on the brink of fragmentation. Such low opacity requires extremely efficient growth and settling of solids in the dead zone. This problem is alleviated by moving the outburst towards 1 AU, as suggested by observations (Zhu *et al.* 2007).

GIs may play an important role in planet formation simply by accelerating the formation of planetesimals, planetary embryos and the solid cores of gas giants.

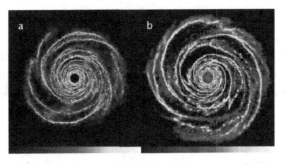

Fig. 14.3. Logarithmic grey scale for the ratio of particle to gas surface density. (a) 50-cm particles without self-gravity. Particles rapidly concentrate into gas-phase spirals by relative factors ~100 (Rice *et al.* 2004). (b) 1.5-m particles with self-gravity. The particles experience GIs and form bound clumps within the spirals (Rice *et al.* 2006).

Because of gas drag, solids in the 10s cm to 1 m size range drift rapidly relative to the gas in the direction of the pressure gradient (Weidenschilling 1977). As shown in Figure 14.3, bodies of this size will be concentrated into the high-pressure regions of GI-induced spiral waves within a few P_{rot} (Rice *et al.* 2004) and could achieve sufficient density to become themselves gravitationally unstable to clumping (Rice *et al.* 2006). GIs can also produce dense gas rings at the boundaries of GI-active and GI-inactive regions (Pickett *et al.* 2003; Pickett & Lim 2004; Durisen *et al.* 2005), where solids may accumulate (Haghighipour & Boss 2003) and promote the growth of gas giant cores. The small-scale physics leading to production of kilometre-sized or larger planetesimals when metre-sized solids gather into high-pressure GI-structures is probably similar to the gravitational and two-stream instabilities expected after midplane settling of solids (Youdin & Shu 2002; Youdin & Goodman 2005; Johansen & Youdin 2007).

Although there might be ways by which gas giant formation by core accretion plus gas capture in the context of a smooth protoplanetary disk seems to work (Lissauer & Stevenson 2007), there are reasons for continued concern, as discussed in Durisen *et al.* (2007). The acceleration of planetesimal and core formation in a GI-active environment could resolve some of these problems. In a similar vein, Klahr & Bodenheimer 2006 show that core accretion plus gas capture is accelerated inside a long-lived vortex. Further exploration of such hybrid scenarios is likely to be fruitful.

14.5 Planet–disk interactions

Once a planet is formed in a protoplanetary disk, it will start to interact gravitationally with the disk. This leads to perturbations in the disk structure and a back-reaction onto the planet, causing planetary orbital migration (see Lin &

Papaloizou 1993; Lin *et al.* 2000; Papaloizou & Terquem 2006; Papaloizou *et al.* 2007 and references therein). The discovery of very short-period extrasolar planets around sun-like stars (Marcy & Butler 1995, 1998; Mayor and Queloz 1995) is supporting evidence that large scale orbital migration occurs in these disks, because it is hard to conceive how such short-period massive planets could have formed in-situ.

14.5.1 The disk response: spiral waves torques and orbital migration

The problem of determining the evolution of a planet orbit involves calculating the disk response to the forcing induced by the gravitational potential of the planet. The gravitational potential ψ of a protoplanet of mass M_p, assumed to be in a circular orbit about a central mass , M_*, with semi-major axis, a, is expressed as a Fourier series of the form

$$\psi(r, \varphi, t) = \sum_{m=0}^{\infty} \psi_m(r) \cos\{m[\varphi - \omega t]\}, \qquad (14.4)$$

where (r, φ) are the usual cylindrical coordinates and ω is the orbital angular velocity of the planet.

An external forcing potential $\psi_m(r, \varphi)$ with azimuthal mode number m and pattern angular velocity, ω, as viewed in an inertial frame, causes density waves to be launched from Lindblad resonances, being disk locations where

$$m(\Omega - \omega) = \pm\sqrt{\kappa^2 + m^2 c_s^2 / r^2}. \qquad (14.5)$$

Here, $\Omega(r)$ is the disk angular velocity, c_s is the local sound speed and for a Keplerian disk, to adequate accuracy, $\kappa \equiv \Omega$ is the epicyclic frequency.

There are two types of Lindblad resonance: the first corresponds to an inner Lindblad resonance located inside the orbit for $\Omega = \omega + \kappa\sqrt{1/m^2 + c_s^2/(r^2\kappa^2)}$ and the second an outer Lindblad resonance located outside the orbit for $\Omega = \omega - \kappa\sqrt{1/m^2 + c_s^2/(r^2\kappa^2)}$. Density waves are launched at Lindblad resonances and subsequently propagate away from the planet. Their travelling wave character results in a spiral form and an angular momentum flux. As a consequence of the latter, a torque acts on the planet. For a Keplerian disk, the torque resulting from waves launched at a Lindblad resonance, through the action of the component of the potential with azimuthal mode number m, can be found in a WKB approximation to be given by

$$\Gamma_m = \frac{\text{sign}(\omega - \Omega)\pi^2 \Sigma}{3\Omega\omega\sqrt{1 + \xi^2}(1 + 4\xi^2)}\Psi^2, \qquad (14.6)$$

with

$$\Psi = r\frac{d\psi_m}{dr} + \frac{2m^2(\Omega - \omega)}{\Omega}\psi_m. \tag{14.7}$$

Here, $\xi = mc_s/(r\Omega)$, and the expression has to be evaluated at the location of the resonance.

The above torque formula was first obtained for a disk with a locally isothermal EOS (Goldreich & Tremaine 1979; Lin & Papaloizou 1979, 1993). It was later generalized to apply for values of ξ of order unity (Artymowicz 1993; Ward 1997; Papaloizou & Larwood 2000).

In a Keplerian disk, waves launched from an outer Lindblad resonance which subsequently move outwards from the planet carry positive angular momentum away, and as a result there is a drag on the planet. On the contrary, waves launched at an inner Lindblad resonance carry negative angular momentum resulting in an acceleration of the planet. The total wave or Lindblad torque must be obtained by summing contributions over m. In addition to this torque, coorbital or corotation torques may be exerted as a result of the disturbance of material close to the planet (see Papaloizou et al. 2007 for a discussion). These are generally a minor effect for smooth surface density profiles and barotropic equations of state.

14.5.2 Type I migration

Type I migration applies to fully embedded planets for which linear perturbation theory may be applied to evaluate the torques. Setting $c_s = H\Omega$, with $H << r$ being a putative semi-thickness of the disk, and $\kappa = \Omega$, one sees that when $m \to \infty$, Lindblad resonances accumulate at

$$r = a \pm \frac{2H}{3}. \tag{14.8}$$

At these locations, the disk flow relative to the planet makes a transition from subsonic to supersonic (Goodman & Rafikov 2001), which is where one indeed expects waves to be launched in a non-rotating homogeneous medium. The important fact that these positions stand off from the location of the planet ensures that when torque contributions are summed over m, they decrease rapidly for $\xi >> 1$, resulting in the so-called torque cutoff.

In most cases, the contribution from the outer resonances is larger resulting in inward migration (Ward 1997). The radius where the flow undergoes the sonic to supersonic transition is more distant from the planet inside the orbit than outside. This makes the outer wake stronger than the inner wake (Figure 14.4) leading to inward migration.

Surface density

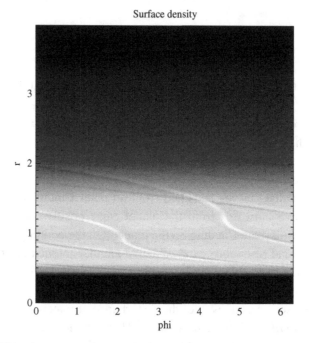

Fig. 14.4. This shows two planets of mass $4M_\oplus$ undergoing type I migration in a region of a 2D disk with uniform surface density profile (the ordinate is dimensionless radius and abscissa is φ). The spiral wakes are clearly visible and pass through each other without significant disturbance. This indicates that the disk response to the protoplanets is in the linear regime. Note that the outer wakes which originate slightly closer to the planet are stronger than the inner wakes which is consistent with stronger torques leading to inward migration.

Linear calculations by Tanaka *et al.* (2002) that take into account 3D effects give

$$\tau_M \equiv a/\dot{a} = (2.7 + 1.1\eta)^{-1} \frac{M_*^2}{M_p \Sigma a^2} h^2 \omega^{-1}, \qquad (14.9)$$

for a surface density profile $\Sigma \propto r^{-\eta}$. For an Earth mass planet around a solar-mass star at $r = 1\,\text{AU}$, in a disk with $\Sigma = 1,700\,\text{g cm}^{-2}$ and $h \equiv H/r = 0.05$, this translates into $\tau_M = 1.6 \cdot 10^5$ years.

This semi-analytic estimate has been verified by means of 3D numerical simulations (Bate *et al.* 2003; D'Angelo *et al.* 2003a). Both find an excellent agreement in the limit of low-mass planets. But note that deviations from the linear estimates (Figure 14.5) set in at a planet mass which varies with the disk model, a phenomenon which is apparently related to the behaviour of the corotation torque. The type I migration timescale is in general significantly shorter than the build-up time of the $M_p \sim 5 - 15 M_\oplus$ solid core of a giant planet (Pollack *et al.* 1996; Papaloizou

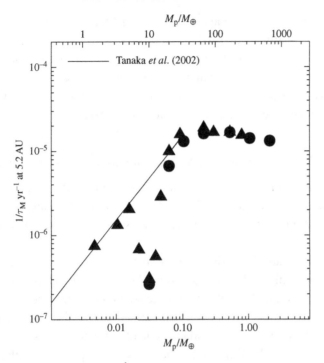

Fig. 14.5. The migration rate, τ_M^{-1}, for different planet mass ratios from 3D fully non-linear nested grid simulations. The symbols denote different approximations (smoothing) for the potential of the planet. The *solid line* refers to linear results for type I migration by Tanaka *et al.* (2002), see Eq. (14.9). The reduction of the migration rate below the linear value for intermediate masses is a non-linear effect related to corotation torques. Adapted from D'Angelo *et al.* (2003a).)

& Nelson 2005). Hence, the existence of type I migration may result in difficulties for the accumulation scenario for such massive cores.

Attempts to resolve this issue note that the inclusion of more detailed physics, such as opacity transitions (Menou & Goodman 2004), the effects of a toroidal magnetic field (Terquem 2003), stochastic migration induced by turbulence (Nelson & Papaloizou 2004; Nelson 2005), the possible effects of a finite orbital eccentricity (Papaloizou & Larwood 2000), vortex production (see Koller *et al.* 2003; Papaloizou *et al.* 2007) or self-shadowing near the planet (Jang-Condell and Sasselov 2005), have led to effective lower estimates of type I migration rates which might help resolve the accretion/migration timescale discrepancy. The above analysis as well as almost all hydrodynamical and MHD simulations adopt a barotropic or locally isothermal EOS without radiation transport. However, hydrodynamic simulations including the latter have been carried out recently by Paardekooper and Mellema (2006). These indicate that type I migration might be reversed for a range of disk optical depths.

An example of two $4M_\oplus$ planets undergoing type I migration in a uniform Σ region of a 2D locally isothermal inviscid disk with $H = 0.05R$ is illustrated in Figure 14.4. These orbit close together and their orbits can undergo resonant locking and joint migration (see Papaloizou & Szuszkiewicz 2005 for details). This may lead to systems of low-mass planets in near resonant orbits just interior to the disk inner boundary (see Terquem & Papaloizou 2007).

14.5.3 Non-linearity, gap formation and type II migration

The physical basis for orbital migration is that the gravitational perturbation from an orbiting planet on the gaseous disk induces frictional torques. Below we give a simplified description of the production of these torques based on a local impulse approximation.

Suppose that a particle of disk matter in circular orbit has a close encounter as it shears past a planet in circular orbit. We make a local approximation such that in the planet frame the particle approaches in a straight line with speed u and impact parameter $x \ll a$ (Lin & Papaloizou 1979, 1993). Interaction with the planet then produces a scattering and an angle of deflection δ. Assumed to be small, this is given by $\delta = 2GM_p/(xu^2)$. The associated specific angular momentum transferred is

$$\Delta j = au(1-\cos\delta) \sim \frac{au\delta^2}{2} = \frac{2aG^2M_p^2}{x^2u^3}. \tag{14.10}$$

This transfer is directed along the original direction of motion as seen by the planet. It is frictional such that in a Keplerian disk, interior matter accelerates the planet, while being torqued down, and exterior matter decelerates the planet, while being accelerated. Note that to adequate approximation $|u| = |a(\Omega - \omega)| \sim 3\omega x/2$.

The fact that the scattering occurs in a rotating frame requires the application of a correction factor (see, e.g., Lin & Papaloizou 1993) $C = 4\left[2K_0(2/3)+ K_1(2/3)\right]^2/9$, where K_0 and K_1 denote modified Bessel functions. Thus,

$$\Delta j = C\frac{2aG^2M_p^2}{x^2u^3}. \tag{14.11}$$

The associated rate of exchange of angular momentum with one side of the disk is

$$\frac{dJ}{dt} = \int_{x_{min}}^{\infty} a\Sigma\Delta j|\Omega - \omega|dx = \frac{8G^2M_p^2a\Sigma C}{27\omega^2x_{min}^3}. \tag{14.12}$$

Here, x_{min} is the minimum distance between disk matter and the protoplanet orbit that should be considered.

Because the interaction is repulsive, a strong enough perturbation will cause a gap to open. In a disk with small viscosity, ν, gap formation starts to occur once the

Hill radius, $r_H = a(M_p/M_*)^{1/3}$, exceeds the disk scale height (the non-linearity condition, see Papaloizou *et al.* 2004, 2007 for more details and references).

In addition, for a gap to be maintained in the presence of viscosity, the angular momentum exchange rate given by Eq. (14.12) should exceed the angular momentum flow rate due to viscosity

$$\left(\frac{dJ}{dt}\right)_{\text{visc}} = 3\pi \nu \Sigma a^2 \omega. \tag{14.13}$$

We note that ν is not a real 'viscosity' but rather a way of representing the flow of angular momentum resulting from turbulent stresses. In practice, time averaging over long periods is required to specify it (Papaloizou & Nelson 2003). Use of Eq. (14.12) gives a condition for gap formation

$$\frac{8M_p^2 a^3 C}{81\pi M_*^2 x_{\min}^3} > \frac{\nu}{a^2 \omega}. \tag{14.14}$$

As a gap is expected to extend beyond a Hill radius from the planet, $x_{\min} \geq r_H$. Taking $x_{\min} = 2r_H$, Eq. (14.14) gives

$$\frac{3M_p C}{81\pi M_*} > \frac{\nu}{a^2 \omega}. \tag{14.15}$$

Taking $C = 81\pi/120$, we obtain

$$\frac{M_p}{M_*} > \frac{40\nu}{a^2 \omega} \tag{14.16}$$

(Lin & Papaloizou 1986, 1993). For more recent discussions and numerical investigations of gap-opening criteria, see Bryden *et al.* (1999) and also Crida *et al.* (2006).

For a typical disk model, $\nu/(a^2\Omega) \sim 10^{-5}$, such that gap formation is expected for planets slightly above the mass of Saturn. Note that for this mass $r_H \sim H$, marginally satisfying the non-linearity condition. Under these conditions, the migration rate is a maximum at this mass (see also D'Angelo *et al.* 2003a). The corresponding minimum time can be found from Eq. (14.9) by setting $r_H = H$ as

$$\tau_r = \frac{M_*^{4/3}}{2.7 \times 3^{2/3} M_p^{1/3} \Sigma a^2 \omega}. \tag{14.17}$$

For $\Sigma = 200\,\text{g cm}^{-2}$ at 5.2 AU, this gives $\sim 10^4$ years. This simply obtained estimate is in reasonable agreement with the results presented in Figure 14.5. However, we note that this migration time may be even shorter in more massive disks because of the action of coorbital torques, which under some circumstances may produce a runaway (Masset & Papaloizou 2003).

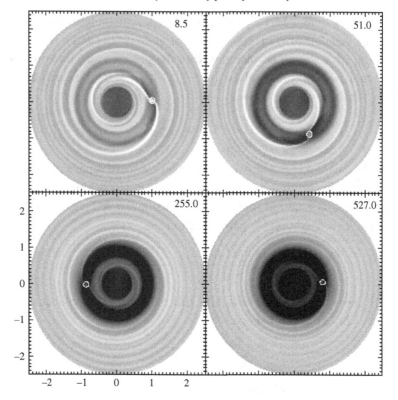

Fig. 14.6. This shows the development of a gap after a Jupiter mass planet is inserted in a disk with $h = 0.05$ and $\nu = 10^{-5}a^2\omega$. The number of orbital periods after insertion is indicated in the upper right corner of each panel. Note the initially strong spiral waves and the development of an inner cavity, which have been found in extensive calculations of this kind (Lubow & D'Angelo 2006).

For larger masses, a gap opens and migration becomes driven by the angular momentum transport in the disk (Lin & Papaloizou 1986; Nelson *et al.* 2000b). When the planet is less massive than the disk mass contained within its orbital length scale, it evolves as a free disk particle, migrating on the viscous timescale of typically $\sim 10^5$ years. But the evolution slows when the planet mass dominates the disk (Ivanov *et al.* 1999). High-resolution numerical computations allow for a detailed computation of the torque exerted by the disk material on the planet, its mass accretion rate and orbital evolution. When the orbital evolution of the planet is followed (Nelson *et al.* 2000b), results are in very good agreement with the estimates obtained from the models using a fixed planet. These simulations show that during inward migration they may grow up to about $5M_{\rm Jup}$.

The thermodynamic effects (viscous heating and radiative cooling) on the gap formation process and type II migration have been studied by D'Angelo *et al.*

(2003b), the effect of self-gravity by Nelson and Benz (2003) and the effect of MHD turbulence producing the effective viscosity by Nelson and Papaloizou (2003) and Winters *et al.* (2003). For expected protostellar disk parameters, these phenomena are robust.

14.5.4 Consequences for evolution in young planetary systems

A number of studies with the object of explaining the existence and distribution of giant planets interior to the 'snow line' at 2 AU, which make use of type II migration, have been performed (see, e.g., Trilling *et al.* 1998, 2002; Armitage *et al.* 2002; Alibert *et al.* 2004). These assume formation beyond the snow line followed by inward type II migration that is stopped by one of the following: disk dispersal, Roche lobe overflow, stellar tides or entering a stellar magneto-spheric cavity interior to which the disk planet interaction is expected to be negligible (Papaloizou 2007). Reasonable agreement with observations is attained. But only if type I migration is suppressed significantly, possibly as a result of one of the effects discussed above.

14.6 Outlook

As we have shown in this chapter, protoplanetary disks are being modelled in various ways, depending on the goals one wishes to reach. For understanding the spectra and images of these disks, typically smooth parameterized disk models are used as input to sophisticated multi-dimensional radiative transfer methods. On the contrary, for studying the hydrodynamics of disks for planet formation and disk–planet interaction, sophisticated multi-dimensional hydrodynamics codes are used, but with simplified radiative transfer. It is evident that one of the goals for next generation models will be to combine the best of both worlds. We will then have hope to have more definite answers to questions such as whether gas giant planets can form from GIs, exactly how planetary migration works and how such models compare to modern protoplanetary disk observations.

Acknowledgements

RHD's research is supported by NASA grant NNG05GN11G. During part of this work, RHD enjoyed the hospitality of the Max Planck Institute for Astronomy in Heidelberg. R.H.D. would also like to thank two of his students: S. Michael, for substantive help with figures, and A.C. Boley, for commenting on an early draft.

References

Adams, F. C., Ruden, S. P., and Shu, F. H. (1989), *ApJ.* **347**, 959–976

Adams, F. C. and Watkins, R. (1995), *ApJ.* **451**, 314–327

Alexander, R. D. and Armitage, P. J. (2006), *ApJ. Lett.* **639**, L83–L86

Alexander, R. D., Clarke, C. J., and Pringle, J. E. (2006a), *MNRAS* **369**, 216–228

Alexander, R. D., Clarke, C. J., and Pringle, J. E. (2006b), *MNRAS* **369**, 229–239

Alibert, Y., Mordasini, C., and Benz, W. (2004), *A&A* **417**, L25–L28

Armitage, P. J., Livio, M., Lubow, S. H., and Pringle, J. E. (2002), *MNRAS* **334**, 248–256

Armitage, P. J., Livio, M., and Pringle, J. E. (2001), *MNRAS* **324**, 705–711

Artymowicz, P. (1993), *ApJ.* **419**, 155–165

Balbus, S. A. (2000), *ApJ.* **534**, 420–427

Balbus, S. A. (2003), *A R A&A* **41**, 555–597

Balbus, S. A. and Papaloizou, J. C. B. (1999), *ApJ.* **521**, 650–658

Banerjee, R., Pudritz, R. E., and Holmes, L. (2004), *MNRAS* **355**, 248–272

Barranco, J. A. and Marcus, P. S. (2005), *ApJ.* **623**, 1157–1170

Bate, M. R., Lubow, S. H., Ogilvie, G. I., and Miller, K. A. (2003), *MNRAS* **341**, 213–229

Bodenheimer, P. (1995), *A R A&A* **33**, 199–238

Boley, A. C. and Durisen, R. H. (2006), *ApJ.* **641**, 534–546

Boley, A. C., Durisen, R. H., Nordlund, A., and Lord, J. (2007a), *ApJ.* **665**, 1254–1267

Boley, A. C., Durisen, R. H., and Pickett, M. K. (2005), in *Chondrites and the Protoplanetary Disk*, eds. A. N. Krot, E. R. D. Scott, and B. Reipurth, Volume 341 of Astronomical Society of the Pacific Conference Series, pp. 839–848

Boley, A. C., Hartquist, T. W., Durisen, R. H., and Michael, S. (2007b), *ApJ.* **656**, L89–L92

Boley, A. C., Mejía, A. C., Durisen, R. H., Cai, K., Pickett, M. K., and D'Alessio, P. (2006), *ApJ.* **651**, 517–534

Boss, A. P. (1997), *Science* **276**, 1836–1839

Boss, A. P. (2000), *ApJ. Lett.* **536**, L101–L104

Boss, A. P. (2001), *ApJ.* **563**, 367–373

Boss, A. P. (2002a), *ApJ. Lett.* **567**, L149–L153

Boss, A. P. (2002b), *ApJ.* **576**, 462–472

Boss, A. P. (2004a), *ApJ.* **610**, 456–463

Boss, A. P. (2004b), *ApJ.* **616**, 1265–1277

Boss, A. P. (2007), *ApJ. Lett.* **661**, L73–L76

Boss, A. P. and Durisen, R. H. (2005a), *ApJ. Lett.* **621**, L137–L140

Boss, A. P. and Durisen, R. H. (2005b), in *Chondrites and the Protoplanetary Disk*, eds. A. N. Krot, E. R. D. Scott, and B. Reipurth. San Francisco: ASP, pp. 821–838

Bouwman, J., de Koter, A., Dominik, C., and Waters, L. B. F. M. (2003), *A&A* **401**, 577–592

Bryden, G., Chen, X., Lin, D. N. C., Nelson, R. P., and Papaloizou, J. C. B. (1999), *ApJ.* **514**, 344–367

Cabot, W. (1996), *ApJ.* **465**, 874–886

Cai, K., Durisen, R. H., Boley, A. C., Pickett, M. K., and Mejía, A. C. (2008), *ApJ.*, **673**, 1138C

Cai, K., Durisen, R. H., Michael, S., Boley, A. C., Mejía, A. C., Pickett, M. K., and D'Alessio, P. (2006), *ApJ. Lett.* **636**, L149–L152

Calvet, N., D'Alessio, P., Hartmann, L., Wilner, D., Walsh, A., and Sitko, M. (2002), *ApJ.* **568**, 1008–1016

Calvet, N., Patino, A., Magris, G. C., and D'Alessio, P. (1991), *ApJ.* **380**, 617–630

Carpenter, J. M., Wolf, S., Schreyer, K., Launhardt, R., and Henning, T. (2005), *AJ* **129**, 1049–1062

Cassen, P. and Moosman, A. (1981), *Icarus* **48**, 353–376

Chiang, E. I. and Goldreich, P. (1997), *ApJ.* **490**, 368

Clarke, C. J., Gendrin, A., and Sotomayor, M. (2001), *MNRAS* **328**, 485–491

Crida, A., Morbidelli, A., and Masset F. S. (2006), *Icarus*, **181**, 587–604

D'Alessio, P., Calvet, N., Hartmann, L., Franco-Hernández, R., and Servín, H. (2006), *ApJ.* **638**, 314–335

D'Alessio, P., Canto, J., Calvet, N., and Lizano, S. (1998), *ApJ.* **500**, 411–427

D'Alessio, P., Hartmann, L., Calvet, N., Franco-Hernández, R., Forrest, W. J., *et al.* (2005), *ApJ.* **621**, 461–472

D'Angelo, G., Kley, W., and Henning, T. (2003a), *ApJ.* **586**, 540–561

D'Angelo, G., Kley W., and Henning T. (2003b), *ApJ.* **599**, 548–576

Davis, S. S. (2007), *ApJ.* **660**, 1580–1587

Desch, S. J. (2004), *ApJ.* **608**, 509–525

Dullemond, C. P. (2002), *A&A* **395**, 853–862

Dullemond, C. P. and Dominik, C. (2004), *A&A* **417**, 159–168

Dullemond, C. P. and Dominik, C. (2005), *A&A* **434**, 971–986

Dullemond, C. P., Natta, A., and Testi, L. (2006), *ApJ. Lett.* **645**, L69–L72

Durisen, R. H., Boss, A. P., Mayer, L., Nelson, A. F., Quinn, T., and Rice, W. K. M. (2007), in *Protostars and Planets V*, eds. B. Reipurth, D. Jewitt, and K. Keil, pp. 607–622

Durisen, R. H., Cai, K., Boley, A., Mejía, A. C., and Pickett, M. K. (2005), *Icarus* **173**, 417–424

Fleming, T. and Stone, J. M. (2003), *ApJ.* **585**, 908–920

Fromang, S. (2005), *A&A* **441**, 1–8

Gammie, C. F. (1996), *ApJ.* **457**, 355–362

Gammie, C. F. (2001), *ApJ.* **553**, 174–183

Gammie, C. F. and Johnson, B. M. (2005), in *Chondrites and the Protoplanetary Disk*, eds. A. N. Krot, E. R. D. Scott, and B. Reipurth. San Francisco: ASP, 145–164

Goodman, J. and Rafikov, R. R. (2001), *ApJ.* **552**, 793–802

Goldreich, P. and Lynden-Bell, D. (1965), *MNRAS* **130**, 125–158

Goldreich, P. and Tremaine, S. (1979), *ApJ.*, **233**, 857–871

Haghighipour, N. and Boss, A. P. (2003), *ApJ.* **598**, 1301–1311

Haisch, K. E., Lada, E. A., and Lada, C. J. (2001), *ApJ. Lett.* **553**, L153–L156

Hartmann, L., Calvet, N., Gullbring, E., and D'Alessio, P. (1998), *ApJ.* **495**, 385–400

Hartmann, L., D'Alessio, P., Calvet, N., and Muzerolle, J. (2006), *ApJ.* **648**, 484–490

Hawley, J. F., Balbus, S. A., and Winters, W. F. (1999), *ApJ.* **518**, 394–404

Hollenbach, D., Johnstone, D., Lizano, S., and Shu, F. (1994), *ApJ.* **428**, 654–669

Hueso, R. and Guillot, T. (2005), *A&A* **442**, 703–725

Ilgner, M. and Nelson, R. P. (2006), *A&A* **445**, 205–222

Ivanov, P. B., Papaloizou, J. C. B., and Polnarev, A. G. (1999), *MNRAS* **307**, 79–90

Jang-Condell, H. and Sasselov, D. D. (2005), *ApJ.* **619**, 1123–1131

Ji, H., Burin, M., Schartman, E., and Goodman, J. (2006), *Nature* **444**, 343–346

Johansen, A. and Youdin, A. (2007), *ApJ.* **662**, 627–641

Johnson, B. M. and Gammie, C. F. (2003), *ApJ.* **597**, 131–141

Johnson, B. M. and Gammie, C. F. (2005), *ApJ.* **635**, 149–156

Johnson, B. M. and Gammie, C. F. (2006), *ApJ.* **636**, 63–74

Jonkheid, B., Faas, F. G. A., van Zadelhoff, G.-J., and van Dishoeck, E. F. (2004), *A&A* **428**, 511–521

Kamp, I. and Dullemond, C. P. (2004), *ApJ.* **615**, 991–999

Kenyon, S. J. and Hartmann, L. (1987), *ApJ.* **323**, 714–733

Klahr, H. and Bodenheimer, P. (2006), *ApJ.* **639**, 432–440

Klahr, H. H. and Bodenheimer, P. (2003), *ApJ.* **582**, 869–892

Koller, J., Li, H., and Lin, D. N. C. (2003), *ApJ.* **596**, L91–L94

Kratter, K. M. and Matzner, C. D. (2006), *MNRAS* **373**, 1563–1576

Krumholz, M. R., Klein, R. I., and McKee, C. F. (2007), *ApJ.* **656**, 959–979

Lachaume, R. (2004), *A&A* **422**, 171–176

Larson, R. B. (1984), *MNRAS* **206**, 197–207

Laughlin, G. and Bodenheimer, P. (1994), *ApJ.* **436**, 335–354

Laughlin, G., Korchagin, V., and Adams, F. C. (1998), *ApJ.* **504**, 945–966

Laughlin, G. and Rozyczka, M. (1996), *ApJ.* **456**, 279–291

Lesur, G. and Longaretti, P.-Y. (2005), *A&A* **444**, 25–44

Li, H., Colgate, S. A., Wendroff, B., and Liska, R. (2001), *ApJ.* **551**, 874–896

Lin, D. N. C. and Papaloizou, J. C. B. (1979), *MNRAS* **186**, 799–830

Lin, D. N. C. and Papaloizou, J. C. B. (1986), *ApJ.* **309**, 846–857

Lin, D. N. C. and Papaloizou, J. C. B. (1993), in *Protostars and Planets III*, eds. E. H. Levy and J. I. Lunine. Tucson: University of Arizona Press, pp. 749–835

Lin, D. N. C., Papaloizou, J. C. B., Terquem, C., Bryden, G., and Ida, S. (2000), in *Protostars and Planets IV*, eds. V. Mannings *et al.* Tucson: University of Arizona Press, pp. 1111–1134

Lissauer, J. J. and Stevenson, D. J. (2007), in *Protostars and Planets V*, eds. B. Reipurth, D. Jewitt, and K. Keil. Tucson: University of Arizona Press, pp. 591–606

Lodato, G. and Rice, W. K. M. (2004), *MNRAS* **351**, 630–642

Lodato, G. and Rice, W. K. M. (2005), *MNRAS* **358**, 1489–1500

Lovelace, R. V. E., Li, H., Colgate, S. A., and Nelson, A. F. (1999), *ApJ.* **513**, 805–810

Lubow, S. H. and D'Angelo, G. (2006), *ApJ.* **641**, 526–533

Lynden-Bell, D. and Pringle, J. E. (1974), *MNRAS* **168**, 603–637

Marcy, G. W. and Butler, R. P. (1995), *187th AAS Meeting BAAS.* **27**, 1379–1384

Marcy, G. W. and Butler, R. P. (1998), *A R A&A* **36**, 57–97

Marcus, P. S. (1993), *A R A&A* **31**, 523–573

Martos, M. A. and Cox, D. P. (1998), *ApJ.* **509**, 703–716

Masset, F. and Papaloizou, J. C. B. (2003), *ApJ.* **588**, 494–508

Matsumoto, T. and Hanawa, T. (2003), *ApJ.* **595**, 913–934

Matzner, C. D. and Levin, Y. (2005), *ApJ.* **628**, 817–831

Mayer, L., Lufkin, G., Quinn, T., and Wadsley, J. (2007), *ApJ. Lett.* **661**, L77–L80

Mayer, L., Quinn, T., Wadsley, J., and Stadel, J. (2002), *Science* **298**, 1756–1759

Mayer, L., Quinn, T., Wadsley, J., and Stadel, J. (2004), *ApJ.* **609**, 1045–1064

Mayor, M. and Queloz, D. (1995), *Nature* **378**, 355–359

Mejía, A. C., Durisen, R. H., Pickett, M. K., and Cai, K. (2005), *ApJ.* **619**, 1098–1113

Menou, K. and Goodman, J. (2004), *ApJ.* **606**, 520–531

Muzerolle, J., Hillenbrand, L., Calvet, N., Briceño, C., and Hartmann, L. (2003), *ApJ.* **592**, 266–281

Nakamoto, T. and Nakagawa, Y. (1994), *ApJ.* **421**, 640–650

Nelson, A. F. (2006), *MNRAS* **373**, 1039–1073

Nelson, A. F. and Benz, W. (2003), *ApJ.* **589**, 578–604

Nelson, A. F., Benz, W., Adams, F. C., and Arnett, D. (1998), *ApJ.* **502**, 342–371

Nelson, A. F., Benz, W., and Ruzmaikina, T. V. (2000a), *ApJ.* **529**, 357–390

Nelson, R. P. (2005), *A& A* **443**, 1067–1085

Nelson, R. P. and Papaloizou, J. C. B. (2003), *MNRAS* **339**, 993–1005

Nelson, R. P. and Papaloizou, J. C. B. (2004), *MNRAS* **350**, 849–864

Nelson, R. P., Papaloizou, J. C. B., Masset, F., and Kley, W. (2000b), *MNRAS* **318**, 18–36

Nomura, H. (2002), *ApJ.* **567**, 587–595

Nomura, H., Aikawa, Y., Tsujimoto, M., Nakagawa, Y., and Millar, T. J. (2007), *ApJ.* **661**, 334–353

Nomura, H. and Millar, T. J. (2005), *A&A* **438**, 923–938

Nomura, H. and Nakagawa, Y. (2006), *ApJ.* **640**, 1099–1109

Osorio, M., D'Alessio, P., Muzerolle, J., Calvet, N., and Hartmann, L. (2003), *ApJ.* **586**, 1148–1161

Paardekooper, S.-J. and Mellema, G. (2006), *A& A*, **459**, L17–L20

Papaloizou, J. C. B. (2007), *A& A* **463**, 775–781

Papaloizou, J. C. B. and Larwood, J. D. (2000), *MNRAS* **315**, 823–833

Papaloizou, J. C. B. and Lin, D. N. C. (1989), *ApJ.* **344**, 645–668

Papaloizou, J. C. B. and Lin, D. N. C. (1995), *A R A&A* **33**, 505–540

Papaloizou, J. C. B. and Nelson, R. P. (2003), *MNRAS* **339**, 983–992

Papaloizou, J. C. B. and Nelson, R. P. (2005), *A& A* **433**, 247–265

Papaloizou, J. C. B., Nelson, R. P., Kley, W., Masset, F. S., and Artymowicz, P. (2007), in *Protostars and Planets* V, eds. V. P. Reipurth , D. Jewitt, and K. Keil. Tucson: University of Arizona, pp. 655–668

Papaloizou, J. C. B., Nelson, R. P., and Snellgrove, M. D. (2004), *MNRAS* **350**, 829–848

Papaloizou, J. C. B. and Pringle, J. E. (1987), *MNRAS* **225**, 267–283

Papaloizou, J. C. B. and Szuszkiewicz, E. (2005), *MNRAS* **363**, 153–176

Papaloizou, J. C. B. and Terquem, C. (2006), *Rep. Prog. Phys.* **69**, 119–180

Petersen, M. R., Julien, K., and Stewart, G. R. (2007a), *ApJ.* **658**, 1236–1251

Petersen, M. R., Stewart, G. R., and Julien, K. (2007b), *ApJ.* **658**, 1252–1263

Pickett, B. K., Cassen, P., Durisen, R. H., and Link, R. (1998), *ApJ.* **504**, 468–491

Pickett, B. K., Cassen, P., Durisen, R. H., and Link, R. (2000), *ApJ.* **529**, 1034–1053

Pickett, B. K., Mejía, A. C., Durisen, R. H., Cassen, P. M., Berry, D. K., and Link, R. P. (2003), *ApJ.* **590**, 1060–1080

Pickett, M. K. and Durisen, R. H. (2007), *ApJ. Lett.* **654**, L155–L158

Pickett, M. K. and Lim, A. J. (2004), *A G* **45**, 1.12–1.17

Pinte, C., Fouchet, L., Ménard, F., Gonzalez, J.-F., and Duchêne, G. (2007), *A&A* **469**, 963–971

Pollack, J. B., Hollenbach, D., Beckwith, S., Simonelli, D. P., Roush, T., and Fong, W. (1994), *ApJ.* **421**, 615–639

Pollack, J. B., Hubickyj, O., Bodenheimer, P., Lissauer, J., Podolak, M. and Greenzweig, Y. (1996), *Icarus* **124**, 62–85

Pontoppidan, K. M., Dullemond, C. P., Blake, G. A., Boogert, A. C. A., van Dishoeck, E. F., Evans, II, N. J., Kessler-Silacci, J., and Lahuis, F. (2007a), *ApJ.* **656**, 980–990

Pontoppidan, K. M., Dullemond, C. P., Blake, G. A., Evans, II, N. J., Geers, V. C., Harvey, P. M., and Spiesman, W. (2007b), *ApJ.* **656**, 991–1000

Pringle, J. E. (1981), *A R A&A* **19**, 137–162

Rafikov, R. R. (2005), *ApJ. Lett.* **621**, L69–L72

Rafikov, R. R. (2007), *ApJ.* **662**, 642–650

Rice, W. K. M., Armitage, P. J., Bate, M. R., and Bonnell, I. A. (2003), *MNRAS* **339**, 1025–1030

Rice, W. K. M., Lodato, G., and Armitage, P. J. (2005), *MNRAS* **364**, L56–L60

Rice, W. K. M., Lodato, G., Pringle, J. E., Armitage, P. J., and Bonnell, I. A. (2004), *MNRAS* **355**, 543–552

Rice, W. K. M., Lodato, G., Pringle, J. E., Armitage, P. J., and Bonnell, I. A. (2006), *MNRAS* **372**, L9–L13

Robitaille, T. P., Whitney, B. A., Indebetouw, R., Wood, K., and Denzmore, P. (2006), *ApJ.s* **167**, 256–285

Sano, T., Miyama, S. M., Umebayashi, T., and Nakano, T. (2000), *ApJ.* **543**, 486–501

Scholz, A., Jayawardhana, R., Wood, K., Meeus, G., Stelzer, B., Walker, C., and O'Sullivan, M. (2007), *ApJ.* **660**, 1517–1531

Semenov, D., Wiebe, D., and Henning, T. (2004), *A&A* **417**, 93–106

Shakura, N. I. and Sunyaev, R. A. (1973), *A&A* **24**, 337–355

Shu, F. H., Tremaine, S., Adams, F. C., and Ruden, S. P. (1990), *ApJ.* **358**, 495–514

Sicilia-Aguilar, A., Hartmann, L., Calvet, N., Megeath, S. T., Muzerolle, J., *et al.* (2006), *ApJ.* **638**, 897–919

Stamatellos, D. and Whitworth, A. P. (2008), *A&A*, **480**, 879–887

Stone, J. M. and Balbus, S. A. (1996), *ApJ.* **464**, 364–372

Stone, J. M., Hawley, J. F., Gammie, C. F., and Balbus, S. A. (1996), *ApJ.* **463**, 656–673

Tanaka, H., Himeno, Y., and Ida, S. (2005), *ApJ.* **625**, 414–426

Tanaka, H., Takeuchi, T., and Ward, W. R. (2002), *ApJ.* **565**, 1257–1274

Terebey, S., Shu, F. H., and Cassen, P. (1984), *ApJ.* **286**, 529–551

Terquem, C. E. J. M. L. J. (2003), *MNRAS* **341**, 1157–1173

Terquem, C. and Papaloizou, J. C. B. (2007), *ApJ.* **654**, 1110–1120

Tomley, L., Cassen, P., and Steiman-Cameron, T. (1991), *ApJ.* **382**, 530–543

Toomre, A. (1981), in *The Structure and Evolution of Normal Galaxies*. Cambridge: Cambridge University Press, pp. 111–136

Trilling, D. E., Benz, W., Guillot, T., Lunine, J. I., Hubbard, W. B., and Burrows, A. (1998), *ApJ.* **500**, 428–439

Trilling, D. E., Lunine, J. I., and Benz, W. (2002), *A&A* **394**, 241–251

Truelove, J. K., Klein, R. I., McKee, C. F., Holliman, II, J. H., Howell, L. H., and Greenough, J. A. (1997), *ApJ. Lett.* **489**, L179–L182

Truelove, J. K., Klein, R. I., McKee, C. F., Holliman, II, J. H., Howell, L. H., Greenough, J. A., and Woods, D. T. (1998), *ApJ.* **495**, 821–852

Tscharnuter, W. M. (1987), *A&A* **188**, 55–73

Ulrich, R. K. (1976), *ApJ.* **210**, 377–391

Visser, R., Geers, V. C., Dullemond, C. P., Augereau, J.-C., Pontoppidan, K. M., and van Dishoeck, E. F. (2007), *A&A* **466**, 229–241

Vorobyov, E. I. and Basu, S. (2005), *ApJ. Lett.* **633**, L137–L140

Vorobyov, E. I. and Basu, S. (2006), *ApJ.* **650**, 956–969

Walker, C., Wood, K., Lada, C. J., Robitaille, T., Bjorkman, J. E., and Whitney, B. (2004), *MNRAS* **351**, 607–616

Ward, W. R. (1997), *Icarus* **126**, 261–281

Watson, A.M., Stapelfeldt, K.R., Wood, K., and Ménard, F. (2007), in *Protostars and Planets V*, eds. V. P. Reipurth, D. Jewitt and K. Keil. Tucson: University of Arizona, pp. 655–668

Weidenschilling, S. J. (1977), *MNRAS* **180**, 57–70

Winters, W. F., Balbus, S. A., and Hawley, J. F. (2003), *ApJ.* **589**, 543–555

Wolf, S., Padgett, D. L., and Stapelfeldt, K. R. (2003), *ApJ.* **588**, 373–386

Yorke, H. W., Bodenheimer, P., and Laughlin, G. (1993), *ApJ.* **411**, 274–284

Youdin, A. N. and Goodman, J. (2005), *ApJ.* **620**, 459–469

Youdin, A. N. and Shu, F. H. (2002), *ApJ.* **580**, 494–505

Zhu, Z., Hartmann, L., Calvet, N., Hernandez, J., Muzerolle, J., and Tannirkulam, A.-K. (2007), *ApJ.*, **669**, 483Z

15

Planet formation and evolution:
theory and observations

Y. Alibert, I. Baraffe, W. Benz, G. Laughlin and S. Udry

15.1 Introduction

This chapter is devoted to planet formation and to the early stages of evolution of low-mass objects, including low-mass stars, brown dwarfs and exoplanets. We first summarize the general properties of current exoplanet observations (Section 15.2) and describe the two main planet formation models based on disk instability and on the core-accretion scenario, respectively (Section 15.3). Recent progress of the latter formation model allows sophisticated population synthesis analyses which provide fully quantitative predictions that can be compared to the observed statistical properties of exoplanets (Section 15.3.5). The last part of this chapter is devoted to the distinction between brown dwarfs and planets, in terms of structure and evolutionary properties. The existence of a mass overlap between these two distinct populations of low-mass objects is highlighted by the increasing discoveries of very massive exoplanets ($M \gtrsim 5M_{\rm J}$) and by the identification of planetary mass brown dwarfs in young clusters ($M \lesssim 10M_{\rm J}$). These discoveries stress the importance to define signatures which could allow to disentangle a brown dwarf from a planet. We first analyse the effect of accretion on the evolution of young brown dwarfs and the resulting uncertainties of evolutionary models at ages of a few million years. We also analyse different specific signatures of brown dwarfs and planets such as their luminosity at young ages, their radii and their atmospheric properties.

15.2 Observations

15.2.1 General overview

The discovery of an extra-solar planet orbiting the solar-type star 51 Peg (Mayor & Queloz 1995) in 1995 has encouraged the launch of numerous new search programmes, leading now to a steadily increasing number of exoplanet detections.

Structure Formation in Astrophysics. ed. G. Chabrier. Published by Cambridge University Press.

Since the first discovery, more than 260 other planetary companions have been found to orbit stars of spectral types from F to M. From this sample, we have learned that giant planets are common and that the planetary formation process may produce an unexpectedly large variety of configurations covering a wide range of planetary masses, orbital shapes and planet–star separations (Figure 15.1). In particular, some massive planets are found very close to their parent stars (a few solar radii). Not very surprisingly, planets also appear in families around stars or orbit components of multiple-star systems.

The very large majority of these objects have been found through the induced Doppler spectroscopic variations of the primary star (the so-called *radial-velocity* (RV) technique). Most of the candidates are giant gaseous planets similar in nature to Jupiter. The past few years have, however, known a new step forward with the detections of lighter ($5–10 M_\oplus$), mainly 'solid' planets (see the latest announcements in Lovis *et al.* 2006; Melo *et al.* 2007; Bonfils *et al.* 2007; Udry *et al.* 2007). These detections were made possible (i) by the development of a new generation of stable and more precise instruments for RV measurements as, e.g., the ESO HARPS spectrograph designed for high-precision planet search and (ii) by the application of a careful observing strategy to reduce at maximum intrinsic stellar RV jitter, possibly hiding the tiny planet signal.

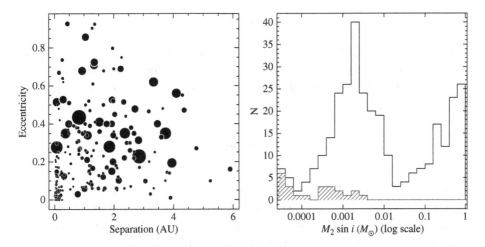

Fig. 15.1. *Left*: eccentricity-separation distribution of the known planetary systems. The size of the dots scales with planet masses. *Right*: mass function of companions to solar-type stars from binaries down to the planet domain (note that the binary and planet populations do not have the same normalization). The lack of companions between planets and binaries (the brown-dwarf desert) is very clear. The left-hand side of the diagram is still observationally strongly biased (the filled histogram indicates the position of HARPS discoveries). Nevertheless, a distinct population of very low-mass planets starts to emerge.

Because of the nature of the method, only the orbital parameters and a lower limit on the planet mass are known from radial velocities alone. Much tighter constraints for planet models are obtained by the observation of a photometric transit of the planet in front of its parent star. Combined with RV measurements, such observations yield the exact mass, radius and mean density of the planet's, providing priceless constraints for the planet's internal structure.

The regularly increasing number of known extra-solar planets lends some confidence to observed features in statistical distributions of the planet and primary star properties. These features are thought to have kept traces of the processes acting during the formation and evolution of the planet systems and thus can help to constrain planet formation models. Understanding the physical reasons for the wide variety in planet-observed characteristics remains a central issue in planet formation theory. Answers to the questions raised by the new discoveries will probably benefit from constraints provided by the statistical distributions of planetary properties. It will in particular help us to point out differences in the formation and evolution processes for gaseous giants and lower-mass rocky/icy planets. In the following sections, we will briefly review the knowledge gained during the past decade from the observation of exoplanet systems. A more detailed description is available in Udry and Santos (2007).

15.2.2 Constraints from statistical results

Along the suite of planet discoveries, our understanding of planetary formation had to integrate several new peculiar characteristics, leading us to continuously re-examine the statistical properties of the derived orbital elements and stellar-host characteristics, in search for constraints to formation and evolution scenarios.

15.2.2.1 General features

15.2.2.1.1 Planets in numbers. A first information available from a planet-search programme is the fraction of stars that host detected planets. For planets more massive than $0.5M_J$ and closer than 4–5 AU, the present surveys find that a minimum average of 6–7% of dwarf stars in the solar neighbourhood harbour giant planets, about 1/6 of which are very close to their parent stars (hot Jupiters).

15.2.2.1.2 Multi-planet systems. An increasing number (25) of multi-planet systems are detected. Some include up to four or five planets (μ Ara, Pepe *et al.* 2007; 55 Cnc, Fischer *et al.* 2008). They represent about 12% of the planet-bearing stars. Some of the multi-planet systems are hierarchically organized with well-separated planetary orbits, while others are resonant systems, with commensurable period ratios. Low-order resonant systems are especially important because planet–planet gravitational interactions are noticeably influencing the system evolution on observable timescale, giving access in the most favourable cases to the orbital-plane inclinations, not otherwise known from the RV technique. Rivera *et al.* (2005) used such an approach for the two-planet system Gl 876 ($P_b = 60.9$ d and $P_c = 30.1$ d) to derive an orbital-plane inclination of $\sim 50°$ and then to detect in the residuals of the model fit the signature of the presence of an inner very light planet of only $\sim 7.5 M_\oplus$. Another useful application of the dynamical analysis of a multi-planet system is the localization of the resonances in the system that shape its overall structure. Stability studies are mandatory as well to ensure the long-term viability of the systems observed now.

15.2.2.1.3 Planets in binaries. Among the extra-solar planets discovered to date by radial velocities, more than 30 are known to orbit one of the members of a double or multiple-star system (see, e.g., Eggenberger *et al.* 2004, 2007, and references therein). Although the sample is not large, some hints for differences between planets in binaries and around single stars are observed. For example, the large majority of the most massive short-period planets are all found in binary or multiple-star systems (Zucker & Mazeh 2002).

15.2.2.2 Planetary mass distribution

15.2.2.2.1 Giant gaseous planets. The strong bimodal aspect of the secondary-mass distribution to solar-type primaries (from stellar companions down to the planetary domain) has generally been considered as the most obvious evidence of different formation mechanisms for stellar binaries and planetary systems. If most of the gaseous giant planets detected have masses less than $5 M_J$, the distribution presents a long tail towards masses larger than $10 M_J$. No real clear limit exists as the distribution goes continuously into the brown-dwarf regime but with a decreasing number of candidates (the so-called brown-dwarf desert, see Figure 15.1).

Towards the low-mass side of the planetary mass distribution, a clear rise is observed. This low-mass edge of the distribution is, however, poorly defined because of observational incompleteness; the lowest mass planets are difficult to detect because the radial-velocity variations are smaller. We thus expect a large population of still unknown sub-Saturn mass planets. This is further supported by accretion-based planet formation models (Ida & Lin 2004; Mordasini *et al.* 2008a,b; see also Section 15.3).

15.2.2.2.2 A new population of 'solid' planets. Most of the detected planets are gaseous giants similar to our own Jupiter, with typical masses of a few hundreds of Earth masses. However, in the past few years, a dozen of planets with masses in the Uranus–Neptune range (5–22 Earth masses) have been detected (Butler *et al.* 2004; McArthur *et al.* 2004; Santos *et al.* 2004; Bonfils *et al.* 2005, 2007; Rivera *et al.* 2005; Vogt *et al.* 2005; Lovis *et al.* 2006, Udry *et al.* 2006, 2007; Melo *et al.* 2007). Because of their small masses and locations in the systems, close to their parent stars, they may well be composed mainly of rocky/icy material. In the planet-mass distribution, they already start to draw a new population at very low masses (Figure 15.1).

Among the most interesting candidates, we can mention the super-Earth $(7.5M_\oplus)$ in the Gl 876 three-planet system (Rivera *et al.* 2005), the *trio of Neptunes* around HD 69830 (Lovis *et al.* 2006) or the three-planet system orbiting Gl 581, two of the planets being super-Earths of 5 and $8M_\oplus$ (Udry *et al.* 2007) residing at the edges of the habitable zone of this M dwarf. A close analysis of the high-precision part of the HARPS planet-search programme provides hints for the existence of many more very low-mass planetary candidates. Among them, an exciting system with three planets with masses ranging from 3.6 to 8 Earth masses, with periods from 4 to 20 days, has been recently discovered (Mayor *et al.* 2008).

The discovery of very low-mass planets so close to the detection threshold of radial-velocity surveys, and over a short period of time, suggests that this kind of objects may be rather common. Moreover, at larger separations (2–3 AU), the microlensing technique is finding similar mass objects (the lightest with a mass of $5.5M_\oplus$, Beaulieu *et al.* 2006) showing that smaller-mass planets can be found over a large range of separations. This is in complete agreement with the latest Monte Carlo simulations of planet formation (see Section 15.3.5).

15.2.2.3 Orbital characteristics

15.2.2.3.1 Period distribution of giant extra-solar planets. The distribution is basically made of two main features: an accumulation around 3 days plus an increasing distribution with period (in log scale). The numerous giant planets orbiting very close to their parent stars (the 'Hot Jupiters' with $P < 10$ days) were completely unexpected before the first exoplanet discoveries. They are supposed to undergo a migration process moving them from their birth place close to the central star (see, e.g., Lin *et al.* 1996; Ward 1997). This scenario requires, however, a stopping mechanism to prevent the planets from falling onto the stars (see, e.g., Udry *et al.* 2003 for a more detailed discussion). A *minimum* flat extrapolation of the distribution to larger distances would approximately double the occurrence rate of planets (Marcy *et al.* 2005). This conservative extrapolation hints that a large population of yet undetected Jupiter-mass planets may exist between 3 and 20 AU.

15.2.2.3.2 Mass–period distributions. A correlation is seen between orbital period and planet mass. The most obvious characteristic is the paucity of massive planets on short-period orbits (Udry *et al.* 2002; Zucker & Mazeh 2002). When neglecting the multiple-star systems, a complete absence of candidate is even observed for masses larger than $\sim 2M_J$ and periods smaller than ~ 100 days.

Another interesting feature of the distribution is the rise in the maximum planet mass with increasing distance from the host star (Udry *et al.* 2003). More massive planets are expected to form further out in the protoplanetary disk (where raw material for accretion is abundant and the longer orbital path provides a larger feeding zone). They are then difficult to move as a larger portion of the disk has to be disturbed to overcome the inertia of the planet.

15.2.2.3.3 Eccentricities of giant planets. Extra-solar planets with orbital periods longer than about 6 days have eccentricities significantly larger than for giant planets in the Solar System, spanning the whole range between 0 and 0.93. Their median eccentricity is $e = 0.29$. A few long-period, low-eccentricity candidates are, however, emerging from the surveys. They form a small subsample of Solar-System *analogs*.

The origin of the eccentricity of extra-solar giant planets has been suggested to arise from several different mechanisms: the gravitational interaction between multiple giant planets, interactions between the giant planets and planetesimals in the early stages of the system formation, or the secular influence of an additional, passing-by or bounded companion in the system. For small periastron distance, giant planets are also likely to undergo tidal circularization. The observed final distribution probably results from multiple processes.

15.2.2.4 Primary star properties

The study of the planet hosts themselves provides additional information to constrain planet formation. In particular, the mass and metallicity of the parent stars seem to be of prime importance for models of planet formation (Ida & Lin 2004, 2005a; Benz *et al.* 2006).

15.2.2.4.1 Primary mass effect. Giant planets are detected around dwarf stars with masses ranging from 0.3 to $1.4M_\odot$ (spectral type from F to M). However, results from on-going surveys indicate that giant gaseous planets are rare around M dwarfs in comparison to FGK primaries (Endl *et al.* 2006, Bonfils *et al.* 2007). In particular, no Hot Jupiter has been detected close to an M dwarf. On the contrary, more than half of the planets found to orbit an M dwarf have masses below $25M_\oplus$ and are probably 'solid' planets. Thus, the occurrence rate for planets around M dwarfs appears to be directly dependent on the domain of planet masses considered.

For more massive primaries, surveys targeting earlier, rotating A–F dwarfs (Galland *et al.* 2005) and programmes surveying G–K giant stars (e.g. Johnson *et al.*

2007; Lovis & Mayor 2007, and references therein) are starting to provide interesting candidates. The detected planets are generally massive ($>5M_J$). The surveys are still strongly observationally biased, but it is already clear that the amount of material trapped into planets is strongly correlated with the mass of the primary star (Lovis & Mayor 2007).

15.2.2.4.2 Metallicity of stars with giant planets. A correlation between the presence of gas giant planets and the high metallicity in the host stars was noted in the early years of extra-solar planet detection. Now, systematic homogeneous studies of the stars on planet-search surveys have confirmed the trend (Santos *et al.* 2001, 2005; Fischer & Valenti 2005). These studies are furthermore showing that the probability of forming a gas giant planet is a steeply increasing function of the primary star metallicity. The mostly accepted explanation for this trend posited that metallicity enhances planet formation because of increased availability of small particle condensates – the building blocks of planetesimals. The other proposed argument suggesting that the high metallicity could result from the pollution of the stellar convective zone, from late-stage accretion of gas-depleted material, is not supported by the fact that metallicity is not observed to increase with decreasing convective zone depth for main sequence stars. A similar conclusion can be drawn from the subsample of subgiant stars (Fischer & Valenti 2005).

The metallicity versus planet occurrence trend seems not to be present neither for the low-mass planets (Udry *et al.* 2006) nor for planets orbiting giant stars (Pasquini *et al.* 2007). If confirmed, these results will bring new strong constraints for planet formation models.

15.2.3 Exoplanet characterization

About 10% of the known planets are actually transiting in front of the disk of their parent star. When transit photometry is combined with high-precision radial-velocity measurements, it is possible to derive an accurate mass and radius as well as the mean planet density (Figure 15.2). These important values constrain planetary interior models as well as the planet-evolution history. In particular, they confirm the gaseous or solid nature of the detected giant planets. However, we are still far from a detailed understanding of the physics of irradiated giant planets. For example, the derived densities of transiting extra-solar planets cover a wide range of values. Some candidates present large radii and low densities, a feature clearly not shared by all very close planets and the origin of which is presently intensively debated.

The detection of the transit of a very low-mass planet (of smaller size) is much more challenging than for a giant planet as the depth of the photometric transit

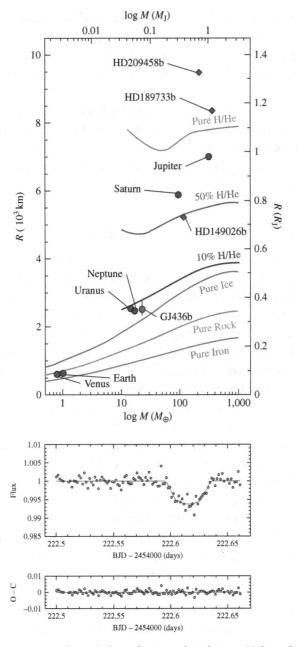

Fig. 15.2. *Top*: mass–radius relation of extra-solar planets. (Adapted from Gillon *et al.* 2007.) These quantities help constraining the planet internal structure. In particular, the position of Gl 436b close to Neptune points towards a composition of the planet principally made of ice on a rocky core, with a probable small atmosphere (15%) of hydrogen and helium. *Bottom*: Photometric transit detection curve of Gl 436b obtained with Euler-cam on the 1.2-m Swiss telescope at La Silla (Chile).

curve scales with the ratio of the surfaces of the planet and star disks. Fortunately, the transit of the $22M_\oplus$ Gl 436b planet was recently observed from the St-Luc (Switzerland), Wise (Israel) and La Silla (Chile) observatories by Gillon *et al.* (2007; Figure 15.2). The radius of the planet is $4R_\oplus$ pointing towards a composition of the planet dominated by ice.

The brightest amongst the 'transited' stars further offers a large number of possibilities for complementary studies, from the ground or from space. For example, with the Spitzer space telescope, anti-transits are now detected in several infrared bands (Deming *et al.* 2005, 2006; Demory *et al.* 2007), probing the thermal emission of the planet and providing first points for an observed planetary spectrum.

15.2.4 The future: A quest for terrestrial planets

The many recent discoveries of planetary systems harbouring Neptune-mass or super-Earth planets clearly indicate that low-mass planets around solar-type stars must be very common. This general feeling is enforced by statistical considerations (since low-mass planets are more difficult to detect) and by results of newly developed formation models. The exciting discovery of the 'trio of Neptunes' around HD 69830 (Lovis *et al.* 2006), including a 'long-period' (6 months) low-mass planet, not only confirmed this tendency but also showed that even lower-mass planets can be discovered by the Doppler technique. Indeed, the achieved $0.2\,\mathrm{ms}^{-1}$ long-term precision over several years (considering run-averaged values) shows that Earth-like planets are now within reach of HARPS-type instruments (for comparison, the Earth produces a radial-velocity wobble of $9\,\mathrm{cm\,s}^{-1}$ on our Sun). The performances demonstrated by HARPS have excited the imagination of astronomer concerning the observational possibilities. New ultra-stable spectrographs are now studied for large telescopes, with the aim of reaching long-term radial-velocity precisions down to $10\,\mathrm{cm\,s}^{-1}$ (ESPRESSO/VLT) or even at the level of a few $\mathrm{cm\,s}^{-1}$ (CODEX/E-ELT). Such an exquisite precision will also be very useful outside of the planet domain, it, for example, gives access to direct observation of the acceleration of the universe through the measurements of precise velocities of remote quasars over decades (Cristiani *et al.* 2008).

The unambiguous identification of the signature of an Earth-like planet from Doppler measurements requires a large number of observations. Moreover, stellar noise source must be averaged out by increasing the exposure time beyond the requirements dictated by pure photon noise. A search for Earth-like planets around a sample of a couple of tens of stars would then be possible but will be expensive in term of telescope time. Fortunately, the transit technique is rapidly developing and is starting to deliver many new planetary candidates. Due to their limited sensitivity and accuracy, the ground-based transit-search programmes are not able to detect the tiny signature produced by Earth-sized objects. The situation is, however, drastically changing in space. The COROT satellite in operation since December 2007, and later the Kepler/NASA and PLATO/ESA space missions, will start to deliver planetary candidates of much smaller size. In particular, the Kepler and PLATO satellites will have Earth-size planet detection capabilities and are indeed expected to deliver several dozen of candidates.

However, the transit method does not provide any information on the planet mass. In order to acquire insight on the mean density and thus the inner structure of the planet, we need the mass, and precise radial-velocity follow-up is required. The two approaches (radial velocities and photometric transits) will then play a key and complementary role in detecting Earth-type planets in a close future.

15.3 Formation of giant planets by core-accretion model

15.3.1 Introduction

The notion of a *natural cosmogony*, the formation of the Sun and planets through natural law rather than via the fiat of a creator, is an important and enduring product of the Age of Enlightenment. In 1742, for example, the French naturalist Georges Louis Leclerc Comte de Buffon proposed that a 'comet' had a near-collision with the Sun and that the planets in our Solar System were the condensation products of the resulting tidal stream. Buffon's theory accounted for the fact that the planets all orbit in the same direction and in a sense can be viewed as a forerunner to the Giant Impact scenario that is now held responsible for the genesis of Earth's moon. Also in the 1740s, the German philosopher Immanuel Kant postulated that the planets arose from a primordial cloud of spinning gas. This concept was more fully developed later, independently, in the 1790s by Pierre-Simon de Laplace, who imagined that the disk contracted as it cooled, leaving behind a succession of rings that fragmented to form the planets (for more detail, see Brush 1978).

These eighteenth-century cosmogonies were couched in quaint language and are not fully correct, but they nevertheless hit surprisingly close to the mark. Today's theory, observations and simulations all support the general picture provided by Laplace's nebular hypothesis. We are now privileged to have a theory of planetary formation which is increasingly complete and self-consistent and which is testable through observations of planetary systems in a wide range of dynamical and evolutionary states. Nevertheless, there are still significant gaps in our understanding, and both observational and theoretical works are proceeding at a very rapid pace.

With the discovery of a large population of extra-solar planets, a scientific consensus is emerging that the so-called core-accretion theory is responsible both for the formation of the gas giant planets in our Solar System and for a majority (but likely not all) of the nearly three hundred extra-solar planets that have been observed. In the context of an overview article, it thus appears most useful to devote a majority of the theoretical discussion to the foundations, the tests and the open issues surrounding core accretion. It is important, however, to point out that some workers support the competing paradigm of gravitational instability (Boss 2000) as the dominant mechanism for forming giant planets.

15.3.2 *Planet formation by disk instability*

At the very least, disk fragmentation provides a useful foil for a discussion of core accretion. Gravitational instability arises when a protostellar disk becomes massive enough and cool enough to fragment directly into planet-mass condensations. In detailed simulations of the process (Boss 2000; Rice *et al.* 2003; Mayer *et al.* 2004), Jovian-mass planet formation occurs on a dynamical timescale of order hundreds of years (2–10 orbital periods at the planet-forming radius in the disk) followed by a longer period of Kelvin–Helmholtz contraction. A planet that forms via gravitational instability may produce a high-entropy initial condition that would be expected to match well with the initial states assumed in calculations that employ the equations of stellar structure to model the contraction of Jovian planets (Baraffe *et al.* 2003). Core accretion, on the contrary, has been recently suggested to produce Jovian mass planets with far smaller internal entropies and which, at early times, are several to one hundred times less luminous (Marley *et al.* 2007). These suggestions will be discussed in Section 15.4.

Two basic requirements must be met in order for gravitational instability to work. First, the so-called Toomre Q parameter, where $Q = c_s \kappa / \pi G \sigma$, must be near unity somewhere within the disk. In essence, the Q parameter measures a disk's local propensity for self-gravity (scaling with the surface density, σ) to compete against the stabilizing influences of differential rotation (measured by the epicyclic frequency, κ) and pressure (which scales with the sound speed, c_s). Second, the cooling time for fragments must be less than half of an orbital period (Gammie 2001; Rafikov 2005).

Planets formed via gravitational instability mechanism will not have massive cores. Gravitational instability also predicts that the occurrence of planets should be independent of the planet–star metallicity and that planet formation should not depend sensitively on the mass of the parent star. While it is likely that some form of gravitational instability is responsible for the presence of objects such as 2MASS-1207b (Chauvin *et al.* 2004) in which a $\sim 5 M_J$ object is observed in 55 AU orbit around a $\sim 25 M_J$ brown dwarf primary, there are a number of strong observation-based objections to disk fragmentation's ability to form the majority of currently known planets.

There are also theoretical difficulties with the gravitational instability picture. The dynamical modes of fragmentation and planet formation that are seen in simulations are based on initial conditions that posit axisymmetric, yet gravitationally unstable configurations. Linear and non-linear stability analyses show that self-gravitating protostellar disks are susceptible to non-axisymmetric (spiral) modes of instability at considerably higher Toomre-Q values than the axisymmetric ($m = 0$) modes that occur near $Q \sim 1$ (Adams *et al.* 1989; Heemskerk *et al.* 1992; Laughlin & Rozyzcka 1996). This leads to the possibility that non-axisymmetric modes appearing at high Q will interact non-linearly, thereby producing mass and angular momentum transport within the disk sufficient to forestall a planet-forming instability (Laughlin *et al.* 1997).

15.3.3 The core-accretion paradigm

In the classical core-accretion models of Pollack *et al.* (1996), giant planet formation proceeds in three distinct phases, and its completion requires timescales of order 10^6 years, roughly a thousand times longer than the fiducial timescale characterizing gravitational instability. In the first phase of the core-accretion picture, solid particles that are small enough to be carried along with the nebular gas at the disk midplane must coagulate into planetesimals through a process that is still subject to uncertainty. In particular, the details of the planetesimal agglomeration process will depend strongly on whether planet-forming disks maintain significant turbulence. In the case of laminar flow, accumulation may occur via a form of gravitational instability that involves direct collapse of a very thin dust layer at the midplane into kilometre-sized planetesimals (Goldreich & Ward 1973). The presence of turbulence in the disk can quench this so-called Goldreich–Ward instability (Weidenschilling & Cuzzi 1993). Planetesimal agglomeration in a turbulent environment may require concentration of solids within eddies or persistent large-scale vortices (Barge & Sommeria 1995; Tanga *et al.* 1996; Klahr & Bodenheimer 2006). In addition, gas drag plays an important role for solid bodies ranging in size from a millimetre to a kilometre, and for boulder-sized objects, gas drag can lead to rapid orbital decay (Weidenschilling 1977). This decay problem has not yet been fully resolved. For a review of the initial stages of planetesimal growth, see Dominik *et al.* (2007).

As Phase I continues, objects in a protoplanetary nebula with radii larger than a kilometre or so begin to exert substantial gravitational perturbations on neighbouring bodies. A planetesimal that is somewhat larger than its neighbours will preferentially accrete smaller bodies, leading to very rapid (runaway) growth once the escape velocity from the protoplanet's surface exceeds the relative velocity of the bodies in the swarm. Planetesimals – often referred to as *oligarchs* – that find themselves in such a situation experience runaway accretion until their entire radial feeding zones are depleted (Wetherill & Stewart 1989; Ohtsuki *et al.* 2002). A practical expression for the width of the feeding zone is $a_f = \sqrt{12 + e_h^2} R_H$, where $R_H = a(M_P/3M_\odot)^{1/3}$ is the planetesimal's Hill radius (Lissauer 1993).

During Phase I, the accumulation of solids occurs much faster than gas accretion. In the standard version of the core-accretion paradigm (as elucidated, for example, by Pollack *et al.* 1996), the mass increase of the planet during the first phase depends on the protoplanet's radius and the ratio of the gravitational to the geometric cross section:

$$\frac{dM_p}{dt} = \pi R_c^2 \sigma \Omega F_g ,$$

where Ω is the orbital frequency, F_g is the gravitational enhancement factor and σ is the surface density in solids at the protoplanet's orbital radius.

Phase II begins with the end of runaway growth, which occurs when the growing core depletes its feeding zone and when the protoplanet contains at least several Earth masses. As the planet grows through this size threshold, its atmospheric envelope becomes increasingly structurally significant. Modelling of this stage requires computation of the radial hydrodynamical equilibrium structure of the gas envelope and is generally accomplished with 1D codes that solve the standard equations of stellar structure (see, e.g., Henyey *et al.* 1964). Within these models, the luminosity of the nascent planet is set by Kelvin–Helmholtz contraction, accretion luminosity from the solid core and energy input into the envelope via dissolution of planetesimals.

In Pollack *et al.*'s (1996) fiducial model, Phase II of core accretion begins after approximately 0.5 Myr, at which time the isolation mass at $r = 5.2$ AU in a $T = 150$ K nebula (containing $\rho_{neb} = 5 \times 10^{-11}$ g cm^{-3} and $\sigma_{solids} = 10$ g cm^{-2}) is approximately $10 M_\oplus$. The planetary mass increases slowly during the next 7 million years, as gas is gradually accreted from the protoplanetary nebula.

It is tempting to imagine that the slow growth of the planet during Phase II is limited by the planet's small mass and an associated inability to rapidly accrete gas from the more massive nebula. The actual reason, however, stems from the inverse dependence of the Kelvin–Helmholtz contraction time on the planetary mass. Throughout Phase II, the planet fully fills its Roche lobe, and growth in mass is limited by the planet's ability to radiate quickly enough. As the gas mass and core mass gradually increase, the contraction time shortens. When the gas mass and core masses are approximately equal, the KH contraction timescale becomes much shorter and evolution becomes very rapid. Phase III is comprised by this period of runaway growth. In a span of a few tens of thousands of years, the planet inflates from a Neptune-like mass to a Jovian mass. Growth is eventually limited by the availability of gas and the tendency of the planet to open up a gap in the disk.

The fiducial model of Pollack *et al.* (1996) has had a tremendous influence in shaping the astronomical community's understanding of Jovian planet formation. As of 1 January 2008, for example, the Pollack *et al.* (1996) paper has accumulated more than 500 ADS citations. Their baseline model is often criticized, however, for the long formation time that it predicts for Jupiter. Such criticism has been used as a point in favour of gravitational instability (Boss 1998) or, more commonly, as a basis for refining the core-accretion model itself. In the past 5 years, a number of significant advances have been made.

15.3.4 Refined core-accretion models

Put simply, an 8 Myr formation time for Jupiter would require an uncomfortably long-lived solar nebula. Observations of nearby star-forming regions show that the fraction of young stars with excess JHKL emission declines rapidly over a 2–3 million year timescale (Haisch *et al.* 2001). In clusters older than 8 Myr, only a small minority of stars still retain significant gas-rich protostellar disks. The final runaway Phase III of core accretion, however, emphatically requires ample gas. A planet that arrives for its final stage after gas in the disk is depleted will end up as an ice-rich object resembling Uranus or Neptune. It will not be a Jupiter-mass planet with ready RV-detectability.

Furthermore, the Pollack *et al.* (1996) baseline model predicts a core mass of order $20M_\oplus$ for Jupiter. This is in apparent conflict with interior models (see, e.g., Saumon & Guillot 2004) that seek to match the observed values of the Jovian R_{eq}, J_2 and J_4. While ambiguity arises from the fact that the hydrogen–helium equation of state is still uncertain, it appears that Jupiter has a small core, perhaps less than $5M_\oplus$, with an overall planetary heavy element enrichment up to $45M_\oplus$.

A number of mechanisms have been studied, which can speed up the formation of Jupiter and reduce its core size. Grain opacities, in particular, are clearly a key issue. The long coasting era represented by Phase II hinges on the opacity of the planetary envelope as it undergoes Kelvin–Helmholtz contraction. A transparent envelope allows for rapid contraction and a reduced timescale, whereas an opaque envelope has the opposite effect. Pollack *et al.*'s (1996) baseline model adopted dust opacities for the envelope, which assume an interstellar size distribution and which provided very efficient absorption. It is evident, however, that material that enters a giant planet envelope will have been modified from its original interstellar grain state by processes of coagulation and fragmentation. Work by authors such as

Podolak (2003) indicates that once grains enter the protoplanetary envelope, they coagulate and settle out quickly into warmer regions where they are destroyed. As a result, Podolak (2003) argues that the true envelope opacities during Phase II are $\sim 50\times$ smaller than interstellar.

Hubickyj *et al.* (2005) have published a comprehensive study of core accretion that serves as a general update to the Pollack *et al.* (1996) benchmark. They find that by adopting Podolak *et al.*'s (2003) estimates for the envelope opacity, Jupiter's formation timescale is decreased to a fully reasonable ~ 2 Myr. Additionally, Hubickyj *et al.* (2005) address the core mass problem by positing that the growth of a giant planet is affected by competing planetary embryos. Competition for solids from additional planets in the system serves to both reduce the envelope luminosity from infalling planetesimals and plausibly limit the final core mass to $M < 10 M_{\oplus}$.

Giant planets being formed in gaseous disks experience momentum transfer with it, leading to migration. The amount and even the direction of migration are still subject to some debate; however, observations of extrasolar planets very close to their central star point towards important *inward* migration. Migration occurring on timescales comparable to typical planet formation timescale must be included in consistent theoretical models, such as the ones of Alibert *et al.* (2004, 2005a). As shown by these authors, by preventing the depletion of the planet-feeding zone, migration suppresses the afore-mentioned Phase II, reducing the formation timescale to values compatible with protoplanetary disk lifetimes. Moreover, taking into account migration and protoplanetary disk evolution allows to develop Jupiter and Saturn formation models – the formed planets having bulk composition and enrichment in volatile species similar to observed values (Alibert *et al.* 2005b). Finally, when comparing formation models with the observed distribution of extra-solar planets, migration during the formation phases plays a key role. Such extended models can be used to provide fully quantitative comparisons between population synthesis simulations of giant planet formation and the observed aggregate of extra-solar planets, as presented in Section 15.3.5.

15.3.5 Population synthesis

The characteristics of the ~ 250 known extra-solar giant planets begin to provide a database to which the predictions of planet formation theories presented in the Section 15.3.4 can be confronted. In order to do this, the expected planet population is synthesized, based on extended core-accretion scenario, taking into account migration of forming planets, as presented above (Alibert *et al.* 2004, 2005a; Mordasini *et al.* 2008b).

15.3.5.1 Initial conditions and detection biases

In order to generate from our models a population of planets that can be directly compared to the observed one, we need to specify not only the initial conditions but also their probability distribution function (PDF). A number of these quantities can be constrained from observations, another set from relatively robust theoretical considerations and finally some are nothing else than educated guesses. We provide here brief indications on how we define the most important parameters; more detailed information can be found in Mordasini *et al.* (2008b).

The disk mass PDF is derived from Beckwith and Sargent (1996) who give histograms for the distribution of total disk gas masses in the Taurus–Auriga and Ophiuchus star-forming regions. Assuming a power-law surface density with an index of $-3/2$, their results correspond to surface densities at 5.2 AU in the range $50\text{--}1000\,\mathrm{g\,cm^{-2}}$. The total disk mass is between 0.004 and $0.09M_\odot$ for the region inside 30 AU and between 0.014 and $0.29M_\odot$ for the region out to 300 AU. We note that because of the unknown dust opacity and gas-to-dust mass ratio, the masses of the protoplanetary disk masses are rather uncertain. We also point out that submillimetre interferometry is now able to provide independent constraints on disk profiles, albeit only at large distances for the moment (Dutrey *et al.* 2007).

Following Murray *et al.* (2001), we relate the dust-to-gas ratio f_{dg} to the metallicity Fe/H of the disk (and the star) by the relation $\mathrm{Fe/H} = \log(f_{\mathrm{dg}}/f_{\mathrm{dg}\odot})$, where $f_{\mathrm{dg}\odot}$ is the dust-to-gas ratio in the planet formation region of the solar nebula at the time planetesimals were formed. Kornet *et al.* (2001) have shown that dust–gas interactions during the early stages can lead to significant re-distribution of dust leading to an enhanced local dust-to-gas ratio by factors of order 2–4 in the inner planet-forming region. In the simulations presented here, we therefore use $f_{\mathrm{dg}\odot} = 0.04$ rather than $f_{\mathrm{dg}\odot} = 0.0149$ (Lodders 2003), which would be the current solar value. Finally, the probability of occurrence of a given content in heavy element is taken from the metallicity distribution of F, G and K stars in the solar neighbourhood (Nordström *et al.* 2004), which is Gaussian with a mean $\mu \approx -0.14$ and a dispersion $\sigma \approx 0.19$.

Haisch *et al.* (2001) have shown that near-infrared observations of hot ($T \sim 900\,\mathrm{K}$) dust indicate that the circumstellar disk fraction in young stellar clusters is a linearly decreasing function of age with a half-life of about 3 Myr and a maximum lifetime of 6–8 Myr. In addition, more recent studies of accretion in young systems also showed that disk lifetimes are probably less than 10 Myr (Jayawardhana *et al.* 2006). In our models, the disk lifetime is given by a combination of viscous evolution as determined by the viscosity parameter α (we use the so-called *alpha*-disk models) and the rate of photoevaporation $\dot{M}_W$. Both parameters are set in such a way that the disk lifetimes nearly follow the decay law found by Haisch *et al.* (2001). Typical values for $\dot{M}_W$ range between 10^{-10} and $1.5 \times 10^{-8} M_\odot$ year^{-1}.

In contrast to the above-mentioned quantities, the start location of the seed embryo cannot be constrained by observations. However, numerical simulations (Kokubo & Ida 2000) have shown that oligarchic growth leads to embryos separated by a few Hills radii. This leads to a probability distribution which is uniform in $\log(a)$ as used by Ida and Lin (2004) and which we adopt here as well.

Planetary embryos grow at a rate that is in part determined by their distance to the central star. Hence, once the seed embryo's position has been chosen, we determine the time required to grow this seed at its location following Thommes *et al.* (2003). This gives us the time lag between the start of the disk evolution and the time at which the seed embryo is allowed to start growing.

Finally, in order to compare statistically our synthetic planet population with the one actually observed, it is necessary to take into account the various detection biases associated with any particular detection method. We focus on the radial-velocity techniques since the majority of extra-solar planets have been discovered by this means.

15.3.5.2 Results

In this section, we present the results of our planet population synthesis. Due to space limitations, we limit ourselves in presenting mass *versus* semi-major axis results for solar-type stars. Additional results such as those related to the host-star metallicity and chemical composition of the planets, as well as formation process around stars of different masses, are left to more detailed papers (Alibert *et al.* 2008; Mordasini *et al.* 2008b).

15.3.5.2.1 Formation tracks. In order to gain insight into the physical processes at work during planet formation, we consider the formation tracks of some of the planets we have synthesized (Figure 15.3).

The different line styles in the figure allow to distinguish three different formation stages: state 1 – type I migration for low-mass planets (*solid lines*). In this phase, planets grow by accreting solids; gas accretion is negligible. The tracks are very steep, as a result of the highly reduced type I migration we consider in our models (see Alibert *et al.* 2005a for details). At small distances, cores quickly grow to the isolation mass before the gas disk disappears, whereas at larger distances some have not reached it at that moment. As we neglect any evolution after the dispersion of the disk, the latter cores stay at intermediate masses.

In some cases, preferentially at locations not far behind the iceline in massive disks, the core grows so large that it can open a gap in the gas disk significantly

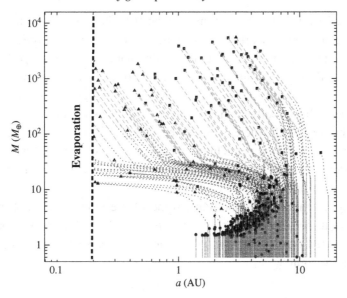

Fig. 15.3. Planetary evolutionary tracks in the mass–distance plane. The tracks lead from the initial position at some $a(0)$ and $M(0) = 0.6M_\oplus$ to the final position reached when the gas disk disappears (*large black symbols*). Planets reaching the inner 0.2 AU may suffer some amount of evaporation and are not used in the comparison with observations. Each track has a line style according to the migration mode: *solid lines* for type I migration, *dotted lines* for ordinary type II migration (when the mass of the planet is small compared to the mass of the local disk) and *dashed lines* for the braking phase (when the mass of the planet is comparable to or larger than the local disk mass). On the tracks, *small black ticks* are plotted every 0.1 Myr to illustrate the temporal evolution along the path. The *short dashed lines* have a slope of $-\pi$ (see text).

before the gas disappears. Note that, during the formation of the planet, the disk itself evolves and its scale height decreases, so that smaller and smaller planets can open up a gap. At this point starts stage 2: the migration mode changes to disk-dominated type II (*dotted lines*). Type I migration being reduced in our models, the switch from stage 1 to stage 2 translates in an increase of migration rate and of the solid accretion rate, as the planet enters regions where there are new planetesimals available. Planets starting inside about 4–7 AU begin to move in disk-dominated type II migration along nearly horizontal tracks at $M \sim 10$–$30M_\oplus$. Planets which underwent such a nearly horizontal phase at some point during their evolution are plotted as triangles at their final position. In some cases, the core grows to a size that triggers runaway gas accretion. Then, the gas accretion rate increases very rapidly, so that it usually reaches the limiting value given by how much the disk can give to the planet ($\sim 10^{-3}$–$10^{-2}M_\oplus$ year^{-1}, see Alibert *et al.* 2005a). With the rapidly growing mass, the planet mass quickly overcomes the local disk mass, so that the migration changes into planet-dominated type II.[1] The planet branches off to the right from the previously nearly horizontal tracks and stage 3 (*dashed lines*) begins.

In this final phase, the planet migrates at the planet-dominated type II rate and accretes at the disk equilibrium accretion rate, so that the planet moves along a straight line with a slope of $-\pi$ (see Mordasini *et al.* 2008b for the derivation of this feature). The velocity of the evolution along the straight line slows down in time as can be seen by the increasing number of small black ticks on the track near the final position of the planet. This is a consequence of the concurrent decrease of the gas accretion rate due to disk evolution and the slowing down of migration as the mass of the planet becomes larger and larger.

Depending on the initial conditions, some of the embryos do not experience these three formation stages and can grow directly to large masses, without a phase of disk-dominated type II migration (*horizontal lines*). Such embryos are plotted as squares in the figure.

15.3.5.2.2 Mass versus semi-major axis diagram. Probably the most straightforward and most important test of population synthesis calculations is to compare the distribution of the semi-major axis a and masses $M \sin i$ of our synthetic planets

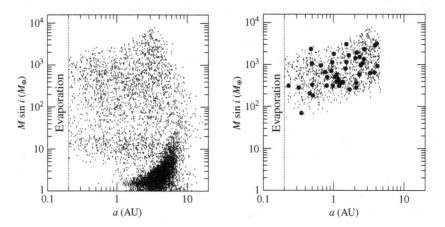

Fig. 15.4. Population synthesis results in a minimum mass versus semi-major axis diagram. The *left panel* shows all the planets formed in the framework of the model, whereas the *right panel* presents planets observable with a $10 \, \text{m s}^{-1}$ radial-velocity survey (black) and an observing period of 10 years. Planets actually detected are indicated by large dots. Also indicated in the plot is the region where subsequent evaporation of the planets during their lifetimes could be significant, thereby preventing a direct comparison with observations.

---

[1] In the planet-dominated regime, the standard type II migration rate of the disk-dominated regime is reduced by a factor proportional to the ratio of the planet mass to the local disk mass, this latter being $\propto \Sigma a^2$, where Σ is the gas surface density at the location of the planet a.

with observations. In Figure 15.4, we show the resulting planet population for our nominal model (*left panel*) as well as the subset of potentially detectable planets with a radial-velocity measurement accuracy of $10 \, \mathrm{m \, s^{-1}}$ (*right panel*). True detections of essentially single planets orbiting solar-like stars ($0.8 M_\odot \leq M \leq 1.2 M_\odot$) on low eccentricity orbits and having $K > 10 \, \mathrm{m \, s^{-1}}$ are indicated by the large dots (Schneider, http://exoplanet.eu).

A radial-velocity precision of $10 \, \mathrm{m \, s^{-1}}$ only allows the detection of a very small subset of all the planets predicted by the model to exist. Not surprisingly, only the most massive planets are detectable, leaving out all the smaller-mass ones. Improving radial-velocity measurement techniques is therefore essential to get access to these smaller-mass planets.

While visually comparing the observed and predicted distributions gives some impression of how different the two can be, statistical tests allow a quantitative and unbiased assessment. To this effect, we apply the Kolmogorov–Smirnov (KS) test (Press *et al.* 1992) in its 1D and 2D versions. We use the 1D version to test the significance of both the distribution of the semi-major a and the masses $M \sin i$ separately and the 2D version to test the significance of the distribution in the $a - M \sin i$ plane. Details on the statistical analysis, as well as their sensitivity to different parameters of the model, are presented in Mordasini *et al.* (2008b). Let us just mention here that applying KS test, the mass distribution of extra-solar planets is well reproduced, whereas the semi-major axis distribution is more difficult to reproduce – the main reason being probably that migration is treated in a very simple way in our models. Future work will address this issue.

15.4 The early evolution of brown dwarfs and planets

15.4.1 Introduction

Despite many efforts devoted to the understanding of the formation processes of low-mass objects (low-mass stars, brown dwarfs and giant planets), their very early stages of evolution are still uncertain. Evolutionary models at young ages still rely on uncertain initial conditions, which depend on the very details of the formation process and in particular on the accretion history. Accretion is a key process during the hydrodynamical phase of gravitational collapse and of disk formation (see Chapter 10 by Megeath, Li and Nordlund in this book). It is also determinant during the quasi-static phase of evolution up to a few million years, where direct comparisons between models and observations can be performed in luminosity-effective temperature or colour-magnitude diagrams. Ideally, the best description of early stages of evolution should combine star/planet formation models, which provide accretion rates as a function of time, and evolutionary models, which describe the structure and evolution of the accreting central object. This is, however, a challenge, which requires to bridge the gap between 3D hydrodynamical time-explicit calculations and 1D time-implicit quasi-static calculations. A multi-dimensional implicit hydrodynamical code would be the appropriate numerical tool to describe this transition. Such a tool still needs to be developed. The effect of accretion on the first stages of evolution is thus currently described by phenomenological or simplified approaches, as described below in Section 15.4.2. The second part of this section (see Section 15.4.3) is devoted to an attempt to define diagnostics, enabling the distinction between a brown dwarf and a planet. The current distinction relies on the deuterium burning minimum mass and has no physical foundation. Given the increasing number of newly discovered objects in the mass range between 1 and 10 Jupiter masses, it is now urgent to find signatures of the different formation processes which characterize respectively, a brown dwarf and a planet. We also critically analyse the recent suggestion that planets formed by the core-accretion model should be fainter than objects formed by gravitational collapse at young ages, as mentioned in Section 15.3.

15.4.2 Effect of accretion on young brown dwarfs

15.4.2.1 Observational constraints

Signatures of accretion on objects belonging to young open clusters and star formation regions of a few million years are commonly observed down to the planetary mass regime (Muzerolle *et al.* 2005). Observations over a wide mass range suggest a trend of decreasing accretion rate with mass, following roughly $\dot{M} \propto M^2$ (Mohanty *et al.* 2005; Natta *et al.* 2006). Accretion rates vary from $\sim 10^{-7} M_\odot \text{year}^{-1}$ for $\sim 1 M_\odot$ to $\sim 10^{-12} M_\odot \text{year}^{-1}$, or less, for brown dwarfs. More importantly, observations show similarities of accretion properties between higher-mass stars and low-mass objects, including brown dwarfs, confirming the idea that stars and brown dwarfs share similar formation histories. Observations also indicate a trend of decreasing accretion rate with time, suggested by a decreasing fraction of accreting objects with age. The determination of accretion rates at stages of evolution earlier than ~ 1 Myr is unfortunately hampered by high extinction surrounding the accreting central object. Much higher rates than currently observed are, however, expected, which will significantly alter the structure and evolution of accreting objects. But could they still affect the position of objects in a luminosity-effective temperature diagram after a few million years? And could accretion explain the large dispersion observed in such diagrams (Figure 15.5)? These questions stress the importance of analysing the sensitivity of evolutionary models used to assign a mass and age to observed objects, to the accretion history.

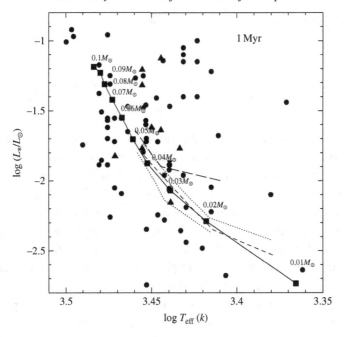

Fig. 15.5. Effect of accretion on the evolution of low-mass stars and brown dwarfs. The curves correspond to isochrones of 1 Myr for masses ranging from 0.01 to $0.1 M_\odot$, with different treatments of accretion (see text). The *solid line* corresponds to the case without accretion: $\dot{M} = 0$. The other curves show the effect of different time-dependent accretion rates and value of ϵ (see text), varied between $\epsilon = 0$ (*dash-dotted line*) and $\epsilon = 0.5$ (*long dashed line*). Symbols indicate observations in different young clusters and star formation region from Comeron *et al.* (2000), Luhman *et al.* (2004), Mohanty *et al.* (2005), Muzerolle *et al.* (2005) and Natta *et al.* (2006).

15.4.2.2 Evolutionary models with accretion

Extending the works by Stahler (1983, 1988) and Hartmann *et al.* (1997) to the brown dwarf regime, Gallardo (2007) and Gallardo *et al.* (2008) recently analyse the effect of accretion based on a phenomenological approach. One main assumption is to consider that accretion proceeds through an accretion disk or is magnetically channelled, disturbing only a small fraction δ of the stellar/substellar surface. The object is thus able to radiate away its luminosity L given by

$$L = 4\pi R^2 (1 - \delta)\sigma T_{\rm eff}^4 \quad \text{with } \delta \to 0. \tag{15.1}$$

One of the main uncertainty inherent to accretion is the determination of the fraction of accreted matter internal energy transferred to the proto-object outer layers. This extra supply of internal energy, per unit mass of accreted matter, is proportional to the gravitational energy, $\epsilon GM/R$, with $\epsilon < 1$ a free parameter. A simple way to explore the impact on the structure of the accreting object is to include an additional energy source to the energy conservation equation, yielding for the luminosity:

$$L = \epsilon \dot{M} \frac{GM}{R} + \int_M \epsilon_d dm - \int_M \left[T \left(\frac{dS}{dt} \right)_m + T \left(\frac{\partial S}{\partial m} \right)_t \dot{m} \right] dm, \qquad (15.2)$$

where ϵ_d is the local nuclear energy generation rate due to deuterium fusion.

Taking into account various accretion rates, constant in time or exponentially decreasing with age, and a variation of ϵ between 0 and 0.5, the results of the Gallardo *et al.* (2008) study are summarized in Figure 15.5. The only constraint on the choice of the accretion rate is to remain consistent with observed rates for a given mass at an age of a few million years. This figure shows observations and isochrones of 1 Myr in a L–T_{eff} diagram for masses ranging from 0.01 to $0.1M_\odot$. As already shown by Hartmann *et al.* (1997), an accreting low-mass star expands less or contracts more than a non-accreting similar object (see also Chabrier *et al.* 2007a). Consequently, an accreting object looks older in an HR diagram, because of its smaller radiating surface for the same internal flux, compared to a non-accreting object of the same mass and age. This is illustrated by the comparison between the isochrone with $\dot{M} = 0$ (*solid line*) and the one with $\epsilon = 0$ and a time-dependent accretion rate (*dash-dotted line*). If a fraction of the thermal energy of the accreted material is added (i.e. $\epsilon \neq 0$) to the object internal energy, the contraction is slowed down. For a significant fraction ($\epsilon > 0.25$), the contraction is significantly slower than for a non-accreting object of same mass, and the corresponding 1 Myr isochrones (dash-dotted and long dashed curves) lie above the isochrone for non-accreting objects ($\dot{M} = 0$).

15.4.2.3 Summary on accretion effects

Evolutionary models accounting for the effect of accretion show that accreting brown dwarfs can look older or younger than their non-accreting counterpart, depending on the amount of accretion energy added to their internal structure. If an object is able to radiate away all the thermal energy released by the accretion process, its contraction is slightly accelerated and it will appear slightly fainter and older than a non-accreting object. For a significant amount of additional accretion energy, the effect is non-negligible, and the accreting object can look significantly younger than its non-accreting counterpart. Even if very simplistic, an exploration of the effect of accretion, as presented in this chapter, illustrates the uncertainties in mass and/or age determinations for young low-mass objects based on current evolutionary models. It also shows that accretion can lead to an important dispersion in a $L - T_{\text{eff}}$ diagram. A better understanding of the accretion process is thus an important task to perform in the future in the field of star/brown dwarf formation. Progress can be achieved with (i) a better determination of the variation of accretion rates with age with the help of numerical simulations and (ii) a better description of the accretion shock on the proto-object and of the feedback of the accretion disk on its inner structure (due to irradiation effects, see Rafikov 2007).

Accretion process, however, cannot explain alone the much larger dispersion displayed by observations. Additional explanations are required, such as (i) an underestimate of the uncertainties in the derivation of L and T_{eff} from observable quantities (colours, magnitude, spectral type) and (ii) missing physics such as magnetic field or rotation in current young low-mass star/brown-dwarf models (see, e.g., Chabrier *et al.* 2007b). These effects still need to be explored.

15.4.3 *Brown dwarfs versus planets*

15.4.3.1 *Definitions*

The IAU defined some years ago a planet as an object with a mass below the deuterium-burning limit. This definition, though still in use in the community, remains controversial and unsatisfactory. Based on physical arguments, the distinction between a brown dwarf and a planet should be based on their different formation mechanisms (see Chabrier *et al.* 2007a). Even if secondary scenarios exist, there is a general consensus that the former primarily forms like stars, e.g. via gravitational collapse of a molecular cloud fragment, and the latter forms in a protoplanetary disc, following the most popular scenario, namely the core-accretion model (see Section 15.3). In this context, two burning questions arise in the star and planet formation community: what is the minimum mass of a brown dwarf and what is the maximum mass of a planet? Answers to these questions are crucial to understand the formation mechanism of both types of objects, but require observational signatures enabling the identification of a brown dwarf or of a planet. A definition purely based on a mass criterion brings utter confusion in the overlapping mass regime where both type of objects exist. The recent study by Caballero *et al.* (2007) of the substellar mass function in σ Orionis stresses again, if necessary, the existence of an overlapping mass regime. They indeed find a continuous mass function from the low-mass star regime down to $\sim 6M_{\text{J}}$, suggesting that such low-mass objects form as an extension of the low-mass star formation process. In parallel, the discovery of the 'super' Jupiter transit of 8–$9M_{\text{J}}$ in a close orbit to a solar-type star (Bakos *et al.* 2007) emphasizes the urgent need for identification criterions. In the following sections, we analyse different properties inherent to brown dwarfs and planets which could provide such criterions.

15.4.3.2 The luminosity of young planets and young brown dwarfs

In a recent paper, Marley *et al.* (2007) suggest that young giant planets should
be fainter and smaller than predicted by standard evolutionary models starting
from arbitrary initial conditions (i.e. assuming a fully convective structure with
very large initial radius and luminosity). To reach this conclusion, the authors
have used initial conditions for planetary evolutionary models derived from the
core-accretion mechanism. Marley *et al.* (2007) find lower initial internal entropy
than usually assumed, as a result of the treatment of the accretion shock through
which the accreted matter is processed. This result is a direct consequence of the
assumption that the accreting gas loses most of its internal entropy through the
shock. Such assumption is derived from the shock boundary conditions of Stahler
et al. (1980), which were initially developed for the study of accretion onto *protostars*. The conclusion of Marley *et al.* (2007) strongly depends on the details of
the accretion shock. But their approach represents an improvement over previous
planetary models based on arbitrary initial conditions.

The work of Marley *et al.* (2007) emphasizes the high sensitivity of early planet
evolution to initial conditions, as previously done by Baraffe *et al.* (2002) for
low-mass stars and brown dwarfs. It has important consequences on the identification of planetary mass objects in young clusters and on detection strategies. This
work also raises the question of whether the faintness of young planetary mass
objects may be used as a criterion to distinguish a brown dwarf from a planet.
The answer is non-trivial, but one must keep in mind that accretion through a disk
(circumstellar or circumplanetary) is a common process of both star and planet
formations. Moreover, as already mentioned, the conclusion that young planets
should be faint relies on an treatment of accretion which can also be applied to
the formation of proto-stars and proto-brown dwarfs (see Section 15.4.2). Thus,
for the same reasons, brown dwarfs may as well be fainter than predicted by current evolutionary models. Multi-dimensional radiative transfer and hydrodynamic
simulations of the accretion process and the resulting shock could provide more rigorous post-shock initial conditions for young *planet* and *stellar* models. Although
still very challenging, this level of complexity seems to be required to improve in
the field.

15.4.3.3 Inner structure diagnostic

Interior and atmosphere structures of brown dwarfs and giant planets can be
described by the same basic physical ingredients (equation of state, opacities,
convection). Some degree of refinements are, however, required for the interior
structure of planets, as shown by the studies of our own giant planets and of transit
exoplanets, which provide valuable information on their mean density and on their
mean composition. Models devoted to the interior of our own giant planets, such

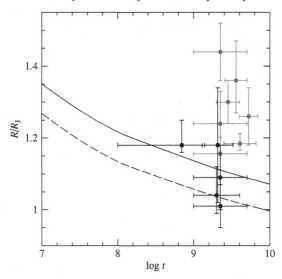

Fig. 15.6. Comparison between the observed radius of transit planets (*solid dots*) and model predictions. The curves show the evolution of a 1 M_J, irradiated by a Sun at 0.05 AU, with solar metallicity (*solid line*) and with a total heavy element enrichment $Z = 13\%$ (*dashed line*).

as Jupiter and Saturn, show that a substantial level of heavy element enrichment is required to fulfil various observational constraints (see Guillot 2005 and references therein). The analysis of transiting planets confirms that the presence of a significant amount of heavy elements is a property also shared by exoplanets. This is illustrated in Figure 15.6, which shows that some of the transit planets recently discovered display a radius significantly smaller than solar metallicity planet models. A heavy element enrichment of at least ~6 times solar (i.e. a metallicity in mass fraction $Z = 13\%$), similar to the amount of heavy elements in Jupiter, is required to explain the observed radius. This enrichment is explained in the framework of the core-accretion model (see Section 15.3) and is thus a signature of the formation process. A planetary mass object with a *significantly* smaller radius than predicted by solar metallicity models can thus be identified as a genuine planet. This criterion to work unambiguously requires very accurate observed radius and a high level of metal enrichment, such as in HD149026b, a Saturn mass planet with ~70 $M_\oplus$ of heavy elements (Sato *et al.* 2005). This criterion, however, becomes useless for moderately heavy element-enriched planets ($Z \lesssim 10\%$).

On the contrary, as illustrated by Figure 15.6, a significant fraction of transiting exoplanets have much larger radii than predicted by current planetary models. Many efforts are devoted to try to explain this puzzling property. It may point to missing physics in current planetary models. This possibility is attractive, in the context of identifying distinct properties between brown dwarfs and planets, since

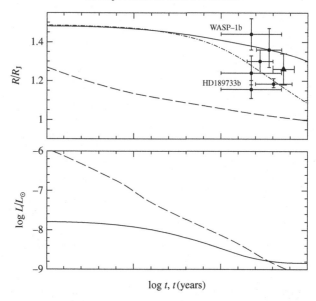

Fig. 15.7. Evolution of the radius and luminosity of a $1\,M_J$ planet orbiting a Sun at 0.05 AU (irradiation effects are included). The models have a central dunite core $M_c = 6\,M_\oplus$ and metal enrichment $Z_{env} \sim 6Z_\odot$ in the envelope. *Long dashed line*, adiabatic interior; *solid line*, model including layered convection with 100 layers; *dash-dotted line*, layered convection with 50 layers. The symbols indicate the observed values of transit planets with abnormally large radii. The triangle corresponds to HD206458b.

abnormally large radii would be the signature of a physical process specific to planets. The recent suggestion of reduced heat transport in planetary interiors due to the presence of a molecular weight gradient, inherited from the formation process, is interesting in this context (Chabrier & Baraffe 2007). These authors show that the conditions in planetary interiors are favourable to the development of double diffusive layered or overstable convection in the presence of a compositional gradient. The efficiency of heat transport is dramatically reduced compared to large-scale adiabatic convection and the planet evolution is slowed down. The effects on the radius and luminosity evolution are displayed in Figure 15.7, indicating that this process can explain the abnormally large radii of transit planets. Another signature of this process is a significantly fainter luminosity compared to the fully convective case at ages younger than 1 Gyr (Figure 15.7, *lower panel*). This property could be used to disentangle a planet from a brown dwarf. The process of double diffusive layered or overstable convection in planetary interiors thus needs to be further investigated, in order to confirm its existence and the later prediction on the luminosity.

15.4.3.4 Atmospheric diagnostic

The signature of the nature of a planetary mass object may be provided by the presence of highly non-solar composition in planet atmospheres, as observed in Jupiter and Saturn. In a preliminary attempt to point out this possibility, Chabrier *et al.* (2007a) highlight the differences in spectra expected between a solar composition spectrum with $T_{eff} = 1200$ K and its counterpart five times enriched in metals (see their Figure 6). The results show a strong difference of around 4.5 μm, due to an increased absorption of CO in the metal-enriched spectrum – a feature reachable by Spitzer. However, metallicity indicators may be strongly affected by non-equilibrium chemistry (Saumon *et al.* 2003), and the problem is far from being solved. A great deal of work must still be done to allow robust predictions on the abundance patterns and non-solar composition signatures in planet atmosphere. This avenue, however, deserves to be explored as it provides one of the most promising identification diagnostic.

15.5 Conclusions

The discovery of extra-solar planetary systems with very different properties from our own Solar System has overturned our theoretical understanding of how planets and planetary systems form. Theorists are therefore faced with an ambitious goal: developing a theoretical understanding of planet formation and evolution that stands up to the observational confrontation with quantitative predictions, guiding the direction of future observations. This effort requires to understand not only how planets form but also how they evolve after formation.

Focusing on the formation of giant planets, the recent years have seen the development of two main theoretical scenarios: one based on the direct collapse formation scenario and the other on the core-accretion model. The former is too numerically demanding to allow a quantitative comparison with current observational data, either for the Solar System or for extra-solar planets. The latter, the core-accretion model, including its recent extensions, has on the opposite reached a sufficient degree of maturity to allow for detailed calculations and explicit comparison not only with our Solar System but also, in a statistical way, with the currently detected population of extra-solar planets. Hence, this formation model has moved from the realm of conjectures to the one of testable hypothesis, opening the era of quantitative planetology.

The knowledge of planet formation will gain in the near future from many observational programmes, which, on the one hand, will improve our knowledge of protoplanetary disk structure and evolution, and, on the other, will enlarge the parameter space of observed extra-solar planets. On the one hand, protoplanetary disk knowledge will be greatly improved, e.g., by the already launched Spitzer Space Observatory, the ALMA facility, the James Webb Space Telescope and the Herschel Space Observatory. On the other hand, extra-solar planets will be discovered and characterized by the already launched CoRoT mission, the imminent KEPLER satellite, the SPHERE VLT-Planet finder, the JWST, ALMA, the GAIA probe, future microlensing surveys and future high-precision radial-velocity surveys (EXPRESSO/VLT, CODEX/E-ELT). Each of these facilities, due to their specific observational technique, will shed light on different parts and evolution epoch of protoplanetary disks and on different populations of extra-solar planets. Understanding the meaning of all these future observations and discoveries requires a unified theoretical framework in which all these pieces of the puzzle can be assembled to bring knowledge on the formation and evolution of planetary systems in general and our own Solar System in particular.

Theorists must therefore develop planetary system formation models, including in a global and unified framework both gas giant planets and terrestrial rocky planets, and combine planet formation models to early inner structure and evolutionary models in order to provide a fully consistent picture between formation and evolution. To reach this goal, they will take advantage of the wealth of observational data that will become available in the near future, both related to protoplanetary disks and to mature planets. Deciphering, in such an unified approach, the complex processes acting during the formation of planetary systems may allow us to understand the origin and degree of uniqueness of our own Solar System. It also represents the very first step towards the understanding of the conditions that led to the apparition of life on Earth (e.g. presence of liquid water) and to answer the exciting question on their ubiquity.

References

Adams, F. C., Ruden, S. P., Shu, F. H. (1989), *ApJ*. **347**, 959

Alibert, Y., Mordasini, C., Benz, W. (2004), *A&A* **417**, L25

Alibert, Y., Mordasini, C., Benz, W., Naef, D. (2008), *A&A*, in preparation

Alibert, Y., Mordasini, C., Benz, W., Winisdoerffer, C. (2005a), *A&A* **434**, 343

Alibert, Y., Mousis, O., Mordasini, C., Benz, W. (2005b), *ApJ*. **626**, L57

Bakos, G.A., Kovacs, G., Torres, G., *et al.* (2007), *Astrophys. J.*, **670**, 826–832 astro-ph/7050126

Baraffe, I., Chabrier, G., Allard, F., Hauschildt, P. H. (2002), *A&A*, **382**, 563

Baraffe, I., Chabrier, G., Barman, T. S., Allard, F., Hauschildt, P. H. (2003), *A&A* **402**, 701

Barge, P., Sommeria, J. (1995), *A&A* **295**, L1

Beaulieu, J.-P., Bennett, D. P., Fouqué, P., Williams, A., Dominik, M., *et al.* (2006), *Nature* **439**, 437

Beckwith, S., Sargent, A. (1996), *Nature* **383**, 139

Benz, W., Mordasini, C., Alibert, Y., Naef, D. (2006), in *Proc. Conf. Tenth Anniversary of 51 Peg-b: Status of and prospects for hot Jupiter studies*, Eds. L. Arnold F. Bouchy, and C. Moutou, Frontier Group, Paris, 24

Bonfils, X., Forveille, T., Delfosse, X., Udry, S., Mayor, M., *et al.* (2005), *A&A* **443**, L15

Bonfils, X., Mayor, M., Delfosse, X., Forveille, T., Gillon, M., *et al.* (2007), *A&A* **474**, 293

Boss, A. (1998), *ApJ.* **503**, 923–937

Boss, A. P. (2000), *ApJ.* **536**, L101

Brush, S. G. (1978), *J. His. Astron.* **9**, 71.

Butler, R. P., Vogt, S. S., Marcy, G. W., Fischer, D. A., Wright, J. T., *et al.* (2004), *ApJ.* **617**, 580

Caballero, J. A., Bjar, V. J. S., Rebolo, R., *et al.* (2007), *A&A* **470**, 903

Chabrier, G., Baraffe, I. (2007), *ApJ.* **661**, L81

Chabrier, G., Baraffe, I., Selsis, F., Barman, T. S., Hennebelle, P., Alibert, Y. (2007a), *Protostars and Planets V*, Eds. B. Reipurth, D. Jewitt, and K. Keil (University of Arizona Press), p. 623

Chabrier, G., Gallardo, J., Baraffe, I. (2007b), *A&A*, **472**, L17

Chauvin, G., Lagrange, A.-M., Dumas, C., Zuckerman, B., Mouillet, D., Song, I., Beuzit, J.-L., Lowrance, P. (2004), *A&A* **425**, L29

Comeron, F., Neuhäuser, R., Kaas, A. A. (2000), *A&A* **359**, 269

Cristiani, S., Avila, G., Bonifacio, P., Bouchy, F., Carswell, B., *et al.* (2008), *Nuovo Cimento B*, **122** (9), 1165–1170, arXiv:0712.4152

Deming, D., Harrington, J., Seager, S., Richardson, L. J. (2006), *ApJ.* **644**, 560

Deming, D., Seager, S., Richardson, L. J., Harrington, J. (2005), *Nature* **434**, 740

Demory, B., Gillon, M., Barman, T., Bonfils, X., Mayor, M., *et al.* (2007), *A&A* **475**, 1125

Dominik, C., Blum, J., Cuzzi, J. N., Wurm, G. (2007), *Protostars and Planets V*, Eds. B. Reipurth, D. Jewitt, and K. Keil (University of Arizona Press)

Dutrey, A., Guilloteau, S., Ho, P. (2007), *Protostars and Planets V*, Eds. B. Reipurth, D. Jewitt, and K. Keil (University of Arizona Press), p. 495

Eggenberger, A., Udry, S., Chauvin, G., Beuzit, J.-L., Lagrange, A.-M., Segransan, D., Mayor, M. (2007), *A&A* **474**, 273

Eggenberger, A., Udry, S., Mayor, M. (2004), *A&A* **417**, 353

Endl, M., Cochran, W. D., Kürster, M., Paulson, D. B., Wittenmyer, R. A., *et al.* (2006), *ApJ.* **649**, 436

Fischer, D., Marcy, G. W., Butler, R. P., Vogt, S., Laughlin, G. (2008), *ApJ.*, **675**, 790–801

Fischer, D., Valenti, J. A. (2005), *ApJ.* **622**, 1102

Galland, F., Lagrange, A.-M., Udry, S., Chelli, A., Pepe, F., *et al.* (2005), *A&A* **444**, L21

Gallardo, J. (2007), Ecole Normale Supérieure de Lyon, PhD thesis Lyon, France

Gallardo, J., Baraffe, I., Chabrier, G. (2008), *A&A*, submitted

Gammie, C. F. (2001), *ApJ.* **553**, 174

Gillon, M., Pont, F., Demory, B.-O., Mallman, F., Mayor, M., *et al.* (2007), *A&A* **472**, L13

Goldreich, P., Ward, W. R. (1973), *ApJ.* **183**, 1051

Guillot, T. (2005), *Ann. Rev. Earth Planet Sci.* **33**, 493

Haisch, K., Lada, E., Lada, C. (2001), *ApJ.* **553**, L153

Hartmann, L., Cassen, P., Kenyon, S. J. (1997), *ApJ.* **475**, 770

Heemskerk, M. H. M., Papaloizou, J. C., Savonije, G. J. (1992), *A&A* **260**, 161

Henyey, L. G., Forbes, J. E., Gould, N. L. (1964), *ApJ.* **139**, 306.

Hubickyj, O., Bodenheimer, P., Lissauer, J. J. (2005), *Icarus* **179**, 415

Ida, S., Lin, D. N. C. (2004), *ApJ.* **604**, 388

Ida, S., Lin, D. N. C. (2005a), *ApJ.* **626**, 1045

Ida, S., Lin, D. N. C. (2005b), *Protostars and Planets V*, Eds. B. Reipurth, D. Jewitt and K. Keil (University of Arizona Press), p. 8141

Jayawardhana, R., Coffey, J., Scholz, A., Brandeker, A., vanKerkwijk, M. H. (2006), *ApJ.* **648**, 1206–1218

Johnson, J. A., Fischer, D. A., Marcy, G. W., Wright, J. T., Driscoll, P., Butler, R. P., Hekker, S., Reffert, S., Vogt, S. S. (2007), *ApJ.* **665**, 785

Klahr, H., Bodenheimer, P. (2006), *ApJ.* **639**, 432

Kokubo, E., Ida, S. (2000), *Icarus* **143**, 15

Kornet, K., Stepinski, T., Rózyczka, M. (2001), *A&A* **378**, 180

Lissauer, J. (1993), *ARAA* 31, 129–174

Laughlin, G., Korchagin, V., Adams, F. C. (1997), *ApJ.* **477**, 410

Laughlin, G., Rozyczka, M. (1996), *ApJ.* **456**, 279

Lin, D.N.C., Bodenheimer, P., Richardson, D.C. (1996), *Nature* **380**, 606

Lodders, K. (2003), *ApJ.* **591**, 1220

Lovis, C., Mayor, M. (2007), *A&A* **472**, 657

Lovis, C., Mayor, M., Pepe, F., Alibert, Y., Benz, W., *et al.* (2006), *Nature* **441**, 305

Luhman, K., Peterson, D. E., Megeath, S. T. (2004), *ApJ.* **617**, 565

Marcy, G.W., Butler R.P., Fischer, D., Vogt, S., Tinney, J.T., *et al.* (2005), *Prog. Theor. Phys. Suppl.* **158**, 24

Marley, M. S., Fortney, J. J., Hubickyj, O., Bodenheimer, P., Lissauer, J. J. (2007), *ApJ.* **655**, 541

Mayer, L., Quinn, T., Wadsley, J., Stadel, J. (2004), *ApJ.* **609**, 1045

Mayor, M., Queloz, D. (1995), *Nature* **378**, 355

Mayor *et al.* (2008), *A&A*, in press, astro-ph/0806.4587

McArthur, B.E., Endl, M., Cochran, W.D., Benedict, G.F., Fischer, D., *et al.* (2004), *ApJ.* **614**, L81

Melo, C., Santos, N.C., Gieren, W., Pietrzynski, G., Ruiz, M.T., *et al.* (2007), *A&A* **467**, 721

Mohanty, S., Jayawardhana, R., Basri, G. (2005), *ApJ.* **626**, 498

Mordasini, C., Alibert, Y., Benz, W., Naef, D. (2008a), *A&A*, submitted

Mordasini, C., Alibert, Y., Benz, W., Naef, W. (2008b), in *'Extreme Solar Systems'* ASP Conference Series, Eds. D. Fischer, F. Rasio, S. Thorsett, and A. Wolszczan, Vol. 398, p. 235

Murray, N., Chaboyer, B., Arras, P., Hansen, B., Noyes, R. (2001), *ApJ.* **555**, 801

Muzerolle, J., Luhman, K., BriceŨo, C., Hartmann, L., Calvet, N. (2005), *ApJ.* **625**, 906

Natta, A., Testi, L., Randich, S. (2006), *A&A* **452**, 245

Nordström, B., Mayor, M., Andersen, J., *et al.* (2004), *A&A* **418**, 989

Ohtsuki, K., Stewart, G. R., Ida, S. (2002), *Icarus* **155**, 436

Pasquini, L., Dollinger, M. P., Weiss, A., Girardi, L., Chavero, C., *et al.* (2007), *A&A* **473**, 979

Pepe, F., Correia, A. C. M., Mayor, M., Tamuz, O., Benz, W., *et al.* (2007), *A&A* **462**, 769

Podolak, M. (2003), *Icarus* **165**, 428

Pollack, J. B., Hubickyj, O., Bodenheimer, P., Lissauer, J. J., Podolak, M., Greenzweig, Y. (1996), *Icarus* **124**, 62

Press, W., Teukolsky, S. A., Vetterling, W. T., Flannery, B. P. (1992). *Numerical Recipes in FORTRAN. The Art of Scientific Computing.* (Cambridge University Press)

Rafikov, R. (2005), *ApJ.* **621**, L69

Rafikov, R. (2007), *Astrophys. J.,* **682,** 527–54, astro-ph/07073636

Rice, W. K. M., Armitage, P. J., Bonnell, I. A., Bate, M. R., Jeffers, S. V., Vine, S. G. (2003), *MNRAS* **346**, L36

Rivera, E. J., Lissauer, J. J., Butler, R. P., Marcy, G. W., Vogt, S. S., *et al.* (2005), *ApJ.* **634**, 625

Santos, N. C., Bouchy, F., Mayor, M., Pepe, F., Queloz, D., *et al.* (2004), *A&A* **426**, L19

Santos, N. C., Israelian, G., Mayor, M. (2001), *A&A* **373**, 1019

Santos, N. C., Israelian, G., Mayor, M., Bento, J. P., Almeida, P. C., *et al.* (2005), *A&A* **437**, 1127

Sato, B., Fischer, D. A., Henry, G. W., *et al.* (2005), *ApJ.* **633**, 465

Saumon, D., Guillot, T. (2004), *ApJ.* **609**, 1170

Saumon, D., Marley, M. S., Lodders, K., Freedman, R. S. (2003), *IAU Symposium* **211**, 34

Stahler, S. W. (1983), *ApJ.* **274**, 822

Stahler, S. W. (1988), *ApJ.* **332**, 804

Stahler, S. W., Shu, F. H., Taam, R. E. (1980), *ApJ.* **241**, 637

Tanga, P., Babiano, A., Dubrulle, B., Provenzale, A. (1996), *Icarus* **121**, 158

Thommes, E., Duncan, M., Levison, H. (2003), *Icarus* **161**, 431

Udry, S., Bonfils, X., Delfosse, X., Forveille, T., Mayor, M., *et al.* (2007), *A&A* **469**, L43

Udry, S., Mayor, M., Benz, W., Bertaux, J.-L., Bouchy, F., *et al.* (2006), *A&A* **447**, 361

Udry, S., Mayor, M., Naef, D., Pepe, F., Queloz, D., *et al.* (2002), *A&A* **390**, 267

Udry, S., Mayor, M., Santos, N. C. (2003), *A&A* **407**, 369

Udry, S., Santos, N. C. (2007), *Ann. Rev. A&A* **45**, 397

Vogt, S. S., Butler, R. P., Marcy, G. W., Fischer, D. A., Henry, G. W., *et al.* (2005), *ApJ.* **632**, 638

Ward, W. (1997), *Icarus* **126**, 261

Weidenschilling, S. J. (1977), *MNRAS* **180**, 57

Weidenschilling, S. J., Cuzzi, J. N. (1993), *Protostars and Planets III* (University of Arizona Press), p. 1031

Wetherill, G. W., Stewart, G. R. (1989), *Icarus* **77**, 330

Zucker, S., Mazeh, T. (2002), *ApJ.* **568**, L113

16

Planet formation: assembling the puzzle

G. Wurm and T. Guillot

16.1 Introduction

Although not precisely in its infancy, the question of building planets and planetary systems still faces many challenges: how are planetesimals assembled from micrometre-sized grains? What does radial transport do to growing dust aggregates? Do solids concentrate close to the star or do (metre size) objects vanish rapidly into the central star? How are planets formed from planetesimals? And how do giant planets form that have to acquire hydrogen and helium before the gas is accreted onto the star or is swept away by stellar winds and photoevaporation? Are many generations of planets formed and then lost? Can we explain the compositions of planets in the structure of our present solar system?

These are but a few of the basic questions which are currently the focus of a highly active research field. Presenting a complete overview of the problem is beyond the scope of this short chapter and would not be a long-lasting one as the field is rapidly evolving. We present here what we believe are some important pieces of the puzzle, in the domain of planetesimal formation and giant planet composition.

16.2 Planetesimals

There are several important steps of structure formation after the solar nebula formed and before full-size planets came into existence. One of them is the formation of planetesimals. There is some ambiguity in the term planetesimal as it is used throughout the literature. We take them as roughly *km-size* objects that were formed from the *dust* in the solar nebula in case of our own solar system or, more generally, that were formed in protoplanetary disks surrounding young stars. This highlights two aspects.

- Planetesimals have to be formed by processes which selectively put dust together. Dust motion as well as the motion of somewhat larger bodies are determined by the interaction

Structure Formation in Astrophysics. ed. G. Chabrier. Published by Cambridge University Press.

with the gas in a gaseous protoplanetary disk. This keeps relative velocities *relatively* small compared to velocities occurring in later stages. For example, bodies on different Keplerian orbits in the current solar system can collide at several tens of kilometres per second. Planetesimal formation *only* has to consider collisions occurring at tens of metres per second, which is a factor of 1000 less.

- Once an object has reached about a kilometre in size, its self-gravity becomes important. The escape velocity is then on the order of $1 \, m \, s^{-1}$, which is on the same order of magnitude as that of fragment velocities after a collision at tens of metres per second. A planetesimal can therefore grow further through collisions with smaller bodies even if a collision is disrupting the smaller body. It just collects the debris afterwards by means of its own gravitational attraction.

This already leads to the heart of planetesimal formation, which are collisions and, supposedly, growth by means of these collisions if sticking of the dusty bodies occurs. As mentioned, sub-planetesimal-size objects collide at speeds lower than $100 \, m \, s^{-1}$, which sounds to be slow in comparison with planetary-scale collisions. But slow and fast are comparative terms, and if particles are supposed to stick together in collisions, then $100 \, m \, s^{-1}$ ($360 \, km \, h^{-1}$) is not a gentle knock and has to be regarded as a high-speed collision for this purpose.

This is a major bottleneck for the formation of planetesimals. While we would like to caution the reader to use everyday experience to judge the outcome of a collision between dusty bodies, it is certainly a valid point of criticism that pre-planetesimals might have a hard time growing at large impact energies. It should be noted that planetesimals are not just a detail in planet formation scenarios. Terrestrial planets have to be built from dust, and it is current wisdom that this encompasses a phase of km-size bodies or planetesimals existing, which then grow further to protoplanets and planets (Weidenschilling & Cuzzi 1993). As we currently do not have a detailed understanding of how planetesimals are formed, we have an undesirable large freedom in speculating when and where planetesimal growth and subsequent planet formation took place. Is planetesimal formation triggered by some events? Does it occur continuously in all parts of protoplanetary disks or does it need certain conditions? But to return to the beginnings, then how far do we understand planetesimal formation?

16.2.1 Gas–grain coupling times

Small dust particles embedded in the gas of a protoplanetary disk couple so well to it that they follow any gas motion almost perfectly. If the gas is turbulent or laminar, accreting onto the star or moving on sub-Keplerian orbit, the particles rapidly adapt to any change in the gas flow. Ideal coupling means that two dust particles would never collide as streamlines cannot penetrate each other. However, collisions occur due to two reasons. On the one hand, there is no perfect coupling. In reality, dust particles are much more massive than gas molecules. To change a dust particle's motion, the gas has to apply a drag force on the particle. A particle reacts to an abrupt change in gas flow with an exponential change of its own velocity on a timescale τ_f. This timescale is the gas–grain coupling time. At low gas pressure, it is given by Blum *et al.* (1996) as

$$\tau_f = \frac{m}{A} \frac{\epsilon}{\rho_g v_m},\tag{16.1}$$

where m is the mass and A the geometrical cross section of the dust particle, $\epsilon = 0.58$ is a constant which has to be determined experimentally, ρ_g is the gas density and v_m is the mean thermal velocity of the gas molecules. A rough value for a micron particle at 10^{-2} mbar pressure (1 AU in a minimum mass solar nebula) is $\tau_f = 1$ s. Assuming for the moment that the radial component of the star's gravity is balanced by a Kepler rotation, a particle at a height z above the midplane of the disk would be subject to a gravitational acceleration towards the midplane of about $a_g = G \frac{M}{r^2} \frac{z}{r}$, with M being one solar mass. At $r = 1$ AU and vertical height $z = 0.1$ AU, this is about $a_g = 6 \cdot 10^{-4}$ m s^{-2}. A micrometre particle therefore settles to the midplane with a typical velocity of $v = a_g \cdot \tau_f \approx 1$ mm s^{-1} (assuming τ_f not to increase too much at 0.1 AU above the disk).

This quantifies the qualitative statement that small dust particles couple well to the gas. A micrometre dust particle would not settle to the midplane of the disk in its lifetime of a few million years. So particles have to grow larger to settle to the midplane. It is clear that any gas motion (e.g. convection and turbulence) can easily disperse dust particles to large heights again. Furthermore, if settling velocities are on the order of millimetres per second, the relative velocities between two dust particles are still an order of magnitude smaller.

16.2.2 First fractal growth phase

Compared to differential sedimentation, Brownian motion of micron-size particles is on the order of millimetres per second. The first collisions of particles in protoplanetary disks are therefore dominantly generated by Brownian motion which only loses importance for 100 micrometre particles. Afterwards sedimentation to the midplane takes over as the dominant source to generate relative velocities. If particles collide at velocities smaller than 1 m s^{-1} – as they initially do – they stick to each other by surface forces (Poppe *et al.* 2000).

It was studied in microgravity experiments that a cloud of particles left to itself in Brownian motion leads to the growth of larger aggregates through sticking collisions (Blum *et al.* 2000) and a rather small size distribution of fractal dust aggregates results. Fractal here refers to the fact that the mass, m, of an aggregate is not proportional to the volume or size, a, to the third power of the aggregate but depends on a not necessarily integer but fractal dimension d_f or $m \sim a^{d_f}$ (Meakin 1991). Wurm and Blum (1998) observed individual collisions between dust aggregates consisting of 0.95-μm (radii) SiO_2 particles at velocities up to 1 cm s^{-1}. In the individual collisions, a hit-and-stick behaviour leading to the formation of larger aggregates could be imaged directly. The aggregates were generated before the experiments in a turbulent environment. Analysis of the turbulent growth shows that typical collision velocities during growth were larger than 7 cm s^{-1} (Wurm & Blum 1998). In these sets of experiments, fractal dimensions $d_f \approx 2$ were found. As seen in Eq. (16.1), the friction time depends on the ratio between particle mass and cross section. For the low fractal dimensions, this does not change significantly during growth. Particle settling is therefore not dependent on particle size and collision velocities stay small. It was found from experiments, computer simulations and moderate scaling that once aggregates of cm-size have formed, the energy is sufficient to lead to a compaction of the fractal aggregates (Dominik & Tielens 1997; Blum & Wurm 2000). Then gas–grain coupling times increase, sedimentation speeds up and aggregates can reach the midplane. As the cross sections of fractal aggregates are large, this does not imply that the initial growth is slow. Particles can and will grow to cm-aggregates in 1000 years (Blum 2004). There is little doubt that this initial growth occurs as it depends little on the disk or particle parameters. The question is rather where the dust particles come from, which are observed in protoplanetary disks for millions of years as they initially vanish rapidly during growth.

16.2.3 *From centimetre to decimetre*

Once particles get compact, the gas–grain coupling time increases, and in general, the velocities with respect to the gas and with respect to each other increase as well. In a disk where the gas density and pressure decrease with radial distance to the star, the disk is supported by this pressure gradient and rotates slightly slower than Keplerian. Solid particles follow the gas motion as long as they are coupled well enough. Quantified, a body of about metre size has a coupling time on the order of one orbit. Smaller objects are coupled to the gas motion in much less than one orbit. They can be regarded as moving on sub-Keplerian orbits with the disk. But as solids are too dense to be supported by the pressure gradient, they feel a residual gravity and radially drift inwards. This leads to relative velocities between different kind of particles of up to 10 m s^{-1} for 10-cm particles colliding with smaller aggregates (Weidenschilling & Cuzzi 1993). Small dust aggregates consisting of micron-size dust particles impinging on a larger dust aggregate no longer stick at these velocities but are fragmented and erosion occurs (Blum & Wurm 2000).

Growth within this size regime is still possible though. One idea here is that the gas within the disk can help reaccreting fragments after a collision with a small dust aggregate (Wurm *et al.* 2001). Even if fragmentation of dust aggregates does occur at about 1 m s⁻¹, a net growth of a target by reaccreting fragments due to gas drag occurs at velocities larger than 10 m s⁻¹.

16.2.4 Metre-size bodies and their problems

The aforementioned reaccretion of gas for small bodies is possible in the regime of large Knudsen numbers (free molecular flow) and the transition regime to low Knudsen numbers. However, once in the hydrodynamic regime, streamlines are usually surrounding a body and fragments are transported around the body rather than back to it. However, the growing bodies are rather porous. In that case, gas flows through the body at the front side facing the gas flow in the protoplanetary disk. As before, this allows fragments to be dragged back to the surface if they are slow enough and small enough to couple well to the gas (Wurm *et al.* 2004). Details are subject to discussion (Sekiya & Takeda 2005).

The search for mechanisms to provide growth at high-collision velocities is based on the premise that these collisions are less suited for growth than slow collisions. However, Wurm *et al.* (2005) found that net growth of a larger dust target in collisions with a projectile several millimetres in size even at 25 m s⁻¹ is possible. In fact, the experiments showed that collisions at higher velocities showed accretion of 50% of the projectile mass by the larger target while collisions slower than 13 m s⁻¹ (for the given parameters) did not show significant accretion (Figure 16.1).

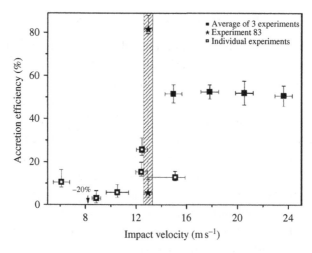

Fig. 16.1. Fraction of a dust projectile accreted onto a dust target during collisions. (Adapted from Wurm *et al.* (2005).)

In current experiments, collisions at velocities larger than $25 \, \mathrm{m \, s^{-1}}$ and at larger impact energies are studied (Wurm, personal communications). Under what conditions growth can still proceed needs to be studied in more detail and will be clearer in the near future. It might be sufficient to have a collisional growth of bodies of several tens of centimetres if gravito-turbulent concentration could proceed from there to build larger objects. This is the focus of current theoretical research (Johansen *et al.* 2007). However, collisional evolution and gas erosion of dense clumps have to be considered in these scenarios in the future as well, and if erosion is the result of a collision and/or gas flow, the growth might not readily produce planetesimals, especially not as dramatic as growing Ceres-like dwarf planets in a small number of orbits, as currently proposed.

Metre-size bodies have yet another problem. Not only is the collision velocity with respect to other smaller particles the largest but they also drift inwards rapidly. Left to itself in a disk where the density decreases with radial distance, a metre-size object can drift about 1 AU in 100 years.

16.2.5 Processing (pre-)planetesimals in the inner disk

There are two more processes related to planetesimals and immediate interaction with the gas within protoplanetary disks which we would like to mention here, as they have not been considered so far. While the interaction of a planetesimal as a whole is decreasing with size, it is very likely that these bodies continue to be composed of loosely bound dust particles for a certain time. At the surface of a planetesimal, this dust still feels gas drag. In wind tunnel experiments, Paraskov *et al.* (2006) studied the effect of gas drag on a dusty body. They found that a dusty body is eroded under certain conditions (Figure 16.2). Applied to protoplanetary disks, planetesimals are subject to potential destruction in the inner AU if they orbit on an (even slightly) eccentric orbit. This leads to a redistribution of matter and

Fig. 16.2. Eolian erosion of a dust pile in a gas flow at 3 mbar at a flow velocity of $63 \, \mathrm{m \, s^{-1}}$. (Adapted from Paraskov *et al.* (2006).)

might favour the formation of larger bodies on circular orbits, but the consequences still have to be worked out in more detail. The same effect of gas erosion might destroy a dense clump of not yet adhesively bound particles much easier and might be an obstacle to the formation of very large gravitationally bound regions of solids as indicated above.

While this process requires a dense gaseous disk to be in place, a second process is important in environments which are optically thin. Recently, Wurm (2007) found that under the combined action of a solid-state greenhouse effect and thermophoresis, a dusty body can be eroded by starlight close to a star up to several tenths of an astronomical unit. Like eolian erosion, erosion by starlight leads to a redistribution of solid matter from large bodies to small dust aggregates. If, as mentioned above, metre-size bodies drift inwards rapidly, they might be destroyed by light-induced erosion, not only to join the dust population within the disk again but also to stay closer to the star. Therefore, the solid matter is not lost to the disk but probably concentrates at the inner edge which might then be a preferred spot for further planetesimal or planet formation.

Currently, models of planet formation usually start with planetesimals distributed homogeneously throughout the protoplanetary disk, accounting for an enhancement beyond the snow line due to ice being present as an abundant solid component. Some recycling of ice and water at the snow line has recently been considered in more detail (Ciesla & Cuzzi 2006). Otherwise, it has not really been considered yet how the inward drifting solid material influences planetesimal and planet formation. While radial transport is recognized as an important mechanism, the fate of drifting solid matter is ignored by unjustified simplifications usually assuming that inward drift is bad for planetesimal formation.

The most widely spread misbelief is that metre-size bodies which rapidly drift inwards have to end up in the central star. This is not so. As discussed above, solids might concentrate at certain distances beyond their sublimation point, which will have profound influence on planetesimal formation.

In the light of the analysis of the giant planets of the solar system and extrasolar systems as outlined below, it is probably not possible to understand the content of heavy elements without considering where planetesimals form and at what rate; also modelling processes at the inner dust edges of protoplanetary disks have to be included in future studies.

16.3 The compositions of giant planets: clues to the formation of planetary systems

Giant planets are important pieces of the puzzle to understand the formation of planetary systems for several reasons: (i) their mass makes them key players for shaping planetary systems, by perturbing the orbits of planets, preventing their formation or even ejecting them; (ii) they must form relatively early, when the gas in the protoplanetary disk is still present, and they are thus direct witnesses of these early times and (iii) their composition, in terms of hydrogen and helium versus other elements ('heavy elements'), as well as the size of their central cores are crucial parameters to constrain formation models.

We can constrain the global compositions and structures of our giant planets, Jupiter, Saturn, Uranus and Neptune, from precise measurements of their gravity field. Their atmospheric composition can also be measured accurately. Further from us, the composition of giant planets orbiting around other stars can be inferred in less detail, but the larger statistics allows further tests for the planet formation models.

16.3.1 Jupiter and Saturn

Jupiter and Saturn have masses of 318 and $95 M_\oplus$, respectively. They are thought to consist of a central dense core (made of refractory material, probably rocks, with possibly ices) and an envelope of hydrogen and helium. Models of their interior structure and composition are based on fits to the gravity fields measured by the Pioneer, Voyager and Cassini-Huygens spacecrafts, as the departure from spherical symmetry caused by the rapid rotation of these planets yields constraints on the internal density profile (see Guillot 2005 and references therein).

Figures 16.3 and 16.4 shows the results of such models for Jupiter and Saturn (Saumon & Guillot 2004). Jupiter is found to possess a central core that is small (less than $12 M_\oplus$, or 4% of its total mass) and possibly inexistant. Saturn appears to have a larger core, in the $10–25 M_\oplus$ range. However, Saturn's structure may be affected by a sedimentation of helium on a very large scale, possibly leading to the formation of a helium shell onto the central core (Fortney & Hubbard 2003). This would certainly lead to a reduction of Saturn's central rock/ice core as needed by the models to fit the data and lead it to values more comparable to those for Jupiter. Note that the most important source of uncertainty remains the equation of state (EOS) of hydrogen, and it is hoped that new high-pressure experiments and simulations will lead to a clearer picture in the nearby future.

The larger than solar (by a factor 1.5–6) compositions of the envelopes of Jupiter and Saturn found from interior models is backed up by measurements in the atmospheres of these planets. Figure 16.5 synthesizes the measurements obtained for the major species in both planets. Unfortunately, water, the dominant species in terms of the mass balance after hydrogen and helium, has escaped a precise abundance measurement because it condenses in the atmospheres of both planets.

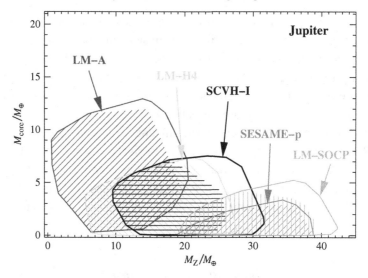

Fig. 16.3. Jupiter's core mass M_{core} and the mass of heavy elements mixed in the H/He envelope M_Z (in Earth masses, $M_\oplus$). The total mass of Jupiter is $318M_\oplus$. Each closed area represents the range of models that satisfy the observational constraints for a given choice of hydrogen EOS. An additional 2% uncertainty was applied to each EOS. (Adapted from Saumon and Guillot (2004).)

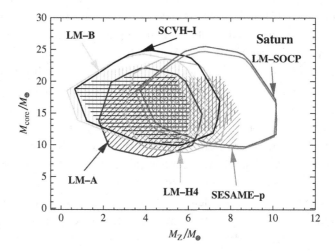

Fig. 16.4. Same as Figure 16.3 for Saturn. Note the difference of scale for M_Z. (Adapted from Saumon and Guillot (2004).)

However, globally, Jupiter's atmosphere appears to be enriched in heavy elements by a factor of 2–4 compared to the Sun (see Wong *et al.* 2004). We have less precise indications for Saturn, except for carbon, which is enriched by a factor of 7–9 (Flasar *et al.* 2005). The real surprise behind these measurements is the larger-than-solar abundance of noble gases (Ar, Kr and Xe) in Jupiter measured by the

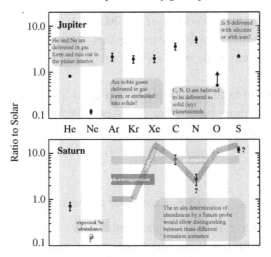

Fig. 16.5. Elemental abundances measured in the tropospheres of Jupiter (*top*) and Saturn (*bottom*) in units of their abundances in the protosolar nebula (in this case from Lodders 2003). The elemental abundances for Jupiter are derived from the in situ measurements of the Galileo probe (see Wong *et al.* 2004). The abundances for Saturn are spectroscopic determinations from Cassini for [He/H] and [C/H] (Flasar *et al.* 2005) and model-dependent ground-based measurements for [N/H] and [S/H] (Briggs & Sackett 1989). The predictions from different formation scenarios are shown as labelled.

Galileo probe, as these species are extremely difficult to trap into solids, raising the question of how their history did not follow that of hydrogen and helium.

16.3.2 Uranus and Neptune

Uranus and Neptune have masses of 14.538 and 17.148$M_\oplus$, respectively, and would be expected to have very similar structures and properties. This is largely the case, except for one mystery: Uranus emits at least ten times less heat in the infrared than Neptune! Apart from that, the characterization of their gravity fields and calculation of interior models show that their density profiles lie close to that of 'ices' (a mixture initially composed of H_2O, CH_4 and NH_3, but which rapidly becomes an ionic fluid of uncertain chemical composition in the planetary interior), except in the outermost layers ($\sim$20% in radius), which have a density closer to that of hydrogen and helium. If we impose the constraint that rock and ices follow a fixed ratio (the protosolar value, $\sim$1/3 and $\sim$2/3, respectively), the overall composition of both Uranus and Neptune is, by mass, about 25% rocks, 60–70% ices and 5–15% hydrogen and helium. Assuming both ices and rocks are present in the envelope, an upper limit to the amount of hydrogen and helium present is $\sim$4.2$M_\oplus$ for Uranus and $\sim 3.2 M_\oplus$ for Neptune (Podolak *et al.* 2000). A lower limit of $\sim 0.5 M_\oplus$ for both planets can be inferred by assuming that hydrogen and helium are only present in the outer envelope at $P < 100$ kbar (see Guillot 2005 and references therein).

The atmospheres of both planets are known from Voyager measurements to be highly enriched in carbon, by a factor of 20–60 over the solar value (with uncertainties coming both from the Uranus/Neptune measurements and from the carbon abundance in the Sun) (Gautier & Owen 1989).

16.3.3 Extrasolar planets

The discovery of extrasolar planets, and particularly of *transiting* extrasolar planets, gives us the possibility to have rough constraints on their compositions and see whether these vary with the different parameters of the problem, in particular the metallicity ([Fe/H]) of the parent star.

An important discovery is that the probability for a star to possess a giant planet companion is directly linked to its metallicity: the probability is around 10% for a solar-type star of solar metallicity and climbs to $\sim 30\%$ for [Fe/H] $= 0.5$ (Santos *et al.* 2004).

With transiting planets, we can now check whether there is a link between the compositions of giant planets and that of their parent star. One has to be reminded, however, of the large uncertainties linked to the EOSs, as discussed previously for Jupiter and Saturn: we cannot hope yet to determine precisely the absolute composition of a planet, but there is a lot to gain by looking at their *relative* compositions (see discussion by Guillot 2005).

This is perfectly illustrated by the first transiting *Pegasid* (aka hot Jupiter), HD209458b, a $0.69 M_J$ planet orbiting at 0.047 AU from its parent star. The radius of HD209458b has been measured from transit photometry to be around $1.35 R_J$, a value that shows that the planet is mostly made of hydrogen and helium but that is also too large to be explained by standard evolution models (Bodenheimer *et al.* 2001; Guillot & Showman 2002; Chabrier *et al.* 2004).

Since that discovery, other planets have been discovered: some are anomalously large, like HD209458b, but some are much smaller than possible for a hydrogen/helium planet, requiring the presence of large masses of heavy elements. The archetype of the last case is HD149026b, a $\sim 120 M_\oplus$ mass planet with a $\sim 70 M_\oplus$ mass core.

There are two possibilities to explain this: either some planets have some peculiarities inherent to their formation or orbital evolution that makes them larger (Bodenheimer *et al.* 2001; Winn & Holman 2005; Chabrier & Baraffe 2007) or there is a missing process (or slightly different physics) that affects all planets in a similar fashion (Guillot 2005; Guillot *et al.* 2006).

With the present sample of 20 transiting planets, Guillot claims that the first solution is unlikely because about 1/3 of the planets in the sample so far are anomalously large: a low-probability mechanism is ruled out.

Figure 16.6 shows the amount of heavy elements necessary to reproduce the observed radii when including an additional source of energy amounting to 0.5% of the incoming stellar flux. This scenario has been proposed on the basis of hydro-dynamical simulations that show that kinetic energy is generated in the atmosphere of Pegasids because of the inhomogeneous heating (i.e. strong winds develop) and is transported downwards, where it may be dissipated by stellar tides (Showman & Guillot 2002). However, the results are very similar when, e.g., arbitrarily increasing the opacities in the envelope.

The results in Figure 16.6 were first shown to imply a positive correlation between the mass of heavy elements in Pegasids and the metallicity of their parent stars on the basis of nine transiting planets known at the time (Guillot *et al.* 2006). The present results strengthen the case for this correlation. They also show that giant planets can possess large masses in heavy elements, a result not expected from formation models.

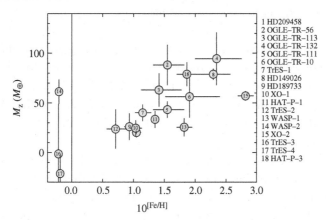

Fig. 16.6. Mass of heavy elements in the planets as a function of the metal content of the parent star relative to the Sun. The mass of heavy elements required to fit the measured radii is calculated on the basis of evolution models including an additional heat source slowing the cooling of the planet. This heat source is assumed equal to 0.5% of the incoming stellar heat flux (Showman & Guillot 2002). Negative core masses correspond to unphysical extrapolations. Horizontal error bars correspond to the 1σ errors on the [Fe/H] determination. Planets with no [Fe/H] determination are to the left of the figure. Vertical error bars are a consequence of the uncertainties on the measured planetary radii and ages. (Figure updated from Guillot *et al.* 2006.)

16.3.4 Consequences for formation models

The above description of the structure of solar and extrasolar giant planets pleads for a formation of these objects by the standard accretion theory (Pollack *et al.* 1996) rather than by a direct gravitational instability in the gas (Boss 2000):

- With maybe the exception of Jupiter, all giant planets in our solar system have massive, dense cores, which would be difficult to accumulate within a gravitational instability scenario.
- Transiting Pegasids also appear to possess large cores, up to almost $100M_\oplus$.
- The enrichment in heavy elements both for our giant planets and for Pegasids requires an early and efficient capture of planetesimals, which is difficult to explain in the gas instability case. Helled *et al.* 2006 discuss such a mechanism; however, this is based on a model in which the contraction of the clump is very slow, and furthermore, this does not include the finding that a forming giant planet opens in a gaseous disk a gap that is almost impermeable to grains (Paardekooper 2007). On the contrary, the enrichments and core masses in Jupiter and Saturn are reproduced by accretion models (Alibert *et al.* 2005b).
- The correlation between frequency of giant planets and metallicity is well understood in the accretion scenario (Ida & Lin 2004; Alibert *et al.* 2005a), not in the gas instability scenario.
- The heavy element mass–metallicity correlation presented here is unexplained but seems to be more naturally linked to an accretion mechanism.

The question of the enrichment of the envelopes of the giant planets remains an important one however. The scenario by Alibert *et al.* (2005b) does indeed reproduce the constraints for Jupiter and Saturn, but it relies on a global migration of the planets coupled to an efficient collection of the planetesimals within their gravitational reach, with almost no shepherding that would limit this process. This should be tested by dedicated simulations.

In any case, this requires an early capture of heavy elements, in a planet that is not already formed. Explaining the observed atmospheric abundances requires an efficient upward mixing and raises the possibility that Jupiter's core was eroded (Guillot *et al.* 2004). This would explain the apparent discrepancy between the present structures of Jupiter and Saturn.

The problem of the capture of noble gases in Jupiter remains. Several explanations have been proposed, involving a capture of very cold, amorphous planetesimals (Owen & Encrenaz 2005), the trapping of noble gases by clathration in crystalline icy planetesimals (Gautier *et al.* 2001) or a photoevaporation of the protosolar disk atmosphere that would lead to a preferential escape of hydrogen, helium and probably neon, species that are not trapped onto grains even at very low temperatures in the outer parts of the disk (Guillot & Hueso 2006). Figure 16.5 shows how different scenarios would affect Saturn, something that could be measured efficiently by a probe sent into Saturn's atmosphere (e.g. the Kronos mission).

16.4 Conclusions

Although a global picture in which collisional growth and gas accretion play a dominant role in the formation of planetary systems seems favourable in general, we still do lack important pieces to complete the puzzle:

- We still do not really know how planetesimals complete their growth to kilometre size.
- It still has to be worked out if it is for good or bad that gas drag brings solid objects close to the central star. If they are lost to the star, this is bad for growth. But if they concentrate at some inner edge, this might well explain certain features in planetary systems.
- Related to this, planetesimals might not form throughout the whole protoplanetary disk but prefer certain locations and times. This is important for the assembly of heavy elements that constitute giant planets and might be where both the small scale of planetesimal formation and the large scale of giant planet accretion might significantly overlap.
- Although it appears that giant planets have to form while the gas disk is disappearing, we do not know how to account for their final masses, orbital distances and eccentricities.
- We still cannot explain the precise composition of planets, in particular the abundance of noble gases.
- The formation of the cores, in particular of Uranus and Neptune, remains difficult.

Hopefully, with CoRoT and the discovery of new transiting extrasolar planets, the Juno mission to Jupiter and many other exciting projects, we should make important progress in characterizing planetary systems. With new laboratory experiments underway and improved numerical models, we should pin down the origin of planetesimals better. And, last but not least, getting a more complete picture of the history of the solar system requires a continued assemblage of the numerous and diverse pieces of the puzzle, such as the making-up of protoplanetary disks, the building of planetesimals, the properties of giant planets and so many others.

References

Alibert, Y., Mordasini, C., Benz, W., Winisdoerffer, C., 2005a, *A&A*, **434**, 343
Alibert, Y., Mousis, O., Mordasini, C., Benz, W., 2005b, *ApJ.*, **626**, L57
Blum, J., 2004, *Astrophysics of Dust*, ed. A. N. Witt, G. C. Clayton, B. T. Draine (ASP Conference Proceedings, Vol. 309. San Francisco: Astronomical Society of the Pacific), 369
Blum, J., Wurm, G., 2000, *Icarus*, **143**, 138
Blum, J., Wurm, G., Kempf, S., Henning, T., 1996, *Icarus*, **124**, 441
Blum, J. *et al.*, 2000, *Phys. Rev. Lett.*, **85**, 2426
Bodenheimer, P., Lin, D. N. C., Mardling R., 2001, *ApJ.*, **548**, 466

Boss, A. P., 2000, *ApJ.*, **536**, L101

Briggs, F. H., Sackett, P. D., 1989, *Icarus*, **80**, 77

Ciesla, F. J., Cuzzi, J. N., 2006, *Icarus*, **181**, 178

Chabrier, G., Baraffe, I., 2007, *ApJ.*, **661**, L81

Chabrier, G., Barman, T., Baraffe, I., Allard, F., Hauschildt, P., 2004, *A&A*, **603**, L53

Dominik, C., Tielens, A. G. G. M., 1997, *ApJ.*, **480**, 647

Flasar, F. M. *et al.*, 2005, *Science*, **307**, 1247

Fortney, J., Hubbard, W. B., 2003, *Icarus*, **164**, 228

Gautier, D., Hersant, F., Mousis, O., Lunine, J. I., 2001, *ApJ.*, **550**, L227

Gautier, D., Owen, T., 1989, in *Origin and Evolution of Planetary and Satellite Atmospheres*, ed. S. K. Atreya, J. B. Pollack, M. S. Matthews (Tucson: University of Arizona Press), 487–512

Guillot, T., 2005, *Annu. Rev. Earth Planet Sci.*, **33**, 493

Guillot, T., Hueso, R., 2006, *MNRAS*, **367**, L47

Guillot, T., Santos, N. C., Pont, F., Iro, N., Melo, C., Ribas, I., 2006, *A&A*, **453**, L21

Guillot, T., Showman, A., 2002, *A&A*, **385**, 156

Guillot, T., Stevenson, D. J., Hubbard, W. B., Saumon, D., 2004, in *Jupiter: The Planet, Satellites, and Magnetosphere*, ed. F. Bagenal, W. McKinnon, T. Dowling (Cambridge, UK: Cambridge University Press)

Helled, R., Podolak, M., Kovetz, A., 2006, *Icarus*, **185**, 64

Ida, S., Lin, D. N. C., 2004, *ApJ.*, **616**, 567

Johansen, A., Oishi, J. S., Low, M.-M. M., Klahr, H., Henning, T., Youdin, A., 2007, *Nature*, **448**, 1022

Lodders, K., 2003, *ApJ.*, **591**, 1220

Meakin, P., 1991, *Rev. Geophys.*, **29**, 317

Owen, T., Encrenaz, T., 2006, *Planet Space Sci.*, **54**, 1188

Paardekooper, S.-J., 2007, *A&A*, **462**, 355

Paraskov, G., Wurm, G., Krauss, O., 2006, *ApJ.*, **648**, 1219

Podolak, M., Podolak, J. I., Marley, M. S., 2000, *Planet Space Sci.*, **48**, 143

Pollack, J. B., Hubickyj, O., Bodenheimer, P., Lissauer, J. J., Podolak, M., Greenzweig, Y., 1996, *Icarus*, **124**, 62

Poppe, T., Blum, J., Henning, T., 2000, *ApJ.*, **533**, 454

Santos, N. C., Israelian, G., Mayor, M., 2004, *A&A*, **415**, 1153

Saumon, D., Guillot, T., 2004, *ApJ.*, **609**, 1170

Sekiya, M., Takeda, H., 2005, *Icarus*, **176**, 220

Showman, A. P., Guillot, T., 2002, *A&A*, **385**, 166

Weidenschilling, S. J., Cuzzi, J. N., 1993, *Protostars and Planets III*, ed. E. H. Levy, J. I. Lunine (University of Arizona Press), 1031–1060.

Winn, J. N., Holman, M. J., 2005, *ApJ.*, **628**, L159

Wong, M. H., Mahaffy, P. R., Atreya, S. K., Niemann, H. B., Owen, T. C., 2004, *Icarus*, **171**, 153

Wurm, G., 2007, *MNRAS*, **380**, 683

Wurm, G., Blum, J., 1988, *Icarus*, **132**, 125

Wurm, G., Blum, J., Colwell, J. E., 2001, *Icarus*, **151**, 318

Wurm, G., Paraskov, G., Krauss, O., 2004, *ApJ.*, **606**, 983

Wurm, G., Paraskov, G., Krauss, O., 2005, *Icarus*, **178**, 253

Part V

Summary

17

Open issues in small- and large-scale structure formation

R. S. Klessen and M.-M. Mac Low

17.1 Introduction

This conference has brought together researchers studying structure formation in astrophysics, at scales ranging from planet and star formation through galaxy formation to cosmic structure formation. Aside from gravity, these fields require knowledge about many further physical processes and phenomena, such as turbulent gas dynamics, magnetic fields, non-equilibrium chemistry and the interaction of radiation with matter. The different communities also all rely on numerical simulations and the same modern, general-purpose, ground-based and space-borne telescopes.

In this proceedings contribution, we attempt to identify some of the major challenges for the future. We furthermore debate whether the physical processes relevant for each field exhibit sufficient overlap to warrant concerted cross-disciplinary efforts or whether the features that define and distinguish these fields prevail and make successful cross-fertilization less likely.

17.2 Planet formation

With the first discovery of a planet around another star in 1995, we have begun to place our solar system in the context of other planetary systems. More than 250 extrasolar planets have been identified, most with characteristics vastly different from our own solar system. Planets around stars such as our Sun may be the rule, rather than the exception, but the observed properties exhibit an enormous spread (see Udry & Santos 2007).

In the core instability model, planet formation begins with the coagulation of dust in protoplanetary disks, forming larger aggregates of solid material through a sequence of collisions and agglomeration. This process continues until solid objects reach the centimetre to metre range. At this point, two effects are important.

Structure Formation in Astrophysics. ed. G. Chabrier. Published by Cambridge University Press.
© Cambridge University Press 2009.

First, the coupling between the gas and the solid objects becomes less effective, as the Stokes number approaches unity, so solid objects start to approach the Keplerian orbital velocity, while the partly pressure-supported gas disk continues to orbit at sub-Keplerian velocity. The resulting headwinds quickly drain angular momentum from objects with Stokes number of order unity, causing them to fall into the star in a few hundred orbits. Second, collisions between solid objects begin to be predominantly destructive in the same size range, making it difficult, though not impossible, to continue to grow to larger sizes by collisional agglomeration. Together, these two effects are known as the metre-size barrier.

Two solutions have been proposed to overcome the metre-sized barrier. One is that sufficiently dense dust disks might be able to assemble large enough planetesimals quickly enough strictly by agglomeration. The effectiveness of collisional fragmentation in preventing growth of such objects remains an important question in this scenario however. This is unfortunately difficult to evaluate experimentally, as it ultimately requires controlled hypervelocity impacts between natural materials. A second solution is an extension of the concept of gravitational instability in a sedimented dust layer originally proposed by Goldreich and Ward (1973). That was, however, understood to require an absolutely quiescent gas disk, which appeared impossible to achieve because the dusty mid-plane layer rotates faster than the rest of the disk, making it susceptible to Kelvin–Helmholtz instabilities. However, curvature terms can help prevent the Kelvin–Helmholtz instability, allowing disks with mildly supersolar metallicities to form planetesimals (Chiang 2008).

Even minimum-mass solar nebulae with solar metallicities have been shown by Johansen *et al.* (2007) to quickly form planetesimals by gravitational instability because of two effects. First, any turbulence present, for example from magnetorotational instability, concentrates boulders in higher pressure regions. Second, the streaming instability further concentrates boulders. These two effects interfere constructively to allow the formation of gravitationally bound boulder clusters the mass of Ceres within seven orbits in a magnetically active region of a disk. However, this study assumed that at least half the disk mass had already assembled into centimetre-to metre-sized objects prior to the beginning of their simulations, a condition that it is not yet clear can be realistically achieved. Further work is still needed on the behaviour of boulders in the dead zones of disks. More details are given in Chapter 15 by Udry and collaborators in this book.

Planetesimals then grow further to planetary cores of up to a few Earth masses by gravitational agglomeration, clearing out their orbits in the protoplanetary disk. This is the formation path of terrestrial planets such as the Earth. If the conditions in the disk are right and if the rocky cores are massive enough, then a phase of runaway growth sets in. The cores accrete appreciable amounts of gas in a very short period of time and turn into giant gas planets. This is the most likely scenario for the origin of the Jovian planets in our solar system.

Explaining the observed diversity of planetary systems is a major challenge. The dynamical interaction of an embedded protoplanet with the ambient gas disk or a population of planetesimals can lead to radial migration of the protoplanet through the disk, resulting in the accretion of terrestrial planets by the central star in less than a million years in the standard case. Understanding the magnitude and direction of this migration under different conditions and the resulting eccentricity evolution of the planet requires continuing efforts. For example, relaxing the usual assumption of an isothermal equation of state can lead to reduced or even reversed direction of migration. Since real disks are usually optically thick at least in regions close to the star and thus have nearly adiabatic mid-planes, this may represent an important effect. Also the effects of magnetorotational turbulence on migration may be profound. This has been demonstrated to result in a random walk rather than inward migration for terrestrial mass planets in the absence of a dead zone. The presence of dead zones reduces the mass of the objects subject to random walks, but does not suppress the effect entirely. Understanding the detailed turbulent structure of protoplanetary disks remains a major challenge, though, because of its dependence on the ionization structure of the disk, which in turn depends on ionizing radiation, cosmic ray irradiation, chemistry, dust properties, mixing and thermal structure. For more detailed discussions, see Chapter 13 by Duchene *et al.* and Chapter 14 by Durisen *et al.* in this book.

17.3 Star formation

Stars are born in the dense, cold cores of supersonically turbulent, magnetized, molecular clouds which become gravitationally unstable and fragment to form single, binary and multiple stars. Turbulence and magnetic fields appear central to understanding the origin of stellar masses and their distribution in different physical environments. Chemistry in the interstellar medium (ISM) is extremely complex. Molecular species play important roles in the thermodynamics of molecular clouds and provide the diagnostic tool to determine the thermal and dynamical structure as well as the evolutionary age of these clouds (see Chapter 9 by Hennebelle *et al.* in this book).

There is a well-established model for the earliest evolutionary stages of solar-type stars, ranging from pre-stellar cores to accreting protostars and, finally, young stars surrounded by disks (see Chapter 10 by Megeath *et al.* and Chapter 11 by André *et al.* in this book). This picture, however, does not yet address what determines the initial stellar mass function (IMF) or how sub-stellar objects, massive stars, binaries and multiple systems form. The IMF regulates chemical enrichment as well as energy and momentum input into the ISM. One can then ask: Is the IMF universal or does this function depend on environmental factors such as gas density, strength of ISM turbulence and metallicity? The answer is a prerequisite for understanding the global star formation history of galaxies at all redshifts.

A number of different approaches to the IMF are currently discussed. The stellar mass spectrum can be related to the properties of supersonic turbulence (Klessen 2001; Padoan & Nordlund 2002) or to competitive accretion from a common gas reservoir in dense clusters (Bonnell & Bate 2006). Numerical simulations support these theoretical approaches although they have been criticized for neglecting heating from the accretion luminosity of the accreting stars (see the Chapter 12 by Krumholz and Bonnell in this book). Radiative heating, protostellar outflows or ionizing radiation have been invoked to determine the IMF either directly by cutting off the accretion onto individual protostars or indirectly by providing support against collapse to the surrounding medium. The IMF has also been explained as the result of the purely random process of collapse in a fractal cloud, invoking the central limit theorem to explain its characteristic shape. Further discussions can be found in Mac Low and Klessen (2004), Bonnell *et al.* (2007), Larson (2007) and McKee and Ostriker (2007).

However, the IMF is not scale-free, and a particularly promising explanation focuses on the thermodynamic properties of the gas. The amount of fragmentation occurring during gravitational collapse depends on the compressibility of the gas (Li *et al.*, 2003). For polytropic index $\gamma < 1$, turbulent compressions cause large density enhancements in which the Jeans mass falls substantially, allowing many fragments to collapse. Only a few massive fragments get compressed strongly enough to collapse in less compressible gas though. In real molecular gas, the compressibility varies as the opacity and radiative heating increase. Larson (2005) noted that the thermal coupling of the gas to the dust at densities above $n_{crit} \sim 10^5$–10^6 cm^{-3} leads to a shift from an adiabatic index of $\gamma \sim 0.7$–1.1 as the density increases above n_{crit}. This sets a mass scale for the peak of the IMF. Jappsen *et al.* (2005) demonstrated numerically that not only does this mechanism set the peak mass, but also appears to produce a power-law distribution of masses at the high-mass end comparable to the observed distribution. Reducing the critical density by increased radiative heating – as occurs in clusters of accreting protostars or in the centres of starburst galaxies – will then lead to a larger characteristic mass.

At the extreme ends of the stellar mass spectrum, our knowledge of the IMF is particularly limited. Massive stars are very rare and rather short-lived. The number of massive stars that are sufficiently near to study in detail and with very high spatial resolution, e.g. to determine multiplicity, therefore is small (for further details, see the review by Zinnecker and Yorke 2007). Low-mass stars and brown dwarfs, on the contrary, are found in abundance but are faint, so they are difficult to study in detail (see review by Burrows *et al.* 2001). Such studies, however, are in great demand, because secondary indicators such as the fraction of binaries and higher-order multiples as function of mass, or the distribution disks around very young stars or possible signatures of accretion during their formation are probably better suited to distinguish between different star formation models than just looking at the IMF.

Also the formation site for brown dwarfs remains controversial. Gravitational instability in protostellar disks followed by ejection from the resulting triple and other small n-systems remains a viable explanation. However, direct collapse has not been ruled out either, in one of two scenarios. Competitive accretion suggests that objects ejected quickly from the gas reservoir will have their accretion truncated early, with a significant fraction of objects never reaching stellar mass. Alternately, it has been suggested that direct collapse from the low-mass end of a gravoturbulent core spectrum is sufficient to explain the observed distribution of brown dwarfs. Although brown dwarf binarity has been proposed as a distinguishing test between the latter two scenarios, it has been recently noted that even competitive accretion followed by ejection produces a significant fraction of brown dwarf binaries.

The other end of the stellar mass range also remains a subject of active research. Massive stars appear to require very high accretion rates (exceeding roughly $10^{-3} M_\odot$ yr^{-1}) to overcome radiation pressure and grow to large sizes. This has led to the suggestion that massive star formation must be triggered by passing pressure waves driven by prior supernova explosions. Three alternative explanations have been proposed. First, that massive stars simply represent the high-mass end of the turbulent cloud core mass spectrum. This would suggest that massive stars are no different from any other type of star in their formation, just rarer. This disagrees with the observation that they are almost always found at the centre of large groups of lower-mass stars. Second, that massive stars are the lucky winners of a process of competitive accretion, where multiple stars accrete from a common massive gas reservoir. However, this model neglects the heating from the accretion luminosity of the stars in computing the fragmentation of the massive core. Third, that massive stars form when accretion luminosity from low-mass stars forming in a core can heat up the core sufficiently to suppress further fragmentation, allowing only one or a few more stars to accrete the rest of the mass of the core. To distinguish between these scenarios clearly, more work is required. In particular, numerical models that consistently take into account all physical aspects of the problem are in high demand. For a further discussion, see Chapter 12 by Krumholz and Bonnell in this book.

Other important open questions concern the various feedback mechanisms that regulate the star formation process and its efficiency. In addition to the heating produced by their accretion luminosity, young low-mass stars show strong X-ray activity at young ages, thereby regulating the local ionization structure of their parent molecular cloud. Young massive stars release enormous amounts of energy and momentum into their immediate environment, first through ionizing radiation and bipolar outflows and then subsequently through line-driven winds and finally supernova explosions. Ionizing radiation carves out giant HII regions that quench further star formation within them although possibly triggering secondary star formation in their shells. Stellar winds act to further expand these cavities into superbubbles, and the supernovae that subsequently occur power their growth into supergiant shells. Although triggered star formation occurs in the compressed shells of all of these structures, it appears unlikely that its efficiency exceeds or even approaches unity. That is, first-generation stars will not trigger anything like their own mass of second-generation stars, so the process winds down rather than being self-sustaining. Although quite a number of numerical studies of these processes have recently been presented, more work is required to fully understand the effects of ionizing radiation and supernova explosions on driving interstellar turbulence. Recent analysis of Spitzer GLIMPSE (Galactic Legacy Infrared Mid-Plane Survey Extraordinaire) data, e.g., revealed more than 300 bubble-like structures produced by young OB stars. Although such structures suggest that triggering may play an important role in shaping the star formation activity on a galactic scale, not much is known yet about the kinematic impact or the change in excitation conditions and chemical properties at such bubble interfaces. We need better theoretical models focusing on predicting the chemical state of the gas. On local scales, molecular outflows from young stars are an important source of energy and momentum. However, their importance for driving turbulence on very small scales remains highly controversial. More effort is needed to understand their relation to the self-regulation of the star formation process on scales of individual clusters (for a further discussion, see Chapter 4 by Pudritz *et al.*).

Star formation in the early universe occurs under different enough conditions that many of the conclusions drawn for modern star formation must be reconsidered. The lack of cooling from dust or molecules other than molecular hydrogen or its isotopic variants seems likely to promote massive star formation, just as trapping of cooling radiation by dust opacity may do so in modern star formation (for recent overviews, see Bromm and Larson 2004 or Glover 2005). The question of how the transition from primordial to modern star formation occurs has recently been shown to be more complex than was thought. It probably does not depend simply on the presence or absence of sufficient atomic fine structure cooling but on some combination of initial conditions and other cooling mechanisms.

One favoured mechanism is dust cooling at high densities and metallicities as low as 10^{-5} solar. Other questions that remain topics of current research include how the metals from the first supernovae are distributed, how important feedback from the first stars is in promoting or suppressing feedback in the surrounding region, and what the actual initial conditions for later generations of stars might be. For example, do they form from warm, ionized gas, or only after it has cooled to low temperatures? For a more detailed discussion, see Chapter 8 by Bromm *et al.* in this book.

17.4 Galaxy formation

Stars, interstellar gas and active nuclei within galaxies are the almost exclusive luminous signposts of structure in the universe and therefore need to be understood both in their own right and as tracers and probes of the dominant but invisible constituents of the universe (Voit 2005). The qualitative framework of galaxy formation within cosmological structure formation is likely to hold in the foreseeable future. Our understanding of how star formation proceeds in galaxies still needs to be improved, however.

One of the most striking observed regularities in galactic star formation is the Schmidt law (Kennicutt 1998). Viable explanations have been proposed that depend on the rate of star formation being limited by the strength of gravitational instability (Kravtsov 2003; Li *et al.* 2005) or by the formation of dense cores within molecular clouds (Krumholz & McKee 2005). It may be that these explanations are compatible, as each theory takes the other as a boundary condition, effectively, but that has not yet been demonstrated. To distinguish between these theories, we must understand the formation of molecular clouds in galactic disks.

While much is known about the internal structure and dynamics of molecular clouds, their origin is still unknown. In particular, the formation mechanism, age and lifetime of molecular cloud complexes are still under considerable debate in the contemporary literature. It is likely that the formation mechanism and the overall lifetime of molecular clouds are closely linked. Models can be divided into two groups: those in which the molecular clouds form rapidly (within a few million years) through dynamical processes and those where the evolution occurs more quiescently (on timescales exceeding 10^7 years). The rapid formation models normally involve some form of large-scale shock or gravitational instability (Ballesteros-Paredes *et al.* 1999) that brings the gas together from a large region of the ISM. The gas accumulated in this model can either be atomic, and undergo rapid chemical evolution to form molecular hydrogen, or even be in the form of pre-existing, but very low density, cloudlets of molecular gas (Elmegreen 1990). Clouds formed in rapid, dynamical processes may or may not be gravitationally bound (for a review, see Ballesteros-Paredes 2004) and need not be in virial equilibrium. In quiescent models, the rate of assembly of gas into clouds is controlled by the rate of some other process, such as the loss of magnetic support from the gas through ambipolar diffusion. Clouds formed in these models are always gravitationally bound (as the re-expansion timescale for an unbound cloud is less than the assumed formation timescale) and are usually also in virial equilibrium.

The debate over cloud formation rates and lifetimes is in a large part due to the complexity of the physics involved. Observations of cold atomic hydrogen (Heiles & Troland 2005) show that it has line widths that are much larger than the inferred thermal line width, and this non-thermal component is generally interpreted as reflecting the presence of supersonic turbulence within the gas. As we have no comprehensive theory of compressible, magnetized turbulence, the effects of this turbulence on the evolution of the gas cannot easily be captured in analytic or semi-analytic models. Instead, large numerical simulations are required. For further details, see Elmegreen and Scalo (2004) and Scalo and Elmegreen (2004), as well as Chapter 1 by Lévêque and Chapter 2 by Schmidt in this book.

How can we approach these problems when modelling galaxy formation and evolution in a cosmological context? In our current view of structure formation, the smallest, lowest-mass objects formed first, while larger structures formed later. Since the largest objects present at any given time are rare, their space density is low compared with smaller objects. These considerations place tight constraints on attempts to numerically model processes related to galaxy formation and evolution. On the one hand, the simulations must be able to resolve the smallest early objects, since they are the building blocks of larger structures that form later. On the other hand, simulations must cover large volumes so that the rarest, latest-forming objects are statistically well represented in the computational domain. Cosmological simulations of structure formation thus present both a multi-scale and a multi-physics challenge (see Chapter 6 by Ellis and Silk and Chapter 7 by Abel *et al.* in this book). For very large-scale simulations, they also present a data management challenge. If one wishes to both resolve the internal structure of galaxies with ten resolution elements, say, and at the same time simulate a volume comparable to the modern surveys ($1 \, \text{Gpc}^3$), one arrives at a spatial dynamic range requirement of six orders of magnitude in three dimensions. This is only achievable at the largest supercomputing facilities available today (see, e.g., Springel *et al.* 2005, 2006).

In galaxies, the energy density of gas, magnetic fields and cosmic rays are on average approximately equal and form the bulk of the total energy density, with very strong variations from galaxy to galaxy and also within individual galaxies. Furthermore, energetic feedback from active galactic nuclei (AGNs) may dominate the energy of certain galaxies throughout distinct periods in time. Thus, in order to understand the formation of galaxies, one has to devise a model that incorporates the complex multi-phase and multi-scale structure of the ISM for star formation, for supernova explosions, for AGN activity, for radiative transport and also for mechanisms with respect to the formation of magnetic fields and the energetics of cosmic rays. Although certain aspects of the problem have been modelled very successfully, a fully self-consistent numerical approach seems quite far away. In the foreseeable future, many of these small-scale physical phenomena can only be included in cosmological galaxy formation calculations in a quasi-phenomenological way. That means future efforts will continue to concentrate on developing better approximations to unresolved or purely resolved sub-grid scale baryonic processes.

17.5 Similarities and dissimilarities

After the above brief and certainly incomplete overview over the current state of planet, star and galaxy formation theory, we want to turn our attention to the question of the similarities and dissimilarities between these three fields and investigate whether theoretical methods and numerical techniques successfully applied to problems in one field can also be useful for the other ones. In other words we ask, are the problems and open issues that we are faced with at the vastly different scales essentially similar, or does each problem require us to think anew about the physical processes relevant for the particular scale of interest, so that the transfer of knowledge and expertise from one field to the other will remain limited?

If we look at the past and current state of the art, then it is evident that most studies so far have focused on a small number of physical processes only. Typical questions are: What happens if we include a particular physical process in our theoretical model or our numerical simulation, and how does it affect the system? How does it modify possible equilibrium states? And how does it influence the dynamical evolution if we apply perturbations? The processes and phenomena taken into account in planet, star or galaxy formation models are hydrodynamics and turbulence, gravitational dynamics, magnetic fields, non-equilibrium chemistry or the interaction of radiation with matter. More sophisticated approaches include several of these items, but none has considered all of them. The challenge in the past was mainly to do justice to the inherent multi-dimensionality of the considered problems. For example, stellar birth in turbulent interstellar gas clouds with highly complex spatial and kinematical structure is an intrinsically three-dimensional problem with one- or two-dimensional approaches at best providing order-of-magnitude estimates (see Chapter 2 by Schmidt in this book). The same holds for the web-like large-scale structure of the universe that provides the framework for galaxy formation and evolution (see Chapter 6 by Ellis and Silk in this book).

This era is coming to an end. Many of today's most challenging problems are multi-physics, in the sense that they require the combination of many (if not all) of the above-mentioned processes, and multi-scale, in the sense that unresolvable microscopic processes can feed back onto macroscopic scales. For example, the coagulation of dust species to larger particles or the interaction of dust with the radiation field from the central stars will eventually feedback into the dynamical behaviour of the gas in protostellar accretion disks. Or, similarly, star formation and baryonic feedback are crucial ingredients of understanding galaxy formation and evolution in our cosmological models. In a realistic description of cosmic phenomena, one is faced with the highly non-linear coupling between quite different kinds of interactions on a variety of scales.

This is not only a challenge, but also a chance, because it may open up new pathways to successful collaborations across astrophysical disciplines. It also reaches out to scientists in neighbouring fields, such as applied mathematics or computer science, as some of the challenges in modern astrophysics are of a technical nature. For example, only a few groups around the world are able to fully benefit from the massively parallel computing architectures that are currently being developed. Peak performances with $\sim$100 teraflops will only be attainable on thousands of CPUs; sustained petaflop computing may require as many as 10^5 CPUs. This asks for a completely new approach to parallel algorithm design, a field where modern computer science is way ahead of the schemes currently used in astrophysics. Regular methodological exchange with applied mathematicians and possibly numerical fluid dynamicists thus holds the promise of both transferring new methods into astrophysics and raising the awareness of mathematicians about numerical challenges in astrophysics.

However, as usual, the devil is in the details. It could very well be that cross-disciplinary exchange of methods and collaborations that appear attractive and promising from a distance can in reality not keep up to the expectations raised. This may have a number of different causes. The way of communicating ideas and organizing collaborative projects can vary significantly across different disciplines. This begins with different communities using disjunct terminologies and acronyms (which is already a problem within the different branches of astronomy and astrophysics), touches on varying procedures to communicate new results and ideas (such as scientific publications versus technical reports) and also concerns fundamental questions of how different communities are organized and structured. For example, the design and development of astrophysical simulation software is often connected to individual people or small groups of researchers. There are quite a number of codes which are available for free under a GNU-type licensing agreement (such as Gadget, Ramses, ENZO, FLASH or Athena to name but a few that were prominently discussed during the conference). Computational fluid dynamics in industry or engineering, on the contrary, mostly relies on a small number of large commercially available software packages. The free exchange of expertise and knowledge may in this case interfere with commercial interests and license restrictions. In conclusion, we expect that in the foreseeable future, the development of astrophysical simulation software will remain mostly driven from within the community and that it will continue to be difficult to generate synergy effects through collaborations across disciplinary borders. However, it should be noted that current efforts in building up a virtual observatory and in grid computing are promising counter examples.

The conference has also left us with mixed feelings about the potential of collaborations amongst the different branches of astronomy and astrophysics itself. The one thing that became very clear during the discussions in Chamonix is that the current challenges of theoretical astrophysics lie in incorporating more physics into our numerical models as well as reaching out for higher spatial and temporal resolution. Only approaches that do justice to the complexity and richness of the astrophysical phenomena associated with the formation and evolution of stars and planetary systems, or of galaxies and the universe as a whole, will be able to reach the precision and predictive power required to interpret the wealth of observational data that will become available soon with the advent of new facilities such as ALMA, LOFAR or JWST. This is a very difficult task and it stands to reason that research groups working on related phenomena collaborate. The physics of accretion disks, e.g., that play a crucial role in star and planet formation may be very similar to the processes that govern formation and evolution of disk galaxies such as our Milky Way or regulate the mass growth of super massive black holes in their centres. The jets and outflows we observe from young stars in the solar neighbourhood may be simply scaled down versions of the highly energetic jet phenomena that are characteristics of AGNs.

While the basic physical processes indeed are very similar, the discussions during the conference revealed that there are also important differences that cannot be neglected. For example, magnetic fields and their ability to generate turbulence via the magnetorotational instability play a major role in the evolution of protostellar disks. The same holds for the coupling between gas and dust, the latter being able to coagulate to build up larger aggregates of solid material which eventually may lead to the runaway growth of planets. The gravitational potential of protostellar disks is determined by the central star and the available reservoir of gas. The potential of galactic disks, on the contrary, is usually dominated by dark matter. The interplay between the stellar component and gas – star formation and stellar feedback – is important in determining the long-term evolution of galactic disks and their observational appearance. Magnetic fields may only play a dominant role outside of the star-forming region of the disk, and dust–gas coupling is likely unimportant. The physics that must be included in realistic disk models thus varies significantly depending on the scale of interest. Consequently, this translates into different requirements in terms of resolution and numerical set-up and renders finding a unified numerical approach very difficult.

Despite these concerns, however, discussions about the similarities and dissimilarities in the theoretical and numerical approaches are very helpful and important. They are essential means of exchanging ideas and information across the various branches of astronomy and astrophysics and help to deepen our knowledge of the physical processes that govern the evolution of the universe and the complex structures that it contains. In this sense, the conference in Chamonix bringing together scientists working in planet formation, star formation and galaxy formation as well as numerical astrophysics was a great success and a memorable event to look back to.

Acknowledgements

We first want to thank Gilles Chabrier. Without his engagement and dedication, this highly interesting and exciting conference would not have been possible. RSK also acknowledges partial support from the Sonderforschungsbereich 439 'Galaxies in the Young Universe' funded by the German Science Foundation (DFG), while M-MML acknowledges the Max Planck Society and the DAAD for stipends in support of a visit to Heidelberg, and partial support from NASA grant no. NNX07AI74G.

References

Ballesteros-Paredes, J. (2004). Molecular Clouds: Formation and Disruption. *Astrophys. Sp. Sci.* **289**, 243–254

Ballesteros-Paredes, J., Hartmann, L. and Vázquez-Semadeni, E. (1999). Turbulent Flow-Driven Molecular Cloud Formation: A Solution to the Post-T Tauri Problem? *ApJ.* **527**, 285–297

Bonnell, I. A. and Bate, M. R. (2006). Star Formation Through Gravitational Collapse and Competitive Accretion. *Mon. Not. R. Astron. Soc.* **370**, 488–494

Bonnell, I. A., Larson, R. B. and Zinnecker, H. (2007). The Origin of the Initial Mass Function, in *Protostars and Planets V*, eds. B. Reipurth, D. Jewitt, and K. Keil (Tucson, Arizona: University of Arizona Press), 149–164

Bromm, V. and Larson, R. B. (2004). The First Stars. *Annu. Rev. Astron. Astrophys.* **42**, 79–118

Burrows, A., Hubbard, W. B., Lunine, J. I. and Liebert, J. (2001). The Theory of Brown Dwarfs and Extrasolar Giant Planets. *Rev. Mod. Phys.* **73**, 719–765

Chiang, E. (2008). Vertical Shearing Instabilities in Radially Shearing Disks: The Dustiest Layers of the Protoplanetary Nebula. *ApJ.* **675** 1549–1558

Elmegreen, B. G. (1990). A Comparison of Cloud Formation Rates in the Gravitational Instability and Random Collisional Buildup Models. *ApJ.* **357**, 125–131

Elmegreen, B. G. and Scalo, J. (2004). Interstellar Turbulence I: Observations and Processes. *Annu. Rev. Astron. Astrophys.* **42**, 211–273

Glover, S. C. O. (2005). The Formation of The First Stars in the Universe. *Space Sci. Rev.* **117**, 445–508

Goldreich, P. and Ward, W. R. (1973). The Formation of Planetesimals. *ApJ.* **183**, 1051–1062

Heiles, C. and Troland, T. H. (2005). The Millennium Arecibo 21 Centimeter Absorption-Line Survey. IV. Statistics of Magnetic Field, Column Density, and Turbulence. *ApJ.* **624**, 773–793

Jappsen, A.-K., Klessen, R. S., Larson, R. B., Li, Y. and Mac Low, M.-M. (2005). The Stellar Mass Spectrum from Non-Isothermal Gravoturbulent Fragmentation. *A&A* **435**, 611–623

Johansen, A., Oishi, J. S., Mac Low, M.-M., Klahr, H. H., Henning, T. and Youdin, A. (2007) Rapid Planetesimal Formation in Turbulent Circumstellar Disks, *Nature* **448**, 1022–1025

Kennicutt, R. C. (1998). Star Formation in Galaxies Along the Hubble Sequence. *Annu. Rev. Astron. Astrophys.* **36**, 189–231

Klessen, R. S. (2001). The Formation of Stellar Clusters: Mass Spectra from Turbulent Molecular Cloud Fragmentation. *ApJ.* **556**, 837–846

Kravtsov, A. V. (2003). On the Origin of the Global Schmidt Law of Star Formation. *Aphys. J.* **590**, L1–L4

Krumholz, M. R. and McKee, C. F. (2005). A General Theory of Turbulence-Regulated Star Formation, from Spirals to Ultraluminous Infrared Galaxies. *ApJ.* **630**, 250–268

Larson, R. B. (2005). Thermal Physics, Cloud Geometry and the Stellar Initial Mass Function. *Mon. Not. R. Astron. Soc.* **359**, 211–222

Larson, R. B. (2007). Insights from Simulations of Star Formation. *Rep. Prog. Phys.* **70**, 337–356

Li, Y., Klessen, R. S. and Mac Low, M.-M. (2003). The Formation of Stellar Clusters in Turbulent Molecular Clouds: Effects of the Equation of State. *ApJ.* **592**, 975–985

Li, Y., Mac Low, M.-M. and Klessen, R. S. (2005). Star Formation in Isolated Disk Galaxies. II. Schmidt Laws and Efficiency of Gravitational Collapse. *ApJ.* **639**, 879–896

Mac Low, M.-M. and Klessen, R. S. (2004). Control of Star Formation by Supersonic Turbulence. *Rev. Mod. Phys.* **76**, 125–194

McKee, C. F. and Ostriker, E. C. (2007). Theory of Star Formation. *Annu. Rev. Astron. Astrophys.* **45**, 565–687

Padoan, P. and Nordlund, Å. (2002). The Stellar Initial Mass Function from Turbulent Fragmentation. *ApJ.* **576**, 870–879

Scalo, J. and Elmegreen, B. G. (2004). Interstellar Turbulence II: Implications and Effects. *Annu. Rev. Astron. Astrophys.* **42**, 275–316

Springel, V., Frenk, C. S. and White, S. D. M. (2006). The Large-Scale Structure of the Universe. *Nature* **440**, 1137–1144

Springel, V. and 16 colleagues (2005). Simulations of the Formation, Evolution and Clustering of Galaxies and Quasars. *Nature* **435**, 629–636

Udry, S. and Santos, N. S. (2007). Statistical Properties of Exoplanets. *Annu. Rev. Astron. Astrophys.* **45**, 339–396

Voit, G. M. (2005). Tracing Cosmic Evolution with Clusters of Galaxies. *Rev. Mod. Phys.* **77**, 207–258

Zinnecker, H. and Yorke, H. W. (2007). Toward Understanding Massive Star Formation. *Annu. Rev. Astron. Astrophys.* **45**, 481–563

18

A final word

E. E. Salpeter

We have heard a lot about probability functions $p(M)$ at this meeting for mass M of planets or stars or clouds or clusters under various conditions. Since we have covered such an enormous range of masses, it is not surprising that power-law distributions close to the scale-invariant power have recurred so often. A power law differing from this distribution in the direction of favouring either low or high masses must of course have a turnover (or termination) towards this end to avoid a divergence. The physical reason for such a turnover is of interest, as is the question of continuity between the various types of objects. Bingelli and Hascher (PASP **119**, 592, 2007) have followed this power-law continuity over 36 orders of magnitude in mass from asteroids to galaxy superclusters. It is instructive to look at similar probability distribution functions in quite different fields. I will give only the examples of two different kinds of human aggregates. One example, which has been discussed for more than a century or so, is the probability distribution for the size (i.e. the number of inhabitants) of a village, town or city. Near the end of the nineteenth century, the deviation from scale invariance was a slight increase towards the bottom end, i.e. overall slightly more people lived in a village of population 100–200 than in a city of 250 000 to 500 000. There was indeed a turnover for the smallest villages, presumably because a village had to be large enough to provide enough of a horseshoe business for at least one blacksmith. In modern times, horseshoes are no longer a consideration and the distribution has in any case shifted towards larger cities. A more gruesome human aggregate is the number n of violent civilian deaths in a single individual incident (from $n = 1$ in a single gunshot to $n = 100$ in a truck bomb or a military manoeuvre) in some large and protracted conflict. Data towards a probability distribution function $p(n)$ have been provided by government sources and media reports in Rwanda, Sri Lanka, Bosnia, various African countries, etc., and most recently (and currently) in Iraq. $p(n)$ from such data already suggests weighting towards small n, but while the official reports underestimate $p(n)$ at all n, they do so especially strongly for small n.

Structure Formation in Astrophysics. ed. G. Chabrier. Published by Cambridge University Press.

The integral of $p(n)\mathrm{d}n$ is underestimated in the official reports by an unknown factor F. In a number of conflicts, this integral, i.e. the total number of violent deaths per year, has been estimated much more reliably by epidemiologists using 'cluster sample surveys', where householders are interviewed about deaths in the family and asked to produce death certificates of killed relatives. The householders do not know the size of the incident involved, although the prevalence of gunshot deaths points towards small n. Such a survey has been done, for each of the first 3 years since the occupation of Iraq started, by an epidemiology team from John Hopkins University (Gilbert Burnham *et al.*, *Lancet*, Oct. 2006). Their estimate of about 500 000 violent deaths of Iraqi civilians up to July 2006 is reliable, giving an underestimation factor F of about 5 or a little larger. However, no such surveys (which are time-consuming and dangerous) have been done since then. I suggest that someone should do the following, purely theoretical, correlation study for all cases where cluster sample surveys are available (separately per year for long conflicts): The official $p(n)$ is likely to give a simple power-law exponent E since, with no $n < 1$, there is probably no turnover at the bottom end. The exponent E is likely to vary appreciably from case to case. The cluster survey gives the factor F by which the integral of $p(n)\mathrm{d}n$ was underestimated in each case. I hope that there is some recognizable correlation between E and F. If so, one can then estimate F from the measured E, in cases without a cluster survey, and hence a better estimate of the total number of violent deaths.

18.1 Science and politics

The likely pitfalls in the correlation study I have just suggested are rather similar to the ones discussed at this meeting. Some of you are therefore well qualified to undertake such a study, even though military conflicts are not similar to astronomy as such and even though your results would have political consequences. I myself have switched from astrophysics to mathematical epidemiology and medical statistics for meta-analysis and have found my new profession quite exhilarating. There is only one proviso, it is advisable not to make a switch all on your own, but to team up with an experienced professional in your new field (in my own case, it is my daughter Shelley who is a doctor in a Stanford Medical School teaching hospital). The usual etiquette at scientific meetings is not to bring up politics, but I feel that it is quite wrong especially at a dangerous time like the present. To say 'I'll think about politics only in my spare time' just does not work; our own work is too time-consuming and so much fun, there is too little spare time. My teacher Hans Bethe lived through the Weimar Republic Germany just before Hitler and the Second World War. He recalled that intellectuals were not for war nor against democracy, but just did not have enough spare time to do anything about it. I feel that scientists and other intellectuals should speak out on political issues, especially on topics where we have some technical expertise or at least have the appropriate background. We are in a privileged position, financially and security-wise, so we can do so with little risk and we are likely to be listened to, at least somewhat. To be able to speak out effectively, one has to be well-informed. For scientists, I recommend reading reports from the Union of Concerned Scientists (www.ucsusa.org), an organization funded by private donations: global warming, energy conservation, ballistic missile defence and new nuclear weapons are some of the topics of interest. UCS gives technical, but readable, analyses of these topics from time to time. On legal matters, the Center for Constitutional Rights (www.ccr-ny.org) similarly does excellent work. One kind of 'speaking out' relates to the harm for human kind, which scientists might cause with their own work. For instance, in 1995, Hans Bethe urged scientists to 'cease and desist' from carrying out any further weapons work. More recently, Lord Martin Rees, in a book and in speeches, has warned of the potential dangers even from non-weapons research in nano-technology, computer advances, genetic engineering, etc. We should all consider whether the time is ripe for a kind of 'Hippocratic oath' to be undertaken by practising scientists. I feel scientists should even speak out publicly on important issues for which they have no technical expertise, but only objectivity. One example is warning politicians of the many disasters which would follow any kind of US attack on Iran. I have given mainly US examples, but much applies to Europe as well. Starting a branch of UCS in Europe would be welcome and a Hippocratic Oath movement might actually succeed first in Europe. Even distant expressions of sanity will have some effect in the United States. Americans do listen to European warnings against the occupation of Iraq.

18.2 Occam's razor

I am of course gratified that we had such a harmonious meeting in spite of the many different sub-fields from planets, stars, gas, galaxies and clusters to the universe. With turbulence, jets and magnetic fields reoccurring, the various experts were close enough to each other that they could ask meaningful questions and learn from the remaining differences. The absence of outright conflicts or irreconcilable differences made for the harmony, but I am worried that without any strife we might be missing out on some things. William of Ockham taught us a long time ago that one should shave off unnecessary appendages from a complex theory. We usually agree that one should adopt the simplest hypothesis, which can explain all the facts in some field or at least explain the most important sounding facts. However, this more relaxed version of Occam's razor can sometimes lead to a self-fulfiling prophecy. I will give one example from the early writings of Sir Arthur

Eddington. In his book, *The Internal Constitution of Stars* (written 1924/1925), he made it clear that there were great puzzles about the structure and evolution of stars, but he also included one chapter 'Diffuse Matter in Space', where he felt there was serene harmony. Although he must have known something about dark nebulae (and certainly lots about rocks and planets), he simply assumed right at the beginning of the chapter that there were *no* solids in the interstellar medium and hence no dust grain opacity. He then made some straightforward derivations: with no dust opacity, there is no obscuration of starlight which must include UV. With all this UV, all ISM atoms (of whatever elements they were made) are at least singly ionized. With all ions positively charged, every pair of ions repel each other and cannot combine to form a diatomic molecule. With no molecules in space, one cannot form solids: QED. Although it is almost as unpopular to talk against Occam's razor as it is to talk about politics, I want to suggest that very occasionally we should consider more complex or far-fetched hypotheses. These may not be correct in themselves, but can help us check whether some simple hypotheses have lulled us into a false sense of security. One such complexity, which has already received some attention in the past but should be reconsidered, is the possibility that some of our dimensionless physical constants are changing with time. Another question is whether some exo-biology has altered (deliberately or inadvertently) its own exo-solar system drastically. Of course, such a civilization would have to be more advanced technologically than we are, but we are fairly close ourselves. For instance, before our first nuclear weapons test during the Second World War, there was some speculation whether this could explode the Earth's atmosphere. One class of anti-Occam examples might involve the combination of two disparate things. The following three questions are just examples: (i) If one accepted *both* Milgrom's MOND alternative to Newtonian dynamics *and* some Dark Matter, could one avoid the need for Dark Energy altogether? (ii) With Cosmic Strings/Branes *plus* multi-universes, should we look for different things in high-redshift observations than we do now? (iii) The abundance of interstellar $H3+$ depends on the intensity of ionizing photons and is not considered to be important for the early universe. However, if dark matter consists of sterile decaying neutrinos, as advocated by Peter Biermann (*ApJ* **654**, 290, 2007) and others, ionization is increased. Could this make the catalytic properties of $H3+$ more important? To put my anti-Occam plea into a different context: I hope that in future meetings, we will have a *few* crackpot talks! However, zero crackpot talks are certainly better than too many such talks. In any case, the interaction between the various disparate groups at this meeting will have done a lot of shaking up, even without going off the deep end.